Springer Proceedings in Physics

Volume 254

Indexed by Scopus

The series Springer Proceedings in Physics, founded in 1984, is devoted to timely reports of state-of-the-art developments in physics and related sciences. Typically based on material presented at conferences, workshops and similar scientific meetings, volumes published in this series will constitute a comprehensive up to date source of reference on a field or subfield of relevance in contemporary physics. Proposals must include the following:

- Name, place and date of the scientific meeting
- A link to the committees (local organization, international advisors etc.)
- Scientific description of the meeting
- List of invited/plenary speakers
- An estimate of the planned proceedings book parameters (number of pages/articles, requested number of bulk copies, submission deadline).

Please contact:

For Americas and Europe: Dr. Zachary Evenson; zachary.evenson@springer.com
For Asia, Australia and New Zealand: Dr. Loyola DSilva; loyola.dsilva@springer.com

Yongdong Li · Anbang Sun · Ming Xu ·
Jiawei Zhang
Editors

Proceedings of the 5th Chinese Conference on High Voltage and Discharge Plasmas

HVDP 2024, 29 Nov–1 Dec, Xi'an, China

Springer

Editors
Yongdong Li
Xi'an Jiaotong University
Xi'an, Shaanxi, China

Anbang Sun
Xi'an Jiaotong University
Xi'an, Shaanxi, China

Ming Xu
Xi'an University of Technology
Xi'an, Shaanxi, China

Jiawei Zhang
Xi'an University of Technology
Xi'an, Shaanxi, China

ISSN 0930-8989 ISSN 1867-4941 (electronic)
Springer Proceedings in Physics
ISBN 978-981-96-5710-0 ISBN 978-981-96-5711-7 (eBook)
https://doi.org/10.1007/978-981-96-5711-7

This Springer imprint is published by the registered company Springer Nature Singapore Pte Ltd.
The registered company address is: 152 Beach Road, #21-01/04 Gateway East, Singapore 189721, Singapore

Contents

Chapter 1
Study on Trap Characteristics of Polypropylene Core-Shell Nanocomposite Dielectrics Based on Constant Temperature Discharge Current

Zeqin Liu, Xuening Wang, Hongxiang Zhao, Yuanyuan Li, and Zhipeng Lei

Abstract In order to investigate the effect of $Al_2O_3@SiO_2$ core-shell nanoparticles addition on trap characteristics of poly-propylene nanocomposite dielectrics, five kinds of polypropylene nanocomposite dielectrics with different core–shell nanoparticles addition amounts were prepared. Their carrier and trap distribution characteristics were analyzed by isothermal discharge current method. The results show that the addition of a small number of core-shell nanoparticles can introduce a large number of new traps, which have a stronger trapping effect on the carrier and reduce the carrier mobility. On the contrary, the addition of a large number of core-shell nanoparticles can weaken the deep trap characteristics, strengthen the shallow trap characteristics, make carrier detachment easier, and significantly increase carrier mobility.

Keywords Polypropylene · Nanocomposite dielectric · Thermostatic discharge current · Trap characteristic

1.1 Introduction

With the rapid development of High Voltage Direct Current (HVDC) cables, polypropylene (PP) insulation materials have gained widespread application in dielectric materials due to their excellent electrical properties, thermal stability, and environmentally friendly, recyclable characteristics. However, traditional polypropylene dielectric materials tend to be brittle and prone to cracking, and their performance under high electric fields is often limited by factors such as dielectric loss, breakdown strength, and thermal stability [1–3]. To further improve the performance of

Z. Liu · X. Wang · H. Zhao · Y. Li · Z. Lei (✉)
Shanxi Key Laboratory of Mining Electrical Equipment and Intelligent Control, College of Electrical and Power Engineering, Taiyuan University of Technology, Taiyuan 030024, China
e-mail: leizhipeng@163.com

Y. Li et al. (eds.), *Proceedings of the 5th Chinese Conference on High Voltage and Discharge Plasmas*, Springer Proceedings in Physics 254,
https://doi.org/10.1007/978-981-96-5711-7_1

polypropylene dielectrics, researchers have introduced elastomers and nanoparticles into the polypropylene matrix [4, 5] to optimize both its mechanical and electrical properties.

Nanoparticles can not only effectively improve the dielectric constant and breakdown strength of polypropylene [6], but also influence its current discharge behavior by modulating the distribution of charge carriers and trap characteristics [7–11]. The dielectric trapping characteristics are closely related to the carrier migration process within the material. Therefore, investigating the isothermal discharge current process and trapping characteristics of polypropylene-based nanocomposite dielectrics is crucial for understanding the mechanisms by which nanoparticle fillers enhance dielectric properties. This, in turn, plays an important role in advancing the design and application of high-performance electrical materials.

This study aims to analyze the effect of adding $Al_2O_3@SiO_2$ core-shell nanoparticles on the trap characteristics of polypropylene-based nanocomposite dielectric materials through isothermal discharge current testing, revealing the optimization mechanisms by which nanoparticle fillers enhance the dielectric properties of the material.

1.2 Arrangement of Experiment

1.2.1 Specimen Preparation

Nanocomposite dielectric samples with varying contents of $Al_2O_3@SiO_2$ core-shell nanoparticles were prepared using melt blending and hot pressing methods. Prior to sample preparation, PPH-T03-S-F401 grade iPP from Yangtze Petrochemical, 6202FL grade polypropylene-based elastomer (PBE) from ExxonMobil, and $Al_2O_3@SiO_2$ core-shell nanoparticles prepared by the Stöber method were dried at 65 °C for 36 h to remove moisture. The temperature of the twin-roll mixer was adjusted to approximately 175 °C, and 8 g of PBE, 12 g of iPP, and varying amounts of core-shell nanoparticles were sequentially added near the active roller end. The PBE and iPP were mixed for 5 min, while the nanoparticles were gradually added to the iPP/PBE mixture in small amounts over 15 min to achieve a homogeneous blend. The prepared nanocomposite dielectric material was then placed with the mold in a hydraulic press, preheated at 195 °C for 5 min, and hot-pressed at 15 MPa for 5 min. The final samples had a thickness of 0.2 mm and a diameter of 100 mm. The mass fractions of core-shell nanoparticles added were 0, 0.5, 1, 2, and 4%.

1.2.2 TEM Test

The structure of the $Al_2O_3@SiO_2$ core-shell nanoparticles was analyzed using a Transmission Electron Microscope (TEM) model JEM-2010.

1.2.3 Isothermal Discharge Current Measurement Method

The isothermal discharge current was measured using a three-electrode system and a KEYSIGHT B2981A picoammeter, with the experimental circuit shown in Fig. 1.1. Under a temperature of 30 °C, the polarization and depolarization currents of the sample were measured sequentially at different applied electric fields (5, 10, 15, 20, 25, 30, 35, 40, 45, 50 kV/mm). The steady-state current density was calculated based on the steady-state current measured 15 min after the applied voltage. After each measurement, the sample was short-circuited for 15 min, and the depolarization current during the short-circuit discharge process was taken as the isothermal discharge current. This procedure was repeated for measurements at the next voltage level.

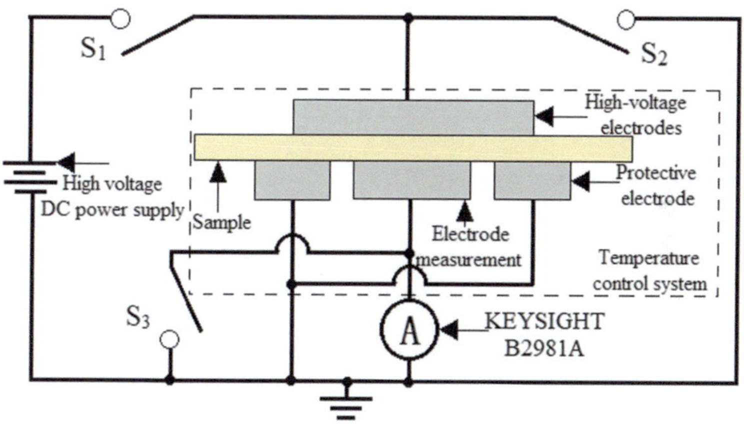

Fig. 1.1 Polarization depolarization current measurement system

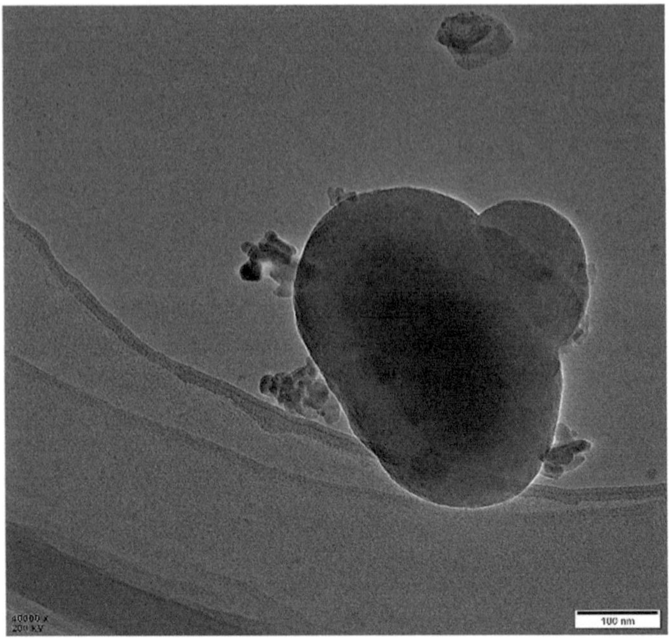

Fig. 1.2 TEM of single core-shell nanoparticles

1.3 Experimental Results and Discussion

1.3.1 Topography of Al_2O_3@SiO_2

Figure 1.2 shows the TEM morphology of Al_2O_3@SiO_2 core-shell nanoparticles. As can be seen from the image, the particle size ranges from approximately 250–450 nm. The Al_2O_3@SiO_2 core-shell nanoparticles exhibit distinct regions with varying grayscale contrast. The darker areas represent the Al_2O_3 core, while the brighter regions correspond to the SiO_2 shell.

1.3.2 Carrier Concentration and Mobility

Based on the quasi-steady-state conductive current in the polarization current, the quasi-steady-state current density J of the sample under various electric field intensities E can be obtained, as follows:

$$J = i_{\infty}(t)/S \tag{1.1}$$

In the equation, $i_\infty(t)$ represents the steady-state conductive current, and S is the area of the measurement electrode.

According to the space charge limited current (SCLC) theory, assuming that all traps are filled, the current density J can be expressed as [12]

$$J = 9\varepsilon_r\varepsilon_0\mu U^2/8d^3 \tag{1.2}$$

In the equation, ε_r is the relative permittivity of the sample; ε_0 is the permittivity of free space, taken as 8.85×10^{-12} F m; U is the applied DC voltage; and d is the thickness of the sample.

The carrier mobility can be estimated from Eq. (1.2). The carrier concentration can also be estimated from the electric field strength E and current density J [12].

$$n = 9E\varepsilon_r\varepsilon_0/8qd \tag{1.3}$$

In the equation, q represents the elementary charge of an electron, which is taken as $1.6 \times 10\,C^{-19}$.

Figure 1.3 shows the carrier concentration and carrier mobility of samples with different contents of core-shell nanoparticles under a field strength of 40 kV/mm.

As shown in the figure, the carrier concentration (n) of all the samples is in the same order of magnitude and exhibits little variation, indicating that the addition of core-shell nanoparticles has a minimal effect on the carrier concentration of the dielectric. The carrier mobility of the samples with 0.5 and 1% nanoparticle addition is significantly lower than that of the sample without core-shell nanoparticles, with the sample containing 1% showing the lowest mobility, approximately 8.82×10^{-18}

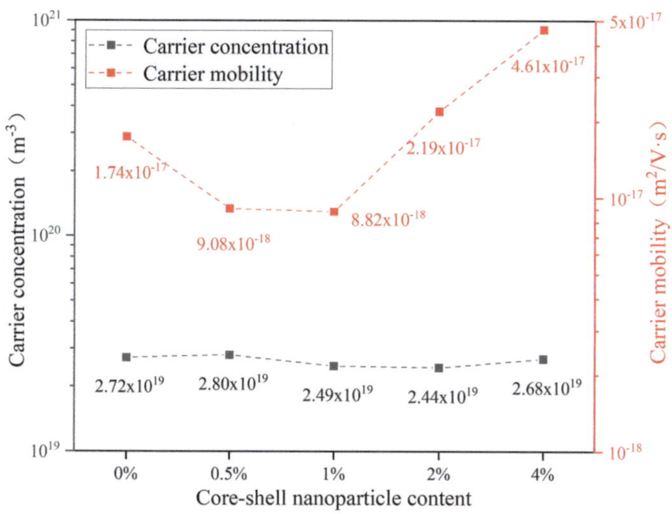

Fig. 1.3 Carrier concentration and carrier mobility

$m^2/(V\ s)$. On the other hand, the carrier mobility of the samples with 2% and 4% nanoparticle addition is higher than that of the sample without nanoparticles. The sample with 4% addition shows the highest carrier mobility, around 4.61×10^{-17} $m^2/(V\ s)$. The addition of a small amount of core-shell nanoparticles can reduce the carrier mobility of polypropylene-based composite dielectrics, while a larger amount increases the carrier mobility. This is because the addition of core-shell nanoparticles introduces new trapping sites.

1.3.3 *Trap Characteristics*

Figure 1.4 shows the isothermal discharge current of samples with different contents of core-shell nanoparticles under a 40 kV/mm DC electric field for the first 600 s. As seen in the figure, the decay rate of the isothermal discharge current slows down after the addition of core-shell nanoparticles. This is because the introduction of core-shell nanoparticles creates new trapping sites, which slow down the discharge rate. For samples with 0.5% and 1% nanoparticle addition, the initial value of the isothermal discharge current decreases, indicating that a small amount of core-shell nanoparticles introduces deep traps, thereby enhancing the deep trap characteristics. For samples with 2 and 4% nanoparticle addition, the initial value of the isothermal discharge current increases, suggesting that a higher concentration of core-shell nanoparticles introduces shallow traps, which weakens the deep trap characteristics and enhances the shallow trap characteristics. This may be due to the fact that high doping concentrations can cause nanoparticles to act as impurities, introducing physical defects, and these defect structures generate more free carriers (electrons or holes).

The isothermal discharge current method is a technique used to study the trap energy level density distribution in materials by measuring the variation of discharge current with time under constant temperature conditions. According to the model proposed by J.G. Simmons in 1973, there is a relationship between the isothermal discharge current I and the trap density-trap energy level distribution function $N_t(E)$ as follows [13]:

$$I = qSdk_B TN_t(E)/2t \tag{1.4}$$

In the equation, t is the depolarization time; k_B is the Boltzmann constant, taken as 8.568×10^{-5} eV/K; and T is the absolute temperature, taken as 303.15 K.

The relationship between the trap energy level depth E_t and the depolarization time t is given by [13]:

$$E_t = k_B T \ln(vt) \tag{1.5}$$

In the equation, v represents the electron vibration frequency, which is taken as 3×10^{12}/s.

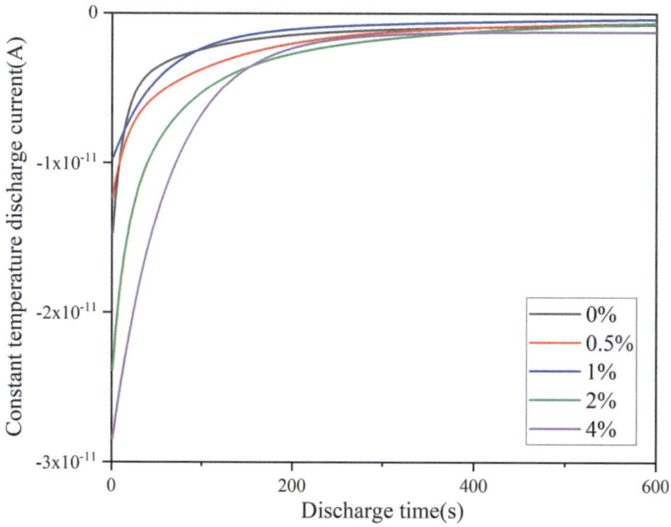

Fig. 1.4 Constant temperature discharge current

Based on Eqs. (1.4) and (1.5), the relationship between trap density and trap energy level can be expressed as:

$$N_t(E)/E_t = 2It/qSdk_B^2T^2\ln(vt) \tag{1.6}$$

Figure 1.5 shows the charge trapping characteristics of samples with different core-shell nanoparticle contents under a 40 kV/mm DC electric field. The trap energy levels between 0.74 and 0.9 eV can be considered shallow traps, while those between 0.9 and 0.94 eV represent deep traps. From the figure, it can be seen that when no core-shell nanoparticles are added (0% content), there is a central energy level at 0.886 eV, with a trap density at the trap center of $4.08 \times 10^{17}/(\text{eV m}^3)$. The addition of core-shell nanoparticles introduces a large number of new traps. When the content is 0.5% and 1%, deep trap centers are introduced at 0.906 eV and 0.916 eV, respectively. Compared to the sample without core-shell nanoparticles, adding 0.5% core-shell nanoparticles increases the trap density at both trap center energy levels, and the trap center energy level also shifts higher. At 1% content, the deep trap energy level reaches approximately 0.916 eV, with a trap density of about $7.58 \times 10^{17}/(\text{eV m}^3)$. At this point, both the trap center energy level and trap density increase, leading to a stronger trapping effect on charge carriers, making it more difficult for them to escape. As a result, the carrier mobility decreases. At 2 and 4% addition, the increase in core-shell nanoparticle content leads to a decrease in the trap center energy levels, and the trap density at the trap centers of both samples remains relatively unchanged. When the addition reaches 4%, the deep and shallow trap energy levels are both the lowest, at approximately 0.843 and 0.853 eV, respectively, with a shallow trap center density

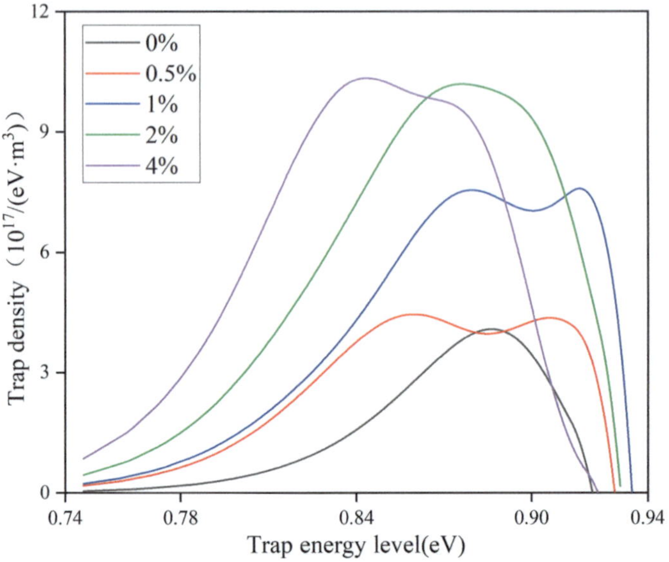

Fig. 1.5 Trap energy level and trap density

of $10.34 \times 10^{17}/(\text{eV m}^3)$. The reduction in both deep and shallow trap energy levels facilitates the escape of charge carriers, leading to an increase in carrier mobility.

1.4 Conclusion

The addition of nanoparticles introduces a large number of new traps, which slows down the decay rate of the isothermal discharge current. The addition of a small amount of core-shell nanoparticles introduces a significant number of deep traps, making the deep trap characteristics of the dielectric more pronounced. This leads to stronger carrier trapping, thereby reducing the carrier mobility. The addition of a large number of core-shell nanoparticles weakens the deep trap characteristics while enhancing the shallow trap characteristics, making it easier for carriers to escape from the traps.

References

1. Tian Fuqiang, Yang Chun, He Lijuan, et al. Recent advances in dielectric properties and mechanisms of polymer/inorganic nanocomposite dielectrics [J]. Transactions of China Electro-technical Society, 2011, 26(03): 1–12.
2. Du Boxue, Li Zhonglei, Zhou Shuofan, et al. Research progress and Prospect of Polypropylene High voltage DC cable insulation [J]. Journal of Electrical Engineering, 2021,16(02): 2–11.

3. Cheng Lu, Liu Hongbo, Liu Biao, et al. Research progress on property control of ternary polypropylene dielectric materials [J]. High Voltage Technology, 2024, 50(06): 2332–234.
4. Huang Bo. Research progress of polypropylene insulation for high voltage direct current cables [J]. Synthetic Resins and Plastics, 2019, 41(03): 63–69.
5. Ouyang Benhong, Huang Kaiwen, Zhao Peng, et al. Research progress of polypropylene materials for power cables [J]. Insulation Materials, 202, 55(08): 6–15.
6. Du Xinyao, Zuo Shengwu. Research progress of modified poly-propylene cable insulation materials [J]. Modern Plastics Processing and Application, 2019, 36(02): 54–57.
7. Xu Hang, Du Boxue, Li Jin, et al. Mechanical and Space Charge Properties of polypropylene/elastomer Composites [J]. High Voltage Technology, 2019, 45(10): 3214–3220.
8. He Qing-Hui, LI Wei, Men Ru-Jia, et al. Effect of tensile ratio on polarization-depolarizing current and trap distribution of ethylene-propylene rubber [J]. Insulation Materials, 2019, 56(06): 28–33. (in Chinese)
9. Lei Z, Fabiani D, Bray T, et al. Space charge behavior of quantum dot-doped polystyrene polymers[J]. IEEE Transactions on Dielectrics and Electrical Insulation, 2021, 28(3): 753–761.
10. Han Kunying, Men Rujia, Guo Meiqing, et al. Al2O3@SiO2 Effect of core-shell nanospheres addition on dielectric and space charge properties of polypropylene [J]. Materials Review, 2018, 38(06): 233–239.
11. Wang Si-Yu, Liu Yang, LEI Zhi-Peng, et al. Study on insulation properties of ethylene propylene rubber for cable under extrusion pressure and thermal stress based on polarization current [J]. Journal of Insulation Materials, 2019, 55(04): 56–61.
12. Kao K C. Electrical conduction and breakdown in insulating polymers[C]//Proceedings of the 6th International Conference on Properties and Applications of Dielectric Materials (Cat. No. 00CH36347). IEEE, 2000, 1: 1–17.
13. Simmons J G, Tam M C. Theory of isothermal currents and the direct determination of trap parameters in semiconductors and insulators containing arbitrary trap distributions[J]. Physical Re-view B, 1973, 7(8): 3706.

Chapter 2
A Review of Research on the Detection Methods of Composite Insulator Decay

Jinghao Chen, Songsong Zhou, Shenghui Wang, Xing Zhang, and Shijie Xu

Abstract With the large-scale use of composite insulators, decay-like fracture failure has occurred multiple times, affecting the normal operation of the power system. Therefore, appropriate detection methods for this failure are quite important. This article starts with the study of the mechanism responsible for decay-like failure, summarizes the commonly used detection methods, and provides several promising detection directions. Offline approaches mainly involve SEM (scanning electron microscopy), FTIR (Fourier transform infrared spectroscopy), TGA (thermogravimetric analysis), and XPS (X-ray photoelectron spectroscopy), while on-site detection approaches mainly rely on infrared detection. However, currently there are few samples of decay-like insulators detected by infrared spectroscopy, resulting to the incomplete algorithm model. Meanwhile, infrared images quality also has high requirements for the shooting environment. Actually, joint detection is a direction to improve diagnostic accuracy. Ultraviolet (UV) pulse, as a low-cost and easy-to-use technology, has the potential to play a certain auxiliary discrimination role in infrared diagnosis. Ultrasonic guided waves can effectively observe the debonding between the core rod and the shell, which is a significant reason for the interface failure of the decay-like composite insulator; Microwaves, due to their unique properties, will bring new and reliable ways for this detection.

Keywords Composite insulators · Decay-like failure · Offline detection · On-site inspection · Ultrasonic guided waves · UV pulse · Microwave

---

J. Chen · S. Wang
North China Electric Power University, Beijing, China

S. Zhou (✉)
China Electric Power Research Institute, Beijing, China
e-mail: z18516979216@163.com

X. Zhang · S. Xu
North China Electric Power University, Baoding, China

Y. Li et al. (eds.), *Proceedings of the 5th Chinese Conference on High Voltage and Discharge Plasmas*, Springer Proceedings in Physics 254,
https://doi.org/10.1007/978-981-96-5711-7_2

2.1 Introduction

Composite insulators are increasingly popularized in transmission lines with voltage levels over 110 kV due to their compact size, light-weight, low cost and superior electrical and mechanical properties [1, 2]. This kind of insulator is mainly composed of the core rod, the shed and the end fitting [3]. Consisting of glass fibers and an epoxy resin matrix, the core rod serves as the mechanical load-bearing component as well as the primary internal insulation of the insulator [4].

Fracture forms of composite insulators mainly include conventional fracture, brittle fracture, and recently discovered decay-like fracture. The mechanism of conventional fracture and brittle fracture have been clearly elaborated [5]. In addition, there are sufficient measures to ensure the composite insulators using new technologies can avoid these two types of fractures. Decay-like fracture is a newly named fracture form in China, precisely reflecting the haptic findings and describing the mechanical and electrical properties of the fractured core rod [6]. As composite insulators are extensively deployed, incidents of decay-like failure are on the rise. From 2010 to 2020, there were 22 decay-like failure cases within the scope of the State Grid, including 4 in Hunan, 4 in Shandong, 5 in Fujian, 3 in Zhejiang, 2 in Jiangsu, and 1 each in Hebei (northern Hebei), Shaanxi, Chongqing, and Henan [7]. Owing to the significant harm caused to the power system, there is an urgent need for effective detection methods to identify the decay-like fracture of composite insulators. This paper reviews the exploration of the decay-like failure in recent years, summarizes the common offline and on-site detection methods for decay-like composite insulators, and proposes potential detection approaches in the future.

2.2 The Development Mechanism of Decay-Like Fracture

Lutz et al. [8] discovered and described a new fracture characteristic that differs from brittle fracture. He believed the reason for the deterioration of composite insulators was the weak adhesion detected between the housing and core rod, namely interface failure. Wang et al. [9] conducted a study on domestic cases and translated the failure as 'decay-like fracture failure'. He concluded that this fracture was more than a mechanical failure and the main causes of this failure may be the internal discharge and the current effects at the interface. Lu et al. [10] found that the formation process of decay and aging is related to partial discharge induced by internal defects in composite insulators. Long term partial discharge will lead to insulation heating, as well as surface corrosion of the core rod glass fiber, oxidation and decomposition of epoxy resin, deterioration of silicone rubber, and generation of nitrate ions, resulting in a degradation in the mechanical properties of the core rod. In terms of defect formation, A. Kutil investigated the defection process at the interface between silicone rubber and epoxy resin under the combined action of stress and partial discharge. Ueta [11] agreed that with rising frequency, the formation of interface defects caused by discharge can be accelerated. Guo [12] simulated the development process of shed corrosion holes based on the spherical growth mode of silicone rubber electric tree branches. He convinced that initial defects appearing at the minimum position of the original axial electric field and close to the high-voltage end are prone to develop into transverse corrosion holes. Additionally, the longitudinal development speed of erosion holes is faster than the transverse development process.

Liang [6] concluded that the degradation of epoxy resin was the direct cause of the decay-like fracture of the core rod in the first paper of their series. He generalized the mechanism as: current erosion of the epoxy resin matrix, production of acidic media due to discharging under damp conditions media and "cutting" of glass fibers by acidic media and mechanical stress. In the second paper, Gao [13] simulated and reproduced the degradation and deterioration of the epoxy resin matrix in the decayed fracture core rod, further verifying that current and discharge are the causes of the degradation and deterioration of the epoxy resin matrix in the decayed fracture core rod. Moreover, after dissecting and analyzing decay-like composite insulator and internal breakdown composite insulators, Lu Ming [14] confirmed that the development process of decaying and aging advanced simultaneously along the axial and radial directions. The deterioration along the axial direction was manifested as a worse electrical performance of the core rod, while the one along the radial direction was described as poor mechanical performance. There must be one that first drops to the critical value of the fault, resulting in corresponding malfunction.

Yuan [15] investigated the degradation characteristics and mechanisms of GRP (glass fiber reinforced epoxy resin polymer) materials in high humidity environments, revealed the failure process of GRP materials under humid and hot conditions. Analyzing the growth characteristics of water branches and electric branches in an electric field under humid and hot conditions, he divided the abnormal heating of composite insulators into four stages: end interface degradation, core rod hydrolysis, epoxy resin oxidation, and core rod pre fracture. He was convinced that the above process would eventually develop into decay-like fracture failure. Xie [16] believed that under humid and hot conditions, interface defects can accelerate the aging process of insulators, leading to interface failure. As interface defects gradually expand, epoxy resin will undergo oxidative decomposition from the interface to the interior, exposing a large amount of glass fibers and bringing about degradation. Along the way, partial discharge and polarization loss of moisture impurities at the interface can make a contribution to abnormal heating faults.

Zhong [17] carried out research on the role of external moisture, alternating loads, and partial discharge in the decay-like failure of core rods. He concluded that the diffusion process of external moisture inside the core rod material conforms to Fick's law. When there is a large amount of wet stress inside the core rod for a long time, it will lead to debonding of the fiber matrix interface, resulting in microcracks and micropores. Under the action of alternating loads, the outer side of the curved arc on the surface of the core rod will first experience matrix fatigue damage and fiber fracture. Due to the cumulative effect of fatigue, the damaged area will expand from the surface of the core rod to the inside. After micro defects appear at the core rod and interface, partial discharge will be triggered, causing degradation of the epoxy resin matrix. As the degradation of the core rod worsens, discharge along the surface of the core rod and electric branches developing towards the interior of the core rod will be further triggered, eventuating partial carbonization.

2.3 Existing Detection Methods

On account of the late start of research on the decay-like composite insulator, answers to key issues such as fracture characteristics, fracture mechanisms, fracture influencing factors, and preventive measures were not comprehensive at first. Thus, offline detection became the main means in the early stage. With deeper understanding, people gradually try online detection in order to timely identify the decay-like composite insulators and reduce potential hazards.

2.3.1 Offline Detection

Scanning Electron Microscope. Scanning Electron Microscope (SEM) uses an electron beam as an illumination source to focus a finely focused electron beam onto a sample in a grating like scanning manner. Secondary electrons and backscattered electrons generated by the interaction between the electrons and the sample are collected and processed to obtain a magnified microstructure image [18]. SEM plays an important role in fracture failure analysis, observation of material microstructure morphology, and composition analysis. Observing the severely deteriorated parts of the decay-like composite insulator core with SEM, Wang [9] found that the glass fibers were not covered by resin and had been exposed to the air. The surface values of the glass fibers in the cracked parts were also severely damaged. Mohammad [19] explained the reason for the large contact angle through surface roughness of aged insulators magnified observed by SEM (Fig. 2.1).

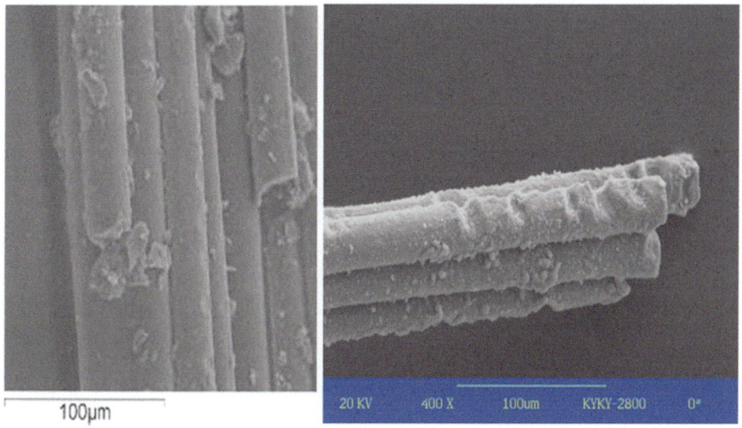

Fig. 2.1 Observation of the microstructure of insulators using SEM

Fourier transform infrared spectroscopy. Fourier transform infrared spectroscopy (FTIR) is a technique based on the absorption characteristics of substances towards infrared radiation, which obtains information on the molecular structure and chemical composition of samples by analyzing their infrared spectra. Different molecules have different vibrational energy levels and absorb specific wavelengths of light in the infrared spectral region (wavelength 10–3 ~ 1 mm). As such, different chemical bonds are supposed to produce different absorption frequencies under infrared light irradiation. Test the characteristic peaks of the substance's absorption infrared spectrum and find the chemical bonds corresponding to absorption peaks at different positions, then deduce the possible functional groups and types of chemical bonds in the sample [20, 21]. The main components of the composite insulator core rod are bisphenol A glycidyl ether epoxy resin, some acid anhydride curing agents, and silane coupling agents containing ester groups. Understanding this information and comparing the infrared spectra of each sample, Zeng [22] analyzed the decomposition of polydimethylsiloxane and flame-retardant aluminum hydroxide during the aging process. Alok Ranjan [23] carried out a comparative analysis and compared FTIR spectroscopy between fresh and aged samples. He found that some peaks were reduced, manifesting loss of the main chain (Si–O–Si) and the reason for loss of material properties (Fig. 2.2).

Thermogravimetric Analysis. Thermogravimetric Analysis (TGA) is a technique for measuring the relationship between the mass of a substance and temperature changes under program-controlled temperature. It is widely used to investigate the thermal stability, composition, and thermal reaction changes of materials. TGA can record the changes in sample weight with rising test temperature, thereby obtaining the relationship between weight loss rate and temperature or time. As for decay-like

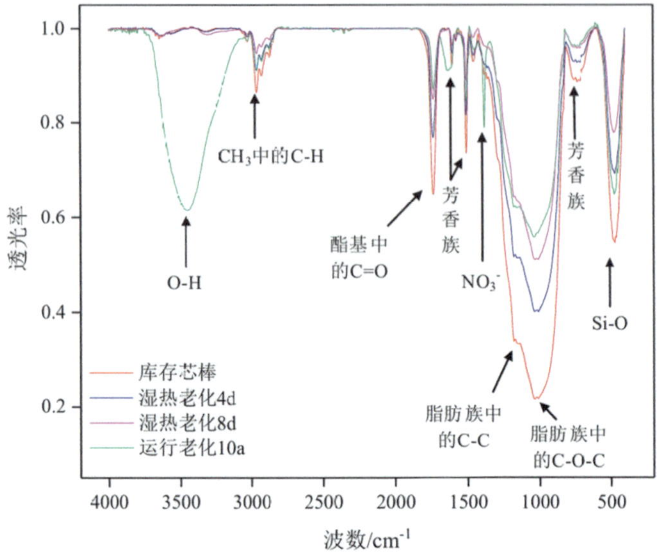

Fig. 2.2 FTIR analysis results of the sample

failure, TGA can be used to study the thermal stability and composition changes of composite insulators. It helps identify quality loss caused by thermal aging and chemical degradation. In composite insulator core rods, glass fiber, as an inorganic fiber, has excellent heat resistance and has the characteristics of non-combustion and non-oxidation. In fact, its softening point is between 550 and 750 °C and melting temperature is 1600 °C. The thermal weight loss process of fiberglass materials below 1000 °C is a result of the thermal weight loss of the epoxy resin matrix. Therefore, TGA can be a method for measuring the glass fiber content in fiberglass core rod [6]. According to the obtained results of TGA, I. Ahmadi-Joneidi et al. [24] figured out the reasons of two weight loss belonged to degradation of LMWS, water, and volatile elements and break of the polymeric chains respectively (Fig. 2.3).

X-ray photoelectron spectroscopy. X-ray photoelectron spectroscopy (XPS) is a surface analysis technique based on the principle of photoelectric effect, which obtains XPS spectra by measuring the kinetic energy of electrons emitted from the material surface. This technology can accurately measure the inner electron binding energy and chemical displacement of atoms and provide information about chemical bonds. The principle of XPS is to utilize the characteristic of X-rays to excite the inner shell electrons of elements. Due to the binding energy of the inner shell electrons being related to their element type, chemical bonding state, etc., the photoelectron peak of the inner shell electrons shifts with the chemical composition and element type of the element. The correspondence between the different photoelectron peaks

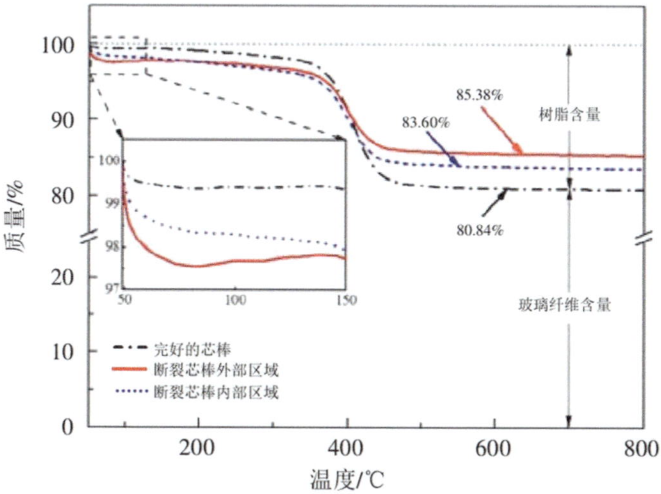

Fig. 2.3 TGA curve of a certain core rod

generated by X-ray excitation of inner layer electrons and the elemental chemical states can be used for qualitative and quantitative analysis of the elemental composition and chemical states of materials [25]. Gao [26] discovered through XPS that the peak areas corresponding to the C–C and C–H chemical bonds in the non-decayed region were larger than those in the decayed region. The envelope area of the C–COO peak measured in the decayed region sample remarkably increased, indicating an increase in oxygen-containing chemical bonds and oxidation processes on the surface of the core rod during the decay process (Fig. 2.4).

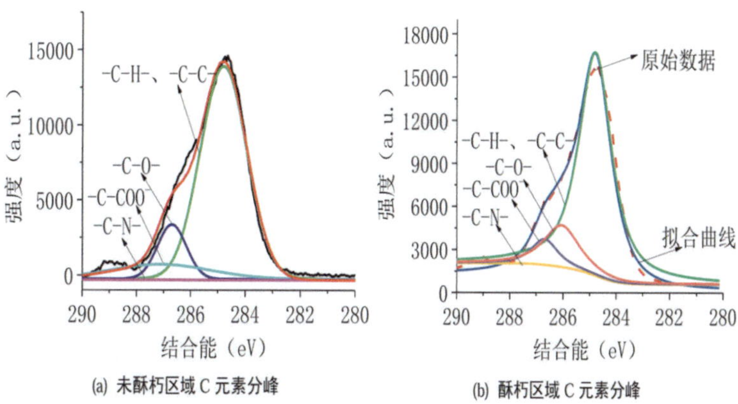

Fig. 2.4 Schematic diagram of XPS analysis results

2.3.2 *Online Detection*

Lu et al. [27] suggests that during the decay process of composite insulators, the thermal and chemical effects of local currents, as well as the polarization loss of moisture under strong electric fields, release a large amount of heat, contributing to the abnormal local heating of composite insulators at the macroscopic level. Partial discharge can also increase the dielectric loss factor of insulation materials, affecting their temperature rise [17]. Thereby, infrared detection of actual operating composite insulators is an effective way of detecting potential composite insulator decay-like fracture accidents.

Infrared thermal imager is an instrument that detects object radiation and establishes a mutual relationship between radiation and surface temperature, even in the completely dark detection environment. Infrared specific band signals of object surface thermal radiation are detected by photoelectric technology and converted into images that can be visually distinguished. Then temperature values are carefully calculated and represented by various colors. Finally, this thermal image can correspond to the thermal distribution field on the surface of the object (Figs. 2.5 and 2.6).

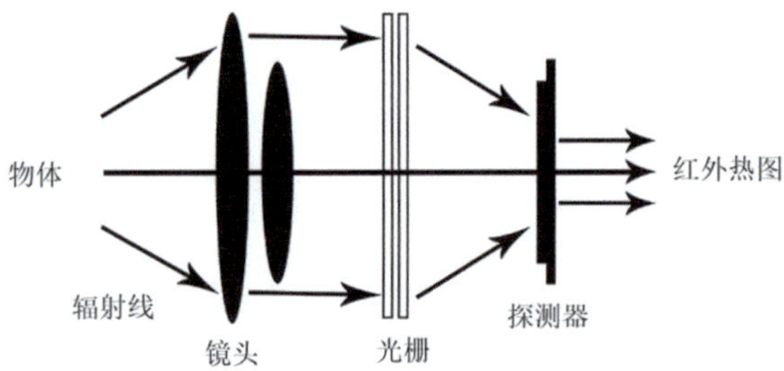

Fig. 2.5 Infrared imager

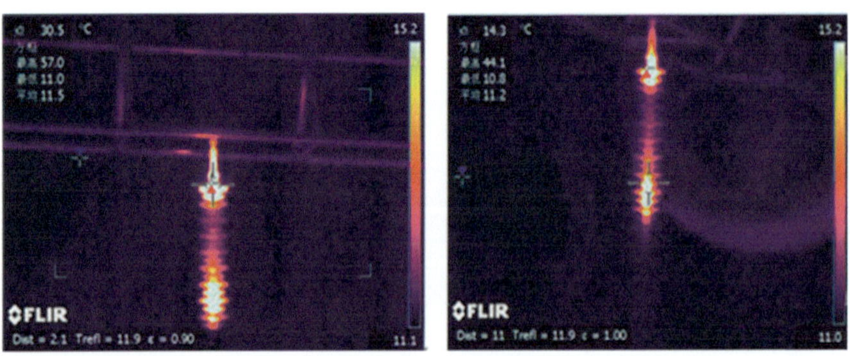

Fig. 2.6 Infrared detection image

Compared with traditional handheld infrared testing, drone infrared testing is more efficient and convenient [28, 29]. the problem of insufficient summary of infrared spectrum features faced by drone infrared testing has been alleviated, and the lack of on-site testing standards for drone infrared has been solved. Since 2022, Jiangsu Electric Power Research Institute has built a comprehensive infrared data sample library for unmanned aerial vehicle and helicopter inspections, collecting over 200,000 infrared samples, covering various types of line components such as insulators, wires, and fittings, Furthermore, they obtained recommended parameters such as infrared detection distance and pixel density on the basis of the unmanned aerial vehicle inspection experience. In 2021, Liu et al. [30] conducted unmanned aerial vehicle (UAV) infrared testing on the abnormal heating defects of composite insulators on 500 kV transmission lines at different testing distances. By normalizing, the variation law of temperature difference test amplitude of heating insulators with testing distance was obtained. A testing distance control method suitable for UAV operation on site was proposed based on the analysis of infrared test lens parameters, testing distance, and the proportion of tested insulators on the infrared test screen. Besides, Zhejiang Electric Power Research Institute led the preparation of the "Guidelines for Infrared Testing of Composite Insulators for Overhead Lines by Unmanned Aerial Vehicles" (T/ZJSEE 0015-2023) in 2023, which standardized the testing standards for on-site infrared testing of composite insulators for overhead lines by unmanned aerial vehicles. The State Grid also issued the "Guidelines for Infrared Detection of Over-head Transmission Lines" in the same year.

However, the abnormal temperature rise of composite insulators is not only caused by the deterioration defects of the core rod, but also by surface fouling, aging and moisture of the sheath, which can contribute to abnormal heating of composite insulators [31]. For example, Cheng [32] conducted experimental research on the heating of air gap defects inside the core rod. They found that when the heating amplitude on the surface of the sheath reached 1.4 K, the temperature difference amplitude generated by the heating inside the sheath had reached 207 K, being sufficient to able to trigger the decomposition of silicone rubber. For internal conductivity defects or large areas of core rod moisture, the temperature difference between the inside and outside of the housing is limited. Infrared is an effective means of detecting defects [33]. Moreover, the heating caused by moisture or mild contamination of the composite insulator shell is limited to the high voltage end, and the heating amplitude decreases with decreasing humidity [34, 35].

Maintenance decisions and emergency response measures are not the same for different types of abnormally heated composite insulators, so it is necessary to distinguish the infrared manifestations of these types. In the "Guidelines for Infrared Detection of Overhead Transmission Lines", the State Grid provides intuitive methods for distinguishing surface heating, surface fouling, and interface defects of the shell in infrared images, such as observing the heating position, amplitude, and other insulator conditions in the same area. Defining temperature characteristic quantities and

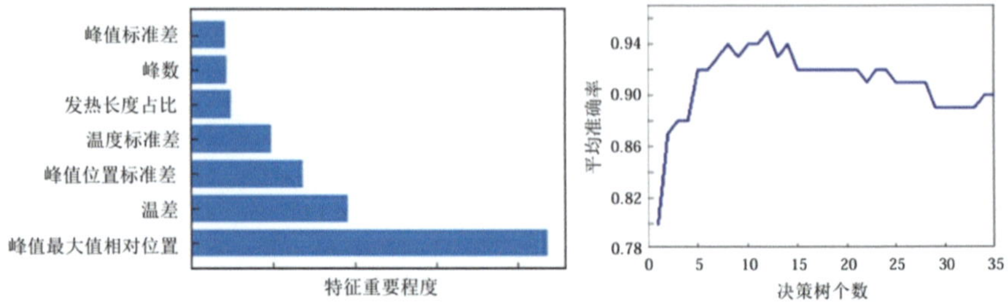

Fig. 2.7 The importance ranking of temperature characteristic quantities and the variation trend of algorithm accuracy with the increase of the number of decision trees

combining them with algorithms to characterize the abnormal state of composite insulators is also a viable option. In 2022, Xu [36] analyzed the temperature curves of the central axis on the surface of composite insulators with different types of abnormal heating. He extracted five characteristic quantities including temperature difference, temperature standard deviation, relative position of maximum temperature, peak number, and proportion of heating length to characterize the heating state of composite insulators and classify the types of abnormal heating. In the same year, Zhong [17] defined temperature gradient characteristic coefficient in the literature and proposed a multi feature quantity infrared diagnosis method for decay-like composite insulators on account of decision trees. This method can effectively distinguish the decay-like failure from aged housing, while these two kinds have similar degree of temperature rise and characteristics of the temperature rise area. In 2024, Hu [37] analyzed the surface central axis temperature curves of composite insulators with surface fouling, shell aging and moisture, and core rod decay and heating. Temperature characteristic quantities were defined to characterize the abnormal heating state of composite insulators, and a classification model for abnormal heating composite insulators based on random forest algorithm was proposed with a classification accuracy of over 85% (Fig. 2.7).

However, the accuracy of directly searching for features in infrared images or combining algorithms to identify temperature feature quantities to characterize abnormal heating states is seriously related to the quality of infrared images. The shooting of infrared images, whether for general testing or retesting, has high requirements for the environment. Background interference will greatly affect its quality and limit reliable shooting time. In addition, there are currently few samples of decay-like insulators for infrared detection, and the various classification models of the algorithm are unified and imperfect. To improve diagnostic accuracy, other detection methods are needed to assist.

2.4 Promising Detection Methods in the Future

2.4.1 Ultrasonic Guided Waves

Interface debonding is the starting point of decay-like failure. And poor adhesion between the core rod and thc sheath is a fundamental reason for poor interface performance, which is usually caused during the production process due to non-compliance with production requirements, such as temperature or improper handling and application of primer. Ultrasonic guided waves can play a certain role in detecting interface adhesion [38].

Due to the characteristics of single ended excitation, long propagation distance, and small attenuation along the propagation path, ultrasonic guided waves are widely used in non-destructive testing of industrial pipelines and anchor rods [39, 40]. Compared with traditional ultrasonic point scanning, ultrasonic guided waves perform line scanning, causing particle vibrations that can spread throughout the interior and surface of the component, thus detecting defects inside and outside the component. Not only does it improve detection efficiency, but it also reduces the risk of missed detections. Meanwhile, due to the presence of defect size information in guided wave signals, quantitative estimation of defect size can be made by extracting their feature quantities [41].

In addition to defect echo signals, the signals collected in ultrasonic non-destructive testing also contain various types of noise signals. It is mainly divided into incoherent noise and coherent noise. Noncoherent noise is mainly caused by the environment and is usually represented by the Gaussian white noise model, while coherent noise is caused by the material itself, which is relatively random and chaotic, and may completely mask the defect echo signal. Therefore, it is necessary to process the collected signal [42]. For example, singular value decomposition or independent component analysis can be used to separate crack signals from background noise [43]. Sparse signal reconstruction method can also be used to iteratively project the signal into a redundant waveform dictionary, select the waveform that best matches the pre analyzed signal, and effectively extract the most meaningful features from the signal [44]. Furthermore, it is proposed that the attenuation coefficients of L-mode guided waves and T-mode guided waves can be used to determine whether composite insulators have debonding defects [45].

However, it is hard for ultrasonic testing to use online. Because it needs for tight contact between the ultrasonic probe and the material, as well as sufficient coupling agents to assist in transmitting ultrasonic waves into the sample. Therefore, ultrasonic guided waves can be used as an offline means to detect whether the interface of composite insulators has failed (Fig. 2.8).

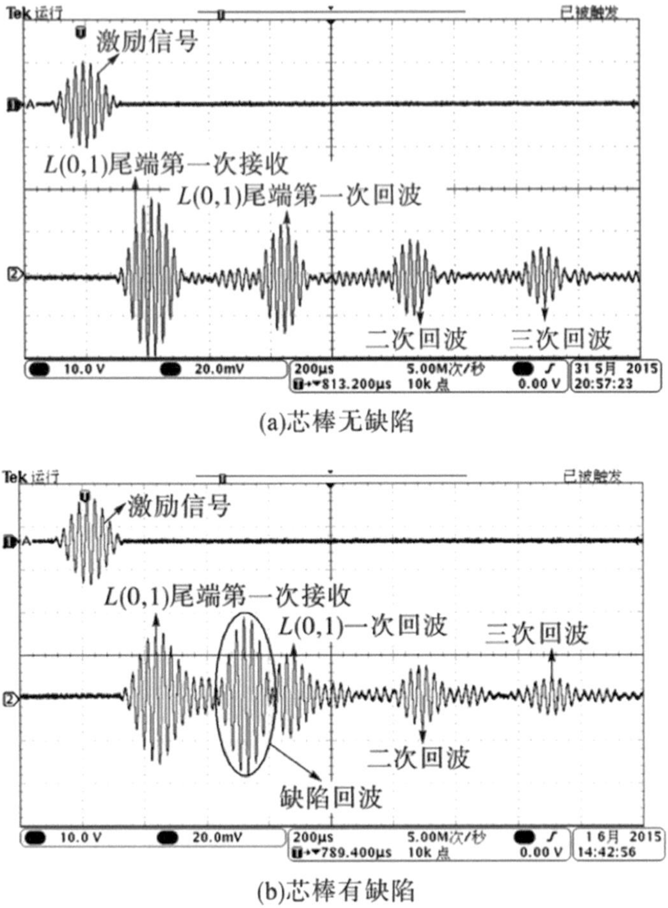

Fig. 2.8 Experimental waveform of guided wave propagation in core rod

2.4.2 *Ultraviolet Pulse*

As a non-contact and high-precision approach, ultraviolet (UV) pulse method can be used to assist infrared detection. During the operation of contaminated composite insulators, the pollution layer will change the electric field distribution on the surface of the insulator, increase the local electric field strength, and lead to partial discharge [46]. This technology is essentially the detection of the energy change of the discharge pulse on the surface of the insulator when the potential distribution of the insulator changes. The ultraviolet pulse detection method uses ultraviolet sensors—ultraviolet phototubes—to collect ultraviolet light emitted during insulator discharge under the

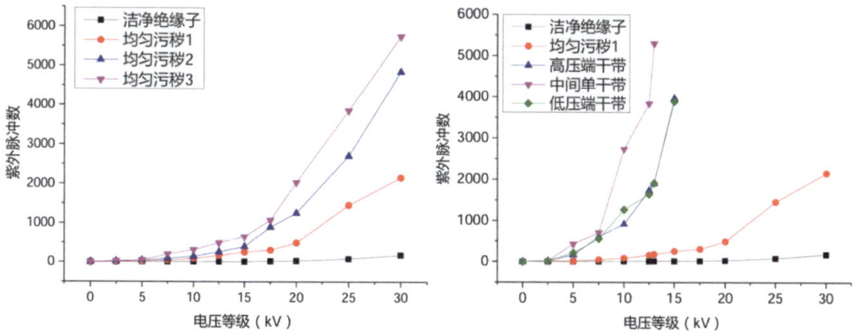

Fig. 2.9 The number of ultraviolet pulses at different levels of pollution and clean dry belt positions under AC voltage

premise of shielding environmental noise interference. Then convert it into current pulse signals, and count the number of effective pulse signals per unit time as the basis for judging the intensity of ultraviolet light, reflecting the pollution status and discharge intensity of insulators [47].

It is confirmed that UV pulse can be used to accurately determine the external insulation condition of high-voltage electrical equipment and achieve non-contact quantitative detection of corona discharge in high-voltage equipment [48, 49]. Wang [50] studied the relationship between the power of corona discharge UV pulses and the surface electric field strength of high-voltage equipment. They measured the number of ultraviolet pulses on the equipment with the UV sensor and achieved the evaluation of the surface electric field of high-voltage equipment. Besides, Wang et al. [51] proposed a method for detecting corona discharge of polluted insulators using UV sensors, and developed an insulator pollution ultraviolet detector to achieve ultraviolet pulse detection of insulator discharge. References [52, 53] studied the relationship between the state of polluted insulators and the number of ultraviolet pulses, and established an insulator discharge UV pulse detection system.

Decay-like composite insulators usually discharge slightly, so the UV pulse method can be used to detect whether there is a significant discharge situation in composite insulators with abnormal heating. If there are many UV pulses, abnormal heating is likely caused by surface contamination of the composite insulator. For cases where both decay and surface fouling co-exist, one can try to distinguish them based on the number of pulses received (Fig. 2.9).

2.4.3 Microwave Detection

Microwave thermal imaging method. For composite insulators with minor internal defects in the early-stage of decay-like failure, their heating and surface discharge

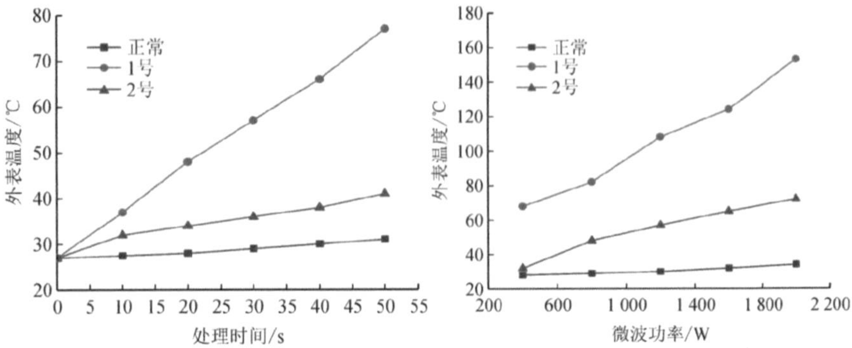

Fig. 2.10 Temperature variation of insulator samples with microwave treatment time/power

phenomena are relatively mild, as such the infrared inspection effect is not significant. Microwave heating can help infrared diagnosis of early decay-like composite insulators. The structure of composite insulators is a double-layer structure of silicone rubber coated fiberglass core rods. When there are decay defects at the interface, their electrical parameters exhibit high dielectric loss and conductivity [54]. Thus, when microwaves are vertically incident, their internal propagation characteristics will change owing to the presence of defects at the interface. Compared with the propagation of microwaves in conductors, the attenuation constant of microwaves propagating in insulating media is small, indicating that the vast majority of microwaves can pass through the insulating media without being absorbed. Therefore, when there are decay defects inside the composite insulator, microwaves can penetrate the sheath with minimal energy loss. The vast majority of microwave energy will be absorbed by the decay defects and converted into internal energy in the form of polarization loss, generating heat. The characteristic that microwaves can heat up decay defects while the temperature rise in other parts is relatively small, can be utilized to clearly observe decay defect location in infrared thermography. Thus, the detection of decay defects can be assisted through microwave heating [55] (Fig. 2.10).

Microwave humidity measurement. Among the several types of abnormal heating, except for decay, surface moisture and sheath surface heating all vary significantly with humidity. In low humidity conditions, the heating amplitude of these two situations is significantly reduced. Microwave humidity measurement can be used to detect the surface humidity of abnormal heating insulators. If the surface humidity is low, the temperature rise is roughly caused by decay-like failure. The invasion of water vapor after interface failure is an important reason for the decay of composite insulators [56], and the polarization loss it brings is manifested as abnormal temperature rise. By detecting the moisture content of composite insulators, it is possible to determine whether there is decay-like failure. The principle behind microwave moisture measurement technology lies in exploiting the correlation between the dielectric constant of materials and changes in moisture content. By transmitting and receiving microwaves of a specific frequency into the material, and measuring physical parameters such as power attenuation, phase change, and resonant frequency, the actual moisture content of the material can be inferred.

Microwave hydrometallurgy can be used to study the distribution patterns of water bodies at different heights and depths from multiple levels. Multi-level scanning provides direct data support for disease mechanisms and their distribution patterns. In civil engineering, the microwave moisture measurement non-destructive testing method is commonly used to evaluate concrete, determine structural degradation and abnormalities. For example, scholars such as Kääriäinen [57] have used microwave moisture meters to measure and study the moisture content in building materials. In the field of cultural relic protection, there are also related applications, such as Scott Allan Orr et al. [58] using micro-wave moisture measurement and radar technology to compare and study moisture monitoring in stone masonry on granite walls and sandstone towers, and obtaining the applicability conditions of the two methods in different situations; Scholars such as Dai [59] analyzed the St. Mary's Snow Hall Church in Macau using microwave moisture measurement and infrared thermal imaging techniques, and confirmed that the main cause of mural lesions was the severe lack of compactness of the outer protective body of the church (Fig. 2.11).

Microwave thermoacoustic imaging. Microwave thermoacoustic imaging is widely used in medicine. Based on the dielectric properties of tissues, microwave absorption and reflection affect the generation and propagation of thermoacoustic signals. When pulsed micro-wave is irradiated onto biological tissue, the absorbing bodies in the biological tissue absorb microwave energy, generate instantaneous temperature rise, then generate sound waves through thermoelastic effect, namely thermoacoustic signals. Through inverting the thermoacoustic signals, the distribution images of different absorbing bodies in biological tissue can be obtained [60].

Different brain tissues have different dielectric properties [61]. The high sensitivity of the microwave thermoacoustic imaging technology to changes in tissue dielectric differences and electromagnetic properties caused by tissue lesions enables it to obtain structural and functional information of tissue related areas, map the electrical properties of tissues, and form complete tomographic imaging of tissues [62]. Therefore, microwave thermoacoustic imaging technology can clearly distinguish various regions of the brain, joints, and adjacent tissues. Chi et al. [63] successfully used microwave thermoacoustic imaging technology to detect rheumatoid arthritis in finger joints, and the imaging results were highly consistent with magnetic resonance imaging (MRI) results, proving that microwave thermoacoustic imaging technology is able to effectively reconstruct the soft and hard tissues of interphalangeal joints. However, there are still certain challenges in terms of imaging quality and technological maturity at present.

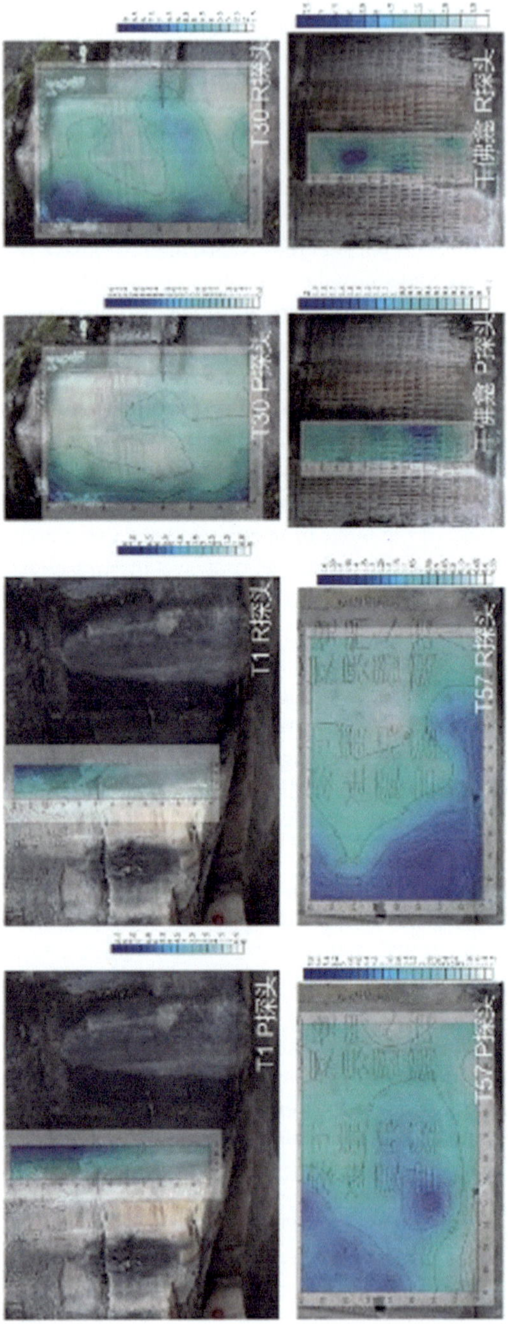

Fig. 2.11 On site detection results of microwave moisture measurement for stone carvings

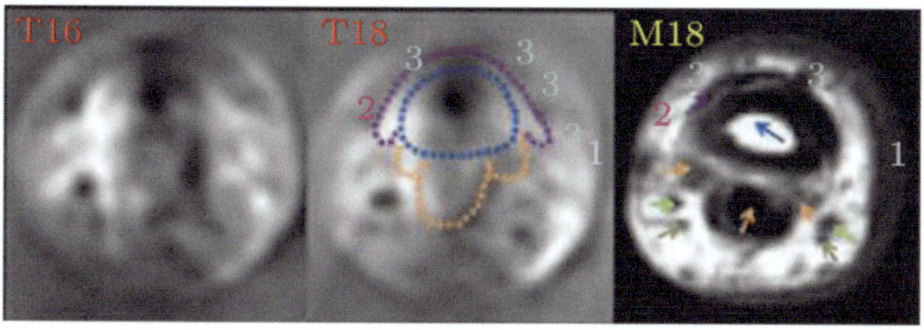

Fig. 2.12 Thermoacoustic image of the tester's middle finger

Microwave thermoacoustic imaging combines the advantages of microwave and ultrasound, and with the addition of coming deep learning algorithms to process received information, internal images can be more clearly reconstructed. In the future, this technology can be a candidate to diagnose whether there is a tendency for decay inside composite insulators (Fig. 2.12).

2.5 Conclusion

This paper starts with the mechanism of decay-like failure in composite insulator, summarizes the commonly used offline and online detection methods, and provides several possible future detection directions.

Offline methods mainly involve SEM, FTIR, TGA, and XPS. These methods directly and deeply reflect the changes before and after decay from the microstructure of the core rod, absorption peaks of chemical groups, epoxy resin content, component mass, and element content. The existing mature on-site detection methods mainly rely on infrared detection, and unmanned aerial vehicle infrared detection technology and feature quantities to characterize abnormal heating and distinguish various types have gradually improved. However, currently there are few samples of decay-like insulators detected by infrared spectroscopy, resulting to the incomplete algorithm model. Meanwhile, infrared images quality also has high requirements for the shooting environment. Actually, joint detection is a direction to improve diagnostic accuracy. Ultraviolet (UV) pulse, as a low-cost and easy-to-use technology, has the potential to play a certain auxiliary discrimination role in infrared diagnosis. Ultrasonic guided waves can effectively observe the debonding between the core rod and the shell, which is a significant reason for the interface failure of the decay-like composite insulator; Microwaves, as a commonly used tool in various fields such as military, cultural relics, civil engineering, and medical, will bring new and reliable ways for this detection due to their unique properties.

Acknowledgements This work is supported by the State Grid Corporation of China Headquarters Management Technology Project Funding (No. 5500-202455305A-1-2-LZ).

References

1. Liang X D, Wang S W: Development of composite insulators in China. IEEE Transactions on Dielectrics and Electrical Insulation 6(5). 586–594 (1999).
2. Power Supply Research Office of the Academy of Railway Sciences, High Voltage Teaching and Research Office of Tsinghua University: Research on the Electromechanical Performance of Composite Insulators. High Voltage Technology 04, 2632 (1983).
3. LIANG Xidong: Research on 500 kV Composite Insulators. Tsinghua University, China (1990).
4. GUAN Zhicheng, LIU Yingyan, ZHOU Yuanxiang: External Insulation of Insulators and Transmission and Transformation Equipment. Tsinghua University Press, Beijing, China (2006).
5. LIANG Xidong, DAI Jianjun: Study on ultrasonic detection of brittle fracture cracks in fiberglass core rods for insulators. Chinese Journal of Electrical Engineering (03), 112–116 (2005).
6. LIANG Xidong, GAO Yanfeng: Study on Decay Fracture of Composite Insulators (Part 1): Main Characteristics, Definitions, and Criteria of Decay Fracture. Chinese Journal of Electrical Engineering 36 (17), 4778–4786 (2016).
7. LU Ming, GAO Chao: Study on the Distribution Pattern of Composite Insulator Decay and Fracture Accidents. High Voltage Apparatus 60 (02), 133–142 (2024).
8. Lutz B, Cheng L, Guan Z C: Analysis of a fractured 500 kV composite insulator-identification of aging mechanisms and their causes. IEEE Transactions on Dielectrics and Electrical Insulation 19(5), 1723–1731 (2012).
9. WANG J F, LIANG X D: Failure Analysis of Decay-like Fracture of Composite Insulator. IEEE Transactions on Dielectrics and Electrical Insulation 21(6), 2503–2511 (2014).
10. LU Ming, ZHANG Zhonghao: Analysis of the Causes of Decay, Heating and Aging of Composite Insulators. Power Grid Technology 42 (04), 1335–1341 (2018).
11. Genyo U, Junichi W: Insulation characteristics of epoxy insulator with internal crack-shaped micro-defects-study on the equivalence of accelerated degradation by frequency acceleration test. IEEE Transactions on Dielectrics and Electrical Insulation 21(3), 1216–1225 (2014).
12. GUO Jun'an, LI Li: Simulation and Analysis of the Formation Process of Electric Corrosion Holes in the Sheath of Decayed Composite Insulators. Electric Porcelain Surge Arrester (03), 140–147+156 (2024).
13. GAO Yanfeng, LIANG Xidong: Study on Brittle Fracture of Composite Insulators (II): Experimental Simulation and Discussion on Preventive Measures for Brittle Fracture. Chinese Journal of Electrical Engineering 36 (18), 5070–5077+5132 (2016).
14. LU Ming, WU Chuan: Correlation analysis of brittle fracture and internal breakdown faults of composite insulators. Electric Porcelain Surge Arrester (01), 157–164 (2022).
15. YUAN Zhikang: Degradation characteristics and mechanism of composite insulator materials under high humidity environment. North China Electric Power University, China (2019).
16. XIE Congzhen, LI Yu: Study on the interface aging mechanism of composite insulators containing core rod sheath interface defects in humid and hot environments. Insulation Materials 57 (04), 72–81 (2024).
17. ZHONG Zheng: Study on the Mechanism and Temperature Rise Characteristics of Core Rod Decay of Composite Insulators under High Humidity Environment. North China Electric Power University, China (2022).
18. YU Lingzhu, LU Jian. Basic Principles and Applications of Scanning Electron Microscopy. Experimental Science and Technology 17 (05), 85—93 (2019).
19. Mohammad G, Leila S K: Ageing evaluation and lifespan estimation of silicone rubber insulators in MV power distribution networks. Engineering Failure Analysis 161, 108289 (2024).

20. XIE Congzhen, ZENG Leilei: Thermal decomposition characteristics of composite insulator core rod based on thermogravimetric infrared spectroscopy. Journal of Electrical Engineering 33 (S1): 227–233 (2018).
21. WENG Shifu. Fourier Transform Infrared Spectroscopy Analysis. 2nd edition. Chemical Industry Press. Beijing, China (2010).
22. ZENG Leilei: Research on Abnormal Heating and Damp Heat Aging Characteristics of Composite Insulators. South China University of Technology, China (2019).
23. Alok R V, Subba R: Aging studies on polymeric insulators under DC stress with controlled climatic conditions. Polymer Testing 68, 185–192 (2018).
24. Iman A J, Amir A S A: Lifetime prediction of 20 kV field-aged silicone rubber insulators via condition assessment. IEEE Transactions on Dielectrics and Electrical Insulation 24(6), 3612–3621 (2017).
25. GAO Yanfeng, LIANG Xidong: Effect of electrical stress on glass fiber reinforced polymer used in high voltage composite insulator under wet environment. Composites Science and Technology 155, 151–159 (2018).
26. Study on Radial Aging Characteristics of Decayed Composite Insulators. Electric Porcelain Surge Arrester, http://kns.cnki.net/kcms/detail/61.1129.TM.20240521.1753.002.html. last accessed 2022/09/10.
27. LU Ming, GAO Chao: Diagnosis of Early Decay Fault and Abnormal Heating Mechanism of Composite Insulators. Electric Porcelain Surge Arrester (02), 205–212 (2022).
28. Wang C, Pei H, Tang G: Pointer meter recognition in UAV inspection of overhead transmission lines, Energy Reports 8(5), 243–250 (2022).
29. Li K, Yan X, Han Y, Multi-mechanism swarm optimization for multi-UAV task assignment and path planning in transmission line inspection under multi-wind field. Applied Soft Computing 150, 111033 (2024).
30. LIU Li, LI Te, ZHAO Haoran: Study on the influence of testing distance on infrared testing of on-site heating composite insulator unmanned aerial vehicles and control suggestions. Electric Porcelain Surge Arrester (06), 184–191 (2021).
31. WANG Xiaojie, XU Jun: Analysis of Temperature Rise Phenomenon and Causes of Composite Insulators in Transmission Lines. Electric Porcelain Surge Arrester (05), 165–171 (2022).
32. CHENG L, LIAO R J, YANG L J: An optimized infrared detection strategy for defective composite insulators according to the law of heat flux propagation considering the environmental factors. IEEE Access (6), 38137—38146 (2018).
33. LI R H, XU K T, ZHANG F Z: A preliminary study on the impact of environment to the heat of insulator. In: 2013 Annual Report Conference on Electrical Insulation and Dielectric Phenomena, pp. 415–417, IEEE, Shenzhen, China. (2013).
34. Huang Zhen, Zhang Zhonghao: Analysis of abnormal heating phenomenon on the surface of composite insulators and its suppression measures. Electric porcelain lightning arrester (6): 171–175, 186 (2020).
35. WANG Liming, ZHANG Zhonghao: The influence of moisture on the abnormal temperature rise at the end of composite insulator sheath. Power Grid Technology 40 (2): 608–613 (2016).
36. XU Xing: Study on Temperature and Physical and Chemical Characteristics of Abnormal Heating Composite Insulators. Chongqing University, China (2022).
37. HU Qin, LIU Li, XU Xing: Classification Model of Abnormal Heating Composite Insulators Based on Random Forest Algorithm. Power Grid Technology 48 (01), 416–425 (2024).
38. SCHMUCK F, BÄR C: Experience with and Application Guide for Composite Line Insulators Overhead Lines (2023)
39. ZHOU Shaoping, ZHANG Pugen: Detection of crack defects in bent pipes based on guided waves. Journal of Mechanical Engineering 51 (6), 58–65 (2015).
40. WU Bin, SUN Yaxin: Application research on high-frequency ultrasonic guided wave detection of full-length bonded anchor rods. Journal of Rock Mechanics and Engineering 26 (2): 397–403 (2007).

41. ZHANG X, TANG Z: Implementing ultrasonic guided wave mode control in hollow cylinders. In: 2015 IEEE Far East. NDT New Technology & Application Forum (2015).
42. DENG Honglei, HE Zhanfeng: Research on L-mode ultrasonic guided wave detection of composite insulator core rod defects based on matching tracking. Electric Porcelain Surge Arrester (02), 168–174 (2019).
43. Wu Biao: Defect localization and sparse signal reconstruction method for ultrasonic non-destructive testing. Harbin Institute of Technology, China (2017).
44. ZHAO Xianya: Research on Rail Crack Identification Method Based on Ultrasonic Guided Waves. Beijing Jiaotong University, China (2021).
45. DENG Honglei, HE Zhanfeng: Ultrasonic guided wave detection of debonding defects in composite insulators. High Voltage Technology 45 (01), 196–202 (2019).
46. Wen Shuangduan: Ultraviolet Imaging Detection for the Discharge of Polluted Insulators Based on Iterative Threshold Segmentation. In: 2019 Photonics & Electromagnetics Research Symposium—Fall (PIERS—Fall), pp. 1423–1427. Xiamen, China (2019).
47. GE Hao, ZHANG Manman: A method for detecting and evaluating the pollution status of ceramic insulators based on ultraviolet pulse method. Guangdong Electric Power 31 (08), 91–97 (2018).
48. LI Zhou, CHENG Lu: Study on Partial Discharge Characteristics of Main Insulation in Power Capacitors. Shaanxi Electric Power 44 (10): 77–81 (2016).
49. YAO Shuyou, XIONG Lan: A monitoring system for corona discharge of polluted insulators based on ultraviolet pulses. High Voltage Technology 35 (04), 844–848 (2009).
50. WANG Jingang, HE Wei: Relationship between surface electric field of insulators and ultraviolet pulses, and their application in detecting degraded insulators. Journal of Electrical Technology (06): 137–142 (2008).
51. WANG Ching, HE Wei: Non-contact-scope for ultra-high-voltage applications based on ultraviolet pulse detection. Transactions of China Electro Technical Society 25 (7), 186–191 (2010).
52. HE Wei, CHEN Tao: Monitoring of insulator pollution status based on ultraviolet pulse method. High Voltage Technology (10), 39–42 (2006).
53. XU Changfu, ZHOU Guangyang: Research on Improved Algorithm for Quantitative Extraction of UV Imaging Atlas Parameters. Smart Power 45 (11), 52–56 (2017).
54. PENG Xiangyang, ZHANG Zhonghao: Degradation characteristics of rod-shaped composite insulators for high-temperature heating lines in Guangdong Province. High Voltage Apparatus 56 (3), 104–110 (2020).
55. LIU Hui, ZHANG Yang,: Microwave Induced Thermal Testing Method for Internal Defects of Composite Insulators. High Voltage Apparatus 59 (03), 179–188 (2023).
56. LI Longlong: Study on the Influence of Solar Radiation and Moisture Immersion on the Temperature Rise of Composite Insulators. Lanzhou University of Technology, China (2023).
57. Kääriäinen H, Rudolph M: Moisture measurements in building materials with microwaves. NDT&E International 34 (6), 389–394 (2001).
58. Scott A O, Lucie F: Moisture monitoring of stone masonry: A comparison of microwave and radar on a granite wall and a sandstone tower. Journal of Cultural Heritage 41, 61–73 (2020).
59. DAI Shibing, TANG Zhong: Research on the Protection of the Murals in the Snow Palace of Our Lady of Macau, a World Heritage Site. Architectural Heritage 3, 114–122 (2016).
60. Rahpeima R, Soltani M: Numerical study of microwave induced thermoacoustic imaging for initial detection of cancer of breast on anatomically realistic breast phantom. Computer Methods and Programs in Biomedicine 196, 105606 (2020).
61. Zhao Y, Shan T: Thermoacoustic tomography of germinal matrix hemorrhage in neonatal mouse cerebrum. Journal of X-Ray Science and Technology 28(1), 83–93 (2020).

62. Zheng Z, Jiang Y: An improved method for quantitative recovery of conductivity using tomographically measured thermoacoustic data. Journal of X-Ray Science and Technology 28(1), 137–145 (2020).
63. Chi Z, Huang L: First assessment of thermoacoustic tomography for in vivo detection of rheumatoid arthritis in the finger joints detection of rheumatoid arthritis in the finger joints. Medical Physics 49(1), 84–92 (2022).

Chapter 3
Study on the Current-Collecting Quality of Pantograph-Catenary System Under Additional Transverse Magnetic Field Conditions

Guoliang Cai, Fengyi Guo, Huazhang Gao, Hongyun Guo, Along Zhang, and Long Xu

Abstract The pantograph-catenary system is a crucial component in the electrification of rail transit, and its current collection quality directly impacts the operation of electrical equipment on high-speed trains. The method of applying a transverse magnetic field is proposed to improve the current receiving quality of the pantograph-catenary system. The variation laws of contact resistance, current stability coefficient, and current carrying efficiency of the pantograph-catenary system under an additional magnetic field were studied using a self-developed sliding electric contact experimental machine. The experimental results show that the additional transverse magnetic field can effectively reduce the contact resistance and improve the current-carrying efficiency. On this basis, the experimental data was quantitatively analyzed using path analysis method. By comparing the standardized path coefficients of the external transverse magnetic field with the contact current, contact pressure, and operating speed on the current carrying efficiency, the order of the influence of the four operating conditions on the current carrying efficiency was determined as follows:

G. Cai · F. Guo (✉) · H. Gao · H. Guo · A. Zhang
College of Electrical and Electronic Engineering, Wenzhou University, Wenzhou, China
e-mail: fyguo64@126.com

G. Cai
e-mail: 1062096503@qq.com

H. Gao
e-mail: 1194251269@qq.com

H. Guo
e-mail: 228609014@qq.com

A. Zhang
e-mail: 2193014825@qq.com

L. Xu
HuaNeng Beijing Cogeneration Co. Ltd, Beijing, China
e-mail: 294179700@qq.com

Y. Li et al. (eds.), *Proceedings of the 5th Chinese Conference on High Voltage and Discharge Plasmas*, Springer Proceedings in Physics 254,
https://doi.org/10.1007/978-981-96-5711-7_3

contact current > transverse magnetic field > contact pressure > operating speed. It has been confirmed that the additional transverse magnetic field has a significant effect on improving the current carrying efficiency of the pantograph, which has certain guiding significance for the design of new pantograph.

Keywords Sliding electrical contact · Current-collecting quality · Additional transverse magnetic field · Path analysis method

3.1 Introduction

The transmission of electrical energy is carried out through sliding electrical contact between the pantograph and the contact wire. The deterioration of the train operating environment will cause the current collection quality of the pantograph-catenary system to decline, which directly affects the stable operation of high-speed trains. Therefore, it is necessary to find a new optimization method to improve the current collection quality of the existing pantograph-catenary system.

In recent years, domestic and foreign scholars have conducted a large number of experiments and simulation studies on the current-collecting quality of the pantograph-catenary system under different operating conditions, and achieved a series of research results. Guo [1] found that the current carrying efficiency increases first and then decreases with the increase of the contact pair roughness, and the current stability coefficient increases with the increase of roughness. Wang [2] introduced the second-order partial correlation coefficient and obtained the ranking of the influence of different roughness parameters on contact resistance. Wang [3] established a mathematical model of the contact resistance of the pantograph-catenary system taking into account the rough surface characteristics. Song [4, 5] found that when the pantograph-catenary system operated at a crosswind speed of 30 m/s, some contact forces exceeded the safety threshold.

The variation laws between the external transverse magnetic field and contact resistance, current stability coefficient, and current carrying efficiency are explained in this article, and the mechanism of the influence of transverse magnetic field on the current quality of the pantograph-catenary system was revealed from a microscopic perspective. The experimental data obtained are analyzed using path analysis to obtain the influence of the applied transverse magnetic field and other macroscopic operating conditions on the current-collecting quality of the pantograph-catenary system.

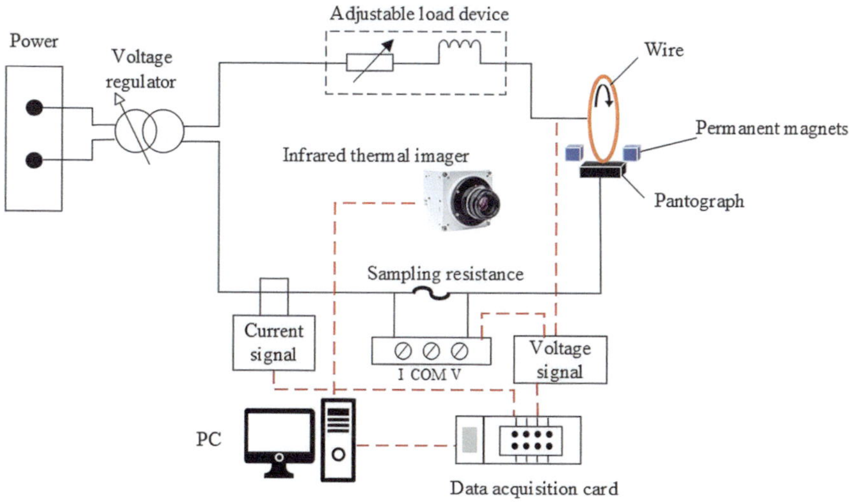

Fig. 3.1 Schematic diagram of the sliding electrical contact experimental platform

3.2 Experimental Platform and Scheme

3.2.1 Experimental Platform

The current-collecting characteristics experiment of the pantograph-catenary system under the transverse magnetic field was carried out using the self-developed sliding electric contact experimental platform, and the structural schematic diagram is shown in Fig. 3.1. A pair of permanent magnets with adjustable spacing are installed on the side of the friction pair. In order to obtain a clear temperature image of the pantograph-catenary, the infrared thermal imager is placed 60cm to the side of the contact center.

3.2.2 Experimental Scheme

Select a pair of 30 mm * 40 mm * 50 mm neodymium iron boron magnets as the source of an external magnetic field, with the magnetic field lines perpendicular to the direction of current conduction. The slide size is 250 mm * 36 mm * 27 mm, and the cross-sectional area of the copper wire is 120 mm^2. The material properties at room temperature are shown in Table 3.1.

The experimental conditions are shown in Table 3.2. There are 135 experiments in total, and each experiment lasts for 30 min. Before each experiment, the slide was polished to the same initial roughness.

Table 3.1 Physical properties of the copper-carbon slide and contact wire

Parameters	Slide	Contact wire
Hardness [N/mm^2]	56.7	96.2
Density [10^3 kg/m^3]	3.4	8.9
Resistivity [μΩ m]	12.32	0.0175
Thermal conductivity [W/(m k)]	15	398
Specific heat [J/(kg K)]	660.2	384

Table 3.2 Experimental scheme of sliding electric contact under transverse magnetic field

Conditions	Values
Operating speed (km/h)	30 50 70
Contact pressure (N)	30 50 70
Traction current (A)	30 50 70
Magnetic field strength (mT)	0 12.5 25 37.5 50

3.3 Experimental Results and Analysis

3.3.1 Effect of Transverse Magnetic Field on Contact Resistance

The experiment was carried out at a running speed of 50 km/h, a contact pressure of 50 N, and traction currents of 30, 50, and 70 A. The obtained average contact resistance changes with the transverse magnetic field as shown in Fig. 3.2.

It can be seen from the figure that the contact resistance decreases with the increase of the transverse magnetic field. The transverse magnetic field will increase the movement speed of the pantograph-catenary arc, accelerate the arc extinction, and reduce the resistance generated by the arc. The transverse magnetic field will accelerate the discharge of wear debris in the pantograph-catenary system, prevent the conductive spots from being covered, and reduce the resistance of wear debris accumulation. In summary, the transverse magnetic field can effectively reduce the contact resistance of the pantograph-catenary system.

3.3.2 Effect of Transverse Magnetic Field on Current Stability Coefficient

Figure 3.3 is a three-dimensional graph showing the current stability coefficient of the pantograph-catenary system as it changes with different operating speeds and magnetic field intensities under the conditions of an operating speed of 50 km/h, a contact pressure of 50 N, and a traction current of 30 A.

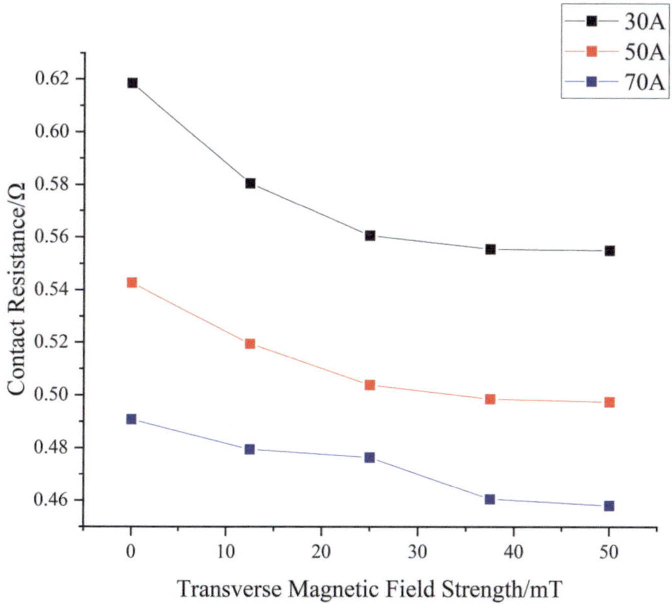

Fig. 3.2 Contact resistance curve with magnetic field strength

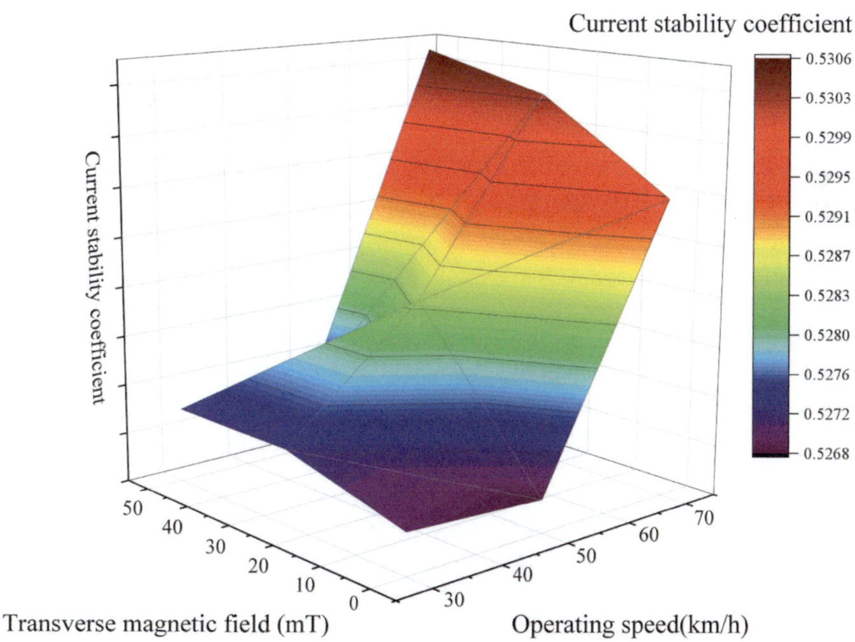

Fig. 3.3 Variation of current stability coefficient under different magnetic field strengths

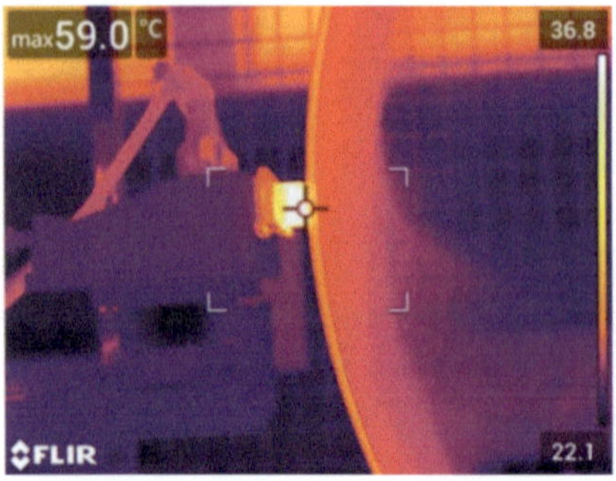

a) Maximum instantaneous temperature without magnetic field

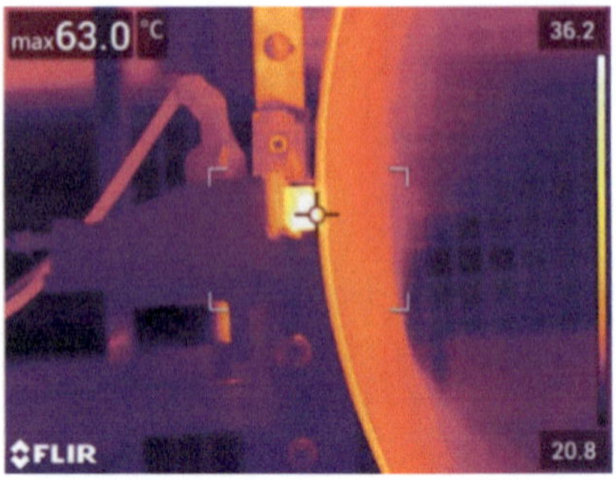

b) Maximum instantaneous temperature at 50mT magnetic field

Fig. 3.4 The temperature of the slide surface with and without magnetic field

In the figure, the current stability coefficient changes approximately in positive correlation with the magnetic induction intensity of the transverse magnetic field. The applied transverse magnetic field reduces the arc stability and causes the current coefficient to increase. The magnetic field will have a significant impact on the power emitted by arc heat convection. Under the action of the transverse magnetic field, the instantaneous temperature emitted by the arc will be higher and the amplitude of temperature change will be larger, which is confirmed by the infrared images taken in the experiment (Fig. 3.4).

3.3.3 Effect of Transverse Magnetic Field on Current Carrying Efficiency

The experiment was carried out at a running speed of 50 km/h, contact pressures of 50 and 70 N, and a traction current of 50 A. A continuous section of sample points was selected from the obtained data to draw a current carrying efficiency variation graph. The results are shown in Fig. 3.5.

The experimental results show that the current carrying efficiency of the pantograph-catenary system is positively correlated with the magnetic induction intensity. When the contact pressure is 50 N, the improvement of the current carrying efficiency of the pantograph-catenary system by the transverse magnetic field is greater than that when the contact pressure is 70 N.

The experiment was carried out at running speeds of 50 and 70 km/h, contact pressure of 50 N, and traction current of 30 A. The current carrying efficiency variation graph was plotted and the results are shown in Fig. 3.6.

The experimental results show that the transverse magnetic field has a more obvious effect on the improvement under 70 km/h conditions. In the actual operation of the pantograph-catenary system, faster speeds and smaller contact pressures

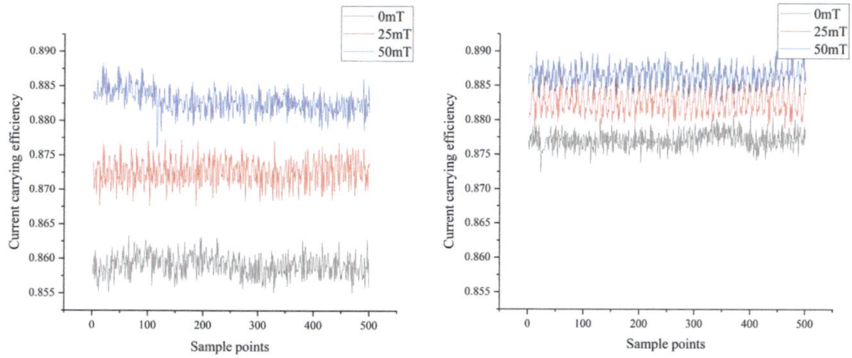

a) Contact pressure 50N b) Contact pressure 70N

Fig. 3.5 Current carrying efficiency at different contact pressures

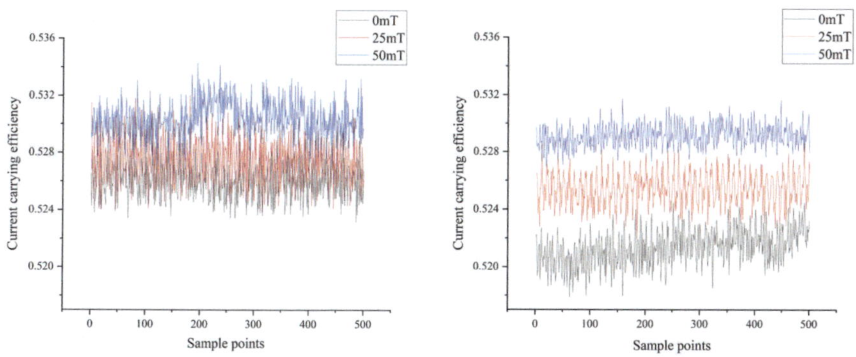

a) Operating speed 50km/h b) Operating speed 70km/h

Fig. 3.6 Current carrying efficiency at different operating speeds

will cause more frequent arcing in the pantograph-catenary system. The arc extinguishing function of the transverse magnetic field enables the pantograph-catenary system to maintain a higher current carrying efficiency when the operating conditions deteriorate.

3.4 Analysis of the Current Collecting Quality of the Pantograph-Catenary System Based on Path Analysis Method

In order to further explore the relationship between multiple operating conditions and multiple current collection quality indicators, path analysis is used to measure the importance of each operating parameter to the current collection quality of the pantograph-catenary system. Figure 3.7 is a flowchart of path analysis.

3.4.1 *Selection of Dependent Variables*

The three dependent variables analyzed are contact resistance R, current stability coefficient ε, and current carrying efficiency η. The current carrying efficiency is selected as the final result variable, assuming that the contact resistance and current stability coefficient are endogenous variables, which are affected by the operating conditions and also affect the current carrying efficiency. A simple correlation analysis of contact resistance, current stability factor and current carrying efficiency was performed before drawing the assumed path model. The calculation results are shown in Table 3.3.

It can be seen from the table that there is no statistical significance between the current stability coefficient and the current carrying efficiency. This path can be deleted when building the mode.

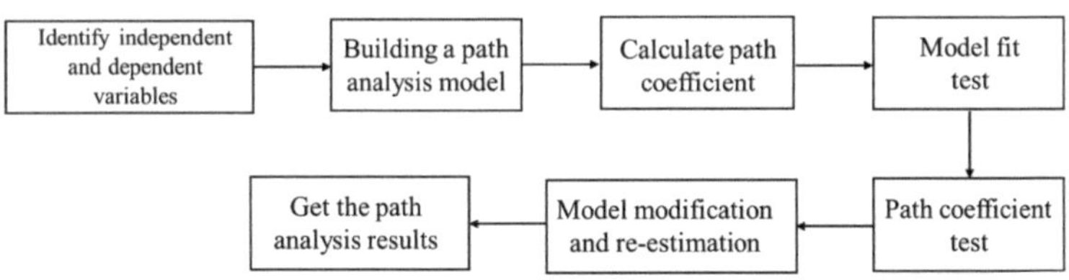

Fig. 3.7 Path analysis flow chart

Table 3.3 Correlation coefficient calculation results

	Sample size	Correlation coefficient	P value
$r\eta$, R	135	0.232	0.062
$r\eta$, ε	135	−0.806	0.000

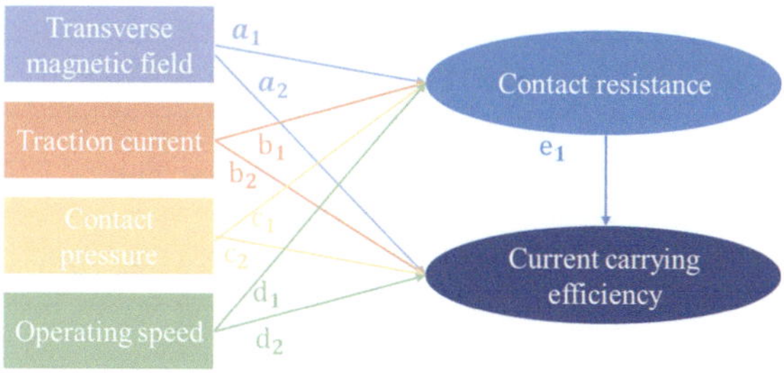

Fig. 3.8 Analysis diagram of the quality path of the pantograph-catenary current collection

Table 3.4 Standardized path coefficient fitted value

Standardized path coefficients	Fit value	Standardized path coefficients	Fit value
a_1	−0.498	a_2	0.666
b_1	−0.608	b_2	0.701
c_1	−0.560	c_2	0.531
d_1	0.469	d_2	−0.201
e_1	−0.349		

3.4.2 Developing Path Analysis Model

According to relevant theories, the path analysis model is drawn as shown in Fig. 3.8. Each path in the figure corresponds to a path coefficient. By comparing the absolute values of the path coefficients, we can determine the influence of different operating conditions on the current receiving indicators. The calculated standardized path coefficients are shown in Table 3.4. After calculation, the fitting indicators of the model meet the requirements.

From the path coefficients a_1 and a_2 of the magnetic field on the contact resistance and current carrying efficiency, it can be seen that the magnetic field has a significant negative impact on the contact resistance, and the transverse magnetic field has a significant positive impact on the current carrying efficiency.

By comparing the absolute values of the standardized path coefficients, the influence of each operating condition on the current carrying efficiency is ranked as follows: traction current > transverse magnetic field > contact pressure > operating speed.

3.5 Conclusion

Through experiments on the current-carrying characteristics of pantograph under transverse magnetic fields, it is found that the transverse magnetic field will reduce the contact resistance of pantograph and increase the current-carrying efficiency.

Through the correlation analysis between contact resistance, current stability coefficient and current carrying efficiency, it is found that the reduction of contact resistance directly increases the current carrying efficiency, and the correlation between the change of current stability coefficient and the current carrying efficiency is not significant. Finally, the constructed path analysis model was used to quantitatively analyze the experimental data. The order of the impact of various operating conditions changes on the current carrying efficiency is traction current > transverse magnetic field > contact pressure > operating speed.

Acknowledgements This research was supported in part by the National Natural Science Foundation of China (52077158), and by the Master's Innovation Foundation of Wenzhou University (3162024004123).

References

1. F. Guo, X. Wang, J. Kou, et al: Impact of Surface Roughness on Pantograph-Catenary Current Collection Quality. In: IEEE Access, 10, 93951–93959 (2022).
2. X. Wang, F. Guo, F. Lil, et al: Influence of Surface Roughness of Sliding Friction Pairs on Pantograph-Catenary Contact Resistance. In: IEEE Transactions on Transportation Electrification, 10, 1109–1119 (2024).
3. Z. Wang, Q. Zhou, F. Guo, et al: Mathematical Model of Contact Resistance in Pantograph-Catenary System Considering Rough Surface Characteristics. In: IEEE Transactions on Transportation Electrification, 8, 455–465 (2022).
4. Y. Song, F. Duan, F. Wu, et al: Assessment of the Current Collection Quality of Pantograph–Catenary with Contact Line Height Variability in Electric Railways. In: IEEE Transactions on Transportation Electrification, 8, 788–798 (2022).
5. Y. Song, F. Duan, S. Gao, et al: Crosswind Effects on Current Collection Quality of Railway Pantograph–Catenary: A Case Study in Chengdu–Chongqing Passenger Special Line. In: IEEE Transactions on Instrumentation and Measurement, 71, 1–13 (2022).

Chapter 4
Study on Lightning Impulse Voltage Test of Multi Gap Arrester for Distribution Lines

Ming Chen, Mingming Xu, Ruirui Xie, Jinshuai Zhang, Bowen Shang, Yanyan Xue, Yunbo Li, Jintian Chen, Caiwei Pan, and Yeqiang Deng

Abstract Distribution arrester failure has accounted for a considerable proportion of the total number of faults in the distribution network, and has become one of the main factors affecting the reliability of power supply in the distribution network. Therefore, from the test point of view of lightning arrester lightning impact withstand capacity is essential. In this paper, for the 10 kV multi-gap arrester specimen, to carry out positive and negative polarity of the lightning impact voltage test, through the recorded discharge process high-speed image, can be obtained multi-gap arrester impact discharge process also includes the process including the initial stage, movement and development stage, dissipation and recession stage, the whole process duration of 0.6 ms. through the "lift method" to determine the specimen's 1.6 ms. Through the "lifting method" to determine the specimen of 1.2/50 μs positive and negative lightning impact 50% discharge voltage were 90.5 and 100.3 kV, and then based on the atmospheric environment at the time of the test on the experimental results of the correction, the positive and negative lightning impact 50% discharge voltage correction value of 89.8 and 99.5 kV, to meet the provisions of the standard.

Keywords Multi gap arrester · Lightning impulse · 50% impulse voltage

M. Chen (✉) · M. Xu · R. Xie · J. Zhang · B. Shang · Y. Xue · Y. Li
State Grid Henan Electric Power Company Electric Power Scientific Research Institute Zhengzhou, Zhengzhou, China
e-mail: 1326789256@qq.com

J. Chen · C. Pan · Y. Deng
Wuhan University, Wuhan, China

Y. Li et al. (eds.), *Proceedings of the 5th Chinese Conference on High Voltage and Discharge Plasmas*, Springer Proceedings in Physics 254,
https://doi.org/10.1007/978-981-96-5711-7_4

4.1 Introduction

4.1.1 A Subsection Sample

As an indispensable and important protective equipment, the performance of lightning arrester is directly related to the safe and reliable operation of lines and equipment. However, statistics show that the distribution arrester failure has accounted for a considerable proportion of the total number of faults in the distribution network, and has become one of the main factors affecting the reliability of power supply in the distribution network [1, 2].

After the operation of the ZnO arrester, due to the excellent electrical characteristics and fast response capability of its internal zinc oxide valve sheet, it can immediately restore the high resistance state after the end of the over-voltage shock, and the time of the IFC renewed current is very short [3]. While the short gap relies on air self-energy arc quenching, the IFC arc shading needs a certain amount of time [4], therefore, it is essential to specifically study the IFC shading capability of the short gap structure from the experimental point of view.

In recent years, researchers for ZnO arrester lightning current impact characteristics of the study, often from the perspective of the proportional unit, carried out the relevant characteristics of the study, such as domestic and foreign scholars carried out different lightning current impact on ZnO arrester impact of the study [5]. On the ZnO arrester resistor resistor sheet in the pulsed current impact under aging, the harmonic characteristics of the leakage current [6]. Distribution network line arrester resistor resistance sheet 90/200 μs inrush current withstand characteristics of experimental research [7]. Natural lightning, ZnO arrester residual voltage characteristics analysis [8] and other research work. However, for the multi-gap arrester, the relevant experimental work carried out is still less.

Therefore, in this paper, for the 10 kV multi-gap arrester specimen, the lightning current impact withstand test is carried out with reference to the standards of electric power system DL/T815-2012, T/CSEE 0082-2018, etc., and the 50% breakdown voltage of the specimen is analyzed by the lifting and lowering method, so as to analyze the difference of the specimen's resistance to the impact lightning current of different polarities.

4.2 Experimental Program

Lightning surge discharge voltage test is used to verify whether the test piece is broken down and discharged under a certain surge voltage.

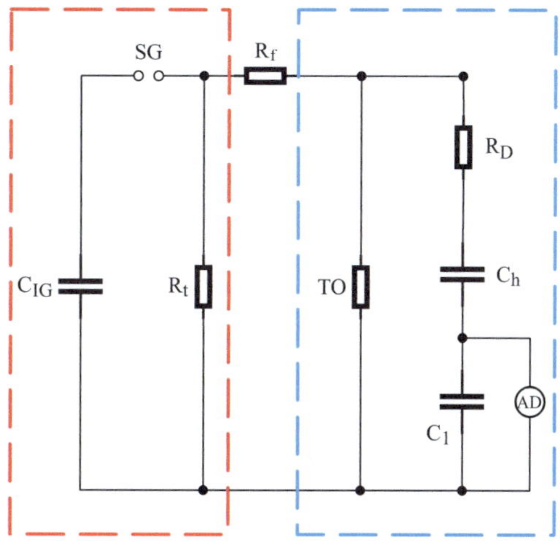

Fig. 4.1 Lightning impact discharge test circuit diagram

4.2.1 Test Circuits and Platforms

The test was conducted using a 300 kV/30 kJ shock voltage generator, the number of levels for 3, can output 1.2 μs ± 30%/50 μs ± 20% standard shock lightning shock voltage full wave and 2 ~ 6 μs standard lightning shock voltage intercept wave, to meet the requirements of this test. The acquisition device is a voltage divider and Tektronic oscilloscope which are matched with the impact voltage generator, and the discharge process is recorded by Phantom v2512 high-speed camera.

The lightning impact discharge test circuit and test platform are shown in Figs. 4.1 and 4.2.

4.2.2 Test Methods

The overvoltage caused by lightning in a power system is a type of impulse voltage with a short duration. Therefore, when tests are conducted in the laboratory, impact voltage generating devices are utilized to generate impact voltages, which are used to simulate overvoltages caused by lightning flash discharges. Due to the decentralized nature of the discharge time delay, the 50% impact breakdown voltage is generally used in engineering to reflect the characteristics of the gap to withstand the impact voltage.

Conduct the test by the "lift method" and record the test voltage at the time of discharge and non-discharge; determine the 50% discharge voltage of 1.2/50 μs lightning impulse of the specimen according to the formula in the "lift method". The specific steps of the "lift method" are as follows: First, estimate the expected value of 50% of the discharge voltage, U'_{50}. If no discharge is induced, the next applied voltage

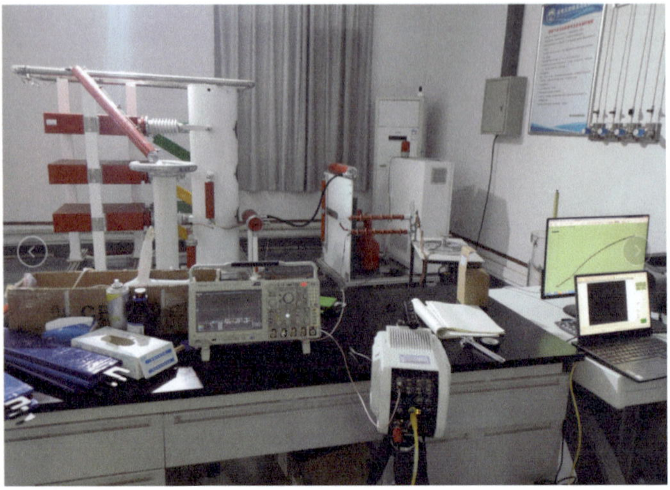

Fig. 4.2 Lightning impact discharge test platform

shall be $U_k + \Delta U$. Here $\Delta U = 1.5\%$ to $3\%\ U'_{50}$. If U_k has caused discharge, then the next applied voltage should be $U_k - \Delta U$. Subsequent pressurization is in accordance with the following law: if the last pressurization has caused discharge, then the applied voltage should be lower than the last voltage ΔU; if the last pressurization has not caused discharge, then the applied voltage should be higher than the last voltage ΔU. This repeatedly lift the pressure to a sufficient number of times, respectively, calculated at all levels of voltage U_i under the pressure number of times ni, the total number of times 20 ~ 40 times, generally take 30 times. Press the following formula to find out the 50% discharge voltage:

$$U_{50} = \frac{\sum U_i n_i}{\sum n_i} \tag{4.1}$$

4.3 Results and Analysis

4.3.1 Positive Lightning Strikes

Figure 4.3 shows the image of the positive lightning impact discharge process of the 10 kV multi-gap arrester obtained by the high-speed camera, and Fig. 4.4 shows the waveform of the positive lightning impact discharge test current of the 10 kV multi-gap arrester. t = 6.6 μs, the lightning impact voltage causes the gap to flashover and form an arc, indicating that the time required for lightning overvoltage breakdown is extremely short, and the reading of the waveform in Fig. 4.4 indicates that breakdown occurs at about 6 μs, which is consistent with the stage of t = 27 ~ 60.3 μs, and the arc develops from the nozzle to enhance the arc cooling effect; t = 73.6 μs after t = 73.6 μs, the arc develops from the nozzle to enhance the arc cooling effect. It is consistent with the discharge process recorded by the high-speed image. t = 27 ~ 60.3 μs stage, the arc is sprayed out from the nozzle and develops, which strengthens the cooling effect of the arc; after t = 73.6 μs, the arc slowly dissipates, and the light intensity of the arc is very weak to t = 600 μs, which indicates that the arc is extinguished. The duration of the whole discharge process was 0.6 ms.

Through the "lift method" on the multi-gap arrester applied positive lightning voltage impact thirty times, each time the size of the applied voltage and the breakdown of the arrester is shown in Table 4.1.

From the table, the number of breakdowns and non-breakdowns during this pressurized test was 15 times, and the positive polarity 50% discharge voltage of 90.5 kV is calculated according to Eq. (4.1).

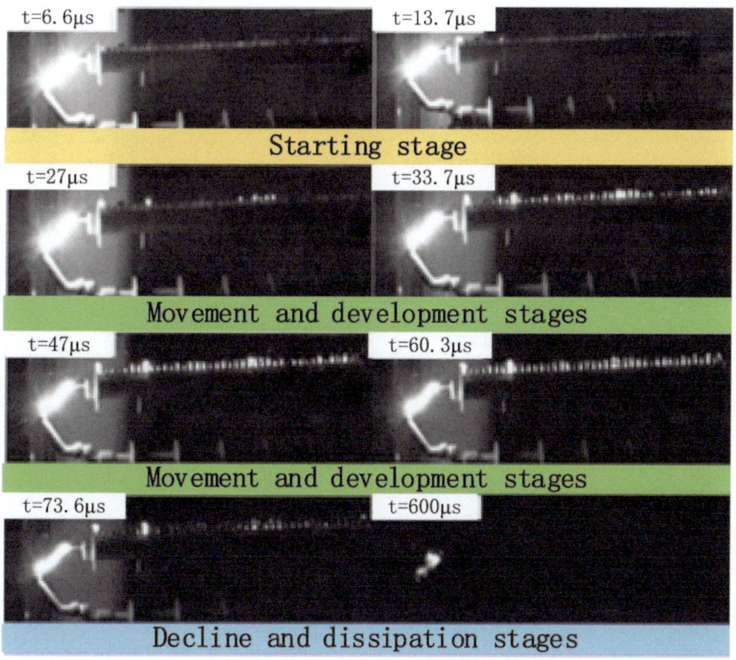

Fig. 4.3 Discharge process of positive lightning impact of 10 kV multi-gap arrester

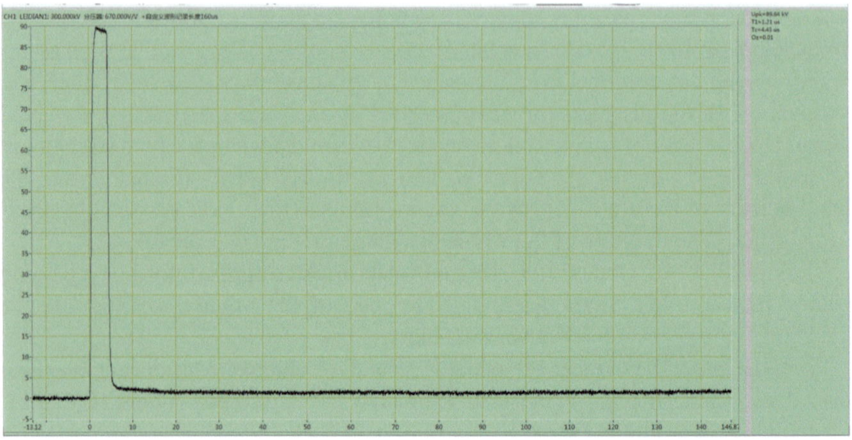

Fig. 4.4 Test current waveform of 10 kV multi-gap arrester positive lightning impact discharge

Table 4.1 Positive polarity lightning impact test results

No	Lightning surge voltage (kV)	Breakdown conditions	No	Lightning surge voltage (kV)	Breakdown conditions
1	90.1	Non-breakdown	16	91.3	Breakdown
2	91.6	Breakdown	17	90.8	Non-breakdown
3	89.6	Breakdown	18	91.3	Breakdown
4	88.7	Non-breakdown	19	90.5	Non-breakdown
5	89.6	Breakdown	20	91.8	Breakdown
6	89.0	Breakdown	21	88.6	Non-breakdown
7	87.7	Non-breakdown	22	90.7	Breakdown
8	89.2	Non-breakdown	23	89.5	Non-breakdown
9	89.5	Breakdown	24	92.7	Non-breakdown
10	89.4	Non-breakdown	25	92.8	Breakdown
11	89.6	Breakdown	26	91.4	Non-breakdown
12	88.6	Non-breakdown	27	92.1	Breakdown
13	89.6	Non-breakdown	28	91.6	Non-breakdown
14	90.8	Non-breakdown	29	92.8	Breakdown
15	91.7	Breakdown	30	92.0	Breakdown

4.3.2 Negative Lightning Strikes

Figure 4.5 is the image of the negative polarity lightning impact discharge process of the 10 kV multi-gap arrester obtained by a high-speed camera, and Fig. 4.6 is the waveform of the negative polarity lightning impact discharge test current of the 10 kV multi-gap arrester. The impact discharge process also includes the process also includes the initial stage, movement and development stage, dissipation and recession stage, similar to the positive polarity, and will not be repeated here.

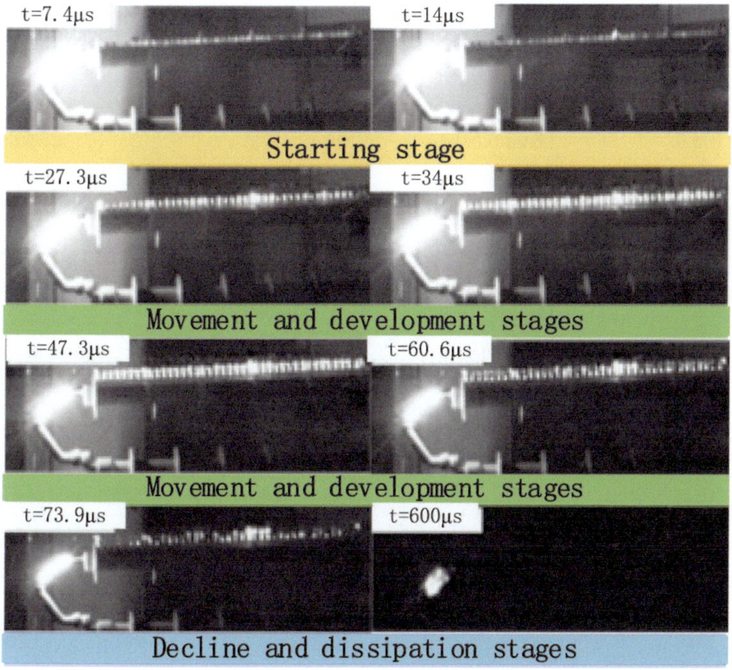

Fig. 4.5 Discharge process of negative lightning impact of 10 kV multi-gap arrester

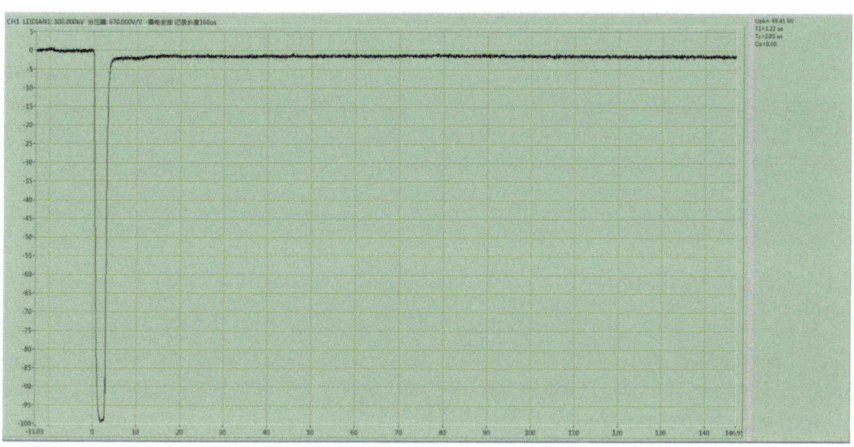

Fig. 4.6 10 kV multi-gap arrester negative lightning impact discharge test current waveforms

Table 4.2 Negative polarity lightning impact test results

No	Lightning surge voltage (kV)	Breakdown conditions	No	Lightning surge voltage (kV)	Breakdown conditions
1	99.4	Breakdown	16	100.2	Breakdown
2	97.5	Non-breakdown	17	99.2	Non-breakdown
3	98.5	Non-breakdown	18	100.1	Breakdown
4	99.6	Non-breakdown	19	98.4	Breakdown
5	100.4	Non-breakdown	20	97.6	Non-breakdown
6	101.0	Breakdown	21	98.8	Non-breakdown
7	99.1	Breakdown	22	99.8	Non-breakdown
8	98.6	Non-breakdown	23	100.6	Non-breakdown
9	99.5	Non-breakdown	24	102.0	Non-breakdown
10	100.5	Non-breakdown	25	103.0	Breakdown
11	101.4	Breakdown	26	101.8	Non-breakdown
12	99.9	Breakdown	27	103.4	Non-breakdown
13	98.7	Non-breakdown	28	104.7	Breakdown
14	99.8	Non-breakdown	29	103.6	Breakdown
15	101.6	Non-breakdown	30	101.6	Breakdown

The same operation was performed as for the positive polarity test and the test results obtained are shown in Table 4.2.

Calculation according to Eq. (4.1) yields a negative polarity 50% discharge voltage of 100.3 kV.

Power system DL/T815-2012, T/CSEE 0082-2018 and other standards stipulate that for line lightning protection devices with series gaps, the lightning impact 50% discharge voltage should be no more than 100 kV.

The atmospheric conditions when this test was carried out were as follows: the atmospheric pressure was 101.3 kPa, the temperature was 22.9 °C, and the relative humidity was 73%. Correcting the above positive and negative polarity test results, the positive and negative 50% lightning impulse voltage of the 10 kV multi-gap arrester is 89.8 and 99.5 kV respectively, which meets the requirements of the standard.

4.4 Introduction

This paper carries out lightning surge voltage withstand test on 10 kV multi-gap arrester, assesses its 50% lightning surge voltage to meet the requirements of the relevant standards, and obtains the following conclusions:

(1) 10 kV multi-gap arrester subjected to 1.2/50 μs lightning impact discharge process similarly includes the process including the initial stage, movement and development stage, dissipation and recession stage, the duration of the discharge is about 0.6 ms.
(2) Through the "elevation method" on the lightning arrester specimen were carried out 30 times positive and negative polarity lightning impact discharge test, the final obtained multi-gap lightning arrester positive and negative polarity 50% lightning impact voltage of 89.8 kV and 99.5 kV respectively.
(3) According to the test results, it can be judged that the test specimen of the multi-gap lightning arrester has a stronger ability to withstand negative lightning strikes than positive lightning strikes.

References

1. Xin, Y., Miaolong, Z., Huanhuan, Z, Haotian, S., Xuehui, D., Weiyu, Z.: Application method of shunt gaps in lightning protection of 10 kV distribution transformers. High Voltage Electrical Appliances 58(04), 179–188 (2022).
2. Huanhuan, Z., Xin, Y., Wei, C., Jie, P., Junhua, Y., Yanfu, W.: Installation of single identical wires for parallel gaps in 10 kV overhead lines. Southern Grid Technology 15(04), 90–98 (2021).
3. Cixuan, C., Shixiong, H., Chunlin, X.: The new development of lightning arrester-zinc oxide gapless arrester. High voltage technology 5(03), 69–80 (1979).
4. Yihao, L., Tingfang, Y., Xinxiu, W., etc.: Self-energizing air cannon protection gap rapid arc extinguishing device. Southern Grid Technology 16(05), 71–78 (2022).
5. Berahim, K.: Efect of multiple lightning strikes onthe performance of Zn0 lightning arrester block. HighVoltage Engineering 37(11), 2763–2771 (2011).
6. Jinbo, W., Jun, H., Shuiming, C., Jinliang, H.: Harmonic characteristics of leakage currents of ZnO varistors under impulse aging. IEEE Transactions on Power Delivery 32(4), 1758–1765 (2016).
7. Qianwei, Q., Mingqiu, D., Ruihai, L., etc.: Characteristic test of 90/200 ms inrush current withstand of distribution line surge arrester resistor sheet. Southern Grid Technology 9(7), 46–51 (2015).
8. Xu, Y., Shaodong, C., Runzhi, J., etc.: Characterization of residual voltage of zinc oxide lightning arrester under natural lightning. China Light and Power 46(7), 72–76 (2013).

Chapter 5
Influence of Ultraviolet Aging on Surface Charge Accumulation Characteristics of Silicone Rubber

Liang Feng, Kezu Wang, Jiao He, Lu Guo, Ling Ma, Dexiong Hu, Feng Wang, and Xingshuo Song

Abstract This paper conducted artificial accelerated aging of silicone rubber by ultraviolet radiation (500, 1000 h), and tested the surface charge accumulation characteristics, SEM, surface resistivity and volume resistivity of the silicone rubber before and after radiation. The test results show that ultraviolet radiation will exacerbate the severity of surface charge accumulation on silicone rubber. Compared with unaged silicone rubber, the severity of surface charge accumulation on silicone rubber with an aging time of 500 h and 1000 h increased by 42.64% and 37.73% respectively. As the ultraviolet aging time increases, holes and cracks appear on the surface of the silicone rubber, resulting in a decrease in both the surface resistivity and the volume resistivity. This work has guiding significance for aging evaluation of composite insulators.

Keywords Silicone rubber · Surface charge · Ultraviolet aging · Degradation

L. Feng · K. Wang · J. He · L. Ma
State Grid Gansu Electric Power Company Material Business Department, Lanzhou, China

L. Guo
State Grid Gansu Electric Power Company Electric Power Science Research Institute, Lanzhou, China

D. Hu (✉) · F. Wang
Hunan University, Changsha, China
e-mail: 1345298273@qq.com

X. Song
State Grid Shandong Electric Power Company Liaocheng Power Supply Company, Liaocheng, China

Y. Li et al. (eds.), *Proceedings of the 5th Chinese Conference on High Voltage and Discharge Plasmas*, Springer Proceedings in Physics 254,
https://doi.org/10.1007/978-981-96-5711-7_5

5.1 Introduction

Composite insulators are widely used in high-voltage power systems, serving as critical components that perform dual functions of insulation and mechanical support for conductors [1]. Typically, composite insulators consist of a core rod, silicone rubber, and metal fittings. The electrical performance of the silicone rubber plays a pivotal role in determining the overall functionality of the insulator. However, during long-term operation, composite insulators are exposed to various environmental factors such as ultraviolet (UV) radiation, humidity, temperature fluctuations, and pollution, all of which contribute to surface aging and degradation of the material's properties. Among these factors, UV-induced aging is a key contributor to the deterioration of composite insulator performance and lifespan.

Recent studies have investigated the physicochemical characteristics of silicone rubber when exposed to UV radiation. For instance, Zuo et al. examined the effects of UV exposure on silicone rubber, reporting that extended UV exposure results in increasingly rough surface textures. Additionally, the initial static water contact angle of the silicone rubber samples exhibited a trend of first increasing and then decreasing with prolonged UV irradiation [2]. Pan et al. conducted accelerated UV aging experiments, demonstrating that surface potential increases post-aging. They summarized the mechanisms of UV-induced aging in silicone rubber and proposed methods to enhance resistance to UV degradation [3]. In a dry environment, Liu et al. explored the impact of UV radiation on the aging behavior of silicone rubber, finding that UV aging hardens the surface and reduces the contact angle by approximately 20°, which diminishes hydrophobicity. Furthermore, they observed an acceleration in surface charge decay with longer UV exposure [4, 5]. UV radiation induces molecular chain scission, cross-linking, and morphological changes in the silicone rubber surface, thereby altering its charge accumulation characteristics. Such changes may lead to uneven electric field distribution, increasing the risk of partial discharge, which can compromise the safety and stability of power systems. However, studies specifically addressing the surface charge accumulation behavior of silicone rubber post-UV aging remain scarce.

Therefore, this paper conducts artificial accelerated UV aging of silicone rubber in a laboratory setting. A surface charge measurement platform was constructed to assess the surface charge accumulation characteristics of silicone rubber samples before and after aging. In addition, the paper examines the surface microstructure, surface resistivity, and volume resistivity of the silicone rubber.

5.2 Materials and Test Platform

5.2.1 Materials

The silicone rubber samples used in this paper were provided by the State Grid Gansu Electric Power Company and are industrial-grade composite insulator materials. The samples were circular discs with a diameter of 100 mm and a thickness of 2 mm. Prior to testing, the surfaces of the samples were cleaned with anhydrous ethanol and then dried in a high-temperature oven for 12 h. Following preparation, the samples were subjected to accelerated UV aging in accordance with the GB/T 14522 standard. The UV aging tests were conducted in a UV aging chamber using UVB-313 light sources, with an irradiance of 0.71 W/m^2 nm. The aging durations were set at 0, 500, and 1000 h.

5.2.2 Test Platform

Surface charge measurement platform. The surface charge measurement platform used in this paper is illustrated in Fig. 5.1, and its components and measurement methodology are consistent with those described in reference [6]. Prior to the experiment, the surface of the silicone rubber sample was cleaned with anhydrous ethanol and dried using a deionized air blower to ensure that the surface was free of any residual charge. The silicone rubber sample was then placed on a grounded aluminum plate, with the distance between the tip of the needle electrode and the top surface of the sample fixed at 20 mm. A DC voltage of 8 kV was applied to the needle electrode for 10 min to induce surface charge on the silicone rubber. After removing the applied voltage, a three-dimensional motion control system was used to move the electrostatic capacitance probe, allowing for precise measurement of the surface charge distribution across the silicone rubber sample.

Microtopography measurement. A scanning electron microscope (SEM, Regulus 8100) was employed to observe the surface microstructure of the silicone rubber samples. The SEM analysis was used to examine the microscopic structural changes in the silicone rubber before and after UV aging.

Resistivity test. An insulation resistance tester (BEST-200A) was used to measure the surface and volume resistivity of the silicone rubber samples. To reduce measurement errors, each parameter was measured three times, and the average value was recorded as the final result.

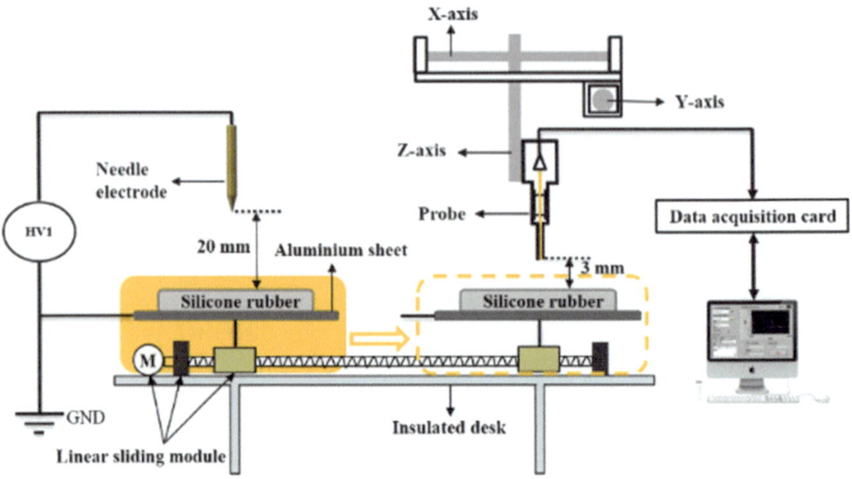

Fig. 5.1 Schematic diagram of a surface charge measurement platform

5.3 Result

5.3.1 SEM

Figure 5.2 shows the surface microstructure of silicone rubber under different UV aging time. As observed, the surface of the unaged silicone rubber remains relatively intact, with no noticeable defects, except for some impurities that may have been introduced during manufacturing. However, after UV aging, distinct defects begin to appear on the surface. The silicone rubber aged for 500 h shows a significant number of raised fragments, while the surface of the sample aged for 1000 h exhibits larger fragments, along with the presence of cracks and holes. These results indicate that the surface damage to silicone rubber becomes increasingly severe as the duration of UV aging increases.

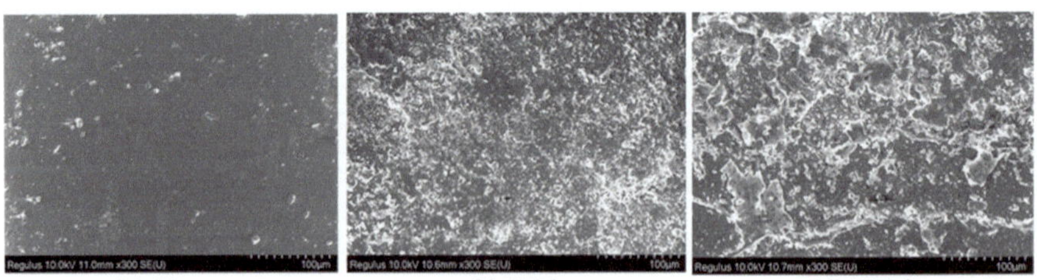

(a) 0 h (b) 500 h (c) 1000 h

Fig. 5.2 Surface microstructure of silicone rubber under different UV aging time

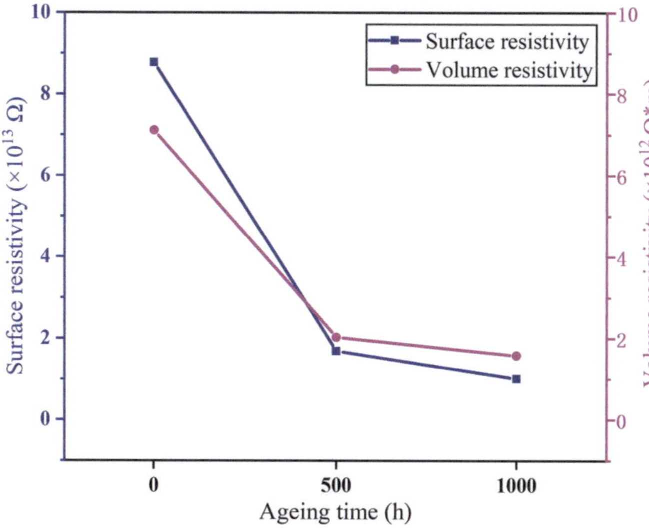

Fig. 5.3 Surface resistivity and volume resistivity of silicone rubber under different UV aging time

5.3.2 *Electrical Resistivity*

It has been reported that the accumulation and dissipation of surface charge are closely related to the surface and volume resistivity of materials [7]. Therefore, we measured the surface and volume resistivity of silicone rubber subjected to different UV aging time, and the results are shown in Fig. 5.3. As the figure illustrates, both surface and volume resistivity of the silicone rubber decrease with increasing UV aging time. The unaged silicone rubber exhibited the highest surface and volume resistivity, measured at $8.78 \times 10^{13}\ \Omega$ and $7.11 \times 10^{12}\ \Omega$ m, respectively. Compared to the unaged sample, the surface resistivity of the silicone rubber aged for 500 and 1000 h decreased by 80.75 and 88.49%, respectively, while the volume resistivity decreased by 71.44 and 77.77%. This decrease can be attributed to the lower bond energy of Si–C and Si–O–C bonds on the silicone rubber side chains, which are prone to break under UV radiation. As a result, the number of methyl and vinyl groups in the silicone rubber decreases, leading to a deterioration of the material's overall performance, manifesting as a reduction in surface and volume resistivity.

5.3.3 *Surface Charge*

Figures 5.4 and 5.5 illustrate the surface charge distribution of silicone rubber under the application of positive and negative DC voltages for various UV aging time. It can be observed that for unaged silicone rubber, the surface charge polarity aligns

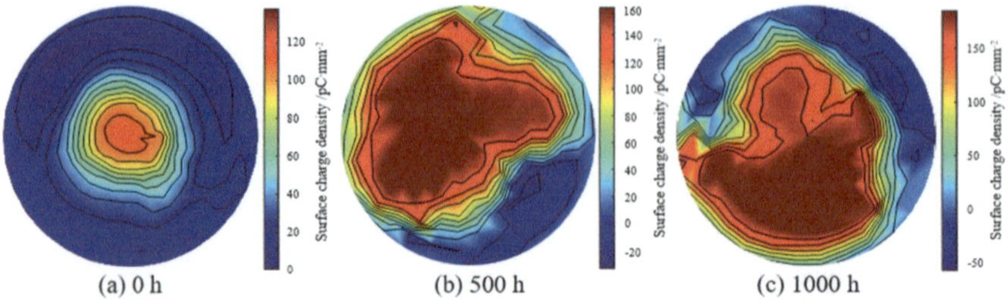

Fig. 5.4 Surface charge distribution of silicone rubber with different UV aging time under positive DC voltage

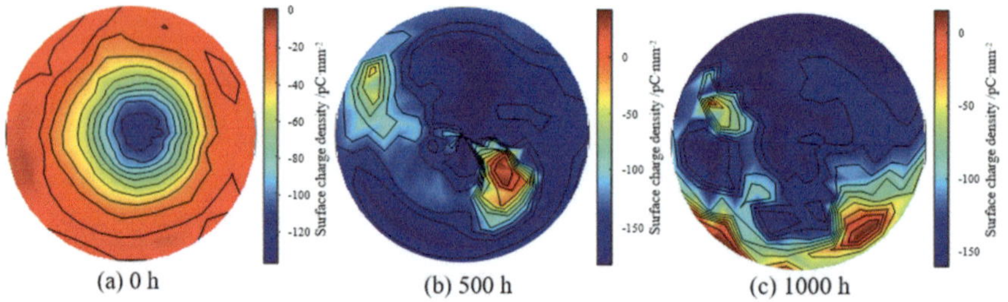

Fig. 5.5 Surface charge distribution of silicone rubber with different UV aging time under negative DC voltage

with the applied voltage, exhibiting a characteristic symmetrical "bell-shaped" distribution. However, after UV aging, the surface charge distribution becomes irregular, with some heteropolar charges appearing at the edges of the silicone rubber. This irregularity may be attributed to localized physical and chemical reactions induced by UV exposure on the surface.

As the UV aging duration increases, the high-potential charge region associated with homopolar charges diminishes after 1000 h of aging, compared to the 500 h condition, and the maximum surface charge quantity also decreases. The variable coefficient (VC), defined as the ratio of the standard deviation to the mean, serves as a measure of data dispersion by normalizing the effects of measurement scale and units. The statistical results for the VC of surface charge density across different aging time are shown in Fig. 5.6. The data reveal that with increasing aging time, the VC initially decreases and then increases. A smaller VC indicates lower data variability and a closer alignment of data points around the mean. Notably, the VC reaches its minimum at 500 h of aging, suggesting that charge accumulation on the silicone rubber surface is most pronounced at this stage. Compared to unaged silicone rubber, the degree of surface charge accumulation increases by 42.64% and 37.73% after 500 h and 1000 h of UV aging, respectively.

When a DC high voltage is applied to the needle electrode, corona discharge occurs at the tip of the needle electrode, generating some charged particles. Among them, the charged particles with the same polarity as the applied voltage will migrate

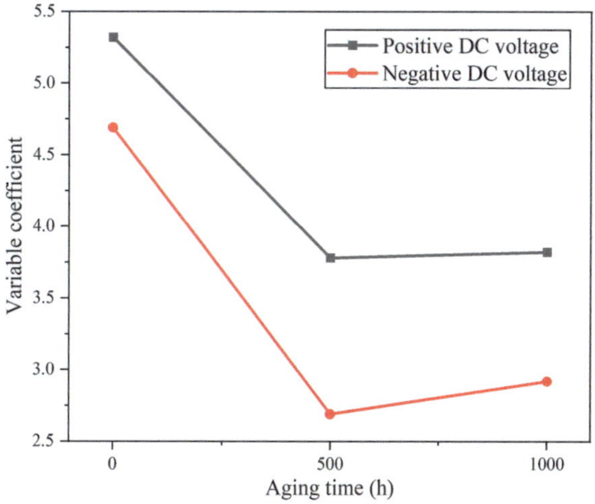

Fig. 5.6 The variable coefficient of surface charge density for silicone rubber subjected to different UV aging time under the application of DC voltage

to the surface of the silicone rubber under the action of the electric field force to form surface charges [8]. Since the electric field lines are the most concentrated below the tip of the needle electrode, most of the charges with the same polarity as the applied voltage will accumulate at the center of the silicone rubber under the action of the electric field force, thus forming a potential distribution characteristic in the shape of a "bell" on the surface of the silicone rubber. When the surface of the silicone rubber is irradiated by ultraviolet rays, combined with the results of SEM, it can be known that the microscopic morphology of the silicone rubber surface has changed. With the increase of ultraviolet irradiation time, more protrusions, holes and other defects will be generated on the surface of the silicone rubber, resulting in an increase in the surface roughness of the silicone rubber, which makes the surface of the silicone rubber more likely to adsorb charges. Therefore, the surface charge accumulation of the ultraviolet-aged silicone rubber is greater, resulting in a larger area of the high—potential charge band generated by the charge with the same polarity as the applied voltage in the central region of the silicone rubber, and the surface charge density distribution no longer presents the characteristics of the regular "bell-shaped" pattern. The accumulation and dissipation of surface charges is a dynamic process. Since the surface conductivity and volume conductivity of the silicone rubber after 1000 h of UV aging are both greater than those after 500 h. Since moderately enhanced conductivity in insulating materials accelerates the dissipation of surface charges, this accounts for the more pronounced charge accumulation observed in silicone rubber aged for 500 h compared to that aged for 1000 h.

5.4 Conclusions

This paper tested the surface charge accumulation characteristics, SEM, surface resistivity, and volume resistivity of silicone rubber before and after ultraviolet radiation. And some important conclusions are obtained as follows: The polarity of the surface charge of unaged silicone rubber is the same as that of the applied voltage, and it shows a regular "bell-shaped" distribution characteristic. However, after ultraviolet aging, the severity of the surface charge accumulation of silicone rubber will be aggravated, and the surface charge no longer shows a regular distribution characteristic. With the increase of ultraviolet aging time, more defects such as protrusions and holes will be generated on the surface of the silicone rubber, resulting in a decrease in both the surface resistivity and the volume resistivity.

Acknowledgements This work was supported by the Science and Technology Project of State Grid Gansu Electric Power Company under grant 522724240001.

References

1. Yuan, C.: Space charge behavior in silicone rubber from in-service aged HVDC composite insulators. IEEE Transactions on Dielectrics and Electrical Insulation 26(3), 843–850 (2019).
2. Zuo, K.: Effect of ultraviolet radiation on properties of room temperature vulcanized silicone rubber. Insulating Materials 53(6), 36–40 (2020).
3. Pan, W.: Research on aging, modification and repair of silicone rubber surface. Surface Technology 51(2), 108–115 (2022).
4. Youn, B.: Surface characterization of silicone polymer used as an outdoor insulator by the measurement of surface voltage decay. Surface and Interface Analysis 33(12), 954–959 (2002).
5. Liu, Y.: Effect of UV Radiation on Aging Performance of HTV-silicone Rubber in Dry Conditions. High Voltage Apparatus 51(4), 129–132 (2015).
6. Fu, Z.: Surface charge dissipation characteristics of Al_2O_3/silicone rubber composites with different weight percentages. IEEE Access 12, 33337–33346 (2024).
7. Yan, Z.: Suppression of surface charge on micro- and nano-structured superhydrophobic silicone rubber. IEEE Transactions on Dielectrics and Electrical Insulation 25(3), 1095–1102 (2018).
8. Wang, F.: Charge carriers on insulating surfaces: an experimental study. IEEE Transactions on Dielectrics and Electrical Insulation 30(6), 2665–2672 (2023).

Chapter 6
Analysis of Guano Flashover Characteristics of Grading Rings with Different Dielectric Insulation Coatings

Hong Wu, Chuanlin Xie, Yuanyi Chen, Bo Wu, Xiao Jin, and Maoqiang Bi

Abstract With the improvement of the ecological environment, bird activities have become more frequent, posing a serious threat to the safe operation of transmission lines. In particular, bird droppings flashover accidents have become one of the important issues affecting the reliability of transmission lines. This paper proposes protective measures using insulating grading rings with different materials to deal with flashover failures caused by bird droppings after installing grading rings on composite insulators of 110 kV transmission lines. First, the influence of grading rings coated with different insulating media (plastic, polyethylene, silicone rubber) on the electric field and potential distribution of composite insulators was analyzed through finite element simulation. It was found that the insulating coating can effectively improve the electric field distribution and reduce the electric field intensity. Secondly, a guano channel flashover test platform was built for testing. The results showed that plastic, polyethylene and silicone rubber insulated grading rings increased the average breakdown voltage by approximately 5.2, 22.1 and 30%, respectively, effectively reducing the probability of guano flashover failures. This study provides an effective solution for improving the safety of transmission lines and has important engineering application value.

Keywords Insulation coating · Guano flashover · Finite element simulation · Composite insulator

---

H. Wu · Y. Chen · B. Wu
Electric Power Research Institute of Ningxia Power Electric Corporation of State Grid, Yinchuan, China

C. Xie · X. Jin · M. Bi (✉)
School of Electrical and Electronic Engineering, Chongqing University of Technology, Chongqing, China
e-mail: 1251784798@qq.com

Y. Li et al. (eds.), *Proceedings of the 5th Chinese Conference on High Voltage and Discharge Plasmas*, Springer Proceedings in Physics 254,
https://doi.org/10.1007/978-981-96-5711-7_6

6.1 Introduction

In recent years, with the continuous improvement of my country's ecological environment, the range of bird activities has been expanding, which has led to overlaps between transmission lines and bird habitats [1, 2]. Birds' excretion, predation, flight, and nesting activities can seriously endanger the safe operation of transmission lines [3]. According to relevant statistics, State Grid Corporation of China suffered a total of 205 bird damages on 500 kV and above transmission lines from 2016 to 2020, accounting for 10.9% of the total number of fault trips. In 2016, a bird-related tripping accident was first discovered on a 750 kV transmission line [4]. According to the analysis of relevant data from State Grid Corporation of China, bird damage has become an important cause of threats to the safe operation of transmission lines, and its degree of harm is second only to external force damage and lightning strikes [5]. Among bird damage faults, bird droppings flashover is the most typical fault type, and it also accounts for the largest proportion, and has occurred in transmission lines of 750 kV and below [6].

Compared with ceramic and glass insulators, composite insulators have the advantages of being lighter, more resistant to pollution, and easier to maintain [7, 8]. They have been widely used in overhead transmission lines. However, due to their structure and materials, composite insulators are prone to electric field distortion at the high-voltage end, which accelerates the aging of insulating materials and seriously affects the safe operation of composite insulators. Therefore, in order to solve the above problems, it is necessary to install grading rings on composite insulators [9, 10]. However, the installation of grading rings shortens the distance between the air gaps at the high and low voltage ends of composite insulators, increasing the probability of bird droppings flashover [11, 12]. According to relevant statistics, among bird-related faults in transmission lines, bird droppings flashover faults caused by bird droppings short-circuiting the grading rings account for more than 70% [13].

Therefore, this paper proposes three different materials for insulating and coating grading rings, so that the pure air gap between the bird droppings and the grading ring during the flashover process is transformed into a gas-solid combined gap, thereby improving the insulation performance of the grading ring and reducing the probability of bird droppings flashover failure caused by the grading ring, thereby reducing bird-related accidents on transmission lines, which is of great significance to improving the safe and stable operation of the power grid.

6.2 Simulation Analysis of Composite Insulator Grading Ring

Based on the actual 110 kV transmission line, 3D modeling and simulation are carried out. This chapter will use AutoCAD to draw the 3D structures of pole towers, composite insulators, grading rings, hardware, etc., and then import the established 3D model into COMSOL finite element simulation software for numerical simulation calculation.

6.2.1 Insulation Coated Grading Ring Principle

The capacitance of an insulator to the ground affects its voltage distribution. The uniformity of the insulator voltage distribution is positively correlated with the size of its capacitance to the ground, that is, the larger the capacitance to the ground, the more uniform the voltage distribution. The capacitance to the ground of porcelain and glass insulators is greater than that of composite insulators, so the uniformity of the electric field distribution of composite insulators is not as good as that of the former two [14]. Adding a grading ring to a composite insulator can increase the capacitance to the ground and improve the uniformity of the electric field distribution. The installation of a silicone rubber-coated grading ring is shown in Fig. 6.1.

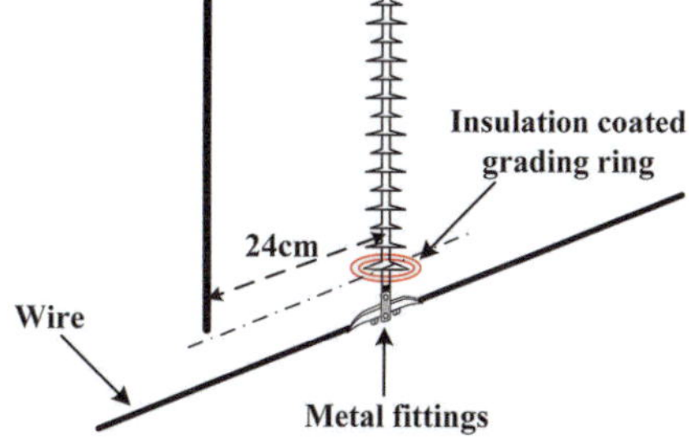

Fig. 6.1 Schematic diagram of the installation of the silicone rubber coated grading ring

6.2.2 *Simulation Modeling of 110 kV Composite Insulator Covered Grading Ring*

In order to improve the calculation efficiency, the model adopts the following simplifications and assumptions [15, 16]: the influence of the ground wire on the surface potential and electric field distribution of the conductor and composite insulator is not considered; the conductor, composite insulator, hardware, etc. are always free of pollution.

(1) Tower model

In this paper, a 1:1 three-dimensional construction of a wine cup tower is carried out in AutoCAD according to the actual line operation. The 110 kV wine cup tower is 28.5 m high and the cross arm is 2.7 m long. The tower model constructed in AutoCAD is shown in Fig. 6.2. After the construction is completed, it is imported into COMSOL Multiphysics for model combination simulation.

(2) Composite insulators

This paper uses the 110 kV composite insulator used in the Ningxia power grid. The sheds of the composite insulator are distributed in "large, medium and small". The structural height of the 110 kV composite insulator is 1438.43 mm, the creepage distance of the insulator is 4263.44 mm, the radius of the large shed is 76.49 mm, the

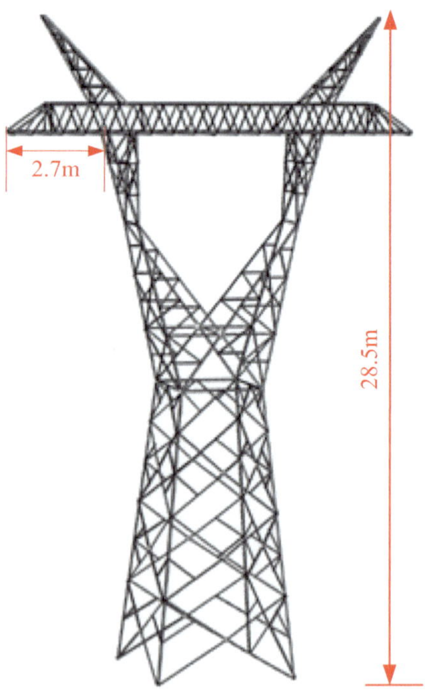

Fig. 6.2 110 kV tower model

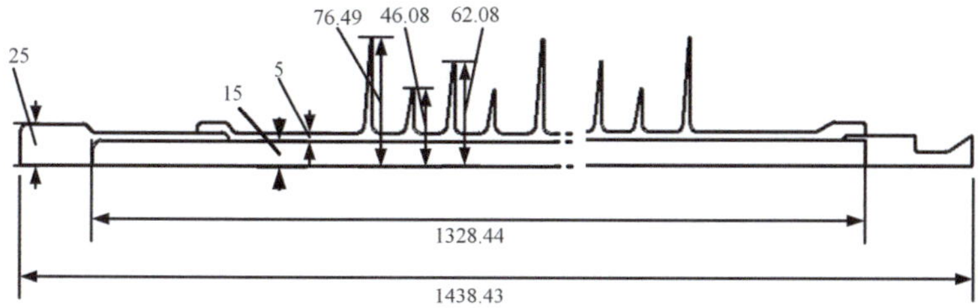

Fig. 6.3 Modeling parameters of 110 kV composite insulator

Fig. 6.4 110 kV composite insulator grading ring

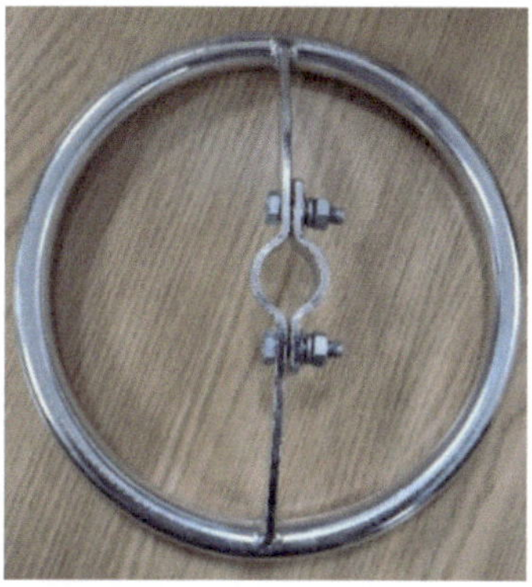

radius of the medium shed is 62.08 mm, the radius of the small shed is 46.08 mm, the shed-to-shed spacing is 20 mm, and the total number of sheds is 41. At the same time, the relevant actual modeling parameters are shown in Fig. 6.3.

(3) Grading ring

The grading ring in this paper is simulated and analyzed based on the grading ring in actual line operation. The diameter of the 110 kV grading ring is 30 mm and the diameter is 260 mm. Figure 6.4 is a real picture of the grading ring of the 110 kV composite insulator.

6.2.3 Simulation Parameter Settings

Electrostatic field simulation can reflect the electric field distribution of the relevant model. The electric field distribution inside the material during the simulation process is mainly related to the relative dielectric constant of the material. The towers, composite insulators, conductors, wire clamps, etc. constructed in the simulation process of this paper mainly use silicone rubber, epoxy resin, metal and other materials. The relative dielectric constants of the materials used are obtained according to the relevant literature [17] as shown in Table 6.1.

Table 6.1 Relative dielectric constants of main insulator materials

Material	Relative dielectric constant ε_{γ}
Silicone rubber	3.2
Epoxy resin	3.6
Metal fittings	100000
Air	1
Steel	100000

Table 6.2 Relative dielectric constant and insulation strength of common insulating media

Insulation medium	Relative dielectric constant	Dielectric strength (kV/mm)
Ordinary aluminum	1e6	\
Plastic	4.8	10 ~ 30
Polyethylene	2.3	20 ~ 40
Silicone rubber	3.2	15 ~ 20

The type of insulating medium used in the insulating coating will also affect the surface distribution of the composite insulator and the grading performance of the grading ring to a certain extent. This paper intends to use three insulating materials, namely plastic, silicone rubber, and polyethylene, for insulation coating. The relative dielectric constant and insulation strength of the insulating medium are shown in Table 6.2.

6.2.4 Simulation Results Analysis

The difference in relative dielectric constant of the insulating medium will have a great impact on the performance of the insulating coated grading ring. At the same time, the electric field and potential distribution of the composite insulator will also change when installing the grading ring coated with different insulating media. This section selects three materials: plastic, polyethylene, and silicone rubber to explore the influence of different insulating medium coated grading rings on the electric field and potential distribution of the composite insulator. The surface potential and electric field distribution of the composite insulator after installing different insulating medium coated grading rings are shown in Fig. 6.5.

As can be seen from Fig. 6.5, the potential distribution of the composite insulator when installing an ordinary metal grading ring is located below the curve. After adding the insulating medium, the surface potential distribution of the composite insulator stretches outward, indicating that the grading ring coated with the insulating medium can improve the potential distribution of the composite insulator, but the change is small, and after the distance along the surface of the composite insulator exceeds 2000 mm, the potential distribution of the grading rings with three different

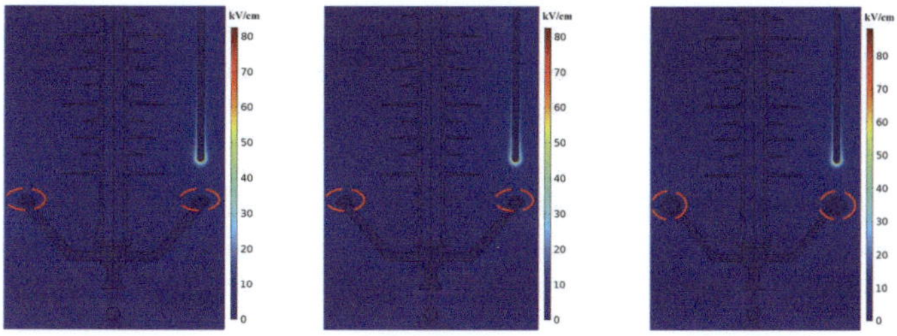

(a) Plastic grading ring (b) Polyethylene grading ring (c) Silicone rubber grading ring

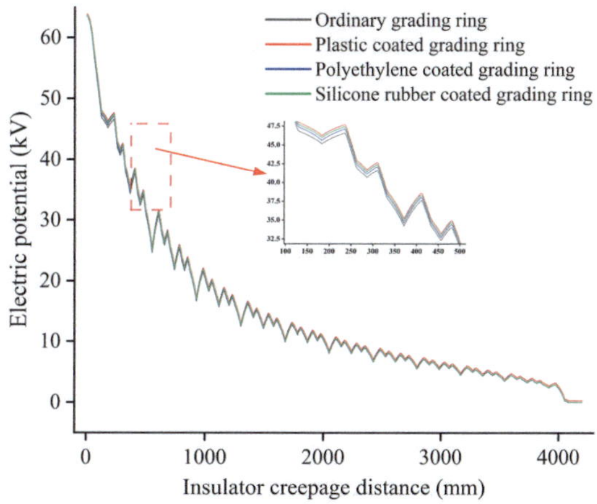

(d) Surface potential distribution of composite insulator

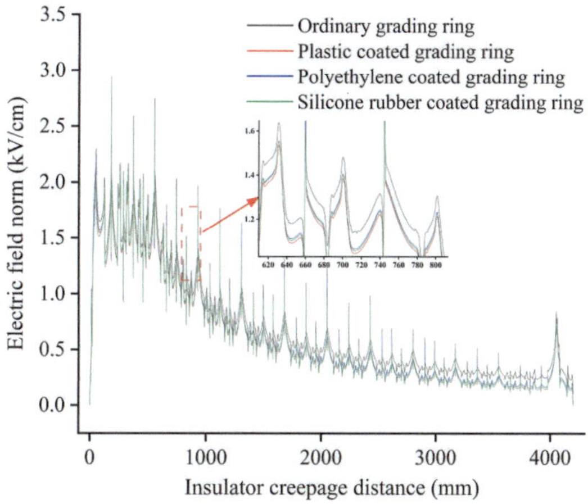

(e) Electric field distribution of composite insulator

Fig. 6.5 Different media coated grading ring

insulating media is basically the same. It can be seen from Fig. 6.5e that the grading ring can reduce the surface electric field of the composite insulator after being coated with the insulating layer, and the electric field strength at the high and low voltage ends has improved to varying degrees. In addition, as the relative dielectric constant of the insulating medium increases, the improvement effect is also improved to a certain extent. The plastic-coated insulating grading ring reduces the electric field to a greater extent than the silicone rubber material and the polyethylene material grading ring.

6.3 Test Results and Analysis

To further verify the validity of the simulation results, this paper built a guano channel flashover test platform for 110 kV composite insulators and grading rings. At the same time, in order to understand the flashover characteristics of the insulation-coated grading ring and guano, this section conducted flashover breakdown tests on ordinary grading rings, plastic-coated insulation grading rings, polyethylene-coated insulation grading rings, and silicone rubber-coated insulation grading rings.

6.3.1 Test Platform

This section mainly builds a 110 kV composite insulator insulation coated grading ring test platform, where the portal frame is 5000 mm high, 5000 mm long and 130 mm wide. The composite insulator is model FXBW-110, and the composite insulator is suspended under the portal frame through a ball head hanging ring. The insulator is connected to the conductor using a suspension clamp. At the same time, the simulated bird droppings have a diameter of 10 mm. The test platform is shown in Fig. 6.6.

The high voltage power supply during the test is a HRYDJW test transformer with a maximum output voltage of 200 kV and a maximum output current of 250 mA. The AC/DC voltage divider used for measurement is NHRC with a rated voltage of 100 kV, a transformation ratio of 1000:1, and an accuracy of 2%. The test transformer and AC/DC voltage divider are shown in Fig. 6.7.

This paper conducts relevant tests on plastic coated grading rings, polyethylene coated grading rings and silicone rubber coated grading rings. The samples are shown in Fig. 6.8.

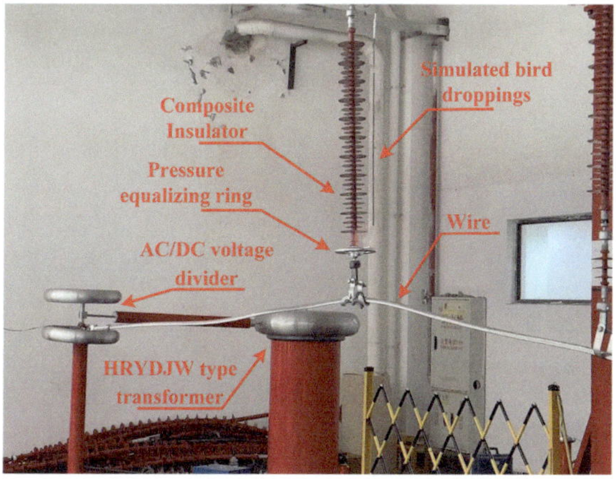

Fig. 6.6 Test platform

(a) Test transformer

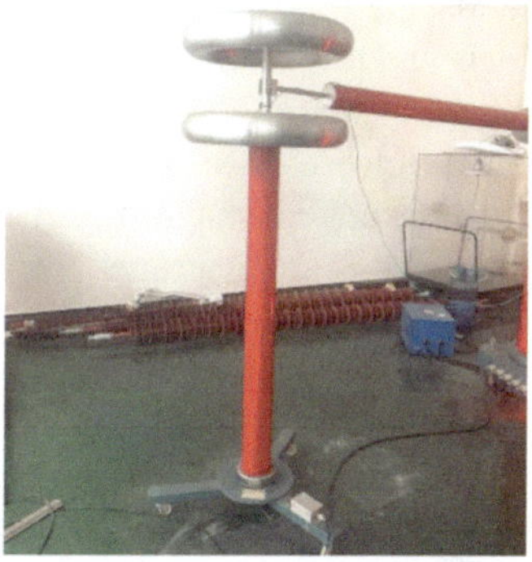

(b) AC/DC voltage divider

Fig. 6.7 High voltage test equipment

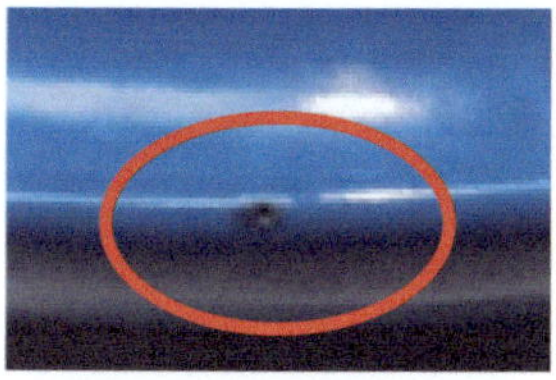

(a) The 7th Test

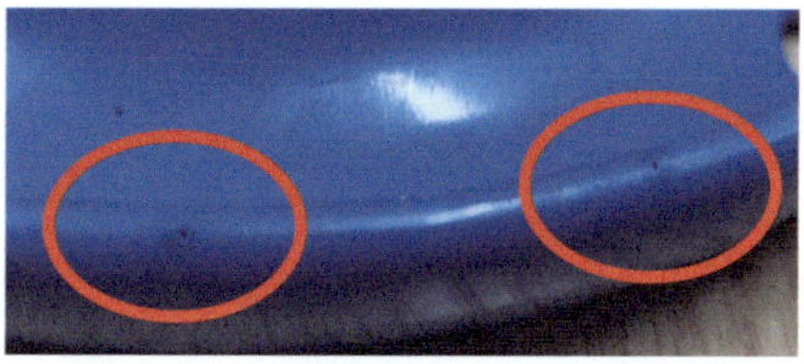

(b) The 13th Test

Fig. 6.8 Four types of grading ring test samples

6.3.2 Common Grading Ring and Plastic Grading Ring Breakdown Test

In this section, plastic material is selected to insulate the grading ring. The thickness of the plastic material is 1.5 mm. The bird droppings are fixed 24 cm away from the center axis of the composite insulator. The end of the bird droppings is parallel to the grading ring to conduct a breakdown test of the grading ring covered with plastic material. The test results are shown in Table 6.3.

It can be seen from Table 6.3 that the average breakdown voltage of the plastic-coated grading ring is 3.2 kV higher than that of the ordinary grading ring, which is 5.2% higher. The breakdown trace of the plastic-coated grading ring is shown in Fig. 6.9.

As can be seen from Fig. 6.9, the plastic-coated grading ring has obvious pinhole-shaped breakdown marks after the breakdown test. The breakdown marks are mostly located at the corners of the semicircular grading ring. At the same time, it can be clearly seen that there are black burn marks around the breakdown marks. According to the test data and test results, the breakdown voltage of the plastic-coated grading ring has a slight increase. The breakdown voltage has increased significantly in the first five tests. However, because the thickness of the plastic is only 1.5 mm, the final average breakdown voltage after 20 breakdown tests is only slightly higher than that of ordinary metal grading rings. The breakdown points are mostly located at the corners of the semicircular grading ring. This may be because the radius of curvature at the corners of the semicircular grading ring is too small, resulting in more serious electric field distortion at the corners than other parts. At the same time, the thickness differentiation treatment was not performed at this place during the insulation coating process, and the corners eventually became weak points for breakdown.

Table 6.3 Breakdown voltage of common grading ring and plastics grading ring

	Breakdown voltage (kV)	
	Grading ring (semicircle)	Plastic grading ring
Average value U_{ave} (kV)	61.5	64.7
Standard deviation σ	0.51	0.79

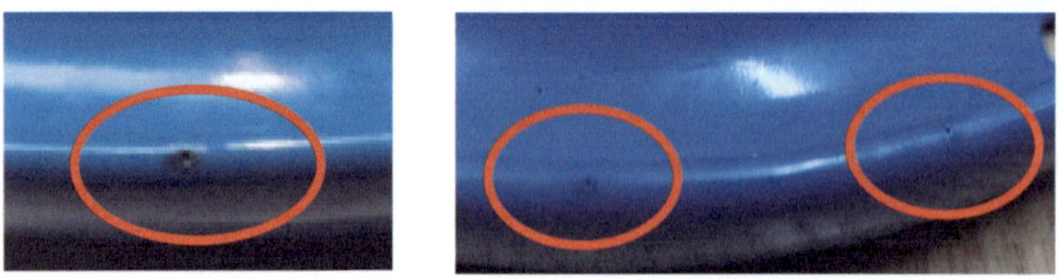

(a) The 7th Test (b) The 13th Test

Fig. 6.9 Breakdown trace of plastic grading ring

6.3.3 Common Grading Ring and Polyethylene Grading Ring Breakdown Test

Polyethylene has good heat resistance, insulation performance, mechanical properties and chemical resistance, so it is widely used in insulated cables. This section uses polyethylene coated grading rings. During the breakdown test, the simulated bird droppings are fixed at 24 cm from the center axis of the composite insulator, and the end of the bird droppings is parallel to the grading ring. The polyethylene coated grading ring breakdown test data is shown in Table 6.4.

As shown in Table 6.4, when ordinary metal grading rings are used for breakdown tests, the average breakdown voltage is 61.5 kV, while the thickness of the insulating coating of the polyethylene-coated grading ring is 4 mm, and the average breakdown voltage is 75.1 kV, an increase of about 22.1%. According to literature data, when the coating thickness increases by 1 mm, the breakdown voltage increases by about 5 kV, indicating that increasing the thickness can increase the breakdown voltage of the insulating-coated grading ring [18]. Combined with the test data, it can be seen that excessive thickness will shorten the air gap, resulting in a lower breakdown voltage than when the coating thickness is thinner. Some of the breakdown test results of the polyethylene-coated grading ring are shown in Fig. 6.10.

As shown in Fig. 6.10, the breakdown marks of the polyethylene-coated grading ring are smaller than those of the plastic grading ring, and the breakdown hole is smaller. The insulation coating layer near the breakdown hole is less damaged and is not easy to be directly observed by the naked eye.

Table 6.4 Breakdown voltage of common grading ring and polyethylene grading ring

	Breakdown voltage (kV)	
	Grading ring (semicircle)	Polyethylene grading ring
Average value U_{ave} (kV)	61.5	75.1
Standard deviation σ	0.11	1.09

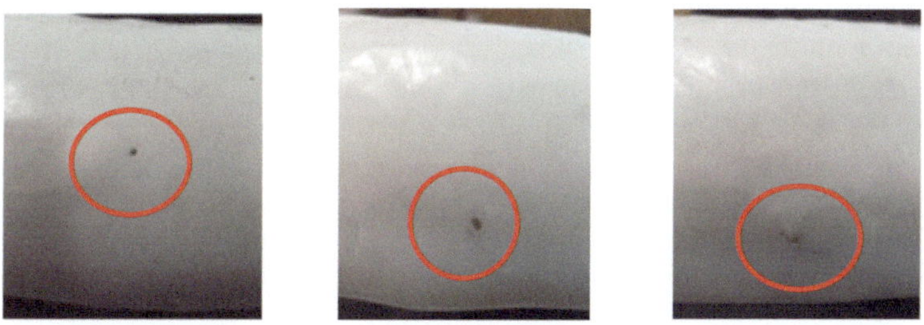

(a) The 5th Test (b) The 9th Test (c) The 14th Test

Fig. 6.10 Breakdown trace of polyethylene grading ring

It can be seen from the test data and breakdown marks that when the coating thickness is thin, the breakdown voltage of the insulating coated grading ring increases slightly, and the breakdown marks are mainly concentrated in the area close to the simulated bird droppings; when the coating thickness increases, the breakdown voltage of the insulating coated grading ring increases, and under the same arrangement, the breakdown location is transferred from the surface of the grading ring to the corners of the grading ring. This is because the electric field at the corners of the grading ring is more distorted than the electric field on the surface of the grading ring, which makes it easier to breakdown here. At the same time, the edges and corners of the polyethylene coated grading ring are not treated differently during the manufacturing process, which leads to insufficient insulation protection here.

6.3.4 Normal Grading Ring and Silicone Rubber Grading Ring Breakdown Test

HTV silicone rubber has the advantages of good high and low temperature resistance, UV resistance, and electrical insulation. In order to avoid the situation where the curvature of the edges is large during the coating process of the semicircular grading ring, the HTV silicone rubber coated grading ring body in this section selects a tubular grading ring. It can be seen from Fig. 6.8e that the surface thickness of the HTV silicone rubber coated grading ring is relatively uniform, and there are no defects such as bubbles inside. During the breakdown test, the simulated bird droppings were fixed at a distance of 24 cm from the center axis of the composite insulator, and the end of the bird droppings was parallel to the grading ring. The breakdown test data of the HTV silicone rubber coated grading ring are shown in Table 6.5, and the flashover breakdown test results of the HTV silicone rubber coated grading ring are shown in Fig. 6.11.

It can be seen from Table 6.5 that the HTV silicone rubber coated grading ring increases the breakdown voltage by about 30%, and no obvious breakdown point is found after the test. It can be seen from Fig. 6.11 that the flashover paths of ordinary metal grading rings and HTV silicone rubber coated grading rings are basically consistent with the breakdown paths of the polyethylene grading rings mentioned above. The flashover breakdown path of ordinary metal grading rings is mainly the air gap and the grading ring body closest to the bird droppings, while the discharge path of the HTV silicone rubber coated grading ring is mainly the breakdown of

Table 6.5 Breakdown voltage of common grading ring and silicone rubber grading ring

	Breakdown voltage (kV)	
	Grading ring (tubular)	Silicone rubber grading ring
Average value U_{ave} (kV)	74.6	96.7
Standard deviation σ	0.37	0.97

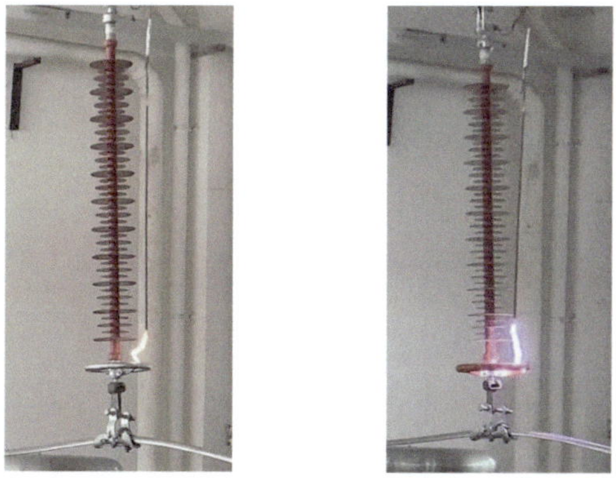

(a) Ordinary metal grading ring (b) Silicone rubber coated grading ring

Fig. 6.11 Phenomenon of grading ring breakdown test

the air gap and the surface flashover of the grading ring surface covered with the bond, that is, the breakdown gap is transformed from a pure gas gap to a gas-solid combined gap.

After summarizing the above breakdown tests, it is recommended that the insulation-coated grading ring be coated with a tubular grading ring. If the insulation-coated grading ring is a semicircular grading ring, the coating thickness at the corners needs to be differentiated because its curvature is large and easy to break through; at the same time, the coating thickness needs to be reasonably considered in terms of voltage level, transmission line, frequency of bird damage, etc.

6.3.5 Comparative Analysis of Breakdown Voltage of Grading Rings Made of Different Materials

Plastic, polyethylene and HTV silicone rubber, the three insulation materials have different insulation properties. Therefore, this section will compare and analyze the three, as shown in Fig. 6.12.

Figure 6.12 shows that the average breakdown voltage of the grading ring coated with plastic, polyethylene and silicone rubber is higher than that of the semi-circular ordinary grading ring by 3.2 kV and 13.6 kV, respectively, and higher than that of the circular ordinary grading ring by 22.1 kV. Among them, the plastic-coated insulating grading ring and the polyethylene-coated insulating grading ring are both semicircular structures, and the silicone rubber-coated insulating grading ring is a circular structure. Therefore, the average breakdown voltage of the plastic and polyethylene-coated insulating grading rings is lower than that of the silicone rubber-coated insulating grading ring. This is mainly because the electric field distortion of

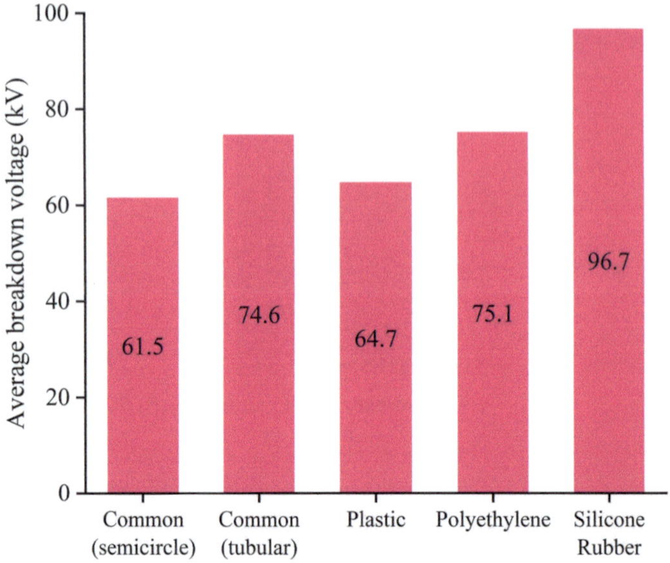

Fig. 6.12 Average breakdown voltage of grading rings with insulation coatings of different materials

the semicircular grading ring is greater at the edge, and the surface of the grading ring is not smoothed, so breakdown is more likely to occur here.

6.4 Conclusion

Aiming at the problem that the grading ring of 110 kV overhead transmission line is easily affected by bird droppings flashover fault, this paper conducts simulation analysis on grading rings coated with different materials of composite insulators, and uses copper rods to simulate bird droppings liquid to explore the relevant performance of insulation-coated grading rings in preventing bird droppings flashover. The main conclusions of this paper are:

(1) The simulation found that adding an insulating coated grading ring to a composite insulator can improve the surface potential and electric field distribution of the composite insulator. At the same time, it was found that the greater the relative dielectric constant of the insulating coating medium, the lower the surface electric field of the composite insulator; the thicker the insulating coating layer of the insulating coated grading ring, the greater the decrease in the surface electric field strength of the composite insulator.
(2) The test results show that the average breakdown voltage of the plastic-coated grading ring and the polyethylene-coated grading ring is increased by 5.2% and 22.1% respectively compared with the semi-circular ordinary grading ring,

and the HTV silicone rubber-coated grading ring is increased by about 30% compared with the tubular ordinary grading ring. At the same time, it was also found that the electric field at the corners of the semicircular grading ring was more distorted than the electric field on the surface of the circular grading ring, making it more prone to flashover breakdown.

Acknowledgements This work is supported by Ningxia Natural Science Foundation Project (2023AAC03849).

References

1. Chen Hong, Wu Hong, Liu Shitao, et al. Analysis and prevention of bird damage faults on 110 kV and above transmission lines in Ningxia power grid [J]. Ningxia Electric Power, 2019, No. 210(06): 16–21.
2. Ma Weilin. Simulation study on the pollution characteristics of transmission line insulators in Shizuishan area of Ningxia [D]. Ningxia University, 2022.
3. Bai Jianfeng. Research on transmission line status assessment and fault prediction based on big data analysis [D]. South China University of Technology, 2019.
4. Yang Guanghua. Study on the characteristics and mechanism of bird droppings flashover of DC overhead transmission lines [D]. North China Electric Power University (Beijing), 2019.
5. Pang Chuntao. Study on the influence of bird droppings on the flashover characteristics of integrated lightning protection insulators [D]. Shandong University of Science and Technology, 2020.
6. Q. Yang, J. Ran, R. Han, S. Sun, Y. Chen and F. Zhou, "Protection radius of bird-preventing installation design for 110 kV suspension insulators based on withstand tests," in IEEE Transactions on Dielectrics and Electrical Insulation, vol. 23, no. 6, pp. 3514–3522, Dec. 2016.
7. Xie Yuncheng. Research and application of bird sound recognition based on deep learning [D]. North China Electric Power University (Beijing), 2021.
8. Wu Hong, Hao Jinpeng, Yang Kai, et al. Research on influencing factors of guano flashover on 110 kV composite type I insulator [J]. Ningxia Electric Power, 2022, No. 225(03):12–18.
9. Bi Maoqiang, Pan Aichuan, Dong Yang, et al. Corona aging characteristics of HTV silicone rubber material under high humidity conditions [J]. China Materials Progress, 2021, 40(01):74–80.
10. Bi Maoqiang, Pan Aichuan, Yang Junwei, et al. Simulation design of multi-needle-plate electrode corona aging device and research on aging characteristics of silicone rubber [J]. Journal of Chongqing University of Technology (Natural Science), 2021, 35(02):177–184.
11. Wu Kun. Research on optimization of grading ring of composite insulator for UHV DC line[D]. Chongqing University, 2008.
12. Huang Xingmin. Research on improvement of electric field at the end of composite insulator[D]. Shandong University, 2011.
13. H. Wu, Z. Xu, X. Li, J. Hao and Y. Hu, "Research on bird droppings flashover of 110kV transmission line composite insulator insulation-coated grading ring," 2022 IEEE International Conference on High Voltage Engineering and Applications (ICHVE), Chongqing, China, 2022, pp. 1–4.
14. Wu Zhengrong. Online detection of disc suspension porcelain insulators [J]. Porcelain Lightning Arrester, 2006 (03):10–13.

15. Chen Qingguo, Qin Yanjun, Shang Nanqiang, et al. Analysis of the influence of temperature on the electric field distribution in the intermediate joint of high-voltage DC cable [J]. High Voltage Technology, 2014, 40(09): 2619–2626.
16. Liu Zhilu, Li Hengzhen, Liu Gang, et al. Analysis of radial temperature field of overhead conductors based on finite element simulation [J]. Insulation Materials, 2021, 54(03):84–90.
17. Liang Xidong, Zhou Yuanxiang, Zeng Rong. High Voltage Engineering [M]. Tsinghua University Press, 2015.
18. Hu Xiulei. Research on insulation technology of composite insulator grading ring of overhead transmission line [D]. Chongqing University of Technology, 2023.

Chapter 7
Comparison of Abnormal Heating Characteristics of the Decay-Like Degraded Composite Insulator Under Field and Laboratory Conditions

Haiyun Tan, Wenhui Li, Shuqi Liu, Yong Zhang, Xiaoyu Zhou, Yuntu Jiang, Xin Lv, Zhou Zuo, and Xidong Liang

Abstract Field infrared inspection has become one of the most effective approaches to screen out the faulted insulators during operation, but the influence factors of the abnormal heating characteristics of the in-service composite insulators still remain unclear. In this paper, 11 composite insulators with abnormal heating behaviors during operation were acquired after ceasing operation, and were re-energized in the high voltage (HV) test hall to acquire the temperature rise (TR) characteristics in different conditions. The result indicates that the ambient condition influences the number of TR segment, but has no influence on the TR location of each segment. More importantly, the TR characteristics of the samples were acquired after removal of silicone rubber (SR) housing to study the abnormal heat source. Similarly, for some samples, the numbers of TR segment were reduced compared to the test results with SR housing, but the TR location is stable. Remarkable TR was recorded for all samples after removal of SR housing, and large leakage current was detected at the HV end of selected samples. These evidences indicate that abnormal heating phenomenon of the tested composite insulators is likely to be ascribed to the dielectric loss of the decay-like degraded core rod.

Keywords Decay-like degradation · Abnormal heating segment · Composite insulator

---

H. Tan · W. Li · Y. Zhang · X. Zhou · Y. Jiang · X. Lv
Taizhou Power Supply Company, State Grid Zhejiang Electric Power Co., Ltd., Taizhou, Zhejiang, China

S. Liu (✉) · Z. Zuo · X. Liang
State Key Laboratory of Power System Operation and Control, Department of Electrical Engineering, Tsinghua University, Beijing, China
e-mail: liu-sq20@mails.tsinghua.edu.cn

Y. Li et al. (eds.), *Proceedings of the 5th Chinese Conference on High Voltage and Discharge Plasmas*, Springer Proceedings in Physics 254,
https://doi.org/10.1007/978-981-96-5711-7_7

7.1 Introduction

In recent years, abnormal fracture accidents of composite insulators induced by internal degradation of the Fiber-Reinforced-Polymer (FRP) core rod have aroused extensive attention from the power operation and maintenance department [1]. The concept of "decay-like" degradation was put forward to describe the crispy and corroded structure of the degraded FRP core rod [2]. Abnormal heating is widely observed for these decay-like degraded composite insulators in service, thus field infrared inspection has become one of the most effective approaches to screen out the faulted insulators during operation, which is seen as a powerful pre-warning technique to prevent power outage of overhead lines [3, 4].

Aiming at improving defect recognition capabilities of infrared detection technique, substantial attentions have been paid to establish the relationship between infrared measurement results and internal defect conditions. Some researchers analyzed the heating source of the composite insulators, and the different heat derivation of composite insulator is reported under different circumstances. For the heat source located close to the HV end, it is believed that the polarization loss of the aged SR sheath under high humidity should be responsible for the observed TR [5]. Some researchers reported another abnormal heating pattern, with several isolated heating points distributed along the composite insulator string, together with partial discharge phenomenon detected by UV device [6]. This type of heat source was attributed to surface contamination under high humidity. The above-mentioned two types of heating source are significantly affected by ambient humidity, and the TR amplitude usually presents obvious decrease when the infrared detection is conducted in dry weather [7]. Moreover, the core rod makes no contribution to the TR of both types of abnormal heating phenomenon, therefore these types of heating phenomenon should not be identified as internal defect problem. For the TR induced by internal defect of FRP core rod, however, it is reported that the heating source will not disappear in dry conditions, with substantial TR amplitude being detected in several test results [8, 9]. Nevertheless, the effect of ambient condition on the TR distribution is still lack of sufficient attention, and the equivalence between laboratory test results and field infrared inspection remains poorly studied. It is thus necessary to compare the abnormal heating characteristics in different conditions, and acquire the relationship between TR distribution and defect location.

In this paper, the heating characteristics of the decay-like degraded composite insulators are studied in different conditions, including field operation environment, laboratory test conditions using composite insulator samples with and without of SR housing. The TR amplitudes different conditions are compared, and the TR distribution characteristics are analyzed. During laboratory re-energization, the leakage current was recorded for selected samples. The visual degradation status of each sample is provided for reference. This work provides valuable information for the study of abnormal heating mechanism of the decay-like degraded composite insulators.

7.2 Samples and Test Methods

7.2.1 Sample Description

The samples acquired in this paper are 500 kV composite insulators that have been operated in two adjacent transmission lines located in southeastern coastal area of China since 2009, and the string type is V-type. In 2023, abnormal heating behaviors are observed for all samples during routine infrared inspection. After ceasing operation, re-energization at the HV test hall was performed, followed by visual inspection before and after peeling off of the SR housing.

Figure 7.1 presents the typical appearance of the samples. As depicted in Fig. 7.1a, most of these samples exhibit severe hydrophobicity loss phenomenon for the SR sheds close to the HV end. Cracks and puncture holes can also be witnessed at the area close to the HV end in Fig. 7.1c, indicating that the moisture can easily ingress into the interface area of the composite insulator. During the SR housing peeling off process, debonding can be seen between the SR housing and the degraded FRP core rod, some of which account for large area, as shown in Fig. 7.1b. Finally, serious decay-like degradation was confirmed for all samples after complete removal of SR housing around the degradation region, where chalking and corroded appearance of the FRP core rod can be observed, as shown in Fig. 7.1d.

7.2.2 Re-energization Test

A 288 kV (RMS), power frequency voltage was applied to all samples in the HV test hall. After 30 min, the TR was recorded with a FLIR handheld infrared device. After manually peeling of the SR housing at the decay-like degraded segment of the samples, these degraded composite insulators were energized again at the same condition, and similar TR recording procedure was performed. Apart from that, sample #1 and #6 were selected for leakage current testing. To be specific, two microammeters were added in series with sample #1. One of them was located between the HV output and the HV end of the sample, while the other was between the grounded side of sample and the ground. The range of the microammeters was 2 mA, and the precision was 1 μA. The leakage current at the grounded end side was also measured by a 10 kΩ sampling resistor in parallel with an oscilloscope, which automatically calculate the sampling voltage based on the voltage waveform signal.

Fig. 7.1 Typical appearance of the aged composite insulators acquired from the operation site

7.3 Results

7.3.1 Comparison of TR Amplitudes in Different Conditions

The samples used in this paper experienced three round of TR measurement in different conditions: the first time for field operation, the second time for laboratory re-energization test with SR housing, and the final round for laboratory re-energization test without SR housing. Figure 7.2 demonstrates the comparison of the TR amplitudes under these conditions. It can be found that the TR amplitudes of most samples during operation are much higher than the values acquired in laboratory. Considering that the humidity usually maintains at a high level in the coastal area where these composite insulators were located during service, it is thus assumed that the contrast may be induced by humidity.

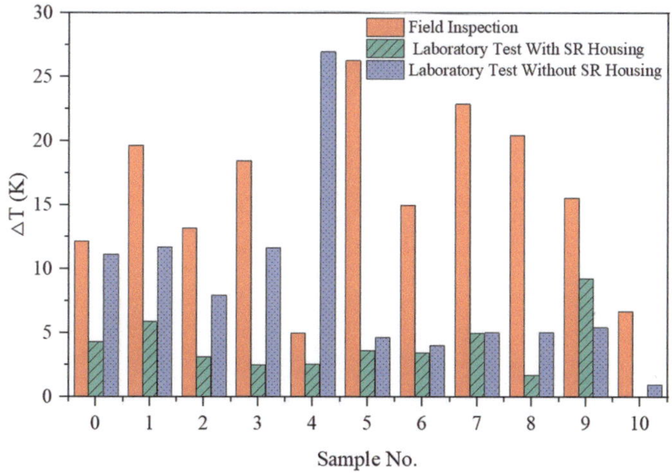

Fig. 7.2 Comparison of TR amplitudes of tested composite insulators in different conditions

Another noticeable fact is that for the majority of samples, the temperature rise amplitudes are higher when tested without SR housing compared with the values when tested with SR housing. This is because of the shielding effect of the SR sheath, since the temperature measured at the SR sheath is lower than the temperature of the heat source due to heat conduction within the SR sheath.

7.3.2 Comparison of TR Distribution Characteristics in Different Conditions

In order to better compare the TR characteristics in different conditions, the temperature distribution curve along the axial direction is extracted for each composite insulator samples. Subsequently, the data is plotted in the form of color-bar as depicted in Fig. 7.3, the color of which stands for the TR values at different locations. To be specific, the left color-bar corresponds to the field infrared inspection result, the middle to the laboratory test result with SR housing, and the right to the laboratory test result with SR housing. In addition, the visible ending location of the decay-like degradation channel on the FRP core rod surface is also marked by the star sign for each sample.

According to Fig. 7.3, a large difference can be observed regarding to the temperature distribution in different conditions. Firstly, for the field inspection results, the TR segment is larger than the adjacent big shed space for all samples, and more than one TR segment can be observed for most of the samples. It indicates that rather lengthy defects may be embedded inside the SR housing for the heating composite insulators, and they are likely to be located in multiple sites. For instance, a major

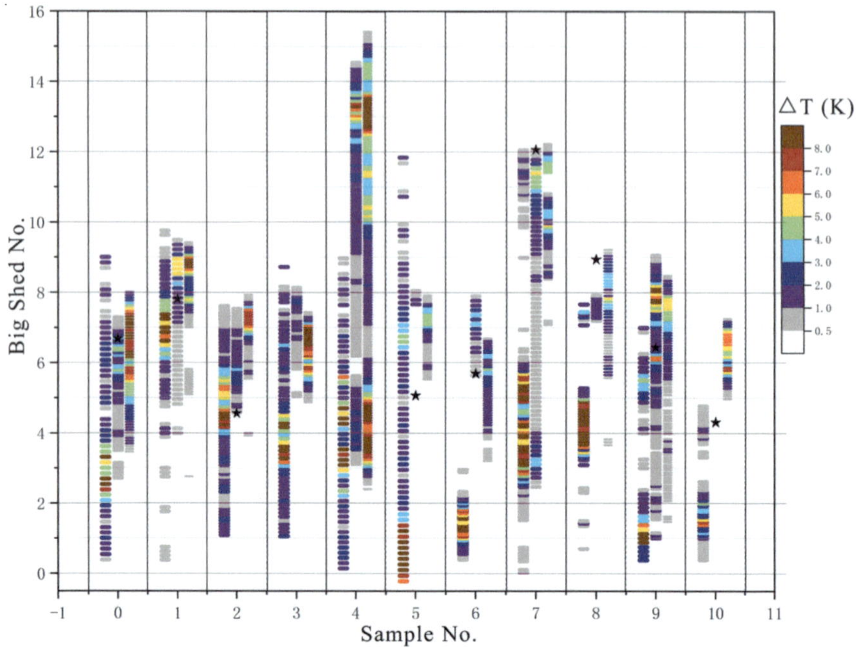

Fig. 7.3 Comparison of TR distribution of tested composite insulators in different conditions. Left to right for each sample: field infrared inspection result, laboratory test result with SR housing, laboratory test result without SR housing. Star sign: ending location of the decay-like degradation channel observed at the corroded FRP core rod surface

TR segment can be found between No. 2 and No. 4 big sheds for sample #0, while another TR segment with lower TR value is located at No. 6 big shed. However, some heating segments disappear when the samples are moved to the HV test hall for re-energization, like the heating segment between No. 2 and No. 4 big sheds for sample #0. In view of the TR amplitude decrease phenomenon shown in Fig. 7.2, it is inferred that the heat power in these segments reduce significantly due to the decrease of ambient humidity, leading to much lower TR values. Although some heating segments disappears, it can be observed that the remaining TR segments in each laboratory test result are located in the same sites as they are in the field inspection result. That is to say, the alteration of ambient condition will not shift the heating locations of the original defects.

7.3.3 Leakage Current of Samples During Laboratory Re-energization Test

The leakage current test for sample No. 1 and No. 6 were conducted successively after removal of SR housing, and the result was presented in Fig. 7.4. It can be found

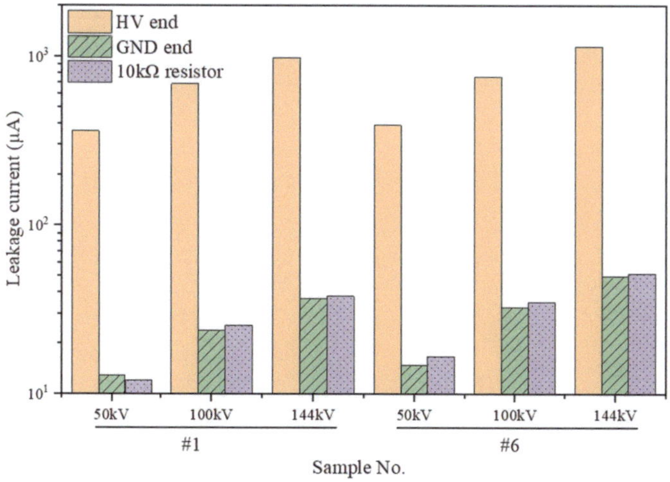

Fig. 7.4 Leakage current of sample #1 and #6 measured at the HV end, the GND end, and at the 10 kΩ sample resistor

that the leakage current measured from the grounded side maintains at a rather low level even when the RMS value of voltage exceeds 100 kV. However, the leakage current measured from the HV side reaches hundreds of μA or 1 mA, forming a significant contrast with the grounded side values. This can be a crucial evidence that accounts for the abnormal heating phenomenon of the composite insulators, because enormous current that flows into the decay-like degraded segment of the core rod is very likely to induce considerable energy dissipation. In contrast, the negligible leakage current measured at the grounded side will not cause detectable TR at the intact region of composite insulators.

7.4 Conclusions

In this paper, and the heating characteristics of the decay-like degraded composite insulators are studied in different conditions. Based on the results obtained in field and laboratory conditions, the following conclusions can be drawn:

(1) Segmental type of TR phenomenon is observed for the decay-like degraded composite insulators in this paper, in both field and laboratory conditions. The TR segments are not limited around the HV end, and more than one heating segments are observed for most samples.
(2) After being transferred to the HV test hall for re-energization, the TR amplitudes are significantly reduced for most samples, and some heating segments disappear compared to field inspection results. Nevertheless, the location of remaining heating segments stay at the same location as the original heat defects

do. Therefore, it is believed that the ambient humidity tends to affect the TR amplitude and range, but will not shift the heat defect location.

(3) The leakage current measured at the HV side is significantly higher than that measured at the grounded side, indicating that the heating energy is very likely to come from the dielectric loss of the degraded FRP core rod. It is assumed that both conduction loss and polarization loss are involved in the heat source.

Acknowledgements This work was supported by the State Grid Zhejiang Electric Power Co., Ltd. science and technology project (grant number: 5211TZ240001).

References

1. Liang, X., Bao, W., Gao, Y.: Decay-like fracture mechanism of silicone rubber composite insulator. IEEE Trans. Dielectr. Electr. Insul. 25(1), 110–119 (2018).
2. Wang, J., et al.: Failure analysis of decay-like fracture of composite insulator. IEEE Trans. Dielectr. Electr. Insul. 21(6), 2503–2511 (2015).
3. Li, Te, et al.: Temperature distribution characteristics and heat defect judgment method based on temperature gradient of suspended composite insulator in operation. IET Gener. Transm. Dis. 15, 2554–2566 (2021).
4. Yuan, Z., Tu, Y., Li, R., Zhang, F., Wang, C.: Review on characteristics, heating sources and evolutionary processes of the operating composite insulators with abnormal temperature rise. CSEE J Power. Energy. Syst. 8(3), 910–921 (2020).
5. Wang, L., et al.: Effect of Damp Sheath on Abnormal Temperature Rise at End of Composite Insulator. Power System Technology. 40(2), 608–613 (2016) (in Chinese).
6. Liu, H., Shen Q., Yong, J., Meng, H.: Detection and processing for ±660 composite insulator abnormal discharge. 2014 International Conference on Power System Technology. 1404–1409 (2014).
7. Li, R., et al.: A preliminary study on the impact of environment to the heat of insulator. 2013 Annual Report Conference on Electrical Insulation and Dielectric Phenomena. 415–417(2013).
8. Zhong, Z., Chen, Y., Liu, Y., Hou, S., Geng, J.: Study on the influence mechanism of ambient humidity on the temperature rise of decay-like composite insulators. High Volt. 7(5), 916–924 (2022).
9. Zhang, Z., Pang, G., Lu, M., Gao, C., Jiang, X.: Study on decay-like fracture of 500 kV composite insulators: Infrared, ultraviolet and electric field distribution detection. IET Gener, Transm & Dis, 16(20), 4132–4141 (2022).

Chapter 8
Cause Diagnosis of Lightning Damage in Transmission Corridor Based on Parameter Normalization Model

Chaoying Fang, Ziqi Lin, Junfeng Zhang, Xiuyi Li, Yeqiang Deng, Yu Wang, Haochen Zhang, Xiao He, and Zhuokai Shao

Abstract Transmission lines have different risks of lightning accidents due to the differences of route topography, line structure, lightning activity intensity and other factors. It is of great significance to study the causes of lightning damage of transmission lines for the safe operation of power grid. Therefore, this paper first establishes parameter reference intervals based on the statistical results of multi-source data and refers to national standards, and then forms a parameter normalization model as the basis for subsequent research. Then, an analysis method of the cause of lightning damage based on parameter state is proposed. A 220 kV line and a typical tower are analyzed. It is found that the ground flash density is the biggest cause of lightning damage, accounting for 27.9%, and the pole tower of the line may have a serious risk of lightning damage.

Keywords Multi-source data · Parameter normalization · Lightning damage cause diagnosis

8.1 Introduction

Transmission lines are the fundamental components of the power grid, and their safe operation is of great significance to the safety of the power grid. The terrain of the transmission corridor is quite complex, with diverse landforms and meteorological conditions, and its safe operation and maintenance are greatly threatened by lightning damage. Related studies have shown that the terrain, line structure, and tower

C. Fang · Z. Lin · J. Zhang
State Grid Fujian Provincial Electric Power Company Limited Electric Power Science Research Institute, Fuzhou Fujian, China

X. Li · Y. Deng · Y. Wang · H. Zhang · X. He · Z. Shao (✉)
School of Electrical Engineering and Automation, Wuhan University, Wuhan, China
e-mail: 2024102070016@whu.edu.cn

Y. Li et al. (eds.), *Proceedings of the 5th Chinese Conference on High Voltage and Discharge Plasmas*, Springer Proceedings in Physics 254,
https://doi.org/10.1007/978-981-96-5711-7_8

parameters of transmission lines, as well as differences in lightning activity intensity, can all lead to varying risks of lightning accidents [1–3]. Therefore, based on the voltage level of transmission lines and the situation of single and double circuits, in-depth research should be conducted on the differences in lightning activity intensity, terrain and landform characteristics, and line structural parameters of the line corridor. Establishing a scientific and standardized lightning damage cause diagnosis model and intelligent recommendation method for optimization measures are of great significance for the safe and stable operation of the power grid.

The current research on the causes of lightning damage and the proposal of lightning protection measures relies heavily on manual labor [4–6], which requires a specific analysis of the line or tower based on environmental parameters, tower parameters, and lightning trip rates. This can lead to lower efficiency and may be influenced by subjective judgments.

In recent years, some scholars have proposed using the Analytic Hierarchy Process to evaluate lightning protection measures. In 2005, Zhang Zhijin [7] et al. constructed a structural model for evaluating lightning protection measures for transmission lines based on the core idea of Analytic Hierarchy Process. In 2018, Zhou Yusheng et al. [8] optimized the Analytic Hierarchy Process (AHP) through extensional methods, using elementary theory, extensional sets, and extensional logic to quantify the degree to which things belong to a certain set and select appropriate lightning protection measures. In 2019, Wang et al. [9] used the 10/10–18/2 scale method in the Analytic Hierarchy Process to establish a comprehensive evaluation model for lightning protection measures from multiple dimensions such as service life, engineering cost, and reducing lightning trip rate.

However, the Analytic Hierarchy Process has its limitations in practical applications. Among them, the scoring process is highly subjective and overly dependent on the personal judgment of experts. In addition, the analysis process is relatively cumbersome and time-consuming, especially when dealing with large amounts of data, making the determination of weights particularly difficult. Although scholars have attempted various optimizations based on the Analytic Hierarchy Process, the subjective color is still difficult to completely eliminate.

Therefore, this article proposes a lightning damage method based on a normalized model. Based on the statistical results of multi-source data and national standards, a parameter reference interval is established to form a parameter normalization model. Finally, based on the normalization results, the cause of lightning damage is diagnosed.

Table 8.1 The power frequency ground resistance of lines with ground wires not connected to the ground wire

Soil resistivity (Ω m)	≤100	100 ~ 500	500 ~ 1 k	1 ~ 2 k	>2 k
Power frequency Ground resistance (Ω)	10	15	20	25	30

8.2 Parameter Reference Interval

8.2.1 Grounding Resistance Interval

According to the Design Code for 110 ~ 750 kV Overhead Transmission Lines (GB50545-2010), the grounding resistance should not exceed the values in Table 8.1 [10].

The standard only specifies the upper limit of each interval and does not provide any explanation for situations below 10 Ω. Based on the statistical data of the tower, 5 −30 Ω can cover the vast majority of cases of the tower's power frequency grounding resistance. 5 Ω is selected as the optimal value of the power frequency grounding resistance interval, and 30 Ω is selected as the worst value of the interval.

8.2.2 Length Rrange of Insulator String Connection

According to the Design Code for 110 ~ 750 kV Overhead Transmission Lines (GB50545-2010), the minimum number of insulators should generally not be less than the specified number in Table 8.2 [10], and the reference interval for the length of the insulator string should be obtained based on this.

According to the Design Code for 110 ~ 750 kV Overhead Transmission Lines (GB50545-2010), when using the creepage distance method [10], the number of insulator pieces should be calculated according to Eq. (8.1).

$$n \geq \frac{\lambda U}{K_{\mathrm{e}} L_{\mathrm{o1}}} \tag{8.1}$$

The Design Code for 110 ~ 750 kV Overhead Transmission Lines (GB50545-2010) [10] also provides the creepage distance under different pollution levels. To eliminate extreme situations, the upper limit of creepage distance at pollution level III

Table 8.2 The minimum number and length of insulator strings

Nominal voltage (kV)	110	220	500
The height of a single insulator (mm)	146	146	155
Number of insulator pieces (Pieces)	7	13	25
Length of insulator string (mm)	1022	1898	3875

Table 8.3 The maximum number and length of insulator strings

Nominal voltage (kV)	110	220	500
The height of a single insulator (mm)	146	146	155
Number of insulator pieces (Pieces)	12	24	52
Length of insulator string (mm)	1752	3504	8060

of 3.20 cm/kV is selected for calculation. In addition, XP-70 insulators are selected for 110 and 220 kV lines; Select XP-160 insulators for 500 kV transmission line towers. According to the calculation results, the optimal reference interval for the length of the insulator string is obtained, as shown in Table 8.3.

As shown above, the optimal length of the insulator string for the 110 kV line is 1752 mm, and the worst value is 1022 mm; The optimal length of the insulator string for a 220 kV transmission line is 3504 mm, and the worst value is 1898 mm; The optimal length of insulator string for 500 kV transmission line is 8060 mm, and the worst value is 3875 mm.

8.2.3 Tower Height Reference Interval

Based on the statistical data of 97 towers in the 110 kV Songxun Round Line, 139 towers in the 220 kV Songyu Line, and 187 towers in the 500 kV Yiquan Yi Line, the height distribution of towers at each voltage level was obtained, as shown in Fig. 8.1. Compared to other lines, the tower data of the above three lines is moderate and can serve as a reference for the normal level of tower height.

The R^2 values of the Gaussian fitting curves for 110, 220, and 500 kV lines are 0.92569, 0.79595, and 0.88473, respectively, indicating that a normal distribution can be used for fitting. Therefore, based on the three sigma principle of normal distribution, the expected value plus or minus standard deviation is adopted, and 1.5

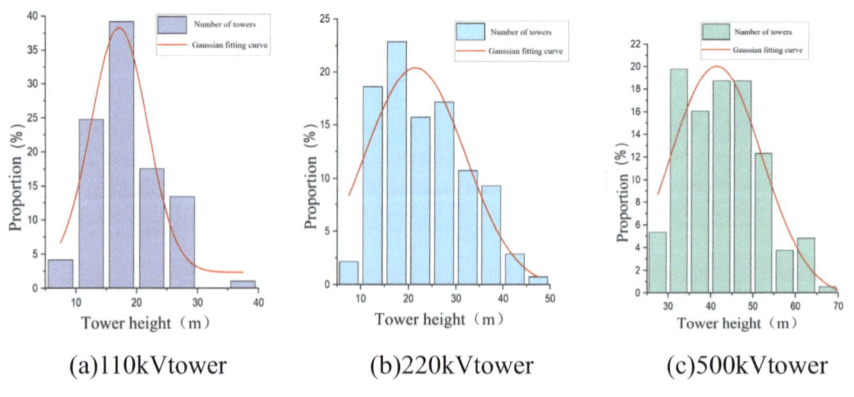

Fig. 8.1 Tower height statistics

Table 8.4 The characteristic value of tower height statistics

Voltage level (kV)	110	220	500
Expected μ	18.50	23.19	42.67
standard error	5.48	8.77	9.07
$\mu - 2\sigma$ (m)	10.28	10.04	29.07
$\mu + 2\sigma$ (m)	26.71	36.35	56.27
Proportion <2σ (%)	4.12	2.25	4.81
Proportion >2σ (%)	10.31	13.48	8.02

standard deviations above and below the expected value are used as the reference interval for tower height. The processing results according to the above requirements are shown in Table 8.4.

As can be seen, the parameter range obtained by the above method covers most of the tower heights and can well represent the reference interval. Therefore, the above parameter range is selected as the reference interval. The optimal value for a 110 kV tower is 10.282 m, and the worst value is 26.710 m; The optimal value for a 220 kV tower is 10.037 m, and the worst value is 36.351 m; The optimal value for a 500 kV tower is 29.070 m, and the worst value is 56.271 m.

8.2.4 *Reference Interval for Lightning Protection Angle of Lightning Protection Wire*

According to the Design Specification for 110 ~ 750 kV Overhead Transmission Lines (GB50545-2010), for single circuit, the protection angle of 110 and 220 kV lines should not exceed 15°, and the protection angle of 500 kV lines should not exceed 10°. For dual circuits, the protection angle of 110 kV lines should not exceed 10°, and the protection angle of 220 and 500 kV lines should not exceed 0° [10]. According to the above standards, the worst protection angle of the lightning protection line can be determined.

The standard has not yet specified the optimal value for the protection angle of lightning conductors. Analyze tower statistical data based on the classification method in GB50545-2010. For a single circuit, based on the 118 single circuit towers in the 110 kV Songxun Round Line and 220 kV Songyu Line, when selecting 5° as the optimal value for the 110 and 220 kV lines, 3.39% of them fall outside the optimal value; According to the 137 single circuit towers in the 500 kV Yiquan Line, when selecting 0° as the optimal value for the 500 kV line, 1.46% of them fall outside the optimal value. For double circuit lines, based on the 70 double circuit towers in the 110 kV Songxun round line, when selecting 0° as the optimal value for the 110 kV line, 1.43% of them fall outside the optimal value; According to the 99 double circuit towers in the 220 kV Songyu Line and 500 kV Yiquanyi Line, when selecting −10° as the optimal value for the 110 kV line, 1.01% falls outside the optimal value.

Based on the above analysis, the optimal value of lightning protection angle can be determined.

In summary, the worst value of the lightning protection angle for 110 kV single circuit lines is 15°, and the optimal value is 5°; The worst value for a dual circuit line is 10°, and the worst value is 0°. The worst value for a 220 kV single circuit line is 15°, and the optimal value is 5°; The worst value for a dual circuit line is 0°, and the optimal value is −10°. The worst value for a 500 kV single circuit line is 10°, and the optimal value is 0°; The worst value for a dual circuit line is 0°, and the optimal value is −10°.

8.2.5 Ground Inclination Reference Range

According to the statistics of 1228 towers in 433 110 kV lines such as Puzheng Line, 523 towers in 201 220 kV lines such as Baizhu Line, and 258 towers in 80 500 kV lines such as Ajin Jia Line, a total of 2000 towers have left and right ground inclination angles, which are fitted with Gaussian curves. The distribution is shown in Fig. 8.2.

The R-squared value of the Gaussian fitting curve is 0.98483, indicating that the normal distribution fitting effect is very good. Therefore, based on the three sigma principle of normal distribution, the processing results of ground inclination are shown in Table 8.5.

Selecting the expected value plus or minus two standard deviations as the range of ground inclination angle, the proportion of data falling outside the range is 4.73%, which belongs to the extreme data that should be filtered out and can well represent

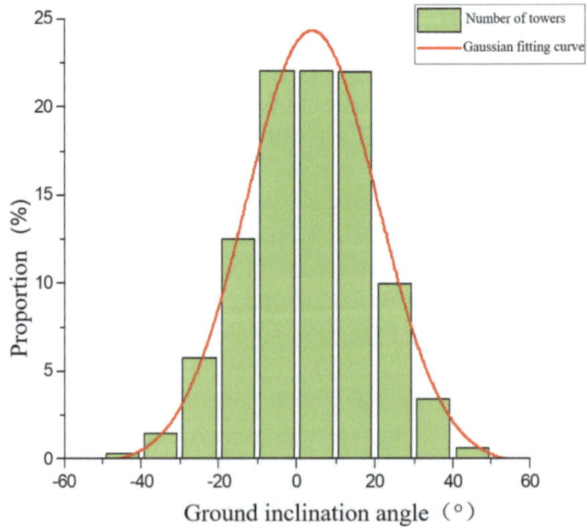

Fig. 8.2 Ground slope angle statistics

Table 8.5 The characteristic value of ground slope angle statistics

Calculate the characteristic values of the results	Angle
Expected μ	3.20°
Standard error	15.72°
μ-2σ	−28.23°
μ + 2σ	34.63°
Proportion <2σ	2.85%
Proportion >2σ	1.88%

the reference interval. Therefore, the optimal ground inclination angle is −28.23° and the worst is 34.63°.

8.2.6 Reference Interval for Ground Flash Density

According to DL/T 1533-2016 "Method for Drawing the Distribution Map of Lightning Zones in Power Systems" [11], ground flash density can be divided into 7 levels.

Based on the statistical data of ground flash density, it is considered that the part included in D2 level belongs to extreme cases. 11.0 times/(km^2 a) is selected as the worst value of ground flash density, and the optimal value of ground flash density is 0 times/(km^2 a).

8.3 Analysis of Thunder Damage Causes Based on Parameter Reference Interval

8.3.1 Normalization Method Based on Parameter Reference Interval

Based on the parameter reference interval obtained in Sect. 8.1, further normalize the parameters to eliminate dimensional and amplitude differences, as the basis for analyzing the causes of lightning damage and optimizing measures. Normalize the insulator string length, tower height, lightning protection angle, ground inclination angle, and ground flash density using Eq. (8.2).

$$\overline{x} = \frac{x - x_{worst}}{x_{best} - x_{worst}} \tag{8.2}$$

In the formula, $\overline{x}$ represents the per unit value of the parameter, x represents the parameter before adopting the per unit value, x_{best} and x_{worst} respectively refer to the optimal and worst values of the parameter in the reference interval under the current voltage level and single/double circuit conditions. Among them, the maximum ground inclination angle is selected to calculate the ground inclination angle per unit.

For grounding resistance, logarithmic calculation followed by normalization is more in line with the actual situation, as shown in Eq. (8.3).

$$\overline{x} = \frac{\ln x - \ln x_{worst}}{\ln x_{best} - \ln x_{worst}} = \log_{\frac{x_{best}}{x_{worst}}} \frac{x}{x_{worst}} \tag{8.3}$$

In the formula, $\overline{x}$ represents the per unit value of the grounding resistance parameter, x represents the grounding resistance parameter before adopting the per unit value, x_{best} and x_{worst} respectively refer to the optimal and worst values of the reference interval of the grounding resistance parameter.

8.3.2 Diagnostic Methods for the Causes of Lightning Damage

The normalization method based on parameter reference intervals in Sect. 2.1 makes it possible to compare the superiority and inferiority of parameters under different dimensions. When the standard value is 0, it corresponds to the worst value of the parameter reference interval; When the standard value is 1, it corresponds to the optimal value of the parameter reference interval. When the unit value is <1, the difference between the current parameter unit value and 1 can represent the distance between the current parameter state and the good state, and can characterize the degree to which the adverse condition of the parameter affects the level of lightning damage risk. For parameters with a standard value greater than or equal to 1, it can be considered that they have reached a good state and are not considered to be the cause of high lightning damage risk. According to the above analysis method, the proportion of lightning damage causes with different parameters can be obtained. Taking the grounding resistance parameter as an example, the proportion of lightning damage causes is shown in Eq. (8.4).

$$C_w = \frac{\max(0, 1 - \overline{w})}{\begin{array}{c}\max(0, 1 - \overline{w}) + \max(0, 1 - \overline{x}) + \max(0, 1 - \overline{y}) + \\ \max(0, 1 - \overline{z}) + \max(0, 1 - \overline{u}) + \max(0, 1 - \overline{v})\end{array}} \tag{8.4}$$

In the formula $\overline{w}$, $\overline{x}$, $\overline{y}$, $\overline{z}$, $\overline{u}$, $\overline{v}$ respectively represent the unit values of grounding resistance, insulator string length, tower height, lightning protection angle, ground inclination angle, and ground flash density parameters. C_w represents the proportion of lightning damage caused by grounding resistance.

8.3.3 *Diagnosis of Lightning Damage Causes in 220 kV Transmission Lines*

According to the analysis method of lightning damage causes in Sect. 2.2, the proportion of lightning damage causes obtained is shown in Fig. 8.3.

It can be seen that there are some differences in the causes of lightning damage among different towers, and the problems of lightning protection angle and grounding resistance only exist in some towers. In addition, the proportion of causes of different parameters varies overall, and the average proportion of causes of 39 high-risk towers is shown in Fig. 8.4.

The ground flash density is the largest cause of lightning damage on the 220 kV line, accounting for up to 27.9%. The line is located in a strong lightning zone with a high frequency of ground flashes, so the cause of this parameter is relatively large. Following closely behind are three parameters: ground inclination angle, insulator string length, and tower height, accounting for 17.7%, 17.2%, and 16.1% respectively, which are also significant contributing factors. The lightning protection angle accounts for 13%. Finally, the grounding resistance, which accounts for a relatively small proportion of 8.1%, is the cause. This indicates that the grounding resistance of the circuit is in a relatively good state. However, considering that ground flash

Fig. 8.3 Percentages of lightning causes of each tower

Fig. 8.4 Average percentages of lightning causes

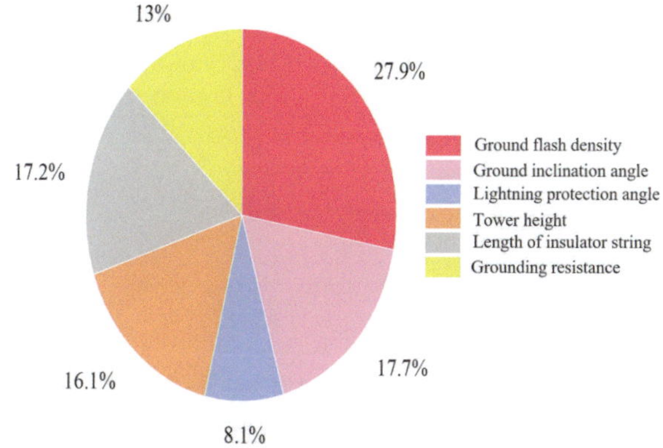

density and ground inclination are unchangeable environmental parameters, further optimization measures can only be proposed from the other four parameters. Firstly, the protection angle of lightning rods can improve the risk of flashover caused by large ground inclination angles. But the protection angle of some tower lightning protection wires has already reached the standard. By reducing the height of the tower, the ground strike distance can better cover the tower conductors, and the problem of excessive ground inclination can also be improved.

In summary, the most serious problem with this line is the high ground flash density, which is an environmental factor that cannot be changed. In addition, the large inclination angle of the ground and the relatively high tower height may cause serious risk of flashover. If the protection angle of the lightning protection line has already met the standard, this can be optimized by reducing the tower height. In addition, the grounding resistance is in a relatively safe state, and the risk of counterattack in this circuit should not be too high.

8.3.4 *Analysis of the Causes of Lightning Damage in Tower 034# and Tower 052#*

According to the ledger query and calculation data, the specific calculation parameters of tower 034# and tower 052# are shown in Table 8.6.

According to the analysis method of lightning damage causes in Sect. 2.2, the proportion of lightning damage causes can be obtained from various characteristic parameter information. The proportion of lightning damage causes for tower 034# and tower 052# is shown in Fig. 8.5.

Among them, as with the overall situation of the line, the main reason for towers 034# and 052# is the high ground flash density. Meanwhile, the heights of these two towers are also relatively high. The ground inclination angle of tower 034# accounts for 16.4%, because this tower has a positive ground inclination angle of over 15°,

Table 8.6 Characteristic parameters of 034# tower and 052# tower

Tower number	Ground flash density	Left ground inclination angle	Right ground inclination angle	Grounding resistance
034#	13.41 time/(km^2 a)	16.62°	−16.00°	15 Ω
052#	12.20 time/(km^2 a)	−23.67°	40.48°	15 Ω
	Insulation string length	Tower height	Mid distance of lightning protection wire	Lightning protection angle
034#	2190.0 mm	36.0 m	7.5 m	0°
052#	2190.0 mm	39.0 m	7.5 m	0°

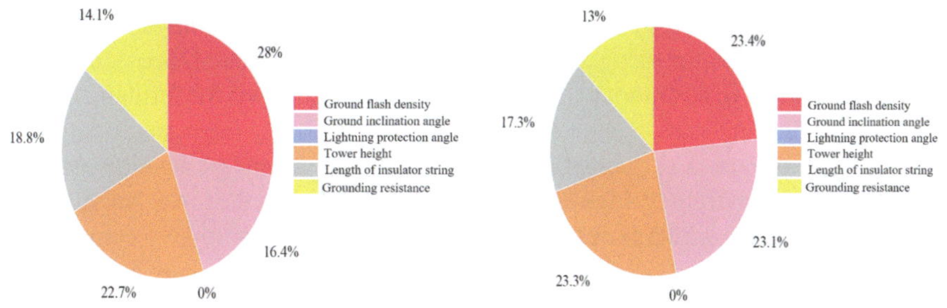

(a) 034#tower (b) 052#tower

Fig. 8.5 Percentages of lightning causes of 034# tower and 052# tower

which may cause a significant risk of derailment. The ground inclination angle of Tower 052 accounts for 23.1%, because the ground inclination angle of this tower exceeds 40°. The environmental conditions are extremely harsh, making it almost impossible to protect the towers and lines within the range of ground strikes, resulting in a great risk of derailment, and it may be difficult to take measures to reduce the risk of derailment.

8.4 Conclusion

At present, the analysis of the causes and optimization measures of lightning damage to transmission towers mostly relies on manual labor and lacks intelligent analysis models. Therefore, this article proposes a parameter normalization based method for diagnosing the causes of lightning damage and an intelligent optimization method for lightning protection measures based on gradient thinking, and analyzes a certain 220 kV line and typical towers. The main achievements and conclusions of this article are as follows.

(1) Based on national standards and parameter statistical data, the parameter reference ranges of six characteristic parameters, including grounding resistance, insulator string length, tower height, lightning protection angle, ground inclination angle, and ground flash density, were obtained through calculation and analysis under different voltage levels and single/double circuit conditions. This range can cover the vast majority of normal data, excluding a few extreme data. Furthermore, based on this range, a normalization method can be formed to characterize the parameter state and serve as the basis for subsequent analysis.

(2) A parameter state based method for diagnosing the causes of lightning damage is proposed based on parameter normalization. This method is based on the state of the tower parameters and provides the proportion of lightning damage causes for each parameter. According to this method, analyze the 220 kV line. The main cause of lightning damage on this line is the ground flash density, accounting

for 27.9%, followed by a large ground inclination angle, and insufficient tower height and insulator string length. There may be a significant risk of derailment on this route. In addition, two typical towers were selected for further analysis. The ground inclination angle of tower 034# is relatively large, exceeding 15°, which may pose a certain risk of derailment. The ground inclination angle of Tower 052 is extremely high, even reaching up to 40°, and the tower height is also relatively high, which poses a serious problem of exposure due to surrounding impact.

Acknowledgements Project Supported by Science and Technology Project of State GridFujian Electric Power Company Limited: Research and Demonstration Application of Digital Key Technologies for Improving Lightning Protection Performance of Overhead Power Lines in Complex Mountainous Areas (52130423004L).

References

1. Chen Jiahong, Lv Jun, Qian Zhiyin, et al. Differentiation Technology and Strategy of Lightning Protection for Transmission Lines[J]. 高电压技术, 2009(12): 12.
2. SARDI J, AB KADIR, M Z A. Investigation on the effects of line parameters to the lightning performance of 132 kV Kuala Krai-Gua Musang transmission line[C]//2013 IEEE 7th International Power Engineering and Optimization Conference (PEOCO 2013), 3–4 June 2013, Langkawi, Malaysia: Institute of Electrical and Electronics Engineers, 2013: 594–599.
3. ZHAO Longbo. Research and Application of Differential Lightning Protection Technology for Transmission Lines[D]. Shandong University of Technology, 2020.
4. Suo Zhen. Research on Lightning Protection Technology of Transmission Line in Handan[D]. North China Electric Power University, 2015.
5. Zhu Shiyang, Deng Yurong, Xie Yaoheng, et al. The Application of Differentiation Technology of Lightning Protection in Guangxi Power Grid[J]. Southern Power System Technology, 2012, 6(z1):80–83.
6. WANG Yewei. Differentiated Lightning Protection Technology Application and Evaluation of Transmission Lines in Jiangxi Province [D]. North China Electric Power University, 2013.
7. Zhang Zhijin, Sun Caixin, Jiang Xingliang, et al. Application of analytic hierarchy process in estimation of compositive transmission line lightning protection measures[J]. Power System Technology, 2005. 29(14): 68–72.
8. ZHOU Yusheng, LIU Chaozhi, ZHOU Shun. Comprehensive Evaluation of Lightning Protection Retrofitting Measures for 500 kV Transmission Line Based on Extension Analytic Hierarchy Process[J]. Insulators and Surge Arresters, 2018(5): 1–6, 12.
9. Wang Yan, Chen Weiwei. Evaluation of lightning protection measures for transmission lines based on 10/10-18/2 scale method[J]. Electronic Measurement Technology, 2019, 42(13): 49–54.
10. GB 50545-2010, 110kV~750kV Code for design of 110kV~750kV overhead transmission line [S]. 2010.
11. Drawing method of lightning zones map for power system: DL/T 1533-2016[S]. 2016.

Chapter 9
Online Measurement Mechanism of Surface Charge Surge on GIL Insulators

Fangwei Liang, Xianhao Fan, Hanhua Luo, Zuodong Liang, Chuanyang Li, and Jinliang He

Abstract Surface charge of insulator has the characteristic of nonuniform distribution. In particular, the electric field can be severely distorted by a surge of surface charges. Thus, the insulation reliability of DC gas-insulated transmission lines (GILs) is threatened. In this paper, the online measurement method and mechanism for surface charge surge is proposed based on the Kelvin electrostatic probe. The online measurement platform for surface charge surge is constructed. The effectiveness of the measurement mechanism is validated. The effects of initial surface charge are analyzed. The results indicate that it is easy to lead to a sudden change in the surface potential when the insulator surface charge sharply increases. Furthermore, there is dissipation of the accumulated surface charge. Once the surface charge is surged, there is a rapid decrease in surface potential. The decay rate of the surface potential is positively related to the initial surface charge. Therefore, the surface potential is affected by the surface charge surge. It is inferred that the insulator surface charge surge can be characterized based on the change in surface potential at the pouring port of basin insulators measured by the Kelvin electrostatic probe. The work in this paper provides a reference for online measurement of the surface charge surge on GIL insulators, improving the reliability of DC GIL insulators by online monitoring to avoid surface flashover.

Keywords Gas-insulated transmission line · Insulator · Surface charge · Surface potential · Surface charge dissipation · Surface charge measurement

---

F. Liang (✉) · X. Fan · H. Luo · Z. Liang · C. Li · J. He
Department of Electrical Engineering, Tsinghua University, Beijing, China
e-mail: liangfangwei@tsinghua.edu.cn

Y. Li et al. (eds.), *Proceedings of the 5th Chinese Conference on High Voltage and Discharge Plasmas*, Springer Proceedings in Physics 254,
https://doi.org/10.1007/978-981-96-5711-7_9

9.1 Introduction

Insulator surface charge accumulation could reduce the surface flashover voltage, which is one of the main threats to the insulation reliability of DC gas-insulated transmission lines (GILs) [1, 2]. The surface charge exhibits uneven distribution due to partial discharge caused by insulation defects such as metal particles and triple junction of the gas-epoxy-conductor [3, 4]. Specifically, the distortion of the electric field on the insulator surface is more severe due to the surge of surface charge in local regions. However, the surface charge distribution of insulators cannot be directly measured. The operation voltage of the conductor in GIL, the insulator surface charge, and the internal space charge of the insulation medium are the main sources of the insulator surface potential. Therefore, the surface charge distribution can be obtained through the surface charge inversion algorithm by measurement of the surface potential distribution [5, 6]. Measurement techniques for the surface potential of insulators mainly include dust map, Pockels method, capacitive electrostatic probe method, and Kelvin electrostatic probe method [7, 8]. The polarity distribution of surface potential in the dust map is reflected through the adsorption of charged colored dust particles. This method is simple but only suitable for qualitative analysis [9, 10]. The Pockels method is based on the electrooptic effect of Pockels crystals, which characterizes the difference of electric field by the change of light intensity, thus enabling the quantitative measurement of surface potential [11]. Insulators are usually epoxy resin doped with aluminum oxide, which is a nontransparent material. Therefore, the Pockels method is not applicable to GIL insulators [12, 13]. It is inferred that the electrostatic probe method is the main method for surface potential measurement of GIL insulators.

The electrostatic probe method is divided into two types: capacitive and Kelvin. For the capacitive electrostatic probe, electrostatic induction occurs through the probe, the induced potential on the sensitive electrode is collected, and then surface potential is calculated based on the division principle of capacitive voltage [14]. This method is often suspected due to the accuracy of capacitive electrostatic probes. Therefore, the Kelvin electrostatic probe method based on the principle of electric field compensation is an excellent measurement method for the surface potential of insulators. During the measurement, the Kelvin probe is adjusted in real time by the position control system. The probe is kept perpendicular to the measured surface and about 3 mm away from each other [15, 16]. The surface potential of the entire insulator surface can be obtained through measuring point by point [17, 18]. However, the Kelvin electrostatic probe method is limited to offline measurement of the insulator surface potential [19]. That is to say, the high voltage electrode must be grounded before measurement. This is mainly because the range of the Kelvin electrostatic probe is only $\pm$ 20 kV, which is much lower than the operation voltage of GIL. For the offline measurement method of Kelvin probes, the surface potential depends on the dissipation of surface charges and the depolarization of space charges inside insulators. In contrast, the change of surface potential for online measurements is influenced by the surface charge surge. It inferred that the change of surface potential at the pouring port of the basin insulator measured by the Kelvin electrostatic probe could characterize the insulator surface charge surge. However, there is still a lack of an online measurement mechanism for the insulator surface charge surge in GIL based on Kelvin probes.

In this paper, corona discharge is used to accumulate an initial surface charge on post insulators. The change of insulator surface potential with time is measured based on the Kelvin electrostatic probe. The differences at different applied voltages, insulator heights, and temperatures are compared to reveal the measurement mechanism of surface charge surge on the GIL insulator.

9.2 Experimental Arrangement

In order to verify the theory of surface potential change, an online measurement platform for the insulator surface charge surge is established, as shown in Fig. 9.1. The Al_2O_3-epoxy post insulator with a diameter of 80 mm are placed in a constant temperature drying chamber, which is grounded. The bottom surface of the insulator is in close contact with the grounded electrode. The corona needle is situated in the middle region on the insulator side. The corona needle is connected to a DC voltage source through a high voltage bushing, which is used to rapidly accumulate initial surface charges during corona discharge to simulate the surface charge surge on insulators. Besides, the Kelvin probe is installed vertically at the center of the upper surfaces on the insulator, about 3 mm away from the tested surface. The surface potential collected by the probe is output from the electrometer to the oscilloscope for real-time storage.

Before each test, anhydrous ethanol is used to clean the residual charges on the insulator surface. After the temperature inside the chamber has stabilized, the corona needle is adjusted by the position control system to a position 2 mm above the insulator surface. Then, the DC voltage is ramped up to the target voltage within 8 s. After 40 s of corona discharge, the applied DC voltage is disconnected and the corona needle is removed. Furthermore, the variation of insulator surface potential with time is collected in real time by the Kelvin probe.

9.3 Results

Comparison of Fig. 9.2a, b shows that the decay curves of surface potential with time are similar. For the applied voltage boosting stage, the surface potential rapidly increases. During the corona discharge, the surface potential gradually increases and tends to saturation as the insulator surface charge rapidly accumulates. Furthermore, the maximum of the insulator surface potential is positively related to the applied voltage. Once the applied voltage is turned off, the surface potential rapidly diminishes. Then, the potential of the insulator surface decreases with the dissipation of

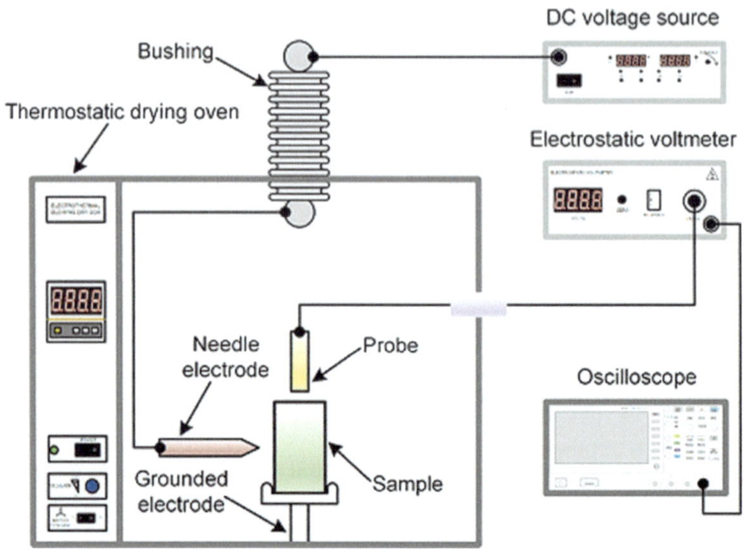

Fig. 9.1 Online measurement platform for surface charge surge on post insulators

surface and space charges. For insulators of the same height at room temperature, when the applied voltage is raised to −6 kV, the decreased percentage of surface potential after 62 s without power increases from 5.8% and 1.6% to 7.4% and 6.8%, respectively. Thus, the decay rate of the surface potential increases with the initial charge. Moreover, the surface potential after 120 s is also positively correlated with the initial surface charge. This indicates that the insulator surface potential varies with the surface charge. Thus, the surface charge surge can be effectively characterized by the measurement of the change in surface potential over time at the pouring port of GIL basin insulators.

9.4 Discussion

The operation voltage on the conductor of GIL is the fundamental source of surface potential on basin insulators. Moreover, the effect of surface charge and space charge on the insulator surface potential cannot be disregarded. As the conductor operation voltage increases, the insulator surface potential rapidly rises at the same rate. This is mainly because the electric field is an essential property of charge, which propagates without a medium and is established at the speed of light. Besides, there is electronic displacement polarization and turning polarization inside the insulator due to the operation voltage, as depicted in Fig. 9.3. During the withstand voltage, the polarization charge at the gas–solid interface and the Al_2O_3-epoxy interface are the main sources of slow changes in the insulator surface potential. The insulator surface

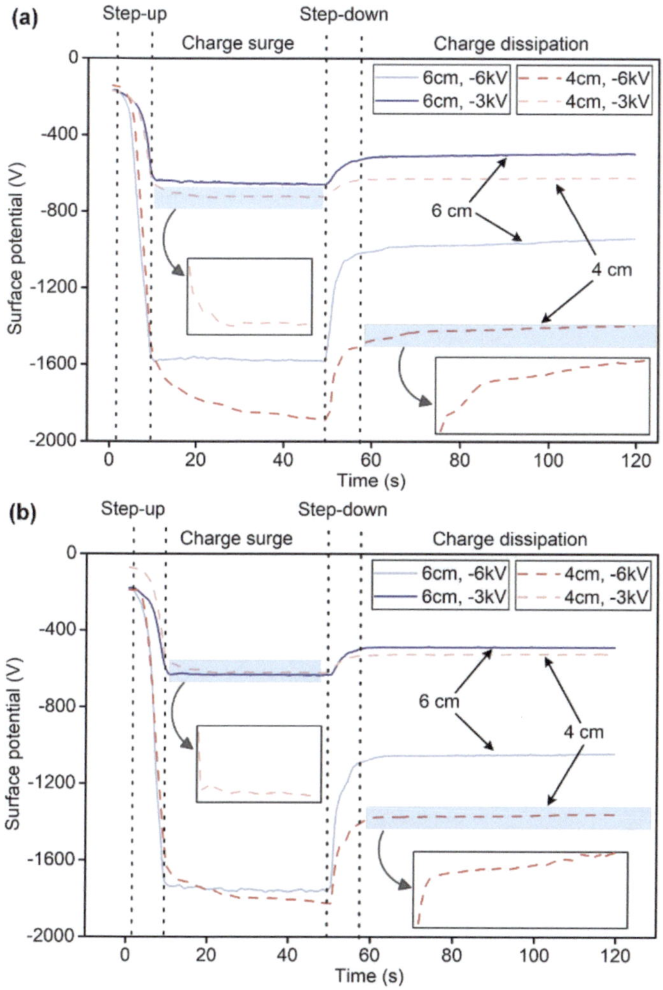

Fig. 9.2 Variation of the surface potential with time. (**a**)–(**b**) Ambient temperature is 26 °C and 66 °C, respectively

potential gradually tends to saturate with the extension of the application time of the DC voltage. However, partial discharges caused by insulation defects are inevitable within the GIL.

Figure 9.4a shows the typical initial potential distribution on the insulator surface after a corona discharge. The center of surface charge distribution is elliptical and is located in the middle area of the insulator directly in front of the corona needle. Furthermore, the amplitude of surface potential is inversely related to the distance to the corona needle. It is inferred that the initial charge on the insulator surface depends on the distribution of the normal electric field. The number of charged particles in

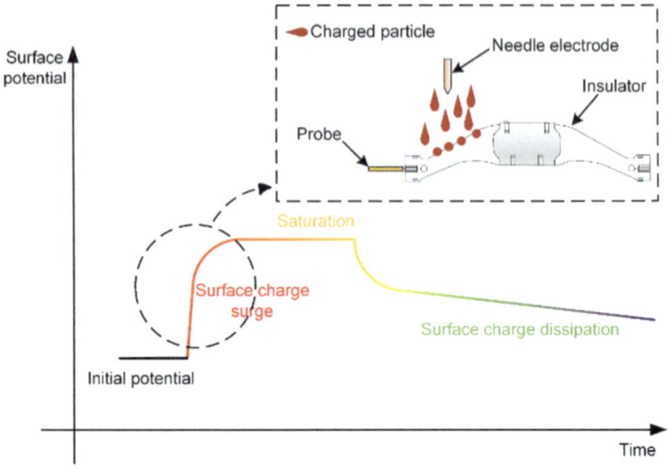

Fig. 9.3 Mechanism of sudden change in the basin insulator surface potential

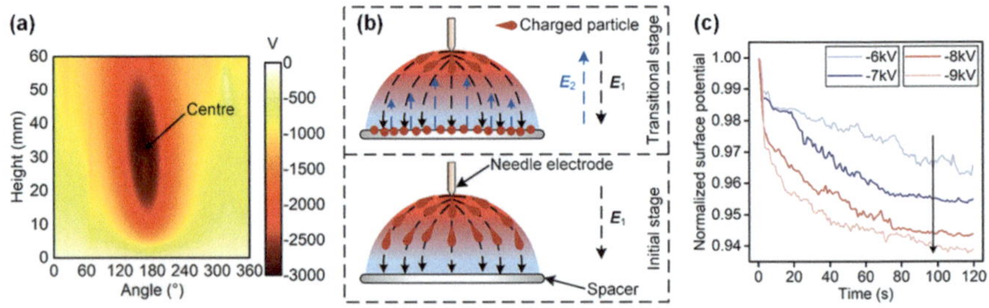

Fig. 9.4 Analysis of factors affecting the insulator surface potential. **a** Initial surface potential distribution (the applied voltage is −6 kV and the discharge time is 40 s). **b** Schematic of surface charge surge. **c** Surface potential dissipation curves under different applied voltages (the ambient temperature is 66 °C)

the gas side increases after corona discharges. Charged particles of the homopolar migrate to the insulator surface along the electric field line due to the presence of an external electric field $\boldsymbol{E}_1$. With the increase of surface charges, the electric field $\boldsymbol{E}_2$ generated by the homopolar surface charge cannot be ignored. Due to the opposite direction between $\boldsymbol{E}_1$ and $\boldsymbol{E}_2$, when $\boldsymbol{E}_2$ is approximately equal to $\boldsymbol{E}_1$, the charged particles generated by discharge are repelled by $\boldsymbol{E}_2$ and cannot migrate to the insulator surface. Thus, the charge accumulation tends to saturate, as shown in Fig. 9.4b. That is to say, the surface charge surge is influenced by the normal electric field on the insulator surface. It is inferred that the partial discharge in the gas side easily leads to surface charge surges and sudden changes in the surface potential. There is a dissipation of surface charge after the charge surge. Thus, the surface potential of insulators rapidly decreases.

Figure 9.4c shows the rate of charge dissipation at different initial charges. The surface potential after 120 s is about 95% of the initial value under different applied voltages. The decreased percentage of the surface potential increases from 3 to 6%. The decay rate of surface potential is positively correlated with the voltage of corona discharges (i.e., the initial surface charge). It is inferred that the surface charge surge can be monitored by measurement the magnitude of the basin insulator surface potential and its rate of change.

9.5 Conclusion

In this paper, an online measurement mechanism is proposed for insulator surface charge surge. Based on post insulators, an online measurement platform for surface potential is constructed. The effectiveness of the online measurement theory is verified. The results indicate that the insulator surface potential gradually tends to saturate with the extension of the application time of the DC voltage. However, the partial discharge in the gas side can easily lead to the insulator surface charge surge and the sudden change of the surface potential due to the influence of the normal electric field. After the surface charge surge, the surface potential of the insulator rapidly drops due to the autonomous dissipation of charges. For insulators of the same height at room temperature, when the applied voltage increases to −6 kV, the surface potential decrease percentage after 62 s without power rises from 5.8% and 1.6% to 7.4% and 6.8%, respectively. This shows that as the applied voltage increases, the surge charge accumulated on the insulator surface grows, and the decline rate of surface potential is accelerated. Therefore, the surface charge surge can be monitored by measurement the magnitude of the insulator surface potential and its rate of change. It is inferred that the surface charge surge can be characterized based on the change in surface potential at the pouring port of basin insulators measured by the Kelvin electrostatic probe.

References

1. An, Z., Xiao, H., Liu, F., Zheng, F., Lei, Q. and Zhang, Y.: Improved resistance of epoxy resin to corona discharge by direct fluorination. IEEE Transactions on Dielectrics and Electrical Insulation 23(4), 2278–2287 (2016).
2. Li, X., Wan, M., Zhang, G. and Lin, X.: Surface charge characteristics of DC-GIL insulator under multiphysics coupled field: Effects of ambient temperature, load current, and gas pressure. IEEE Transactions on Dielectrics and Electrical Insulation 29(4), 1530–1539 (2022).
3. Zhuang, W., et al.: Particle behavior and trap design for ±320 kV gas-insulated power transmission line (GIL). Journal of Physics D: Applied Physics 56(32), 325501 (2023).
4. Xing, Y., et al.: Metal particle induced spacer surface charging phenomena in direct current gas-insulated transmission lines. Journal of Physics D: Applied Physics 54(34), 34lt03 (2021).
5. Li, C., et al.: Field-dependent charging phenomenon of HVDC spacers based on dominant charge behaviors. Applied Physics Letters 114(20), 202904 (2019).

6. Pan, Z., Pan, C., Tang, J., Hu, B., Luo, Y. and Serdyuk, Y. V.: Influence of SiC/epoxy coating on surface charging phenomenon at DC voltage—Part I: Charge accumulation. IEEE Transactions on Dielectrics and Electrical Insulation 29(4), 1333–1342 (2022).
7. Li, C., et al.: Dust figures as a way for mapping surface charge distribution — A review. IEEE Transactions on Dielectrics and Electrical Insulation 28(3), 853–863 (2021).
8. Qi, B., Gao, C., Liu, S., Zhao, L. and Li, C.: Surface charge distribution on GIS insulator under DC/AC voltage. IEEE Transactions on Dielectrics and Electrical Insulation 24(5), 3173–3181 (2017).
9. Kumara, S., Hoque, I. R., Alam, S., Serdyuk, Y. V. and Gubanski, S. M.: Surface charges on cylindrical polymeric insulators. IEEE Transactions on Dielectrics and Electrical Insulation 19(3), 1076–1083 (2012).
10. Li, C., et al.: China's 10-year progress in DC gas-insulated equipment: From basic research to industry perspective. iEnergy 1(4), 400–433 (2022).
11. Zhou, H., et al.: Review of charge accumulation on spacer of gas insulated equipment at DC stress. CSEE Journal of Power and Energy Systems 6(3), 496–517 (2019).
12. Xue, J., et al.: Surface charge distribution patterns of a truncated cone-type spacer for high-voltage direct current gas-insulated metal-enclosed transmission line/gas-insulated metal-enclosed switchgear. IET Science, Measurement & Technology 12(4), 436–442 (2018).
13. Li, D., Zhang, G., Hou, Y. and Zhang, B.: Charge distribution on polymer insulator surface under AC voltage. IEEE Transactions on Dielectrics and Electrical Insulation 26(5), 1709–1715 (2019).
14. Du, Q., Zhang, Q., Wu, Z., Ma, J. and Zhao, J.: Surface charge distribution on DC basin-type insulator. IEEE Transactions on Dielectrics and Electrical Insulation 26(1), 17–25 (2019).
15. Gao, Y., Li, Z., Wang, M. and Du, B.: Magnetic field induced variation in surface charge accumulation behavior on epoxy/Al_2O_3 nanocomposites under DC stress. IEEE Transactions on Dielectrics and Electrical Insulation 26(3), 859–867 (2019).
16. Tu, Y., et al.: The control mechanism of micron and nano SiO_2/epoxy composite coating on surface charge in epoxy resin. IEEE Transactions on Dielectrics and Electrical Insulation 25(4), 1275–1284 (2018).
17. Kindersberger, J. and Lederle, C.: Surface charge decay on insulators in air and sulfurhexafluorid - part II: Measurements. IEEE Transactions on Dielectrics and Electrical Insulation 15(4), 949–957 (2008).
18. Shao, T., Kong, F., Lin, H., Ma, Y., Xie, Q. and Zhang, C.: Correlation between surface charge and DC surface flashover of plasma treated epoxy resin. IEEE Transactions on Dielectrics and Electrical Insulation 25(4), 1267–1274 (2018).
19. Wang, F., Liang, F., Chen, S., Tan, Y., Zhong, L. and Sun, Q.: Surface charge inversion method on cylindrical insulators based on surface potentials measured online. IEEE Transactions on Dielectrics and Electrical Insulation 28(1), 192–197 (2021).

Chapter 10
Research on Temperature Distribution Law of High-Voltage Cable Tunnel Fire

Yinglong Liu, Li Chen, Fenglang Shi, Bingzhou Lu, Ke Wang, Jiaxin Lin, Jian Guo, Pan Zhang, and Zhidu Huang

Abstract As high-voltage cable tunnels are mostly closed underground spaces, a fire in the cable will cause great threat to personnel and equipment. For this reason, this paper constructs a high-voltage cable tunnel model through FDS numerical simulation, and simulates and studies the temperature distribution law when a fire occurs in a high-voltage cable tunnel. The study shows that the temperature in the high-voltage cable tunnel rises slowly before 25 s, rises rapidly after 25 s, and gradually stabilizes after 100 s. The closer the location of the fire source is to the top of the tunnel, the significant increase in the temperature difference in the horizontal direction, at the same time, the fire source power of 6 MW, the temperature of up to 450 °C, and the larger the power of the fire source, the temperature will also increase. The research results can provide a reference for the safe design and fire protection of high-voltage cable tunnels.

Keywords High-voltage cable tunnel · Cable fire · Numerical simulation · Temperature distribution

10.1 Introduction

10.1.1 A Subsection Sample

With the rapid development of social economy, the use of cables is increasing. Due to the cable line short circuit, cable leakage, cable cabin maintenance is not thoughtful and other reasons, may make the cable temperature rise and cause fire. In the high-voltage cable tunnels in each layer are laid a large number of cables, and cable tunnels within the space closed, if any layer of cables in the event of a fire, the fire will spread to all the surrounding cables, which in turn will pose a serious threat to

Y. Liu (✉) · L. Chen · F. Shi · B. Lu · K. Wang · J. Lin · J. Guo · P. Zhang · Z. Huang
Nanning Power Supply Bureau of Guangxi Power Grid Co., Ltd., Nanning, China
e-mail: 1095473499@qq.com

Y. Li et al. (eds.), *Proceedings of the 5th Chinese Conference on High Voltage and Discharge Plasmas*, Springer Proceedings in Physics 254,
https://doi.org/10.1007/978-981-96-5711-7_10

the safe operation of the urban power grid [1–3]. In addition, burning cables usually also cause large-scale power failures and communication interruptions, seriously affecting the normal life of urban residents to protect [4–6]. Nowadays, cable fires account for a relatively high proportion of electrical fire accidents in China, bringing immeasurable losses to the country and society [7, 8]. Therefore, high-voltage cable tunnel fires need to attract people's attention.

Researchers have begun to explore and study the behavioral characteristics of cable fires. Tang et al. [9] numerically simulated the temperature distribution and CO diffusion caused by cable fires in an L-type integrated tube corridor, and found that the vertical maximum temperature increased with the increase of the power of the fire source and the ventilation rate under more conditions. Wang et al. [10] analyzed the data generated by the combustion of high-voltage cables by substituting the data into the FDS software, and explored the changing patterns of the fire development and smoke diffusion at the end of cable tunnels using FDS. They investigated the development of fire at the end of the cable tunnel and the changing rules of smoke dispersion, etc. Esther Kim et al. [11] used FDS to simulate tunnel fires including forced, transient and longitudinal ventilation flows, analyzed the grid size and parameters in the model, and came up with an improved FDS smoke layer prediction model. Liu et al. [12] investigated the different combination modes of the installation of a ventilation system, a fire door, and a sprinkler system by using FDS software to analyze these modes. FDS software was used to analyze the effective-ness of these modes in cable corridor fires, and temperature and visibility cloud maps were drawn to compare the performance under different switching modes. Liang et al. [13] simulated T-type cable corridor fires by using FDS software, and analyzed the temperature changes after the occurrence of the fire, as well as the tendency of the corridor fire spread. At present, many scholars have done extensive research on the behavioral characteristics of cable fires, but there is still less research on the temperature distribution law after a fire occurs in a high-voltage cable tunnel under different working conditions, and this negligence may lead to serious safety hazards for high-voltage cables, which may aggravate the hazards posed by fire accidents.

This paper investigates the behavioral characteristics of high-voltage cable tunnel fires under different fire source conditions. Using the FDS numerical simulation method, a fire model of high-voltage cable tunnels is established in the actual application scenario, and the temperature distribution caused by cable fires is analyzed by numerical simulation considering different working conditions of fire source location and power. The temperature distribution law of high-voltage cable tunnel fire under different working conditions is investigated, and this study can provide reference for the fire hazard evaluation and safety design of high-voltage cable tunnel.

10.2 Modeling

10.2.1 Basic Equations

Fire development is a complex process, but the process of using FDS calculations will follow the laws of conservation of momentum, conservation of energy, conservation of components, etc., in which the basic control equations solved by FDS calculations [14] are:

The continuity equation:

$$\frac{\partial \rho}{\partial t} + \left[\frac{\partial(\rho u_x)}{\partial x} + \frac{\partial(\rho u_y)}{\partial y} + \frac{\partial(\rho u_z)}{\partial z} \right] = 0 \tag{10.1}$$

where ρ is the density (kg/m^3); t is the time (s); u_i is the velocity component (m/s).

Momentum conservation equation:

$$\frac{\partial(\rho \vec{u}\,\vec{u})}{\partial t} + \nabla \cdot (\vec{u}\,\vec{u}) + \nabla p = \rho g + \vec{f} + \nabla \cdot \tau \tag{10.2}$$

where g is the acceleration of gravity (m/s^2); $\vec{f}$ is the external force vector (kg/s^2/m); p is the pressure (Pa); t is the time (s); $\vec{u}\,\vec{u}$ is the second-order tensor of velocity; and ∇ is the Hamiltonian operator.

The energy conservation equation:

$$\begin{aligned} \frac{\partial}{\partial t}(\rho h) + \nabla \cdot (\rho h \vec{u}) = {} & \frac{\partial \rho}{\partial t} + \vec{u} \cdot \nabla p - \nabla \cdot \vec{q_r} + \nabla \cdot (k \nabla T) \\ & + \sum\nolimits_i \nabla \cdot (h_i \rho D_i \nabla Y_i) \end{aligned} \tag{10.3}$$

where h is the enthalpy (J/kg); $\vec{q_r}$ is the thermal radiation flux; T is the temperature; and k is the thermal conductivity (W/(m–K)).

Component conservation equations:

$$\frac{\partial}{\partial t}(\rho Y_i) + \nabla \cdot (\rho Y_i \vec{u}) = \nabla \cdot (\rho D_i \nabla Y_i) + m_i \tag{10.4}$$

where Y_i is the mass fraction of the component i; D_i is the diffusion coefficient of the component i (m^2/s); m_i^m is the mass production rate of the component i (kg/s/m^3).

10.2.2 High-Voltage Cable Tunnel Modeling

Based on the 110 kV outgoing cable tunnel in a 220 kV substation, the internal dimensions of the high-voltage cable tunnel are set to be 8.3 m in length, 1.9 m in width, and 2.3 m in height. By setting up temperature measurement points and 3D slices to explore the fire development inside the high-voltage cable tunnel, and at the same time, according to the internal dimensions of the high-voltage cable tunnel, set up five groups of temperature sensors (from right to left in order of A, B, C, D, E), three in each group, the sensor horizontal interval of 2 m, the longitudinal interval of 0.5 m, and the distance from the bottom of each group of sensors were 1.2, 1.7, 2.3 m, and the computational model is shown in Fig. 10.1.

10.2.3 Boundary Condition Setting

High-voltage cable tunnel mainly consider the initial ambient temperature, initial speed, pressure and other factors, the cable symmetrically arranged on both sides of the high-voltage cable tunnel, each side of the 4 groups of cables, the upper and lower spacing of the two layers of 0.4 m, each group of three zigzag arrangement. The characteristics of each material as shown in Table 10.1.

The rate of heat release from unstable fire sources varies with time, and in order to be consistent with fire development in real situations, the t^2 model is used in the FDS numerical simulation calculations to determine the way in which the rate of heat release from the material grows in the form of equation:

$$\dot{Q} = \alpha t^2 \tag{10.5}$$

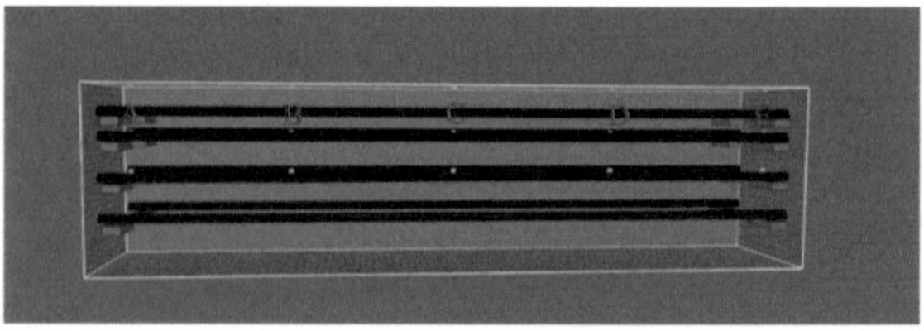

Fig. 10.1 Schematic diagram of high voltage cable tunnel model

Table 10.1 Cable material parameters [15]

Materials	Density/(kg/m^3)	Specific heat/(kJ/kg·K)	Heat conductivity/(w/mK)	Combustion heat/(kJ/kg)	Ignition temperature/°C
Copper	8940	0.38	387	/	/
XLPE	1500	1.5	0.2	2500	330

Table 10.2 Fire growth factor [16]

Fire spread rating	Time/s to reach 1 MW	α/(kW s^{-2})
Slow fire	584	0.0029
Medium speed fire	292	0.01127
High speed fire	146	0.04689
Ultra rapid fire	73	0.01876

Table 10.3 Cable size parameters

/	Materials	Size/mm
Cable diameter	/	98
Cable core diameter	Copper	30
Insulation thickness	PVC	28
Sheath layer thickness	PVC	6

where $\dot{Q}$ is the rate of heat release from the fire source (kW) and α is the fire development factor (kW/s^2) (Table 10.2).

The heat release rate of polyethylene is 265 KW. In this paper, the fire spreading level of cable protection jacket material is rapid fire, α is 0.04689, and the time to reach the maximum heat release rate is calculated to be 75 s.

Cable joints are the high incidence of fire, so the parameters of the insulation layer at the joints set the fire source, the length of the fire source is 0.28 m, and the area of the fire source is 0.028 m^2 (Table 10.3).

10.2.4 Design of Working Conditions

In order to explore the influence of fire source location and fire source power, this paper sets up different locations as well as different power sources to analyze their temperature distribution laws (Tables 10.4 and 10.5).

Table 10.4 Fire source location working conditions

Working condition number	Location of fire source
Working condition 1	Bottom end of cable
Working condition 2	Middle of cable bottom
Working condition 3	Cable high level end
Working condition 4	Cable high level central

Table 10.5 Fire source power conditions

Working condition number	Location of fire source (MW)
Working condition 5	1.5
Working condition 6	3
Working condition 7	6

10.3 Simulation Study of Fire in High Voltage Cable Tunnels

10.3.1 Temperature Distribution in High Voltage Cable Tunnels at Different Ignition Source Locations

Figure 10.2 are working conditions (fire source location in the bottom end) 3 groups of measurement point longitudinal section temperature changes, and working conditions of a horizontal (y = 1.2 m, y = 1.7 m, y = 2.3 m) on the three layers of the measurement point temperature changes, as can be seen from the figure, the temperature of the high-voltage cable tunnel in the 50 s before the slow rise, 50 s after the rapid rise, 100 s after the gradual leveling off. High-voltage cable tunnel closer to the fire source location the higher the temperature, high-voltage cable tunnel left group E measurement point temperature up to 125 °C. The temperature at the top of the high-voltage cable tunnel rises the fastest, the closer to the fire source directly above the location, the greater the fluctuation of the temperature measurement point. From the condition of a horizontal (y = 1.2 m, y = 1.7 m, y = 2.3 m) on the three levels of temperature changes in the measurement point can be seen, y = 1.2 m, y = 1.7 m in the horizontal direction (A-E) temperature changes in the temperature trend is the same, the temperature difference is not significant, y = 2.3 m at the measurement of the temperature of the point of the horizontal direction of the temperature stratification is obvious, the closer the top of the high-voltage cable tunnel, the temperature distribution of the horizontal direction of the temperature distribution shows obvious stratification phenomenon. The high temperature of the smoke is mainly concentrated in the area above 1.7 m, while the temperature change in the position below 1.7 m is relatively small, keeping below 100 °C.

Figure 10.3 are working conditions two (fire source location in the bottom of the middle) three groups of measurement point longitudinal section temperature changes, and working conditions two horizontal (y = 1.2 m, y = 1.7 m, y = 2.3 m) on the three layers of the measurement point temperature changes, as can be seen from the figure, the temperature in the high-voltage cable tunnel in the 50 s before the slow rise, 50 s after the rapid rise, 100 s after the gradual leveling off. High-voltage cable tunnel closer to the fire source location the higher the temperature, high-voltage cable tunnel left group C measurement point temperature up to 125 °C. From the condition of a horizontal (y = 1.2 m, y = 1.7 m, y = 2.3 m) on the three levels of temperature changes in the measurement point can be seen, y = 1.2 m, y = 1.7 m in the horizontal direction (A-E) temperature changes in the temperature trend is

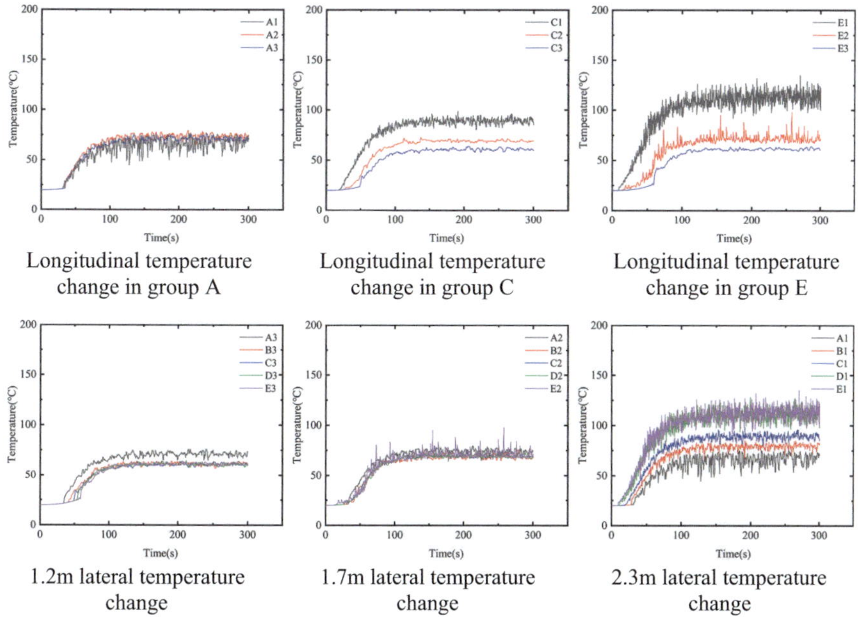

Fig. 10.2 Temperature change of working condition 1

the same, the temperature difference is not very large, y = 2.3 m at the point of the level of the temperature stratification is not as obvious as in the case of a condition, this is due to the high-temperature smoke rises to the top of the tunnel, the tunnel is blocked by the top of the tunnel to the sides of the diffusion, the temperature is not very high. This is due to the high temperature smoke rises to the top of the tunnel, by the tunnel top block to both sides of the diffusion, so that the temperature from the middle of the high-voltage cable tunnel to both sides of the attenuation, at this time, the highest temperature at C1, A1, E1 temperature is the smallest.

Figure 10.4 shows the temperature changes in the longitudinal section of three groups of measurement points in Case 3 (the fire source is located at the end of the high level) and the temperature changes in the three levels of measurement points in the horizontal direction (y = 1.2 m, y = 1.7 m, y = 2.3 m) of Case 3. It can be seen from the figure that the temperature of Group E on the left side of the high-voltage cable tunnel reaches up to 225 °C compared with that of Case 1. The temperature changes in the horizontal direction of Case 3 are more stratified compared with that in the horizontal direction of Case 1. From the condition five horizontal (y = 1.2 m, y = 1.7 m, y = 2.3 m) on the three layers of the measurement point temperature change can be seen, compared with the condition three horizontal direction temperature stratification is more obvious.

Figure 10.5 shows the temperature changes of three groups of measurement points in the longitudinal section of Case 4 (the location of the fire source is in the middle of the middle and high floors) and the temperature changes of the measurement

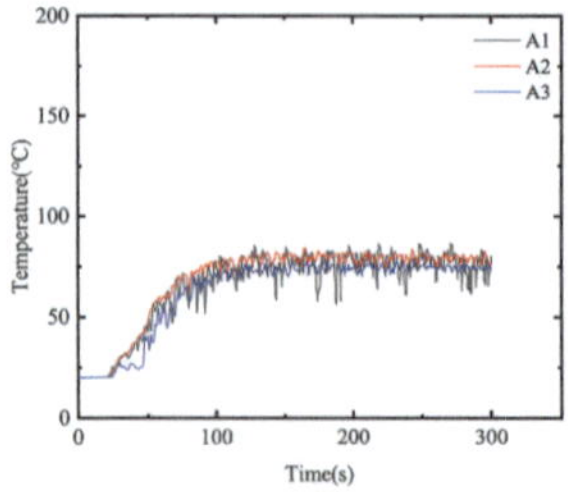

Longitudinal temperature change in group A

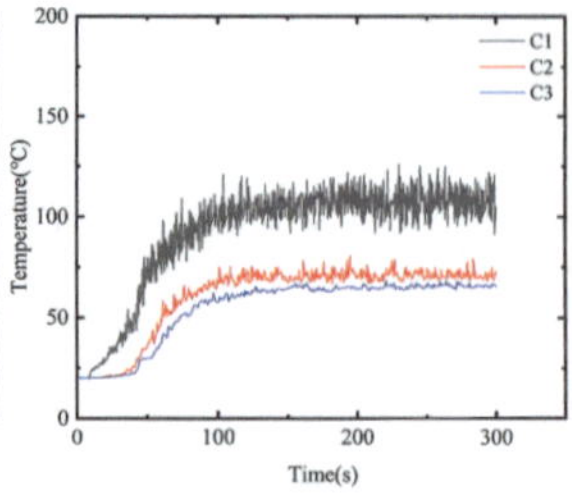

Longitudinal temperature change in group C

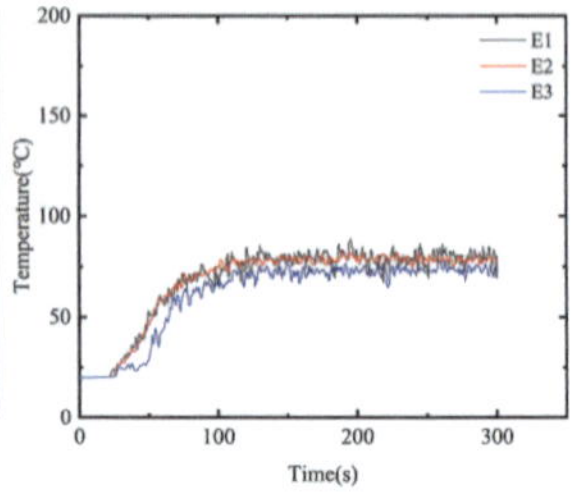

Longitudinal temperature change in group E

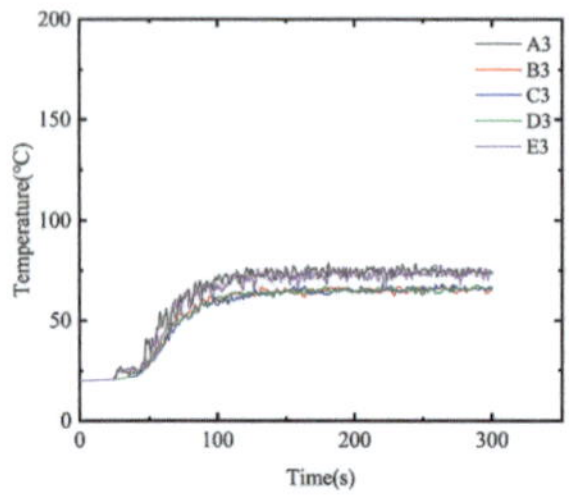

1.2m lateral temperature change

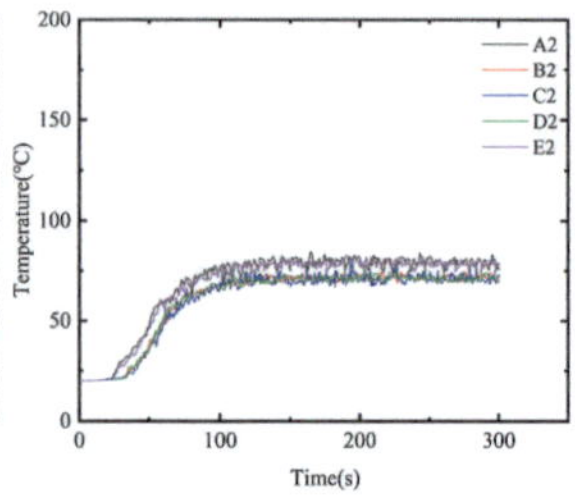

1.7m lateral temperature change

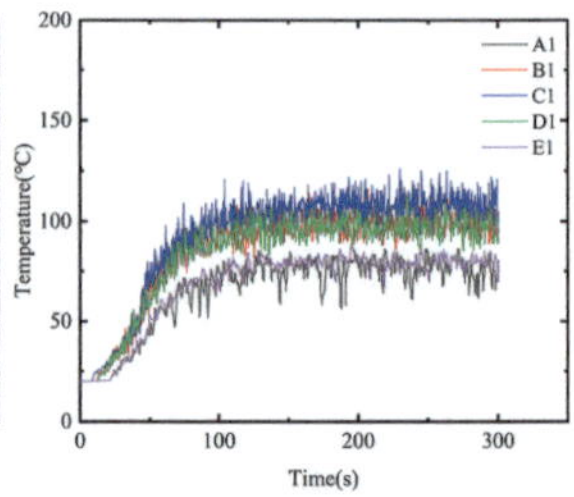

2.3m lateral temperature change

Fig. 10.3 Temperature change of working condition 2

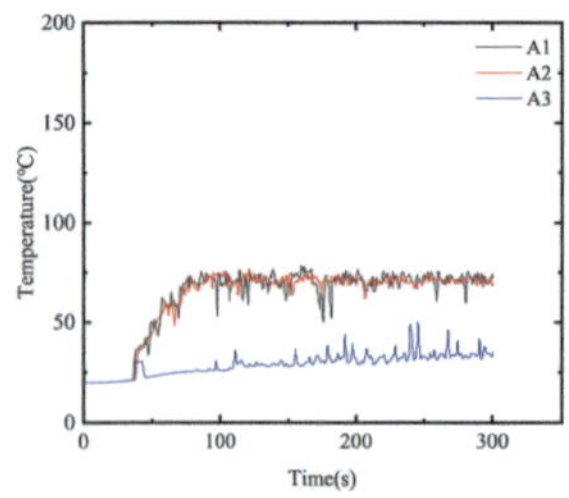

Longitudinal temperature change in group A

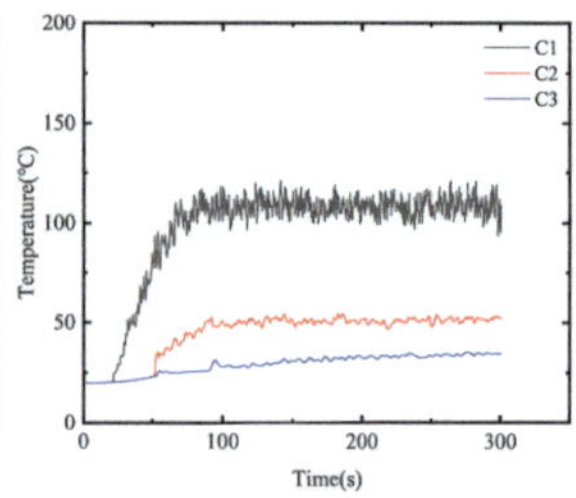

Longitudinal temperature change in group C

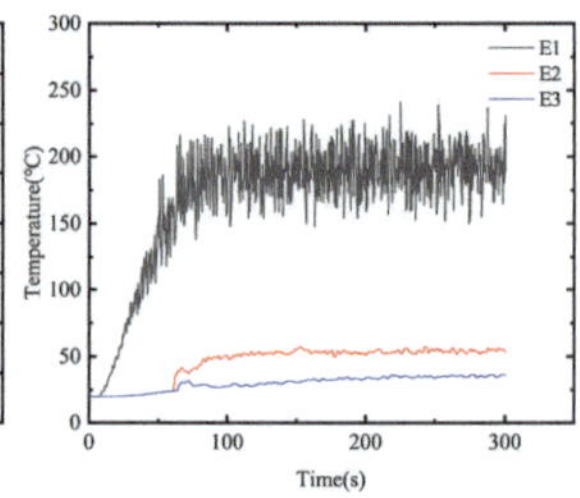

Longitudinal temperature change in group E

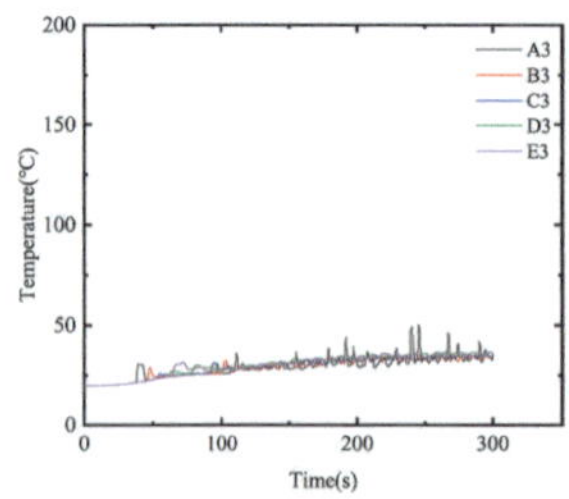

1.2m lateral temperature change

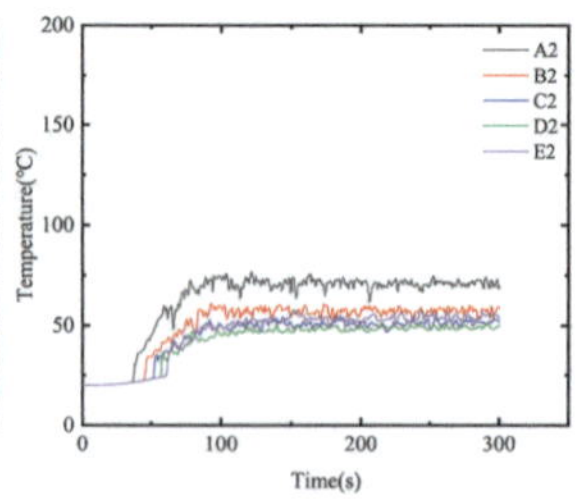

1.7m lateral temperature change

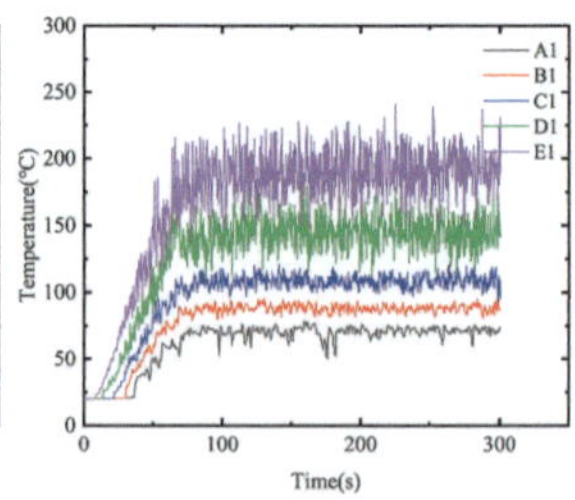

2.3m lateral temperature change

Fig. 10.4 Temperature change of working condition 3

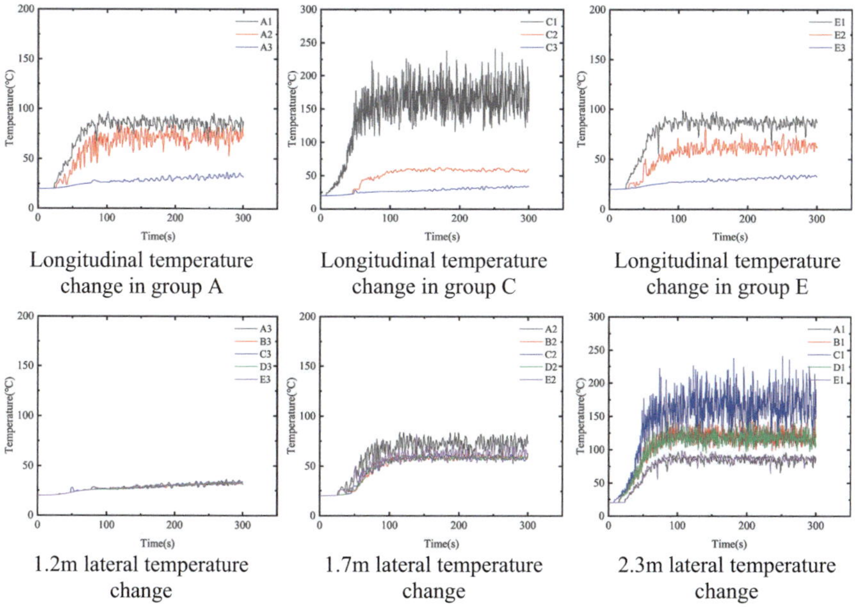

Fig. 10.5 Temperature change of working condition 4

points in the horizontal direction of Case 4 (y = 1.2 m, y = 1.7 m, y = 2.3 m) in the upper three floors, and it can be seen that the temperature of the measurement points in Group C reaches up to 250 °C compared with that in Case 2. From the condition six horizontal (y = 1.2 m, y = 1.7 m, y = 2.3 m) on the three layers of the measurement point temperature change can be seen, compared with the condition four horizontal direction temperature disorder phenomenon is more serious, the temperature stratification is more obvious.

10.3.2 Temperature Distribution in High Voltage Cable Tunnels at Different Ignition Source Locations

Figure 10.6 are working condition 5 (fire power 1.5 MW) temperature changes at the measurement point, it can be seen from the figure, the temperature in the high-voltage cable tunnel in 50 s before the slow rise, 50 s after the rapid rise, 100 s after the gradual stabilization. High-voltage cable tunnel closer to the fire source location the higher the temperature, high-voltage cable tunnel in the middle of the C group of measurement point temperature up to 200 °C. The temperature stratification in the vertical direction of the high-voltage cable tunnel is more obvious, compared with Case 2, the temperature of the measurement point at 2.3 m is higher, and the

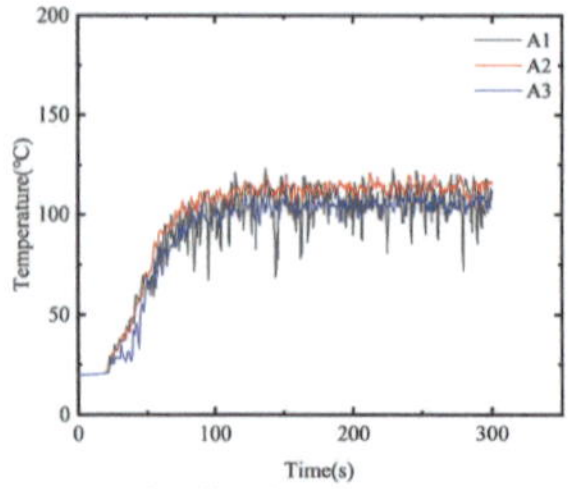

Longitudinal temperature change in group A

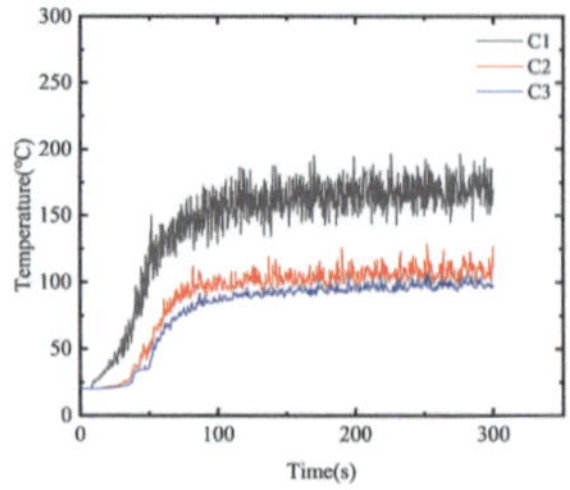

Longitudinal temperature change in group C

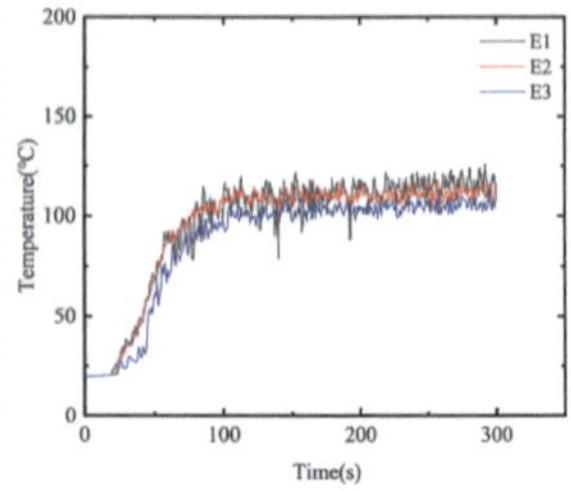

Longitudinal temperature change in group E

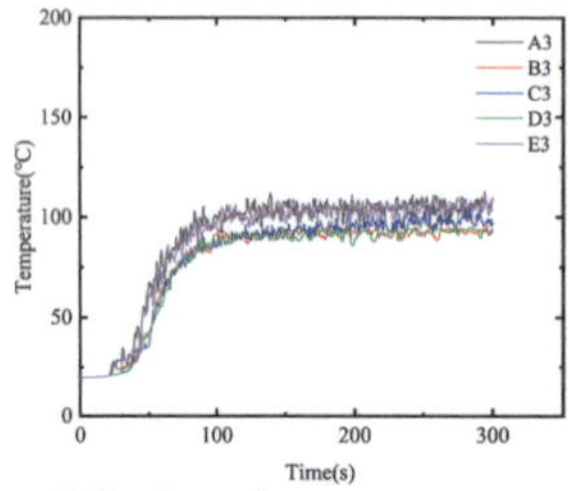

1.2m lateral temperature change

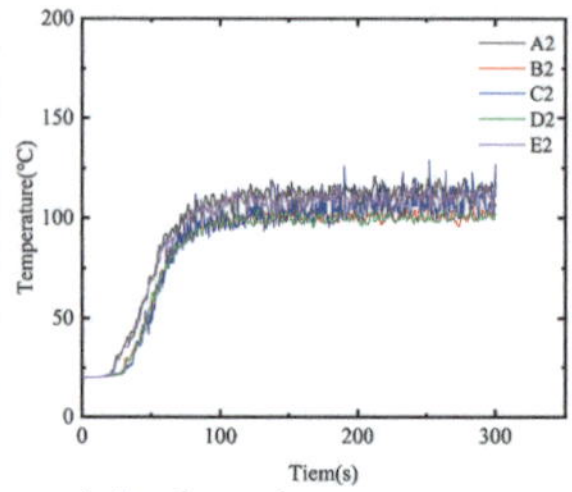

1.7m lateral temperature change

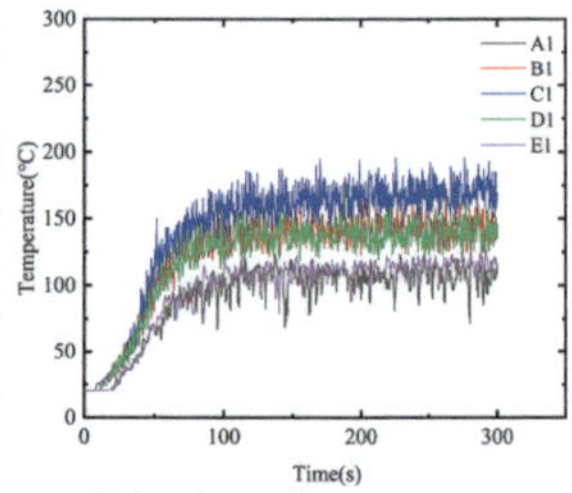

2.3m lateral temperature change

Fig. 10.6 Temperature change of working condition 5

temperature turbulence is more serious, which is due to the fact that with the increase of the fire power, the smoke thermal plume effect is stronger.

Figure 10.7 shows the temperature change of the measurement point in Case 6 (fire source power of 3 MW), and it can be seen from the figure that the temperature in the high-voltage cable tunnel rises slowly before 25 s, rises rapidly after 25 s, and tends to stabilize gradually after 100 s. High-voltage cable tunnel, the closer the location of the fire source, the higher the temperature, high-voltage cable tunnel in the middle of the C group of measurement point temperature up to 300 °C, compared with the working conditions of seven, the temperature fluctuations of the measurement point is greater, the temperature rise rate is higher. The temperature stratification in the vertical direction of the high-voltage cable tunnel is more obvious, compared with Case 5, 1.2 m, 1.7 m at the C group measurement point fluctuations, 2.3 m at the measurement point temperature is higher, the temperature disorder phenomenon is more serious.

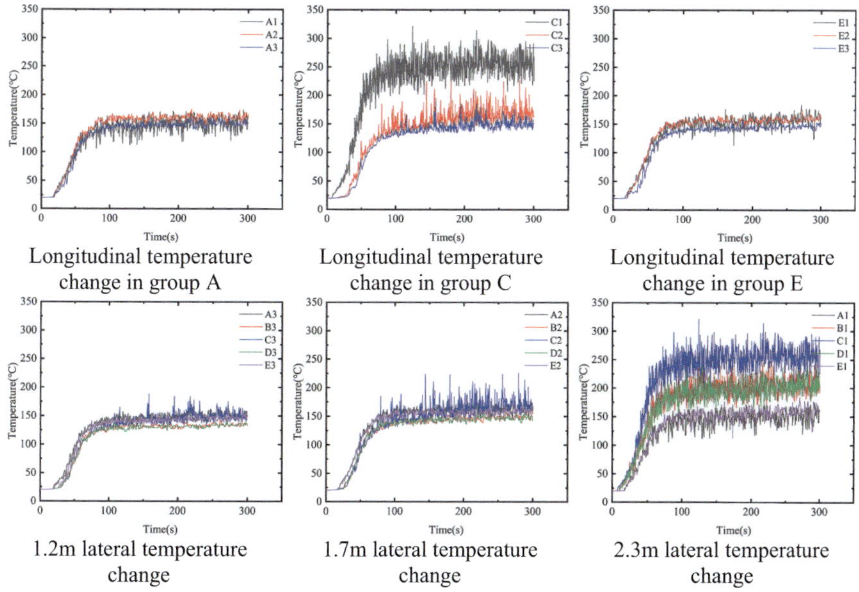

Longitudinal temperature change in group A

Longitudinal temperature change in group C

Longitudinal temperature change in group E

1.2m lateral temperature change

1.7m lateral temperature change

2.3m lateral temperature change

Fig. 10.7 Temperature change of working condition 6

Figure 10.8 shows the temperature change of the measurement point for Case 7 (fire source power 6 MW), respectively. From the figure, it can be seen that compared with Case 6, the temperature of the measurement point in the middle of the high-voltage cable tunnel, Group C, was up to 450 °C, with the largest temperature fluctuation at the measurement point, the highest temperature increase rate, and the most serious temperature disorder. Inside the high-voltage cable tunnel, it is observed that the temperature in the area above 1.7 m is higher and accompanied by significant temperature fluctuations, the increase in source power, the maximum temperature of the high-voltage cable tunnel increases, the more obvious the temperature fluctuations, the closer the location of the fire source is to the top of the tunnel, and the temperature difference in the horizontal direction increases significantly.

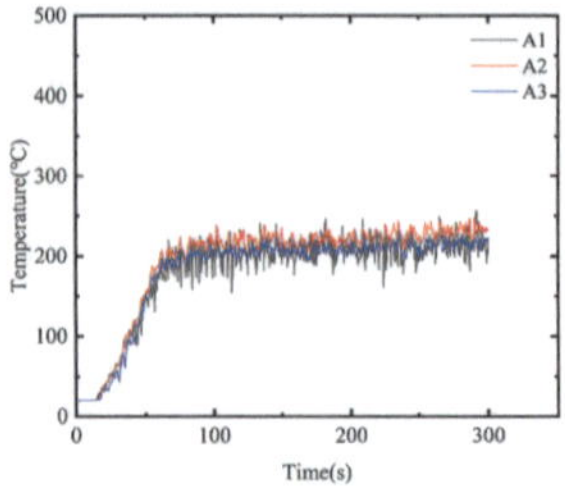

Longitudinal temperature change in group A

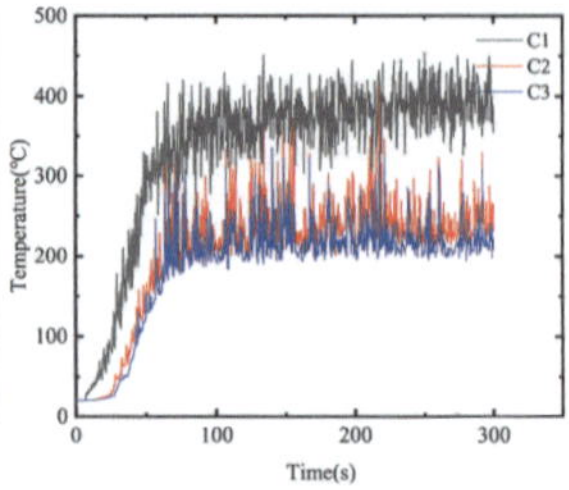

Longitudinal temperature change in group C

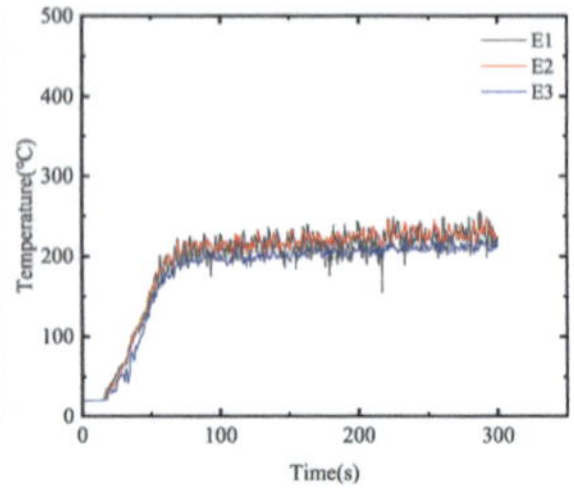

Longitudinal temperature change in group E

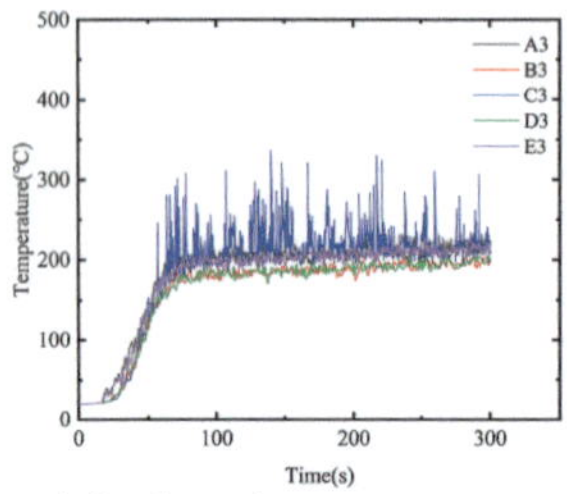

1.2m lateral temperature change

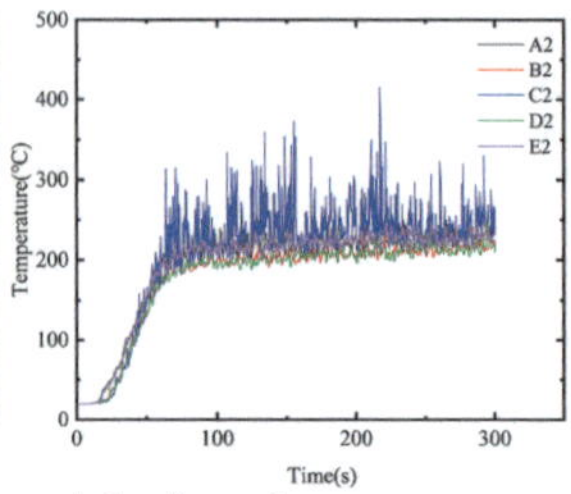

1.7m lateral temperature change

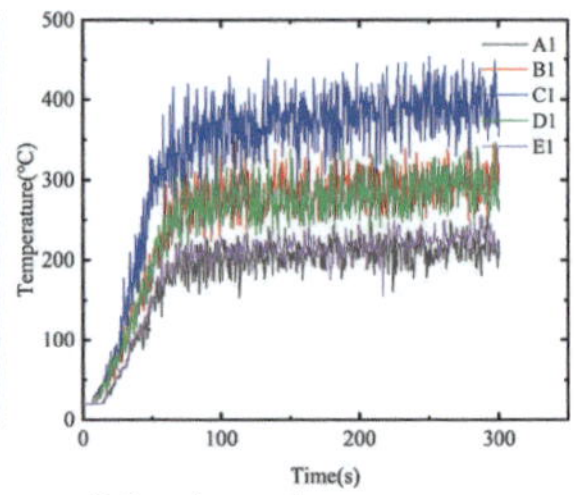

2.3m lateral temperature change

Fig. 10.8 Temperature change of working condition 7

10.4 Conclusion

This paper constructs a high-voltage cable tunnel fire model based on the actual high-voltage cable tunnel parameters. Multiple temperature measurement points are set, and by changing the fire source parameters, the temperature change rule of the fire source parameters on the high-voltage cable tunnel fire is analyzed. According to the results of the study, it can be seen that when the fire occurs in the high-voltage cable tunnel, the temperature in the high-voltage cable tunnel rises slowly before 25 s, rises sharply after 25 s, and tends to stabilize gradually after 100 s. The closer the location of the fire source in the high-voltage cable tunnel, the higher the temperature, and the temperature of the high-voltage cable tunnel up to 450 °C. Inside the high-voltage cable tunnel, it was observed that the temperature was higher in the area above 1.7 m, accompanied by significant temperature fluctuations, and with the increase of the fire source power, the temperature increased, and the temperature fluctuations were more pronounced, and the temperature was up to 450 °C when the fire source power was 6 MW, and the closer the fire source location was to the top of the tunnel, the temperature difference in the horizontal direction increased significantly.

References

1. Zhao Le. An introduction to the application research of urban underground comprehensive pipeline corridor[J]. Science and technology innovation,2017(32):117-118.
2. He L, Ma G, Hu Q, Cai Q, Bai Y, Tang S, Tan J. A novel method for risk assessment of cable fires in utility tunnel. Mathematical Problems in Engineering. 2019. 2019(1):2563012.
3. XU Haozhong, LI Yaozhuang, XU Zhisheng. Numerical simulation study of fire at chamber intersection of urban under-ground comprehensive pipeline corridor[J]. Science and Technology Bulletin, 2018,34(10):264-268.
4. XU, Zhonglin, Haijiang DONG, Yongpeng YYANG, YANG Feng, J. Xu, A. Guo et al. "Design of fire and explosion proof device for high voltage cable joint." (2022): 324–328.
5. Bai Z, Yao H. Experimental Study on Fire Characteristics of Cable Compartment in Utility Tunnel with Fire Source at Shaft Side. Engineering. Letters. 2022 Jun 1;30(2).
6. BOWEN D. Influence of Air Curtain and Mechanical Smoke Extraction on Smoke Diffusion in Tunnel Fire[J]. Journal of Donghua University (Natural Science), 2019.
7. Wang Fangshun. Research on CFD numerical simulation of cable fire in urban comprehensive pipeline corridor[J]. Journal of Armed Police College,2017,33(08):14-18.
8. Roy Gong, Guo Shou, Cheng Yang, et al. Research on the effect of fire source location on high voltage cable trench fire based on FDS[J]. Industrial Safety and Environmental Protection,2023,49(02):21-25.
9. An, Weiguang, Yanhua Tang, Kai Liang, Minglun Cai, Tao Wang, and Zhe Wang. "Study on temperature distribution and CO diffusion induced by cable fire in L- shaped utility tunnel." Sustainable Cities and Society 62 (2020): 102407.
10. Wang, Shuhan, Yongkang Zhang, Ling Wang, Qi Yuan, Chen-bin Wu, and Jianbin Zang. "Fire simulation at the end of cable tunnel based on high voltage cable burning experiment." In Civil Engineering and Disaster Prevention, pp. 701–712. CRC Press, 2023.
11. Kim, Esther, John P. Woycheese, and Nicholas A. Dembsey. "Fire dynamics simulator (version4.0) simulation for tunnel fire scenarios with forced, transient, longitudinal ventilation flows." Fire Technology 44 (2008): 137-166.
12. Mi, Hongfu, Yaling Liu, et al. "A numerical study on the optimization of ventilation mode during emergency of cable fire in utility tunnel." Tunnelling and Underground Space Technology 100 (2020): 103403.
13. Liang, Kai, Xiongfei Hao, Weiguang An, Yanhua Tang, and Yuzhou Cong. "Study on cable fire spread and smoke temperature distribution in T-shaped utility tunnel." Case Studies in Thermal Engineering 14 (2019): 100433.
14. WU Feng, LU Kunlun, XIAO Yang. Numerical simulation study on smoke flow law of cable fire in subway tunnel[J]. Fire Science and Technology,2015,34(7):863-865.
15. Enaru, I, et al. "Numerical Simulation Evaluation of Fire Spreading in a Building Using Fire Dynamics Simulator (FDS)." Journal of Applied Engineering Sciences 13.1 (2023): 65-72.
16. Zheng Yuan. Research on fire smoke movement law and its ventilation optimization in comprehensive pipeline corridor[D]. Hefei: University of Science and Technology of China,2020.

Chapter 11
Simulation Study of Magnetic Force Characteristics of Transformer with Single Helical Windings Under Short-Circuit Conditions

Jun Liu, Kun Li, Peilong Chen, Kui Xu, Ran Zhuo, Meng Gao, Wenhao Ai, Hang Guo, and She Chen

Abstract For large-capacity transformers, the low-voltage winding often adopts single helical windings design. In order to reduce additional losses, transposed wires are widely used. However, further research is needed to investigate the impact of transposition structure of single helical windings on electromagnetic force characteristics. Therefore, this study constructs a finite element model of a transformer with single helical windings based on actual prototype parameters. The field-circuit coupling model is employed to simulate the magnetic field distribution under short-circuit conditions. The results show that the transposition structure of single helical windings distort the magnetic flux density, primarily affecting the transposition area and the adjacent wire cakes. In terms of short-circuit electromagnetic force, there is a significant difference in the radial direction between the two columns of wires in single helical windings, with the outer side of the wire on the right experiencing the greatest radial electromagnetic force. As the distance from the iron core increases, the radial short-circuit electromagnetic force of the wire cake at the transposition structure gradually increases, further exacerbating the non-uniformity of the forces acting on the winding. However, the axial electromagnetic forces are relatively similar. This study is helpful in predicting the mechanical behavior of single helical windings in transformers under multiple short-circuit impacts, and provide guidance for the short-circuit electromagnetic stability verification and operation evaluation of large capacity transformers.

J. Liu · K. Li · P. Chen · K. Xu
Electric Power Research Institute of Guizhou Power Grid Co., Ltd., Guiyang, China

R. Zhuo · M. Gao
State Key Laboratory of HVDC, Electric Power Research Institute, China Southern Power Grid, Guangzhou, China

W. Ai · H. Guo (✉) · S. Chen
College of Electrical and Information Engineering, Hunan University, Changsha, China
e-mail: guohang@hnu.edu.cn

Y. Li et al. (eds.), *Proceedings of the 5th Chinese Conference on High Voltage and Discharge Plasmas*, Springer Proceedings in Physics 254,
https://doi.org/10.1007/978-981-96-5711-7_11

Keywords Single helical winding · Short-circuit magnetic characteristics · Finite element simulation · "Filed-circuit" coupling

11.1 Introduction

With the gradual expansion of the power grid scale and the continuous growth of power demand, power transformers are facing unprecedented operation challenges [1], When a short-circuit current strikes a transformer, its windings are subjected to tremendous stress, and this stress may lead to loosening and deformation of the windings, or even cause permanent damage. According to the statistics of the International Conference on Large Electricity Grids (CIGRE), among the main types of transformer faults, winding and core faults account for the highest percentage, which reaches 28.5% [2]. Especially in transformers that have been in service for 15–25 years, deformation defects in transformer cores and windings are common [3]. These defects not only weaken the mechanical stability of the transformer, but can also lead to electrical faults such as turn-to-turn short circuits and open circuits [4, 5]. Differences in the structure, materials, and processes of different transformer models may lead to significant differences in their response characteristics under short-circuit shocks, which undoubtedly increases the complexity and challenge of the study.

With the wide application of transposed wires in single helical windings of transformers, related research work has been gradually carried out. In the past, scholars generally adopted the assumption of uniform current distribution in the research process. However, the applicability of this assumption is somewhat limited when dealing with helical windings with multiple wires in parallel winding structures [6–8]. Guo et al. established a field-circuit coupling mathematical model to calculate the branch currents of parallel wires, and found that the different transposition forms have significant effects on the branch current distribution, and the results show that the "212" transposition is better in terms of uniform current distribution and loss suppression [9]. Zhao et al. For the low-voltage helical windings of 63000 kVA and 50000 kVA transformers, the results of the study by Zhao et al. studied the current distribution characteristics under the transposition structure, and analyzed its impact on the short-circuit electromagnetic force of the winding, which provides an important reference for the optimization of the transformer design [10]. In view of the complex characteristics of the transformer winding structure, especially the low-voltage winding of the distribution transformer is designed with multiple wires wound in parallel, and this structure may lead to uneven current distribution. As a result, complex magnetic characteristics may arise during a short circuit. Based on the above analysis, it is necessary to further study the magnetic characteristics of transformers with this structure.

In this study, a single-phase transformer prototype model D-400/6.3 is selected as the research object. Based on the actual parameters of the prototype transformer, the short-circuit finite element model of the low-voltage single helical winding is

Table 11.1 Parameters of the transformer prototype

Parameters	Numerical value
Rated capacity/kVA	400
Rated voltage/kV	6.3/0.4
Rated current/A	31.75/500
Core window height/mm	605
Reactance height/mm	455
Column A low voltage/high voltage/Column B low voltage coil forms	Continuous/continuous/single screw

constructed in COMSOL Multiphysics software. In view of the differences in the winding design and the influence of different structures on the current distribution, the magnetic properties of the single helical winding under the short-circuit condition are analyzed and calculated by using the "field-circuit" coupled multi-physics field calculation method, which reveals the mechanism of its influence on the performance of the winding, and provides a reference for engineering applications.

11.2 Finite Element Simulation Model of a Single Helical Winding Transformer

11.2.1 Transformer Prototype Model Parameters

A single-phase transformer prototype of model D-400/6.3 is selected for the study. Table 11.1 lists the related parameters of this transformer prototype.

During the construction of the geometric model, the transformer's manufacturing specifications are strictly followed. This ensures that key parameters, such as the size of the wire cake, the size of the oil passages, and the number of coil turns, match the actual specifications. To accurately simulate the electromagnetic field distribution and magnetic properties, the transformer windings and the airways between them are finely meshed. A relatively sparse mesh is applied to other parts to balance accuracy and computational efficiency. The electromagnetic and structural properties of each component are then defined within the model. Table 11.2 lists the material parameters involved in the model.

11.2.2 Field-Circuit Coupling Model

The geometric model construction involved employing an equivalent simulation approach, where each winding's width was divided based on the radial width of

Table 11.2 Material parameters

Material parameters	Transformer oil	Oil tank wall	Iron core	Coil	Pads
Density (kg/m^3)	895	7850	7650	8900	1300
Relative permeability	1	1	20000	0.9999	0.9999
Conductivity (S/m)	1×10^{-15}	4×10^6	1.12×10^7	5.8×10^7	1×10^{-15}
Relative permittivity	2.3	1	1	1	2.3
Poisson's ratio	0.5	0.3	0.26	0.325	0.28
Young's modulus (Pa)	0	2×10^{11}	2.06×10^{12}	1.15×10^{11}	1.3×10^{11}

the winding and the number of parallel-wound wires to accurately represent the winding structure. At the same time, the height of the wire is set strictly according to the actual height of the wire to accurately reflect the electromagnetic characteristics. Figure 11.1 shows the finite element calculation model and the schematic diagram of the transposition structure of the helical low-voltage winding transformer, where LV1 represents low-voltage winding wire 1 and LV2 represents low-voltage winding wire 2.

Figure 11.2 shows the external circuit wiring diagram of the field-circuit coupling model. In the simulation circuit setup, an AC voltage source is applied at both ends of the high-voltage windings with the rated phase voltage expression of

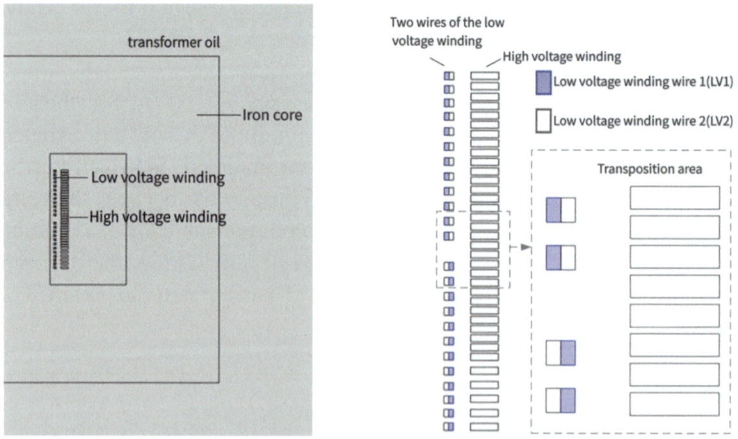

(a) Finite element modeling (b) Schematic diagram of the transposition structure

Fig. 11.1 Schematic diagram of finite element calculation model and transposition structure of single spiral low voltage winding transformer

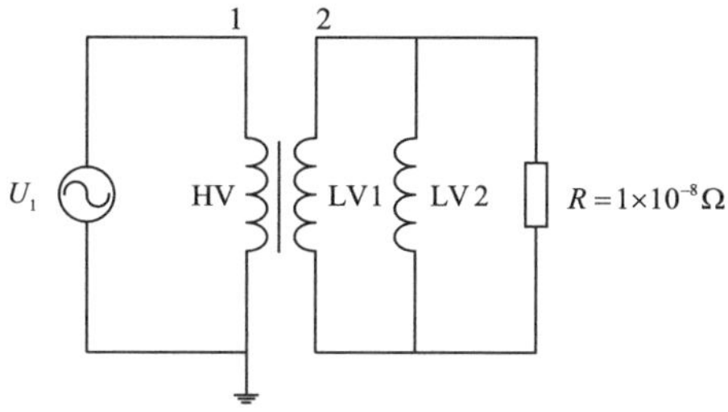

Fig. 11.2 External circuit wiring for "field-circuit" coupling model

$U_1 = 6300 \times \sqrt{2}\sin(100t)$. To simulate a short-circuit condition, the low voltage winding is connected to a small resistor with a resistance value of $1 \times 10^{-8}\Omega$. Each wire of the low-voltage winding acts as a circuit element, and together with the applied voltage on the high-voltage side and the small resistance on the low-voltage side, they form a complete circuit system.

11.3 Magnetic-Force Characteristics of Winding Under Short Circuit Condition

11.3.1 Analysis of Short-Circuit Magnetic Field Results

In this section, we will analyze in detail the specific effect of the low voltage winding transposition on the leakage magnetic field distribution at the moment of $t = 0.0085$ s, and focus on the electromagnetic distribution characteristics of the low voltage winding after the transposition operation. As Fig. 11.3 shown, the overall distribution of the magnetic flux density and the details of the magnetic lines of force around the transposition region are plotted in order to demonstrate the magnetic field distribution characteristics more intuitively.

In the transposition region of the middle of the low-voltage winding, the magnetic lines of force undergo significant bending and spreading, resulting in significant distortion of the magnetic flux density in localized areas. Because the magnetic lines of force always tend to close along the path with the lowest reluctance, the transposition operation causes part of the magnetic flux to crosslink directly over the core in the middle of the low-voltage winding, thus changing the original path of the magnetic flux. The magnetic lines of force in the transposition region are no longer distributed only along the axial direction, but show both axial and radial components, and this change in the distribution pattern exacerbates the inhomogeneity of the magnetic potential distribution along the radial direction of the winding in this region. Due to the transposition region is located in the middle of the low-voltage winding, this inhomogeneity also has an effect on the magnetic flux density distribution in the middle of the high- and low-voltage windings, showing an overall decreasing trend.

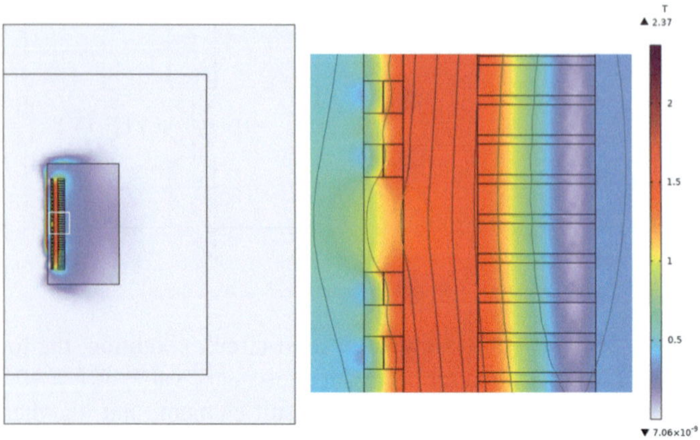

Fig. 11.3 Plot of the overall distribution of magnetic flux density and details of the magnetic lines of force around the transposition region

In addition, the transposition operation results in a widening of the oil channel gap, providing additional circulation paths for the magnetic lines of force. This change complicates the distribution of magnetic flux density, impacting not only the transposition region but also its adjacent wire cakes. To comprehensively understand magnetic field distribution characteristics, this study analyzes the average radial and axial flux densities for each line cake of the left and right column wires. Left-column wires are near the core, while right-column wires are near the main gaps of high- and low-voltage windings, resulting in distinct magnetic field distribution characteristics. As shown in Fig. 11.4, the radial flux densities of both the left- and right-column wires reach extreme values at the line cakes in the transposition region, i.e., at the low-voltage 12 and 13 cakes. This indicates that the transposition operation has a significant effect on the strength and direction of the magnetic field. Meanwhile, the effect of axial leakage field distribution is relatively small, while the radial flux density is relatively high, and the change of radial leakage field is especially obvious.

It is worth noting that the transposition structure not only affects the magnetic field distribution in the transposition region itself, but also has a significant effect on the radial leakage field distribution in the upper and lower neighboring triple pancakes (9, 10, 11 and 14, 15, 16). Although the transposition operation had a significant effect on the radial wise flux density, its effect was relatively small in terms of the axial flux density. This is mainly due to the fact that the axial magnetic field is mainly affected by the winding structure and the iron core, whereas the transposition operation mainly changes the current distribution of the windings and the oil channel gap, which has a limited effect on the axial magnetic field. Compared with the low-voltage winding, the magnetic field distribution of the high-voltage winding exhibits relatively smooth characteristics. Both the axial magnetic field and the radial magnetic field change more gently in the high-voltage winding area. This is mainly due to the structural design of the high-voltage winding itself and

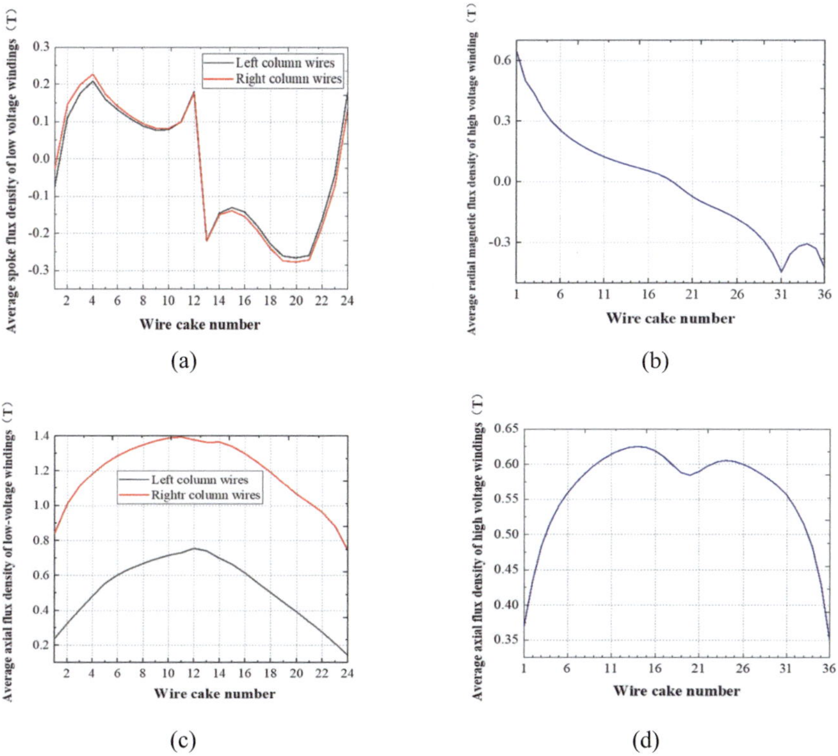

Fig. 11.4 Average magnetic flux density of the left and right columns of wires of the LV winding and each wire cake of the HV winding

the uniformity of the current distribution. Therefore, the effect of the transposition operation on the magnetic field distribution is relatively small in the high-voltage winding.

In summary, the transposition operation has a significant effect on the magnetic field distribution in the transformer windings, especially in the transposition region and its neighboring line cakes. This effect is not only reflected in the size and direction of the magnetic flux density, but also involves the propagation and diffusion effects of the magnetic field. Therefore, the effect of the transposition operation on the magnetic field distribution needs to be fully considered during the design and optimization of the transformer in order to achieve more efficient electromagnetic performance.

11.3.2 Analysis of Short-Circuit Electromagnetic Force

This section delves into the detailed analysis of the impact of low-voltage winding transposition on the short-circuit electromagnetic force distribution of its two

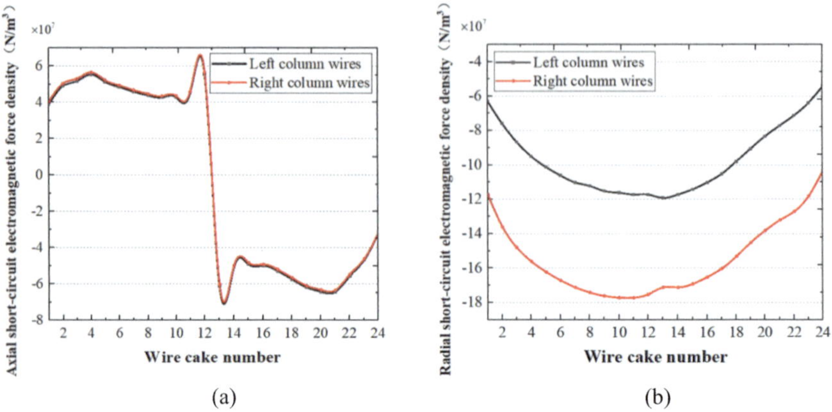

Fig. 11.5 Short-circuit electromagnetic force density of the left and right columns of wires of the LV winding

columns of wires and the high-voltage winding at $t = 0.0085$ s. First, the characteristics of the electromagnetic distribution of the two columns of wires of each line cake of the low-voltage winding are analyzed.

Figure 11.5 shows the electromagnetic distribution curve of the two columns of wires of each line cake of the low-voltage winding of the prototype transformer. Observation of the figure reveals notable numerical disparities in the short-circuit electromagnetic forces of the two columns of wires in the radial direction, largely influenced by the relative positions of the wire and the core. In the radial direction of the winding, the electromagnetic force density gradually increases as it gets farther and farther away from the center of the winding.

In terms of radial electromagnetic force, the most stressed line cakes are the 13th line cake of the left column of wires and the 11th line cake of the right column of wires, whose electromagnetic force densities are respectively $-1.19 \times 10^8 \text{N/m}^3$ and $-1.77 \times 10^8 \text{N/m}^3$. Although the transposition process resulted in a difference in the distribution of wire currents, this difference did not significantly affect the values of axial electromagnetic forces for the two columns of wires. However, at the middle transposition, there is a significant change in the direction of the axial electromagnetic force of the two columns of wires. In particular, the axial electromagnetic force density reaches its highest at the 12th wire cake position, with the left and right column wires being respectively $5.57 \times 10^7 \text{N/m}^3$ and $-5.66 \times 10^7 \text{N/m}^3$, both of which exhibit high values of electromagnetic density.

In order to further investigate the distribution law of the radial electromagnetic force of each column of wires, this paper defines a top-down vertical path on the inner and outer surfaces of the left and right columns of wires in the low-voltage winding. As Fig. 11.6 shows, the radial electromagnetic force on each column of wires in each line cake shows a trend of increasing along the radius outward, among which the outer side of the right wire receives the largest radial electromagnetic force.

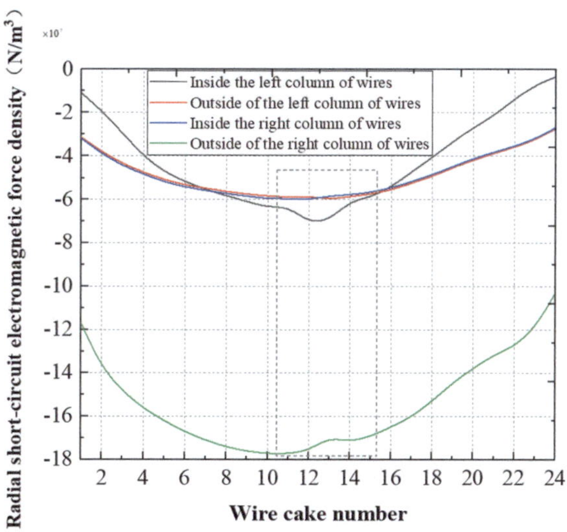

Fig. 11.6 Distribution of radial force density of each wire cake on the inside and outside of the left and right columns wires of the LV winding

The position of the wire cake in the magnetic field at the transposition structure of the left column of wire changes, and the radial short-circuit electromagnetic force shows an increasing trend as it gets farther away from the core. In addition, the non-uniformity of the current distribution also leads to a significant wire change in the radial electromagnetic force at the transposition structure, which further exacerbates the non-uniformity of the winding force. This variation is not only closely related to the difference in current distribution, but may also be affected by the combined effect of the gap of the transposition structure on the local leakage field and current distribution.

In summary, the introduction of the transposition structure not only changes the spatial layout of the wire, leading to the distortion of the leakage magnetic field, but also significantly affects the electromagnetic force distribution characteristics of the winding. Especially under the short-circuit condition, the sharp increase of current further aggravates the magnitude of the leakage field and electromagnetic force, which makes this effect more significant. The presence of the transposition structure as a transition section leads to a non-uniform trend in the force distribution of the winding.

11.4 Conclusion

This study develops a single helical winding transformer model with a transposition structure. The field-circuit coupled simulation method is used to model the uneven current distribution and its impact on the electromagnetic field. The results show that the two columns of wires show a significant difference in the radial-to-radial short-circuit electromagnetic force during the low-voltage winding transposition process, while the axial electromagnetic forces are similar. The difference is due to the varying relative positions between the wires and the core. The non-uniformity of the wire current distribution does not significantly affect the overall distribution of the axial electromagnetic force in most cases. Observing the distribution of the radial electromagnetic force of each column, it can be found that the outer side of the right-hand wire experiences the largest radial electromagnetic force. As the distance from the core increases, the radial short-circuit electromagnetic force of the wire cake at the transposition structure gradually increases, further aggravating the non-uniformity of the winding force. This effect intensifies as the difference in current distribution becomes significant, making the difference in radial electromagnetic force distribution of the parallel wire at the permutation becoming particularly prominent.

Acknowledgements This work is supported by the China Southern Power Grid project 0666002023030103GY00017.

References

1. Li peng, Bi Jiangang, Yu Hao, et al. Intelligent sensing and state sensing technology and application for substation equipment[J]. High Voltage Technology, 2020, 46(09): 3097–113.
2. Tenbohlen S, Jagers J, Vahidi F. Standardized survey of transformer reliability: on behalf of CIGRE WG A2. 37[C]//2017 International Symposium on Electrical Insulating Materials (ISEIM). IEEE, 2017, 2: 593–596.
3. Mortazavi M, Moravej Z, Gharehpetian G B. Detection and localization of LV winding radial deformation in transformers using electromagnetic waves-a feasibility study[J]. International Journal of Electrical Power & Energy Systems, 2024, 155: 109602.
4. Hashemnia N, Abu-Siada A, Islam S. Improved power transformer winding fault detection using FRA diagnostics-part 1: axial displacement simulation[J]. IEEE Transactions on Dielectrics and Electrical Insulation, 2015, 22(1): 556-63.
5. Zhang Bingqian, Xian Richang, Yu Yang, et al. Physical characterization of power transformer windings under inter-turn short circuit fault[J]. High Voltage Technology, 2021, 47(06): 2177–85.
6. Medeiros L H, Maschio G, Oliveira M M, et al. Finite element analysis for calculation of total and distributed electromagnetic forces on power transformers windings arrangements[J]. Electrical Engineering, 2024, 106(1): 931-40.
7. Li Yichun, Liu Wenli, Li Jun, et al. Calculation and analysis of combined short-circuit force in the radial direction of spiral winding of power transformer[J]. Heilongjiang Power, 2016, 38(02): 135–40.
8. Wang Luliang, Liu Wenli, Li Yangyang, et al. Calculation of radial-to-radial short-circuit forces in power transformer windings[J]. Transformer, 2012, 49(02): 1–5.
9. Guo Jian, Lin Heyun, Xu Zihong, et al. Calculation and analysis of field-circuit coupling of branch currents in single spiral-wound transformers[J]. Journal of Electrotechnology, 2010, 25(04): 65–70.
10. Zhao Yi, Wen Tao, Ni Heli, et al. Characteristics of current distribution in power transformer with low-voltage spiral winding wire transposition structure[J]. High Voltage Technology, 2021, 47(09): 3218–25.

Chapter 12
Quantitative Analysis Method of Insulation Strength Based on 35 kV Switchgear Discharge Model

Fei Li, Peng Sun, Zhaoliang Han, Xinxin Bian, Xiaodong Jiang, Kaiji Li, Kai Nie, and Qingquan Li

Abstract Switchgear is widely used in transmission and distribution networks, and its operation status is directly related to the reliability and safety of power supply. This paper takes the 35 kV switchgear as the research object and explores its insulation characteristics. Firstly, the partial discharge simulation calculation of the switchgear is analyzed and the electric field calculation model of the switchgear is established; secondly, different kinds of local discharge faults are set up in the model to investigate the influence on the insulation strength of the switchgear; finally, the insulation performance of the switchgear specimen is characterized by four parameters, namely, partial discharge, change of insulation resistance, relative permittivity and current leakage through the setup of the simulation experiment. The results of the experiment proved that the above four parameters have been analyzed by the simulation experiment. The experimental results prove that the above four parameters can effectively respond to the real-time operation of the switchgear insulation and prevent the switchgear from insulation degradation or failure. The research in this paper provides certain data support for the optimization strategy of switchgear insulation performance.

Keyword High-voltage switchgear · Electric field distribution · Numerical simulation · Insulation optimization

---

F. Li (✉) · P. Sun · Z. Han · X. Bian · X. Jiang · K. Li
State Grid Shandong Electric Power Company Zibo Power Supply Company, Zibo, China
e-mail: liyif1988@163.com

K. Nie · Q. Li
School of Electrical Engineering, Shandong University, Jinan, China

Y. Li et al. (eds.), *Proceedings of the 5th Chinese Conference on High Voltage and Discharge Plasmas*, Springer Proceedings in Physics 254,
https://doi.org/10.1007/978-981-96-5711-7_12

12.1 Introduction

High-voltage switchgear is widely used in power supply systems, and the safe operation of switchgear is closely related to reliable power supply. Data show that the switchgear occurs in all the faults, insulation failure occupies the largest proportion, up to 31.9%. In practice, the switchgear internal insulation degradation and other problems inevitably exist, over time, will cause the failure of the insulation of electrical equipment, which in turn led to a series of failures [1–4]. Therefore, it is crucial to quantify the strength of switchgear insulation in electrical equipment.

Regarding the insulation performance of electrical equipment, there are scholars who have conducted corresponding studies. In the paper [5], based on the finite element electric field simulation analysis, the minimum insulation distance based on circular leads is proposed by analyzing and verifying the insulation distance of circular leads with different structures; Paper [6] put forward the ultrasonic—Transient earth voltage (TEV) fusion sensing technology and designed the fusion sensor, achieving the synchronous measurement of AE signals and TEV signals in space and time, but no analysis was carried out on the frequency domain of the signals and the waveform characteristics of the signals. In paper [7], the partial discharge mechanism of composite insulation materials under DC is analyzed, and the significant characteristics of different partial discharge stages under various defects are obtained. The application of neural network in this paper shows a good recognition accuracy rate. In summary, it can be seen that among the existing studies, there have been a lot of studies on the insulation performance of electrical equipment, but relatively few studies on the insulation characteristics of air switchgear.

Based on this, in order to discuss and study the insulation performance of the switchgear and promote the safe and stable operation of electrical equipment, the paper takes the 35 kV air switchgear as the research object and studies its insulation performance based on electric field calculation and electric aging, aiming to improve the insulation characteristics of the air switchgear. It also provides some reference for the safe and stable operation of the power system.

12.2 Electric Field Simulation of Air Insulated Switchgear

For the calculation of three-dimensional electric field, the Laplace equation is satisfied on the whole, which can be expressed as:

$$\begin{cases} \nabla\varphi^2 = 0 \\ \varphi|_{b_1} = f_1(p) \\ \varphi|_{b_2} = f_2(p) \\ \frac{d\varphi}{dn}|_{b_3} = f_3(p) \end{cases} \tag{12.1}$$

In formula (12.1): φ represents the potential at the node; b_1 indicates the conductive busbar inside the air switch cabinet; $\varphi|_{b1}$ represents the outer boundary and potential of the conductor. b_2 indicates the shell of the air switch cabinet; $\varphi|_{b2}$ indicates the outer boundary and potential corresponding to the ground switch. Further, the corresponding equivalent variational problem can be expressed as:

$$F(\varphi) = \int v \frac{\varepsilon}{2} \begin{bmatrix} \left[\frac{\partial \varphi}{\partial x}\right]^2 + \left[\frac{\partial \varphi}{\partial y}\right]^2 \\ + \left[\frac{\partial \varphi}{\partial z}\right]^2 \end{bmatrix} \mathrm{d}x\mathrm{d}y\mathrm{d}z = \min \tag{12.2}$$

In Eq. (12.2), ε represents the relative dielectric constant.

The discrete functional can be expressed as:

$$F(\varphi) = \int\int\int \frac{\varepsilon}{2} \begin{bmatrix} \frac{\partial^2 \varphi_e}{\partial x^2} + \frac{\partial^2 \varphi_e}{\partial y^2} \\ + \frac{\partial^2 \varphi_e}{\partial z^2} \end{bmatrix} \mathrm{d}x\mathrm{d}y\mathrm{d}z = \frac{1}{2}[\phi]_e[K]_e[\phi]_e \tag{12.3}$$

In formula (12.3), *e* represents the global stiffness matrix, which can be expressed as:

$$K_{i,j}^e = \int\int\int e^{\varepsilon} \begin{bmatrix} \frac{\partial^2 N_i^e}{\partial x}\frac{\partial^2 N_j^e}{\partial x} + \frac{\partial^2 N_i^e}{\partial y}\frac{\partial^2 N_j^e}{\partial y} \\ + \frac{\partial^2 N_i^e}{\partial z}\frac{\partial^2 N_j^e}{\partial z} \end{bmatrix} \mathrm{d}x\mathrm{d}y\mathrm{d}z_e \tag{12.4}$$

In formula (12.4): φ corresponds to the potential value of the electric field; N_i and N_j correspond to the corresponding coordinates in Cartesian coordinates.

In the electric field calculation of air switchgear, the initial model can be simplified by means of grid division, and then the electric field can be analyzed by finite element method more conveniently and accurately. Considering the characteristic that tetrahedron is easy to realize, the finite element mesh of tetrahedron is chosen in the calculation of the electric field of the air switch cabinet. The number of grid elements is 1 388 808, the number of nodes is 1 897 413, and the corresponding calculation time is 7 min.

12.3 Experimental Design

In the actual operation process, the structure and environment of the switch cabinet are very complex, and the same type of office discharge may have multiple forms of expression. According to its causes, the form of partial discharge of high voltage switchgear can be summarized into the following four categories:

(1) Corona discharge caused by metal burrs on the switchgear box and the internal busbar and metal components;
(2) Suspension discharge due to the high voltage isolation switch contact is not closed in place and the distance between the secondary wiring head and the charged body in the switch cabinet is too close;
(3) Due to the air gap inside the insulation material, the air gap discharge caused by the external electric field;
(4) Due to the pollution on the surface of the medium, the metal conductor produces surface discharge along the dirty medium surface.

Therefore, according to GB/T 20,635–2006 standard, four partial discharge defect models including tip discharge, suspended discharge, surface discharge and gap-discharge are set in 35 kV air switchgear, and the insulation performance of the switchgear is tested through short-time power frequency voltage withstand test. The four partial discharge equivalent models are shown in Fig. 12.1.

On this basis, the electrical aging test for power frequency voltage resistance is carried out under different voltage conditions, and the insulation performance of 35 kV air insulated switchgear is investigated by comparing the changes of partial discharge, insulation resistance and leakage current data during the electrical aging process.

12.4 Insulation Performance Analysis Based on Electrical Aging

In the process of electrical aging test, the data distribution and change of local discharge amount, insulation resistance, relative dielectric constant and leakage current of electrode specimen materials are respectively shown in Fig. 12.2.

It can be found from the figure that in the process of frequency electric aging of the air insulated switchgear, with the increase of aging time, the local discharge energy generally shows a change process from stability to growth and then to deterioration; It can be found from the change process of insulation resistance and leakage current of the test product of air insulated switchgear that the corresponding insulation resistance and leakage current of the test product fluctuate within a certain range in the initial stage of electrical aging, and fluctuate to a greater extent when approaching the insulation breakdown. The relative dielectric constant of the test products increases

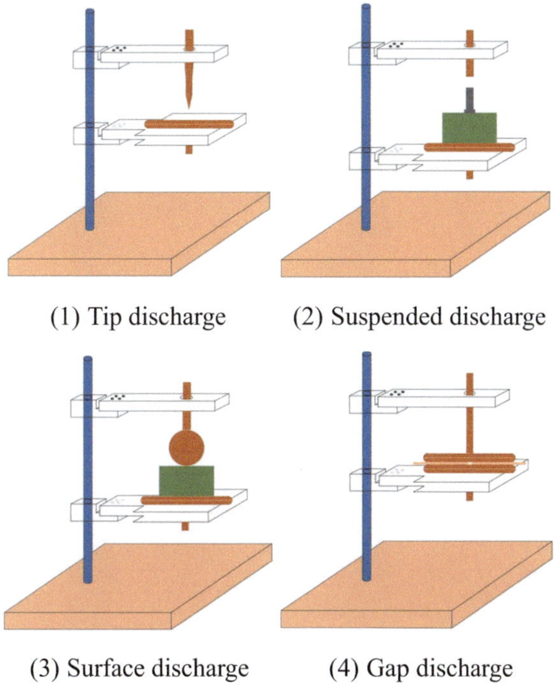

Fig. 12.1 Equivalent model diagrams of four partial discharges

uniformly, but the relative increment of the relative dielectric constant is at a low level throughout the whole electric aging process.

12.5 Conclusion

The safe and stable operation of electrical equipment is affected by multiple factors, but as an important component of electrical equipment, the safe and stable operation of air switchgear is also a direct impact on electrical equipment and even the entire power system. In this paper, 35 kV air-insulated switchgear is taken as the research object, and the distribution changes of partial discharge, insulation resistance and leakage current are found through electrical aging test, which can reflect the working or running state inside the air-insulated switchgear in real time.

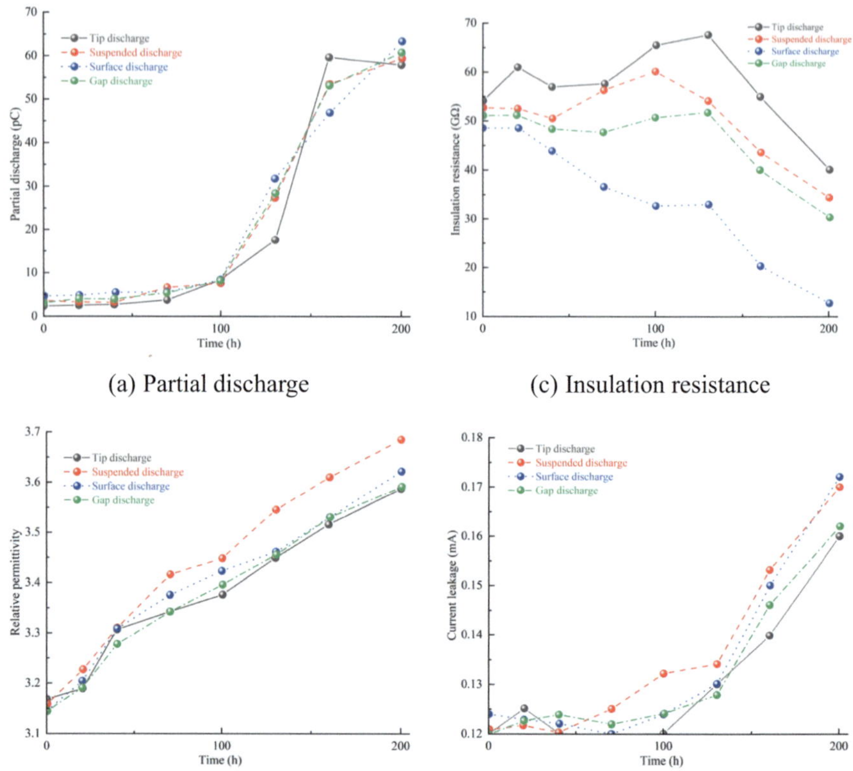

(a) Partial discharge

(c) Insulation resistance

(b) Relative permittivity

(d) Current leakage

Fig. 12.2 Distribution and change of insulation parameters during electrical aging test

Acknowledgements This work was supported by the Science and Technology Project of the State Grid Shandong Electric Power Company: Study on partial discharge characteristics and insulation optimization of 35kV switchgear (52060324000F). The project code is 2024A-024.

References

1. PAN Minglong, FANG Chunhua, YI Mancheng, et al. Influence of cable buckling on bushing force in cable chamber of ring main units[J]. High Voltage Apparatus, 2019, 55(5):94–99.
2. YANG Fang, PAN Qize. Key technologies investigation of damp-proofing and anti-dewingfor 12kV high-voltage switch- gear[J]. High Voltage Apparatus, 2018, 54(8):40–47.
3. LI Yangwei, REN Chengyan, KONG Fei, et al. Research progress on the measurement of secondary electron emis-sion yield of insulation materials and influencing factors[J]. High Voltage Apparatus, 2019, 55(5):1–9.
4. LUAN Lan, XU Benmao, YU Qinxue, et al. Design of reducing the insulation distance of the round lead wire of 10 kV transformer[J]. High Voltage Apparatus, 2019, 55 (2):189–194.

5. XUAN Yaowei, HU Kai, YANG Fengyuan, et al. Research on partial discharge characteristics of typical defects in XLPE cables under DC voltage[J]. High Voltage Apparatus, 2019, 55(7):173–180.
6. SUN Yuan, SONG Yanfeng, WANG Haotian, et al. Ultrasonic-TEV combined detection technology of partial discharge in switchgear[J]. High Voltage Engineering, 2023, 49(12):5052–5060.
7. BAI Lu, LI Xiaojing, DONG Like, et al. Study on the partial discharge types of ring network cabinets based on ultrasonic method[J]. Electrical Measurement & Instrumentation, 2020, 57(11):42–48.

Chapter 13
A Review of the Aging Mechanism and Life Prediction of Silicone Rubber Composite Insulators

Xing Zhang, Songsong Zhou, Hechen Liu, Shijie Xu, and Jinghao Chen

Abstract Silicone rubber insulation materials, with their advantages of weather resistance, water repellency, light weight, and low cost, are widely used in composite insulators in power systems. However, as the number and duration of composite insulator operations increase, and under the long-term exposure to electric fields, mechanical loads, and complex and variable environmental factors, the aging issue of silicone rubber sheds has become increasingly prominent, posing a threat to the safe and stable operation of the power grid. This paper comprehensively reviews the aging characteristics and mechanisms of high-temperature vulcanized silicone rubber materials under the influence of corona discharge, ozone, ultraviolet radiation, acid and alkali, high temperature, surface arcing, and multiple factors simultaneously. Furthermore, various models for predicting the lifespan of these materials are analyzed. Finally, the paper provides an outlook on the aging and lifespan prediction of composite insulators.

Keyword Silicon rubber · Composite · Aging · Life prediction

13.1 Introduction

Composite In the 1950s, insulators made from advanced composite materials emerged, utilizing ethylene propylene diene monomer (EPDM) rubber, polytetrafluoroethylene (PTFE), and blended rubbers for their sheds, marking the beginning of rapid development for composite insulators. Subsequent extensive research has

X. Zhang · H. Liu · S. Xu
North China Electric Power University, Baoding 071003, China

S. Zhou (✉)
China Electric Power Research Institute, Beijing 100192, China
e-mail: z18516979216@163.com

S. Zhou · H. Liu · J. Chen
North China Electric Power University, Beijing 102206, China

Y. Li et al. (eds.), *Proceedings of the 5th Chinese Conference on High Voltage and Discharge Plasmas*, Springer Proceedings in Physics 254,
https://doi.org/10.1007/978-981-96-5711-7_13

shown that composite insulators offer numerous advantages, including light weight, high mechanical strength, strong hydrophobicity and hydrophobic migration, and high pollution flashover voltage, leading to their widespread application. Currently, China has over 7 million composite insulators in operation, making it the country with the largest number of composite insulators in use worldwide [1, 2]. Composite insulators are subjected to a combination of electrical, mechanical, and environmental factors during operation. As the operating time increases, silicone rubber composite insulators may exhibit phenomena such as decreased hydrophobicity, discoloration and hardening of sheds, brittleness, pulverization, cracking, tracking or electrical erosion, failure of end seals, and unexplained flashovers. The aging of composite insulators under long-term electrical stress and environmental factors has garnered significant attention from insulator manufacturers, operators, and research institutions.

Silicone rubber is a polymer with a main chain composed primarily of repeated Si–O bonds, where silicon atoms are directly connected to organic groups. The organic macromolecules of silicone rubber materials are held together by covalent bonds, with the Si–O bond in the silicone rubber main chain having a bond energy of 447 kJ/mol. Under the influence of factors such as electron flows from surface discharges, ultraviolet radiation, mechanical forces, and chemical agents, covalent bonds can break, particularly considering the presence of numerous carbon-hydrogen chemical bonds within silicone rubber materials [3]. In actual operating conditions, factors like high and low temperatures, high humidity, ultraviolet radiation, high-voltage discharges, ozone, salt density, and contamination accelerate the aging of silicone rubber materials. Aging of silicone rubber leads to chain scission within the molecular chains, causing hydrophilic molecules within the chains to precipitate onto the material surface, resulting in a decrease in hydrophobicity and irreversible physical and chemical changes.

China is accelerating the construction of a backbone power grid system centered on ultra-high voltage (UHV) and extra-high voltage (EHV) AC/DC transmission, which is accompanied by severe challenges posed by extreme environments such as high altitudes, heavy pollution, intense ultraviolet radiation, and corona discharges to the external insulation and aging of silicone rubber materials in composite insulators. Addressing these critical issues, it has become an urgent task to deeply analyze the multi-factor mechanism of aging in composite insulators during operation. This paper aims to review the latest advancements in prediction technologies for the service life of silicone rubber insulators by dissecting these influencing factors, thereby providing robust support for effective operation, maintenance, and technical management strategies of composite insulators.

13.2 Aging Types and Mechanisms of Composite Insulators

13.2.1 The Effect of Ultraviolet Radiation on the Properties of High Temperature Vulcanized Silicone Rubber

Ultraviolet (UV) radiation in the upper layers of the Earth's atmosphere accounts for 8% of the total radiation. Near the ground, UV radiation is primarily composed of long-wave UV. Long-wave UV has an energy range of 299 to 380 kJ/mol, with a main peak at 365 nm corresponding to a photon energy of 328 kJ/mol. In High-Temperature Vulcanizing (HTV) silicone rubber, the bond energies of the main chemical bonds are as follows: Si–O has a bond energy of 447 kJ/mol, Si–C has a bond energy of 318 kJ/mol, C-H has a bond energy of 413 kJ/mol, and the bond energies of the less common C–C and O–H bonds are 345 kJ/mol and 463 kJ/mol, respectively. The energy of long-wave UV exceeds the bond energies of certain chemical bonds in HTV, such as Si–C and C–C.

Based on the UV energy spectrum and radiation intensity of sunlight in Yunnan region, Li Guofang [4] conducted 500-h and 1000-h UV radiation accelerated aging experiments on HTV silicone rubber using a self-built UV radiation accelerated aging test chamber. The results indicated that after UV radiation, the tensile strength and elongation at break of HTV silicone rubber decreased, while its hardness increased; the volume resistivity and surface resistivity decreased, and both the breakdown strength and hydrophobicity declined. Microscopically, UV radiation caused the breakage of some Si–C and C–C bonds in HTV silicone rubber, leading to the formation of new free radicals and subsequent crosslinking and oxidation. This resulted in an increase in the surface strength of the silicone rubber, a decrease in elasticity and toughness, a reduction in tensile strength and elongation at break, and an increase in hardness. The breakage of some Si–C bonds within the crosslinking network led to a reduction in crosslinking points and a decrease in breakdown strength. Additionally, partial detachment and oxidation of non-polar methyl groups on the surface of the silicone rubber weakened their shielding effect on the strongly polar main chain, thereby reducing surface hydrophobicity. At the same time, the weak ionization of water itself increased the concentration of charge carriers, leading to a decrease in both volume resistivity and surface resistivity.

Li Jingjing [5] conducted accelerated aging experiments using a self-designed adjustable UV aging test chamber on high-temperature vulcanized (HTV) silicone rubber samples from two manufacturers, subjected to UV irradiation (0, 500 h, 1000 h) with a wavelength range of 320–750 nm. The results indicated that UV-A (UVA) aging of HTV silicone rubber slightly decreased its hydrophobicity and volume resistivity, while increasing the surface roughness. The intensities of the Si-CH3 and Si-(CH3)2 reflection peaks gradually decreased with increasing irradiation time, whereas the intensity of the Si–O–Si reflection peak initially decreased and then increased. Partial detachment of methyl groups on the silicone rubber surface weakened the shielding effect on the strong polarity of the main chain. The cleavage produced gases such as CH4, resulting in pores and unevenness on the material surface, thereby reducing its hydrophobicity. This finding is consistent with the conclusion obtained through the water spray classification method that UV irradiation decreases the hydrophobicity of HTV silicone rubber. It also aligns with the results of SEM surface morphology analysis, which showed the appearance of large particles, pores, and increased roughness on the surface of the irradiated samples.

Li Yawei [6] conducted a 4000-h ultraviolet radiation experiment on silicone rubber using three types of ultraviolet light sources: UVA, UVB, and UVC. After prolonged ultraviolet radiation, changes occurred in the surface morphology of the insulator samples. The surface of the unirradiated silicone rubber sample was smooth, uniform, and without noticeable cracks or damage. After radiation with UVA ultraviolet light, the surface flatness of the sample decreased, and some small protrusions and damages appeared. The number of damages on the surface significantly increased, and the damaged area enlarged after radiation with UVB ultraviolet light. The surface aging of the sample was most severe after radiation with UVC ultraviolet light, becoming extremely rough with textured and laminated microscopic characteristics. Further investigation into the causes revealed that after the surface material of the silicone rubber sample was damaged by ultraviolet radiation, some of the $Al(OH)_3$ filler, which was originally hidden within the composite insulator material, leaked out, making the surface microstructure significantly rougher.

13.2.2 The Effect of Ozone on the Properties of High-Temperature Vulcanized Silicone Rubber

During the operation of high-voltage lines, ozone generated by discharges can react with polymer materials, causing oxidation. Tu Youping [7, 8] conducted ozone aging tests on HTV (High-Temperature Vulcanized) silicone rubber samples under different ozone concentrations and exposure times. The results indicated that under the influence of an ozone environment, the smooth and dense physical structure of the material surface was disrupted, leading to defects such as pores. When the intensity of the ozone environment reached a certain level, the damage to the material's physical structure became more severe. Simultaneously, changes occurred in the content of chemical groups, with a decrease in Si–O–Si structures and an increase in hydroxyl and Si–O bonds, resulting in the breakage of the main and side chains of the sample. B. Ma [9] studied the aging characteristics of HTV silicone rubber materials under AC and DC corona/ozone exposure using two sets of HTV samples with different resistivity and tensile strength. The intensity of DC corona discharge was weaker than that of AC, and the ozone concentration was also much lower. Compared to DC treatment, AC corona exposure had a higher discharge intensity, producing a higher dose of ozone and leading to more severe surface oxidation. The results of tensile strength tests did not show a clear pattern in terms of the observed slight decrease in mechanical properties. A reasonable explanation for this observation is that corona-ozone treatment may lead to surface degradation, while the bulk matrix of the polymer remains essentially unchanged, even under stronger AC corona exposure doses. Finally, FTIR (Fourier Transform Infrared Spectroscopy) and XPS (X-ray Photoelectron Spectroscopy) analyses of the corona-treated sample surfaces confirmed the appearance of surface oxidation, whose intensity was, in turn, correlated with the ozone concentration during treatment.

13.2.3 The Effect of Acids and Bases on the Properties of High-Temperature Vulcanized Silicone Rubber

Composite insulators are mostly operated in moderately or even heavily polluted areas, where acids, bases, and other pollutants in the environment can cause significant damage to the power supply system. Zhang Ni [10] immersed silicone rubber in nitric acid solution with a pH of 1 for different durations and analyzed the aging mechanism of nitric acid on silicone rubber using methods such as GC–MS (Gas Chromatography-Mass Spectrometry) and Thermogravimetric Analysis (TGA). The results showed that polydimethylsiloxane (PDMS) in silicone rubber degraded under the action of nitric acid, leading to an increase in small cyclic siloxane molecules mainly consisting of D4-D7. Fourier Transform Infrared (FTIR) Spectroscopy and X-ray Photoelectron Spectroscopy (XPS) results indicated that after aging with nitric acid, the Si–C bonds on the surface of silicone rubber broke and were oxidized, forming inorganic siloxane structures Si-O3 and Si-O4, resulting in a decrease in surface carbon content. Meanwhile, the breakage of C-H bonds in the methyl groups and the introduction of polar hydroxyl groups led to a decrease in the barrier properties and hydrophobicity of silicone rubber. Peng Xiangyang [11] simulated the aging process of silicone rubber under different environmental conditions using solution immersion tests and systematically investigated the change in surface hydrophobicity with immersion time in acidic or alkaline solutions. In a strongly acidic environment, the contact angle of silicone rubber surface significantly decreased. In an alkaline environment, silicone rubber showed more noticeable aging behavior when the pH was greater than 12. The mechanisms of aging in both acidic and alkaline conditions involved the attack of hydrogen ions or hydroxyl ions on the main chain Si–O bonds, causing them to break and form silanols. Strong bases could also react with reinforcing agents such as silica to generate pores, facilitating water diffusion.

Zheng Hanyu [12] artificially simulated the insulation failure process of composite insulator silicone rubber materials during energized operation in a corrosive environment and analyzed the tolerance gradient of silicone rubber under different operating conditions to reveal the failure characteristics of composite insulator operation. The experimental results showed that the flashover voltage decreased with the increase of applied AC voltage, fog conductivity, and energized operation time. Moreover, as the energized duration increased, the rate of decrease in flashover voltage slowed down. Under the same conditions, the AC flashover voltage of silicone rubber samples was the lowest in an alkaline fog corrosion environment.

Ma Xiaodan [13] selected silicone rubber sheets with the same formula as those used in actual operating insulators and tested their physicochemical and electromechanical properties under different pH values of nitric acid fog, different voltage levels, and different operating times. The results showed that under the continuous effect of an acidic environment, as the pH decreased, the surface roughness increased, resulting in the appearance of pores, cracks, etc. The water absorption increased, leading to an increase in the relative permittivity and dielectric loss. The static contact angle decreased, indicating a reduction in hydrophobicity. The polymer chains underwent cleavage and oxidative cross-linking reactions, resulting in an increase in high-oxidation state Si, a decrease in the relative content of C elements, an increase in the relative content of O elements, and surface discharges that exacerbated these changes.

13.2.4 The Effect of High Temperature on the Properties of High-Temperature Vulcanized Silicone Rubber

Thermal aging of silicone rubber often accelerates the migration of low-molecular-weight substances to the surface and triggers chemical reactions such as cross-linking and thermal degradation within the material. Cheng Yang [14] selected the HTV (High-Temperature Vulcanized) silicone rubber composite insulator sheds that had been in operation for over 10 years as the research objects and conducted accelerated thermal aging tests on them using hot air in a constant temperature oven. The results showed that the aging degree of the HTV silicone rubber composite insulator sheds was greater at the high-voltage end and low-voltage end compared to the middle section. This was mainly manifested by higher hardness, lower -OH peak values, and smaller initial surface flashover voltages in the samples from the high-voltage and low-voltage ends. Under the effect of hot air aging, there were no significant changes in the characteristic groups of silicone rubber, but continuous thermal aging intensified the thermal decomposition of siloxanes, leading to the appearance of local cracks. As the thermal aging time increased, the surface flashover voltage of the samples first increased and then decreased. The change in flashover voltage was related to the surface morphology, hydrophilicity, and initial aging state of the samples. Rong Yiping [15] conducted artificial accelerated thermal aging tests on silicone rubber insulation samples. The results indicated that during accelerated thermal aging, the main reaction occurring in silicone rubber was the oxidation of side-chain methyl groups, which led to further cross-linking reactions. This inhibited the migration of charge carriers, resulting in a decrease in volume conductivity. Simultaneously, the average free path of electrons decreased, which was unfavorable for electrons to gain acceleration through free volume, reducing the destructive energy of electrons and thus lowering the possibility of breakdown. Zhou Yuanxiang [16] carried out combined thermal–mechanical aging tests on silicone rubber to study the aging process before and after. The results demonstrated that the tensile strength and elongation at break of silicone rubber decreased with increasing aging time, while the hardness increased. In the early stage of aging, cross-linking reactions occurred in silicone rubber, leading to an increase in cross-link density, a decrease in free volume and charge carrier mobility, resulting in an increase in breakdown strength and a decrease in dielectric loss. In the later stage of aging, the cross-linking system was disrupted, and degradation reactions intensified, leading to an increase in charge carrier concentration and mobility, an increase in free volume, resulting in a decrease in breakdown strength and an increase in dielectric loss.

13.2.5 The Effect of Electrical Corona on the Properties of High-Temperature Vulcanized Silicone Rubber

Silicone rubber composite insulators are not only subject to the influence of natural conditions during actual use but also operate under the effect of high-voltage electric fields. Electric field factors have a more severe and faster aging effect on silicone rubber. The main reason for corona formation on the surface of insulators is the unreasonable design of the insulator shape or grading ring, which can lead to corona discharge near the metal fittings of composite insulators. Additionally, water droplets on the surface of composite insulators can distort the local electric field, causing corona discharge on the insulator surface. Zhicheng Guan [17], based on a model of two parallel electrodes, studied the impact of droplet distance, droplet volume, and contact angle on corona. The experimental results showed that water droplets significantly reduced the ICV (Initial Corona Voltage) of silicone rubber samples. The electric field intensity under ICV increased with increasing droplet spacing. When the droplet distance was fixed, as the droplet volume decreased, both the ICV and the electric field intensity under ICV increased. Water droplets can enhance the electric field. The electric field intensity of a single water droplet with a contact angle of 90° can be three times that of normal conditions.

To gain a deeper understanding of the mechanism of electrical aging and its impact on insulator performance, systematic studies on corona aging have been conducted in laboratories through various artificial corona accelerated aging tests. Liang Xidong [18, 19] investigated the effects of DC corona and AC corona on the hydrophobicity of silicone rubber materials. By applying DC and AC voltages to silicone rubber using a needle-plate electrode, they found that the chemical reaction induced by DC corona on the surface of silicone rubber was relatively weak and did not cause damage to its hydrophobicity, while under AC corona, the loss of hydrophobicity for both clean and contaminated samples was relatively slow. Guo Xumin [20] conducted corona accelerated aging tests on HTV silicone rubber materials under both AC and DC voltage forms and discovered significant differences in the effects of AC and DC corona on HTV silicone rubber. AC corona aging had the most severe impact on silicone rubber, followed by negative DC corona aging, while positive DC corona aging had the weakest impact. Gao Yanfeng [21] studied the effects of AC and DC corona on the properties of high-temperature vulcanized silicone rubber using a multi-needle electrode system. The results showed that after corona treatment, the hydrophobicity, surface resistivity, tensile strength, and elongation at break of silicone rubber all decreased. Corona reduced the carbon element content and increased the oxygen element content on the surface of silicone rubber, damaging the Si-CH3 on the surface and generating Si–OH. Oxidative crosslinking increased the content of high-oxidation state silicon elements, and a certain number of free radicals were produced on the surface of silicone rubber during the corona process. Xiong Yuanyuan [22] conducted a detailed study on the microscopic mechanism of corona aging characteristics of silicone rubber. The results indicated that corona aging caused the long molecular chains of silicone rubber to break into smaller chains, reducing the flexibility of its molecular chains and forming various defects. The electron and hole trap energy level centers showed a trend of first slowly decreasing and then rapidly decreasing with the increase in corona aging time. Moreover, corona aging led to the gradual transformation of deep traps into shallow traps, weakening the binding effect on electrons and ultimately resulting in surface flashover of silicone rubber.

Tu Youping [23] investigated the influence of three factors—corona voltage, aging time, and relative humidity—on the electrical aging characteristics of silicone rubber materials. They found that as the corona voltage and aging time increased, the energy level traps of the samples deepened and the amount of trapped charge gradually increased. The impact of relative humidity on the corona aging characteristics of silicone rubber was not significant. Liang Ying [24] studied the effects of air pressure, humidity, and aging time on the corona aging process of silicone rubber composite insulators. They discovered that changes in air pressure had the greatest impact on the amount of trapped charge, followed by humidity and aging time. At low air pressures, higher humidity led to a greater amount of trapped charge in the samples, and the influence of humidity became more significant with longer aging times. As air pressure increased, the contribution of humidity to the amount of trapped charge gradually decreased. Moreover, the amount of trapped charge in the samples increased with longer aging times.

13.2.6 The Impact of Surface Arcing on the Properties of High-Temperature Vulcanized Silicone Rubber

Although silicone rubber materials exhibit excellent hydrophobicity and hydrophobic migration properties, when both high field strength and surface contamination combined with moisture are present, arcing may still occur on the surface of silicone rubber, leading to burning or even thermal decomposition of the material. Therefore, arc resistance is one of the crucial properties of silicone rubber materials. The main methods for measuring the arc resistance of silicone rubber materials include the inclined plane method, the flat plane method, and the rotating wheel method.

Wang Ziyang [25] used a DC inclined plane apparatus to study the changes in surface characteristics of silicone rubber materials under the effect of DC arc ablation. The experiment found that the process of the DC inclined plane test can be divided into four stages: initial arcing, concentrated arc ablation, formation of hydrophobic substances, and termination of arc ablation. The appearance of white substances under the action of DC arcs was found to affect the generation and development of arcs. Wang Xiujuan [26] conducted a study on the characteristics of surface arcs along silicone rubber in a DC high-voltage electric field, using silicone rubber plates as the research object and employing a triangular model. The results indicated that the voltage-current characteristics of local arcs on the contaminated silicone rubber surface exhibited a decreasing trend, and the field strength distribution and E-I characteristics of arcs differed when the arc length varied. S. Kumagai [27] studied arc residues on the surface of silicone rubber materials using the inclined plane method in both air and nitrogen environments. The research results showed that the main components of the residues after arc ablation on the silicone rubber surface were SiO2, SiC, etc. Further analysis revealed that under the action of arcs, if oxygen was present, the carbon element in the silicone rubber material mainly escaped in the form of CO and CO2; in the absence of oxygen, C would react with Si to form SiC.

13.2.7 RubberThe Impact of Simultaneous Multifactor Effects on the Properties of High-Temperature Vulcanized Silicone Rubber

Considering the actual operational conditions of silicone rubber, it is often subjected to multiple factors that contribute to its aging. Single-factor and dual-factor aging conditions are often too simplistic, whereas the aging patterns and mechanisms under the influence of multiple factors are more aligned with practical applications. Currently, there is relatively little research on aging under the combined effects of multiple factors, with most studies focusing on the coupling of three factors.

In terms of the coupling effects of electricity-contamination-humidity, Jin Zhe [28] utilized a corona discharge accelerated aging electrode and environmental simulation system to simulate the environmental conditions that contaminated silicone rubber may face during actual operation. They conducted research on how changes in the three influencing factors—contamination concentration, aging time, and environmental humidity—affect the flashover voltage of silicone rubber contaminated with different types of pollutants after corona aging. The interaction between contamination concentration and humidity can cause the contamination layer on the surface of contaminated silicone rubber to be dispersed by electrons generated by corona discharge, hindering the movement of electrons during flashover and resulting in an increase in flashover voltage. At the same time, a loose contamination layer facilitates the migration of silicone rubber's hydrophobicity, leading to an increase in the hydrophobicity recovery angle.

In the aspect of combined electrical-wet-thermal aging, Xu H.S. [29] established a novel aging platform that integrates electrical-wet-thermal environments. They studied the changes in physicochemical, dielectric, and trap characteristics of HTV before and after electrical-wet-thermal aging. The results showed that combined electrical-wet-thermal aging caused more microcracks and microholes on the surface of silicone rubber materials compared to wet-thermal aging. The breakage of PDMS main and side chains produced more LMW substances, while the tensile strength and break yield point of HTV were significantly reduced. The emergence of various physicochemical defects on the surface and inside the material led to a surge in its dielectric constant and trap density, as well as a decrease in withstand voltage strength.

In terms of electrical–mechanical stress aging, Du B.X. [30] investigated the influence of electrical tree growth characteristics during the electrical aging process of HTV silicone rubber insulation materials under combined electrical–mechanical stress. They found that under tensile stress, the probability of electrical tree initiation was higher, and it spread faster. The length, fractal dimension, and cumulative damage of electrical trees were positively correlated with the magnitude of tensile stress. The results under compressive stress were the opposite. Therefore, the combination of tensile stress and high-voltage discharge accelerates the electrical tree aging process of the material.

Regarding the effects of salt density-wet-thermal aging, Mansab Ali [31] studied the changes in hydrophobicity of HTV silicone rubber after aging tests in salt solutions of different concentrations and temperatures. The research showed that the lower the conductivity of the salt solution, the greater the decrease in the contact angle of silicone rubber.

In summary, under the influence of multiple factors, the aging of silicone rubber is more consistent with actual conditions. Research on the environmental adaptability of silicone rubber under the coupling effects of multiple factors helps further explore the aging patterns and mechanisms of the material, accurately assess the aging state, and improve the reliability of equipment operation.

13.3 Research on Life Prediction of Composite Insulators

Currently, many researchers have conducted extensive work on the life prediction of composite insulators. There are mainly three methods for predicting the lifespan of composite insulators. The first method involves using the polymer electrical aging life formula proposed based on the trap theory. The second method establishes an aging model through accelerated aging tests and predicts the lifespan using the Arrhenius equation. The third method begins by analyzing the aging characteristics of composite insulator samples using methods such as the static contact angle method and hardness testing. Based on the test results, life prediction parameters that have a significant correlation with the operational lifespan are obtained through statistical methods such as Fisher Score. Finally, prediction algorithms such as BP neural network algorithm and grey model are used to predict the lifespan. The following is a detailed introduction to these three methods.

13.3.1 Life Prediction Method Based on Electrical Aging

The trap theory posits that there are numerous traps within polymers, which can capture carriers, leading to the accumulation of local space charges. This accumulation subsequently triggers local electric field distortion and may ultimately result in breakdown. Based on this, Tu Demin [32] proposed a formula for the electrical aging lifespan of polymers by studying the correlation between trap theory and the breakdown mechanism of polymers:

$$CE^{b}t^{1/2} = K \tag{13.1}$$

In Eq. (13.1): On the left side of the equation, C and b are constants related to voltage and electrode material; E represents the electric field intensity in kV/mm; t represents the aging time. On the right side of the equation, K represents the number of free radicals generated, which directly reflects the degree of corona aging.

Lei Lan [33] and others conducted accelerated corona aging tests on silicone rubber materials and performed infrared spectroscopy on the aged samples. Based on the polymer electrical aging lifespan formula and experimental results, they determined the undetermined coefficients in the lifespan formula. For corona aging under a highly inhomogeneous pin-plate electric field, the aging lifespan formula for high-temperature vulcanized silicone rubber can be initially determined as follows:

$$3.4E^{0.436}t^{1/2} = K \tag{13.2}$$

13.3.2 Life Prediction Method Based on Accelerated Aging Tests

The Arrhenius equation describes the exponential relationship between the rate constant of a chemical reaction and temperature. The principle of this equation is that as the temperature increases, the average kinetic energy of the reactant molecules increases, enabling more molecules to acquire enough energy to overcome the activation energy of the reaction, thereby converting into activated molecules and participating in the reaction. Therefore, an aging model can be established based on accelerated aging tests, and then the lifespan can be predicted using the Arrhenius equation. In chemical reaction kinetics, by combining the reaction rate equation with the Arrhenius equation, the thermal aging equation for polymer materials can be obtained, as shown in Eq. (13.3).

$$lg\tau = a + b/T \quad (13.3)$$

In Eq. (13.3): τ represents the lifespan of the material; T represents the aging temperature; a is a constant related to the specified failure performance; b = 0.401 × E/R, which is a constant related to the activation energy E, where R is the gas constant with a value of 8.314 J/(mol·K).

13.3.3 Life Prediction Method Based on Mathematical Methods

Currently, composite insulators are widely used in China, and the numerous aged composite insulators provide analytical samples for assessing their service life. BP neural networks are extensively applied in engineering fields such as material performance prediction. Tao Fengbo [34] took composite insulators with different years of operation in coastal industrial areas as the research object, and used methods such as the static contact angle method and hardness testing to test the life prediction parameters of the composite insulator samples. Subsequently, Pearson correlation coefficient and T-test were used to obtain coefficients that were significantly correlated with lifespan. Finally, a BP neural network system was employed to predict the operational lifespan of composite insulators in coastal industrial areas, with the predicted operational lifespan differing from the actual lifespan by only about 5%. Qiu Zhimin [35] believed that BP neural networks have two major defects: inappropriate selection of initial thresholds and weights, and the tendency to fall into local minima during training, which result in insufficient nonlinear learning and poor network performance. Therefore, they adopted the global search capability of the genetic algorithm (GA) to optimize the neural network, making its lifespan predictions more accurate.

However, due to differences between artificial simulations and natural conditions, as well as difficulties in sampling from natural environments, there are certain challenges in establishing a predictive model for the aging lifespan of composite insulators. The grey prediction model, which requires a small sample size and offers high accuracy, is relatively precise in predicting the lifespan of electrical equipment. Therefore, Wang Congmin [36] took normally operating composite insulators with different years of service and varying degrees of degradation in a certain eastern region as the research object, and analyzed their aging characteristics using methods such as the static contact angle method and hardness testing. Subsequently, based on the test results, a series of lifespan prediction parameters with significant correlation to the years of operation were proposed, and a method for calculating the remaining lifespan was introduced using the grey model, resulting in relatively accurate lifespan predictions.

13.4 Conclusion

This paper provides a detailed review of the influencing factors and mechanisms of silicone rubber aging in composite insulators, and summarizes the prediction methods for the service life of composite insulators as well as related research progress in recent years. The main conclusions and prospects are as follows:

The aging of silicone rubber composite insulators is the result of the combined action of various environmental factors. Corona discharge, ozone, ultraviolet radiation, acid and alkali corrosion, high temperatures, and surface arcs can all have significant impacts on the performance of silicone rubber materials. Among them, corona aging has a particularly severe impact on the performance of insulators due to its accompanying physical and chemical aging processes.

During the aging process, the molecular chains on the surface of silicone rubber will break, and the number of hydrophilic groups will increase, leading to a decrease in hydrophobicity. This, in turn, triggers the degradation of the mechanical and electrical properties of the insulator, including a decrease in tensile strength and elongation at break, a reduction in volume resistivity and surface resistivity, and a decrease in breakdown strength and tracking resistance.

The current methods for predicting the lifespan of composite insulators mainly include the electrical aging life formula based on trap theory, the Arrhenius equation prediction method based on accelerated aging tests, and predictions based on mathematical statistical methods (such as BP neural networks, grey models, etc.). These methods have their respective advantages and disadvantages and are suitable for different application scenarios.

In terms of aging mechanisms, future research should focus on deepening the understanding of the aging mechanisms of silicone rubber composite insulators, particularly the aging processes and their microscopic mechanisms under the coupling effects of multiple factors. This will help to more accurately assess the aging state of insulators and provide a basis for formulating scientific and reasonable operation and maintenance strategies. In terms of lifespan prediction, combining modern computing technology and artificial intelligence algorithms to develop more accurate and efficient lifespan prediction models is a promising approach. For example, a database on composite insulators could be established, and machine learning algorithms could be considered for integration with physical models to fit large amounts of data in the database, enabling dynamic simulation and lifespan prediction of the aging process of insulators.

Acknowledgements This work is supported by the State Grid Corporation of China Headquarters Management Technology Project Funding (No. 5500-202455305A-1-2-LZ).

References

1. Zhang Guanjun, Zhao Lin, Zhou Rundong: Current Status and Progress of Aging Characterization Evaluation of Silicone Rubber Composite Insulators. High Voltage Apparatus, 52 (04), 1–15 (2016).
2. Moreno, V.M., and R.S. Gorur: "Impact of Corona on the Long-Term Performance of Nonceramic Insulators." IEEE transactions on dielectrics and electrical insulation 80–95, 10.1 (2003).
3. Yan Bin, Wang Zhihui: Aging Performance Analysis of Silicone Rubber Materials for Composite Insulators . Insulation Materials, 42 (04), 57–59+63 (2009).
4. Li Guofang, Yu Lan, Qin Yongxiong: Study on UV Radiation of High Temperature Vulcanized Silicone Rubber Composite Insulators. Journal of Radiation Research and Radiation Technology, 33 (04), 20–27 (2015).
5. Li Jingjing, Zhang Huihui, Li Guofang: Infrared spectroscopic analysis and aging mechanism of high-temperature vulcanized silicone rubber under ultraviolet irradiation. Spectroscopy and Spectral Analysis, 40 (04), 1063–1070 (2020).
6. Li Yawei, Chen Hongbo, Hu Yuxiao: Effects of UV radiation and corona discharge on the surface morphology and composition of composite insulators in high-altitude areas. Electric Porcelain Surge Arrester, (05), 237–241+247 (2020).
7. Tu Youping, Tong Yuliang, Wang Qian: The Effect of Humidity on the Ozone Aging Characteristics of HTV Silicone Rubber. High Voltage Technology, 37 (04): 841–847 (2011).
8. Tu Youping, Wang Qian, Li Min: Aging effect of ozone concentration on HTV silicone rubber materials. Journal of Electrical Engineering, 28 (01): 21–28 (2013).
9. Ma, Bin, Stanislaw Gubanski, and Henrik Hillborg: "AC and DC Corona/Ozone-Induced Ageing of HTV Silicone Rubber." IEEE transactions on dielectrics and electrical insulation, 1984 18.6 (2011).
10. Zhang Ni, Wang Zheng, Yang Yue: Study on the aging mechanism of silicone rubber by nitric acid. Journal of Electrical Engineering, 36 (22), 4820–4828 (2021).
11. Peng Xiangyang, Wang Zheng, Xu Zhihai: Effects of acid-base aging on the hydrophobicity of silicone rubber composite insulators . Guangdong Electric Power, 30 (06), 110–114 (2017).
12. Zheng Hanyu: Characteristics and characterization methods of AC operation failure of silicone rubber in acidic and alkaline environments. Chongqing University, (2022).
13. Ma Xiaodan: Study on the physicochemical/electromechanical properties of composite silicone rubber under acidic corrosion environment under communication. Chongqing University, (2021).
14. Cheng Yang, Lv Li, Xia Lingzhi:Thermal aging characteristics analysis of HTV silicone rubber composite insulator umbrella skirt. High Voltage Apparatus, 60 (04): 49–55 (2024).
15. Rong Yiping, Song Liang, Wang Yang: The Effect of Thermal Aging on the Electrical Properties of Silicone Rubber with Different Crosslinking Degrees. Electrical Technology, (21): 189–192+195 (2022).
16. Zhou Yuanxiang, Zhang Yunxiao, Zhang Xu: The Effect of Thermal Aging Time on the Initial Characteristics of Silicone Rubber Electric Branches. High Voltage Technology, 40 (04): 979–986 (2014).
17. Guan, Z: "Electric Field Analysis of Water Drop Corona." IEEE transactions on power delivery, 964–969,20.2 (2005).
18. Liang Xidong, Li Zhenyu, Zhou Yuanxiang: Effect of AC corona on hydrophobicity of silicone rubber materials. Chinese Journal of Electrical Engineering, (27), 19–23 (2007).
19. Li Zhenyu, Liang Xidong, Zhou Yuanxiang: Effect of DC corona on hydrophobicity of silicone rubber materials . Chinese Journal of Electrical Engineering, (24), 30–34 (2007).
20. Guo Xumin, Yu Lichao, Liu Hechen: Research on AC/DC Corona Aging of HTV Silicone Rubber Materials. Insulation Materials, 49 (07): 31–35 (2016).
21. Gao Yanfeng, Wang Jiafu, Liang Xidong: The Effect of AC/DC Corona on the Properties of High Temperature Vulcanized Silicone Rubber. Chinese Journal of Electrical Engineering,36 (01): 274–284 (2016).

22. Xiong Yuanyuan, Li Zhiwei, Gong Rui: Microscopic mechanism study on corona aging characteristics of silicone rubber. Insulation Materials, 55 (06), 64–70 (2022).
23. Tu Youping, Luo Meixin, Ying Gaofeng: Orthogonal experimental study on thermal stimulation current characteristics of silicone rubber during corona aging. Chinese Journal of Electrical Engineering, 32 (07), 139–144+202 (2012).
24. Liang Ying, Dong Pingping, Gao Lijuan: Corona aging characteristics of composite insulators based on orthogonal experiments. High Voltage Technology, 44 (12), 4083–4089 (2018).
25. Wang Ziyang, Zhu Zhengyi, Jia Zhidong: Process of surface characteristic damage of silicone rubber by DC pollution discharge. High Voltage Technology, 38 (10), 2611–2616 (2012).
26. Wang Xiujuan, Wang Gang, Xu Aoran: Study on DC Arc Characteristics of Silicone Rubber Surface under Stained Conditions. Electric Porcelain Surge Arrester, 34–38 (06) (2015).
27. Kumagai, S., X. Wang, and N. Yoshimura: "Solid Residue Formation of RTV Silicone Rubber Due to Dry-Band Arcing and Thermal Decomposition." IEEE transactions on dielectrics and electrical insulation 5.2, 281–289 (1998).
28. Jin Zhe: The influence of polluted and humid environment on the surface flashover voltage of silicone rubber during corona aging. North China Electric Power University, (2017).
29. Xu, Huasong: "Effects of Electrical-Hydrothermal Aging Degradation on Dielectric and Trap Properties of High Temperature Vulcanized Silicone Rubber Materials." RSC advances 1.7, 385–3816 (2020).
30. Du, B. X: "Effects of Mechanical Stress on Treeing Growth Characteristics in HTV Silicone Rubber." IEEE transactions on dielectrics and electrical insulation 24.3, 1547–1556 (2017).
31. Ali, M., and R. Hackam: "Effects of Saline Water and Temperature on Surface Properties of HTV Silicone Rubber." IEEE transactions on dielectrics and electrical insulation 15.5, 1368–1378 (2008).
32. Tu Demin, Wang Xinsheng, Liu Fude: Trap theory of polymer breakdown and its verification on polypropylene. Journal of Electrical Engineering, (03), 47–51 (1993).
33. Lan Lei, Wang Hanliang, Wen Xishan: Analysis of Accelerated Corona Aging Test and Life Prediction of High Temperature Vulcanized Silicone Rubber. High Voltage Technology, 38 (04), 782–789 (2012).
34. Tao Fengbo, Bi Xiaotian, Chen Jie: Study on Aging Life Prediction of Composite Insulators in Coastal Industrial Areas. Electric Porcelain Surge Arrester, (01), 182–188 (2022).
35. Qiu Zhimin, Kang Bing, Yan Xia: Study on Aging Characteristics and Operating Life Prediction of Silicone Rubber Composite Insulators under Large Temperature Difference. Insulation Materials, 57 (09), 69–79 (2024).
36. Wang Congmin, Li Minglei, Zhang Ping: Study on the Remaining Life of Silicone Rubber Materials for Operating Composite Insulators Based on Grey Model GM (1, N). Electric Porcelain Surge Arrester, (05), 137–144 (2023).

Chapter 14
Study on the Effect of Strong Wind Erosion on the Properties of Silicone Rubber Composites

Rui Zhang, Huanlong Qu, Jing Zhou, Jingwei Li, Jianghai Geng, Hao Wang, Yang Wang, and Yang Lu

Abstract Silicone rubber, with its excellent mechanical properties such as tear resistance, has become the preferred material for preparing composite insulators. However, the transmission lines in Northwest China pass through regions with high-hardness, irregularly-edged sand and dust particles, which can easily cause erosion on silicone rubber composite materials during high-speed movement, leading to a decline in their mechanical properties and affecting the service life of insulators. This paper takes silicone rubber test pieces as the research object, conducts wind-sand erosion simulation tests on silicone rubber composite materials based on actual measured results of sand parameters and wind-sand erosion simulation systems, compares and analyzes the changes in the mechanical properties of the materials before and after erosion, and systematically studies the impact of wind-sand erosion on the mechanical properties of silicone rubber composite materials. The findings are as follows: wind-sand erosion has a certain abrasive effect on silicone rubber materials; the elongation at break of the test pieces decreases from 218.42% to 156.43%, the tear resistance performance drops by 6.57%; the hardness of the test pieces decreases, and the surface roughness increases by 190.36%; the erosion damage on the material surface mainly manifests as furrows and transverse cracks caused by plowing and micro-cutting. The research results can provide a reference for improving the performance and optimizing the design of silicone rubber composite materials in wind-sand environments.

Keywords Sand and dust erosion · Composite materials · Erosion mechanism · Silicone rubber

R. Zhang · J. Zhou
China Electric Power Research Institute, Wuhan, Hubei, China

H. Qu (✉) · J. Li · J. Geng · H. Wang · Y. Wang · Y. Lu
Hebei Provincial Key Laboratory of Power Equipment Security Defense, North China Electric Power University, Baoding, Hebei, China
e-mail: long2095171127@163.com

Y. Li et al. (eds.), *Proceedings of the 5th Chinese Conference on High Voltage and Discharge Plasmas*, Springer Proceedings in Physics 254,
https://doi.org/10.1007/978-981-96-5711-7_14

14.1 Introduction

Currently, China considers the construction of large-scale clean energy bases in the western desert, gobi, and barren areas (The desert region) as an important direction for future new energy development. However, the transmission equipment in these regions faces unique environmental conditions characterized by strong winds and thick dust [1, 2]. As shown in Fig. 14.1, the dangers posed by dust to power transmission and transformation equipment have become increasingly prominent. In April 2024, southern and eastern Xinjiang experienced sandstorms with wind speeds reaching 20 m/s, resulting in power outages in some areas.

Silicone rubber is the preferred material for manufacturing composite insulators due to its excellent tear resistance and mechanical properties. In the northwest region, the dust particles along transmission lines have high hardness and irregular edges. For silicone rubber composite materials operating in areas with strong wind and sand, the irregular sand particles carried by the wind cause erosive wear on the surface of the material. This wear can damage the surface morphology and internal structure of the composite material, leading to increased surface roughness and even the formation of micro-cracks. As a result, the insulation performance of the insulator deteriorates, and its operational lifespan is reduced, leading to severe consequences. This greatly decreases the electrical safety performance of the insulator and poses serious risks to the safe and stable operation of power transmission lines [3].

Erosion wear of silicone rubber composites refers to the process of material loss caused by a large number of solid sand and dust particles impacting the material's

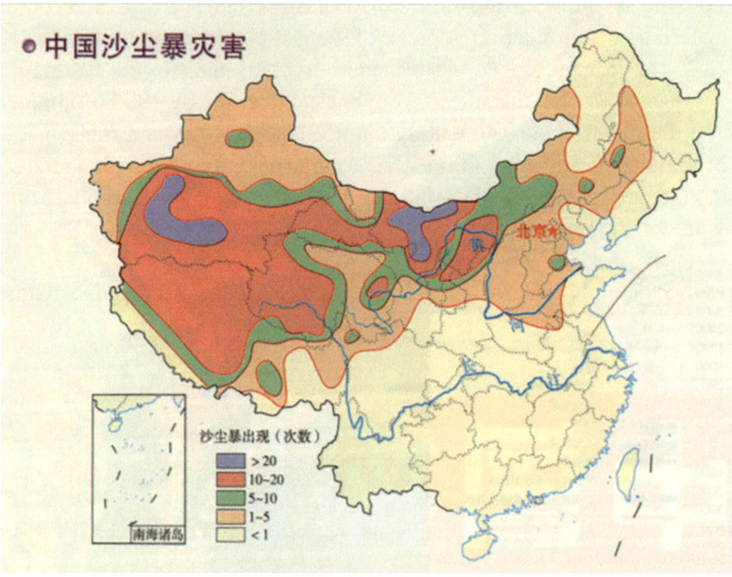

Fig. 14.1 A map of sandstorm distribution

surface at a certain angle and velocity [4, 5]. In 1985, Finnie I first proposed the micro-cutting theory of rigid particles against plastic metals. Subsequent scholars both domestically and internationally have proposed theories such as deformation wear, forging and extrusion theory, brittle material erosion theory, and secondary erosion theory. While the research on erosion mechanisms has become relatively mature, no single erosion principle can fully explain the erosion characteristics of silicone rubber composites in a sandstorm environment.

Most scholars have utilized simulation and experimental methods to investigate the erosion behavior of solid particles on metals and coatings in fields such as natural gas transportation, ore mining, construction material maintenance, and railway transportation. Deng Kuanhai et al. studied the erosion wear characteristics of Al_2O_3 particles on natural gas pipeline material (20G steel) under different angles of incidence and velocities using an air–solid nozzle test method and investigated the erosion behavior of abrasive particles at varying angles [6]. Shu Kang et al. researched the wear characteristics of different rail materials under sand erosion using an airflow sandblasting erosion tester, analyzing the effects of ductility and hardness on sand erosion results under the same experimental conditions [7].

Currently, studies on the performance of silicone rubber composites in a sandstorm environment mainly focus on the issues of pollution accumulation, flashover, and tearing at the skirt root of insulators. Research on the impact of sand erosion on the performance of composite materials is relatively scarce. Zhu Yong et al. systematically studied the effects of sandstorm factors on the mechanical and hydrophobic properties of silicone rubber materials for composite insulators through sandstorm tests [8]. Cao Gui et al. investigated the impact of sand and dust on the flashover characteristics of composite insulators in a high-speed airflow environment, measuring flashover voltage of composite insulators under sand and dust conditions, comparing results with and without sand and dust, revealing the trend of flashover voltage changes in sandy environments [9]. In reference [10], Li Jing et al. established a model for the accumulation of pollution on insulators based on FXBW4-10/100 type composite insulators, studying surface pollution characteristics under various wind speeds, particle sizes, and concentrations. The results showed that wind speed, particle size, and particle concentration all influence the pollution accumulation of composite insulators. Huang Zhicheng from North China Electric Power University studied the collision and rebound motion characteristics of particles with electrical silicone rubber, proposing a collision adhesion model for the surface of composite insulators, and determining the impact mechanism of umbrella structure and parameter on the pollution resistance of composite insulators [11]. Wang Xilin et al. from Tsinghua University used wind tunnel tests and simulation to reveal the interaction process between fluid and the skirt of composite insulators under strong winds, explaining the damage mechanism of the skirt [12].

In response to the issues identified in the aforementioned studies, this paper establishes experimental conditions based on actual meteorological parameters of sand and dust in Xinjiang province, building a sand erosion simulation device based on the principle of airflow sandblasting and conducting erosion wear tests on silicone rubber composites. The mechanical properties and micro-morphologies of silicone rubber samples before and after sand erosion were tested and compared. The study found that sand erosion has a certain abrasive effect on silicone rubber materials, leading to a reduction in the quality of silicone rubber samples. The elongation at break decreased from 218.42% to 156.43%, and tear resistance decreased by 6.57%; hardness also decreased, and surface roughness increased by 190.36%. The erosion damage to the material's surface was mainly manifested as plowing grooves, transverse cracks, and other effects caused by plowing and micro-cutting. This research reveals the erosion wear characteristics of silicone rubber composites in strong sandstorm environments, aiding in the development of optimization measures for improving the anti-sand performance of composite outdoor insulator silicone rubber materials and providing technical support for enhancing the safety of power transmission equipment.

14.2 Based on Dust Parameter Measurement: Wind-Blown Sand Erosion Simulation Experiment

Taking the dust parameters of the Xinjiang Uygur Autonomous Region as the research object, this paper determines the erosive mechanical parameters for the wind-blown sand erosion simulation experiment, constructs the wind-blown sand erosion simulation system, and conducts wind-blown sand erosion simulation experiments using silicone rubber test pieces.

14.2.1 Determination of Erosive Mechanical Parameters and Preparation of Test Specimens

The erosive mechanical parameters required for the experiments in this paper mainly include the particle size of dust particles and the velocity of wind-blown sand flow. Clarifying the mechanical parameters of dust erosion helps to more accurately simulate the real wind-blown sand natural environment in Xinjiang region, obtaining experimental data that aligns with actual conditions [13].

Sand sampling points were set up in the regions of southern Xinjiang and Hami in western Xinjiang, where there are frequent days of sandstorms. A self-designed sand collector was used to collect surface sand, sand at heights of 7–8 m and 12–15 m from the ground, as shown in Fig. 14.2.

The laser diffraction method has advantages such as a wide testing range and high analysis accuracy. This method was chosen to analyze the particle size of collected sand and dust samples, and the results are shown in Fig. 14.2. It was found that nearly 90% of the sand particles in the Shago and Huang regions are within 500 μm, with the main sand particle diameter distribute in 200–500 μm range. Therefore, a particle size of 125–250 μm was selected for the experimental sand and dust particle size.

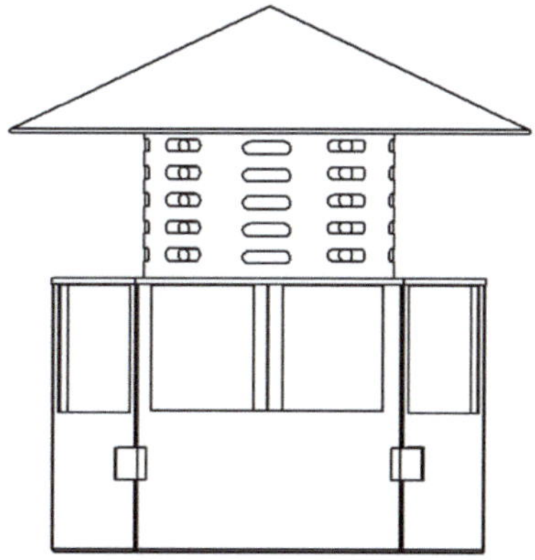

Fig. 14.2 A map of sandstorm distribution

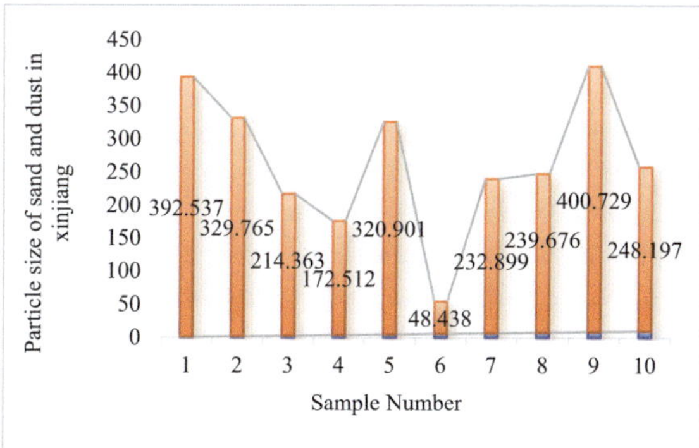

Fig. 14.3 Sand particle size test results in Xinjiang region

The speed of wind-blown sand determines the impact speed of sand particles on silicone rubber, and the higher the wind speed, the more likely it is to cause wear on the silicone rubber material [14]. Considering that maximum wind speeds can reach 28 m/s during some extreme sandstorm events, a wind speed of 30 m/s was chosen to simulate an extremely harsh sandstorm environment (Fig. 14.3).

Silicone rubber composite material samples, identical to those used in FXBT-110 composite insulators, were selected for wind-blown sand erosion simulation experiments, and the mechanical performance changes of the samples before and after erosion were tested. The appearance of the samples is shown in Fig. 14.4.

14.2.2 Construction of Wind and Sand Erosion Simulation Device

Erosion wear test is one of the main methods to study the erosion wear characteristics of materials and also an important means to verify numerical simulations. The erosion

Fig. 14.4 Appearance of silicone rubber test piece

wear characteristics of composite insulators and test pieces under different wind speeds and suspension methods were studied through orthogonal experiments [15].

The erosion wear test device simulates the erosion process by achieving relative motion between particle two-phase flow and the target material. Currently, common types of test devices include wind tunnel type, air jet sandblasting type, and centrifugal acceleration type, etc. [16]. Among them, the air jet sandblasting type test device can control a wide range of wind speed and sand concentration, and can arbitrarily adjust the particle impact angle, which can better simulate the wind speed environment. Therefore, based on the principle of the air jet sandblasting type device, this paper designs an erosion wear test device to restore the aeolian sand environmental conditions, as shown in Fig. 14.5.

The entire wind-sand erosion simulation test device consists of a gas supply system, a sandblasting system, and an erosion system. The gas supply system is composed of an air compressor and a gas storage tank. The air compressor compresses the air and directs it into the gas storage tank to provide gas source power, and the wind speed is controlled by adjusting the air pressure. The sandblasting system consists of

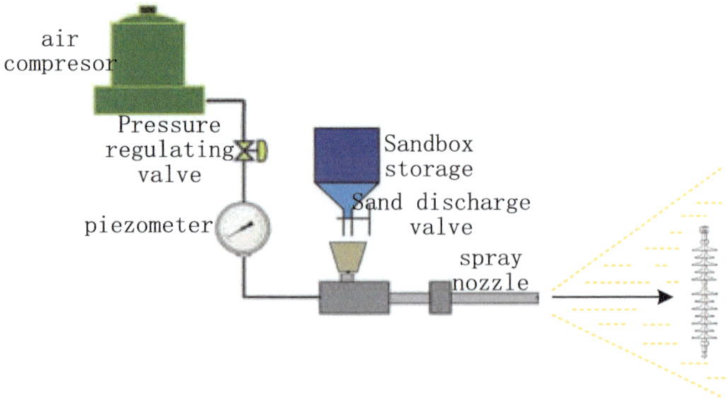

Fig. 14.5 Schematic diagram of wind-sand erosion simulation test device

a sandbox and a fan. When the sand outlet valve is opened, sand particles enter the fan's sand inlet from the sandbox, and the sand flow rate is controlled by adjusting the fan frequency, ensuring that the speed of the gas carrying the sand particles is uniform and adjustable. The erosion system consists of a nozzle and a holding part. The holding part is composed of a door-type bracket and fixed hardware, which can adjust parameters such as the windward angle of the silicone rubber composite material and the distance between the test piece and the nozzle. Sand particles are ejected from the nozzle, and the insulator test piece is fixed by the holding tool and subjected to the erosion wear of the wind-sand flow from the nozzle. The entire erosion process can be directly observed.

During the experiment, the air compressor is first activated to direct air into the gas storage tank, ensuring a steady and continuous airflow during the erosion process. When the gas pressure in the storage tank reaches a certain pressure range, the fan and the sand outlet valve are sequentially opened, allowing sand particles to mix with the airflow and form a wind-sand flow with a certain speed ejected from the nozzle. The silicone rubber material sample is fixed on the door-type bracket.

Methods for Simulating Wind and Sand Erosion Tests

The wind and sand flow diffuses in a fan shape after being ejected from the nozzle, with different wind and sand flow velocities at various distances from the nozzle. Due to the long adjustment time and limited adjustable range required for controlling wind speed using an air compressor, the experiment adopts a method of adjusting wind speed by controlling the distance between the silicone rubber sample and the nozzle, with wind speed measured using a handheld anemometer. Measurements indicate that when the pressure in the air storage tank is 0.27 MPa, the wind speed at a distance of 70 cm from the nozzle is 30 m/s. Based on the actual survey of wind and sand parameters in the Xinjiang region, sand particle sizes of 125–250 μm were selected for the experiment, with a wind speed set at 30 m/s to carry out the wind and sand erosion simulation test, lasting 10 min.

The operation procedure for the erosion test is as follows:

(1) Build the erosion test apparatus and check if devices such as the air compressor, air storage tank, and fan are operating normally. Fix the insulator sample or test piece and check the stability of the holding device. Add sufficient sand particles to the sand box to ensure stable and uniform wind and sand flow during the test. Adjust the layout angle of the insulator or test piece and the distance from the nozzle's centerline to reach the predefined values for each group of tests.
(2) Turn on the air compressor, and when the air pressure reaches 0.27 MPa, switch on the fan power and open the sand discharge valve, controlling the sand flow to a fixed range. At this point, the wind and sand flow with a fixed particle size, sand concentration, and wind speed will strike the surface of the insulator sample or test piece at a certain angle, achieving an erosion wear effect.
(3) After the test time has elapsed, close the sand discharge valve, switch off the fan power and air compressor, remove the sample, and analyze its erosion wear condition.

After the erosion test, mechanical performance and micro-morphology tests of the composite external insulation material after wind and sand erosion will be carried out. The specific operations are as follows:

(1) Wear Amount Inspection
Use an electronic scale to measure the silicone rubber test piece before and after erosion, recording the changes in wear amount.
(2) Micro-Morphology Inspection
Clean the surface of the silicone rubber test piece with anhydrous ethanol, then rinse with deionized water. Use a custom knife to prepare square samples of size 10 mm × 10 mm. Observe and compare the surface micro-morphology of the test piece before and after erosion using a Scanning Electron Microscope (SEM), analyzing the surface changes of the material from a micro perspective.
(3) Roughness Check
Use a TR210 high-precision surface roughness tester to measure the surface roughness of the silicone rubber test pieces before and after erosion. Select five different positions on the same surface with significant erosion to conduct tests and take the average value as the roughness characterization of that surface.
(4) Hardness Test
Use a Shore A durometer to test the Shore hardness of the silicone rubber test piece surface before and after erosion, averaging the data collected from five points to represent the hardness of that surface.
(5) Tear Strength Test
According to GB/T 529 standard, use a cutting knife to prepare rectangular samples before and after erosion. Fix both ends in a tensile testing machine and conduct the tear strength test according to the national standard test method preset in the control software. Test three samples under the same conditions, taking the average value as the tear strength of the sample.
(6) Breaking Strength Test and Elongation at Break Test
Both the breaking strength and elongation at break tests are conducted according to GB/T 528 standard. Use a cutting knife to prepare dumbbell-shaped samples before and after erosion, selecting type 1 dumbbell-shaped samples as specified in the standard for this experiment. Sample Heading (Third Level). Only two levels of headings should be numbered. Lower level headings remain unnumbered; they are formatted as run-in headings (Figs. 14.6, 14.7 and 14.8).

Fig. 14.6 High precision surface roughness tester

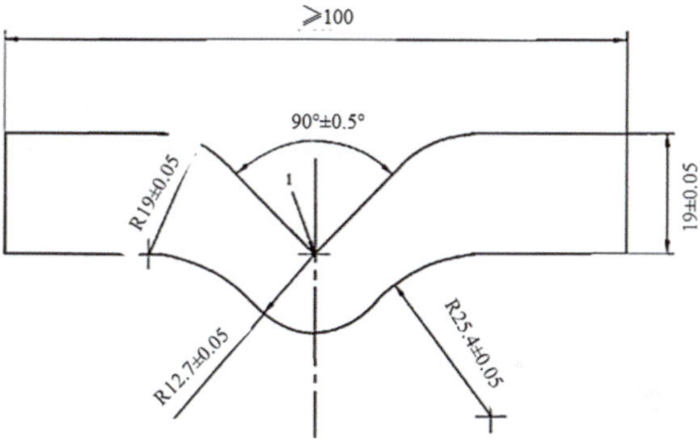

Fig. 14.7 Rectangular specimen

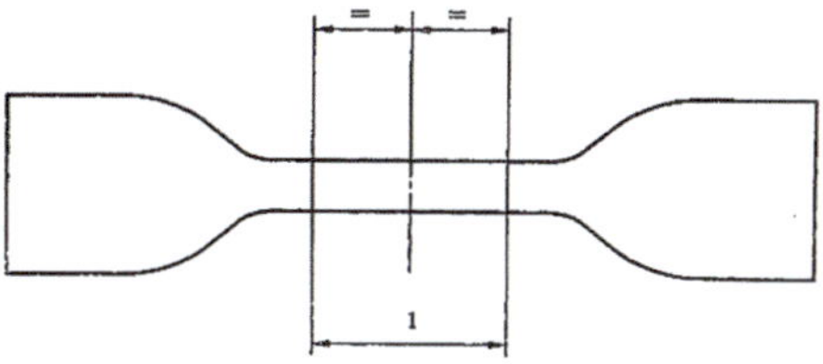

Fig. 14.8 Dumbbell sample

14.3 Erosion Wear Performance Test Results and Analysis of Silicone Rubber Material

14.3.1 Effect of Erosion Wear on the Microstructure of Silicone Rubber Material

To understand the mechanism by which sand erosion affects the mechanical properties of silicone rubber composites, this paper uses SEM to analyze the microstructure of silicone rubber samples before and after sand erosion. The figure shows the microstructure of the silicone rubber sample surface before sand erosion (Fig. 14.9).

As can be seen from the figure, before sand erosion, the material surface is relatively smooth, with only a few scratches and some small particles attached, and the overall surface undulations are minor. This is considered to be caused by the process defects during sample preparation.

Figure 14.10 presents the microscopic morphology of sand erosion and plowing grooves of silicone rubber material observed under an electron microscope at 300 and 3000 times magnification. After sand erosion, many sand particles adhere to the surface of the silicone rubber sample. Due to the irregular shape and angularity of sand particles, the high-speed moving sand particles are repeatedly pressed into the silicone rubber material under vertical stress, creating the pits and some fine cracks shown in the figure. The angular parts of the irregular sand particles attached to the material surface produce a cutting effect on the surface due to the strong wind force impacting the particles, increasing the number of pits and creating fine cracks. This is because the silicone rubber sample undergoes deformation due to repeated impacts from sand particles, and when the material can no longer withstand the stress generated by the sand particle impacts, some cracks appear. When sand particles hit the material surface at a certain speed, they form some grooves. The edges of these grooves form some protrusions due to material displacement. When high-speed sand particles continue to impact these protrusions, the sand particles have a removal effect on the material surface, as shown in Fig. 14.10. The plowing grooves formed

(a). Micromorphology of silicone rubber test piece **(b).** Surface scratch micromorphology

Fig. 14.9 Surface microstructure of silicone rubber test piece before sand erosion

(a). 4x magnification

(b). 300x magnification

(c). 3000x magnification

(d). Micromorphology of plow grooves

Fig. 14.10 Surface microstructure of silicone rubber test piece before sand erosion

are of varying lengths due to the different incidence angles of the particles hitting the material surface, resulting in inconsistent normal impact velocity components during particle incidence.

14.3.2 Effect of Erosion Wear on the Microstructure of Silicone Rubber Material

14.3.2.1 Determination of the Erosion Rate of Silicone Rubber Samples

From the study on the microscopic morphology of material erosion by wind and sand in the previous section, it is evident that sand particles, while moving at high speeds, continuously impact the surface of silicone rubber material, causing numerous pits to form on the surface. These pits further expand to develop cracks, and may even lead to material detachment, ultimately resulting in a loss of material quality. When the wind speed is 30 m/s, the initial kinetic energy of the dust particles is relatively high, which makes the erosion wear phenomenon quite severe. Therefore, the macroscopic weighing results are used to indicate the mass loss of the samples, and the erosion rate is used to measure the extent of the sample's loss [17, 18].

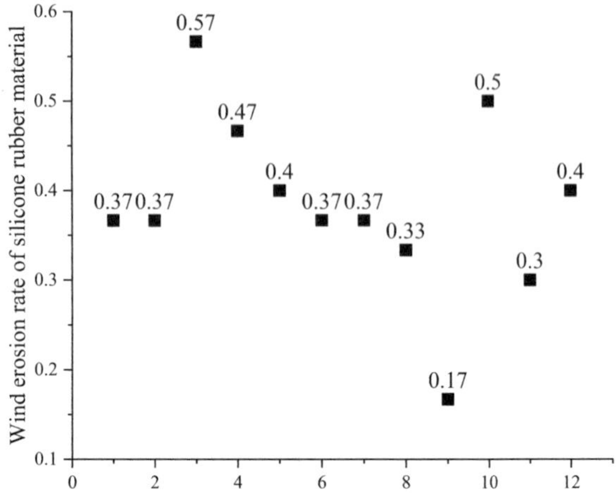

Fig. 14.11 Measurement of silicone rubber roughness before and after sand erosion

$$E = \frac{m_o - m_c}{M_s * t} \tag{14.1}$$

In the formula, E represents the erosion rate of the silicone rubber samples in mg/g; mo denotes the mass of the sample before erosion by wind and sand in mg; mc denotes the mass of the silicone rubber sample after erosion in mg; Ms is the mass flow rate of the sand particles in g/min; and t represents the duration of the erosion test, which is 10 min (Fig. 14.11).

14.3.2.2 The Effect of Erosion Wear on the Surface Roughness of Silicone Rubber Material

The surface roughness of silicone rubber material is an important indicator that affects its pollution resistance and other properties. An increase in surface roughness can influence its insulation performance, as rough surfaces can lead to localized increases in electric field strength, potentially causing discharge phenomena and reducing the insulating capability of the insulator. Using a TR200 roughness testing instrument, the roughness of the silicone rubber material surface was measured before and after sand erosion, with five measurements taken to calculate the average value. The roughness test values before and after erosion are shown in Fig. 14.12. The surface roughness of the specimen before erosion was 0.55, while after erosion it increased to 1.597, indicating that the sand erosion effect caused the surface roughness of the specimen to increase by 190.36%.

Due to the physical impact and abrasive action of sand particles on the surface of the insulator, the roughness measurement of the silicone rubber specimen increased after sand erosion. The sand particles, propelled by the wind, have a certain velocity that causes collisions and cutting against the surface of the silicone rubber specimen,

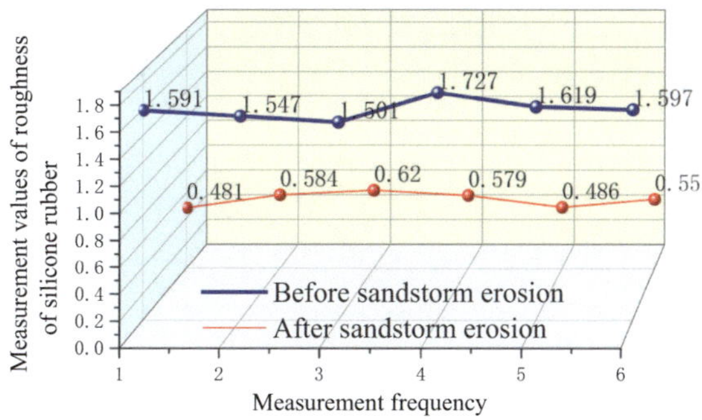

Fig. 14.12 Measurement of silicone rubber roughness before and after sand erosion

Table 14.1 Hardness measurement of silicone rubber before and after sand erosion

Test wind speed	Before sand erosion	After sand erosion
30 m/s	81.80	81.30

resulting in pits, cracks, and other imperfections on what was originally a smooth surface, thereby increasing the roughness measurement.

14.3.2.3 The Effect of Erosion Wear on the Hardness of Silicone Rubber Material

The hardness of the insulator skirt affects its mechanical properties. An increase in skirt hardness is often accompanied by a decrease in mechanical performance, posing significant safety hazards. Therefore, deionized water was first used to clean the insulator after sand erosion, followed by measurements of the hardness changes of the insulator skirt before and after erosion using a Shore durometer, with five measurements taken to calculate the average value. The hardness measurement results are shown in Table 14.1.

From the data in the table, it can be observed that the hardness of the insulator shows a slight decrease before and after sand erosion, but the change in value is not very significant. This may be due to the relatively short erosion time, which did not have a major impact on the hardness of the insulator.

14.3.2.4 The Influence of Erosion Wear on the Mechanical Properties of Silicone Rubber Materials

Assessing the mechanical properties of silicone rubber composites can ensure that they maintain sufficient mechanical strength under various environmental conditions, thereby guaranteeing the safe and stable operation of power systems and effectively preventing damages caused by external forces and reducing the occurrence of power

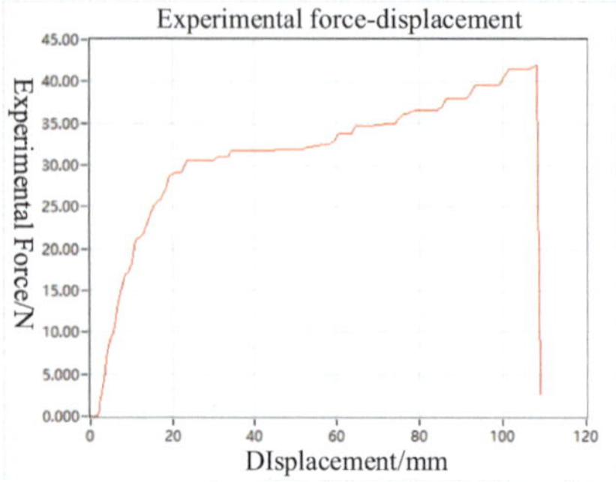

(a). Tensile performance test curve

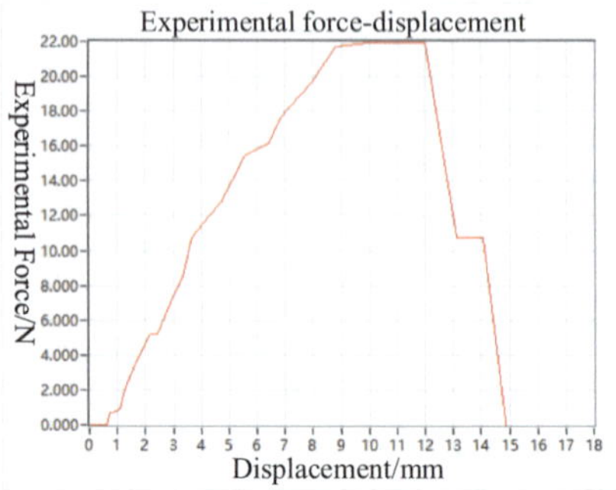

(b).Tear resistance test curve

Fig. 14.13 Test curve of mechanical properties of silicone rubber test piece before sand erosion

outages and other related incidents. Specifically, insulators made primarily of silicone rubber bear various mechanical loads in power systems, including wind pressure, their own weight, and wire tension. Insufficient mechanical properties of insulators may lead to cracking or detachment during operation, causing power failures [19]. Therefore, by testing the tensile strength, compressive strength, and bending strength of the insulators, it can be ensured that they can withstand the anticipated mechanical loads during their design life.

Using a tensile testing machine to test the mechanical properties of the silicone rubber samples before and after sand erosion, the mechanical performance testing curve of the samples before erosion is shown in Fig. 14.13, with specific data results listed in Table 14.2, all of which comply with the specified values in the national standards.

The fracture of the silicone rubber specimen, tested by a tensile testing machine under the conditions of tension and tearing, always initiated from the crack source. From the results of microscopic morphology analysis, it can be seen that after wind-sand erosion, the specimen surface exhibits tiny pits, scratches, and cracks. These damages reduce the smoothness and integrity of the silicone rubber surface, leading to increased surface roughness, more surface cracks, and enlarged existing cracks, making the specimen more susceptible to tearing. As a result, there is a slight decrease in various mechanical properties, among which the elongation at break decreased from 222.08% to 159.01% (Fig. 14.4 and Table 14.3).

Table 14.2 Measurement of silicone rubber before and after sand erosion

Type of test	1	2	3	Measure the mean
Tensile Elongation	207.14%	218.42%	240.68%	222.08%
Tensile Strength	4	4	4	/
Breaking Strength	4	4	2	/
Tear Strength	10.95	10.95	11.13	11.01

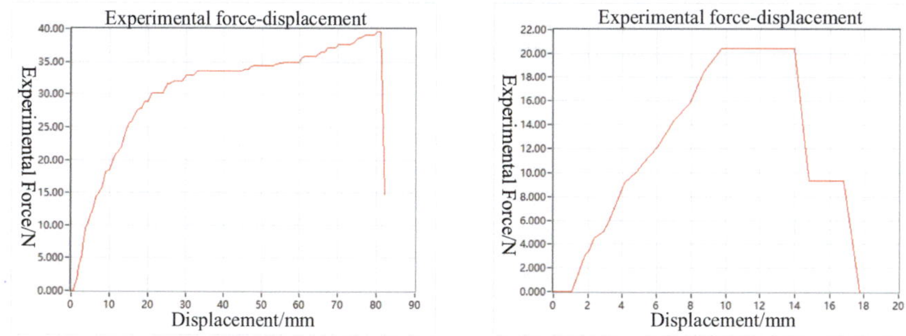

(a). Tensile performance test curve

(b).Tear resistance test curve

Fig. 14.14 Test curve of mechanical properties of silicone rubber test piece after sand erosion

Table 14.3 Measurement of mechanical properties of silicone rubber after sand erosion

Type of test	1	2	3	Measure the mean
Tensile Elongation	156.43%	152.40%	168.20%	159.01%
Tensile Strength	3	3	3	/
Breaking Strength	3	3	1	/
Tear Strength	10.23	10.95	10.14	10.44

14.4 Summary

Strong wind and sandstorms carrying sand and dust particles can degrade the performance of silicone rubber composites operating in the strong wind and sandy regions of Xinjiang, posing threats to the safe and stable operation of power transmission equipment. Based on the actual measurement and analysis results of parameters such as dust particle size and wind speed in Xinjiang, this paper establishes a wind-sand erosion simulation test device to study the impact of erosion wear on the microstructure and mechanical properties of silicone rubber composites. The conclusions are as follows:

(1) Nearly 90% of the sand and dust particles in the Shagehuang region have particle sizes within 500 μm, with the main sand particle diameters distributed in the range of 200–500 μm. Therefore, sand with particle sizes ranging from 125–250 μm was selected for the test.
(2) Wind-sand erosion causes a certain degree of wear on silicone rubber composites. Under the conditions of a wind speed of 30 m/s and an erosion time of 10 min, the erosion rate is approximately 0.3–0.4 mg/g.
(3) The sand particles, under the action of wind, possess a certain velocity, which leads to collisions and cutting on the surface of the silicone rubber specimens. This causes the originally flat surface to develop depressions, cracks, etc., resulting in a 190.36% increase in surface roughness.

(4) Erosion wear leads to an increase in surface roughness, more surface cracks, and the expansion of existing cracks, thereby causing a slight decline in various mechanical properties. Among these, the elongation at break decreased from 222.08% to 159.01%, and the tear strength decreased by 5.2%.

Reference

The research work presented in this article was completed under the funding of the project "Wind-sand Test Service for UHV Insulators and Fittings" (Project Number.: SGZB0000TGJ52301929). The authors would like to express their heartfelt gratitude to the staff members at the Wuhan Branch of China Electric Power Research Institute Co., Ltd., including Mr. Zhang Rui, and to Professors Geng Jianghai and Wang Ping from North China Electric Power University (Baoding).

References

1. Yong, H.: Study on erosion wear and insulation characteristics of EMU roof insulators under wind sand environment. M.D. thesis, North China Electric Power University, 2018.
2. Fochi, W., Yong, H., Hongyu, L., et al.: Numerical simulation of particle impact erosion on insulator of locomotives under wind sand environment. Insulators and Surge Arresters, (06), 198–203 (2017).
3. Wenxia, S.M., Liang, W., Qing., Y.: Effect of sand dust environment on external insulation of electrical power system. High Voltage Engineering, (01), 16–20+52 (2008).
4. Gang, D., Jiuyuan Z.: Effect of sand dust environment on external insulation of electrical power system. Journal of Materials Science and Engineering, (02), 307–312 (2003).
5. Sedrez, A.T, Shirazi, A.S, Rajkumar, R.Y, et al.: Experiments and CFD simulations of erosion of a 90°elbow in liquid-dominated liquid-solid and dispersed-bubble-solid flows. Wear, 426-427(Pt A), 570–580 (2019).
6. Kuanbhai, D., Jinliang, C., Yuanhua, L., et al.: Erosion mechanism of 20G steel based on gas-solid two-phase flow Nozzle experiment. Surface Technology, 53(17), 50–61 (2024).
7. Kang, S., Liang, Z., Wenjian, W., et al.: Windblown sand erosion wear and damage behaviors of different rail steels. Tribology, 42(01), 74–84 (2022).
8. Yong, Z., Xiaorong, Z., Jie, C., et al.: Study on influence of wind sand environment on performance of silicone rubber used for composite insulator umbrella skirt. Electrotechnics Electric, (01), 56–60 (2019).
9. Gui, C., Tonglai, W., Baojiang, C., et al.: Influence of sand dust in high speed airflow on the flashover characteristics of composite insulators. High Voltage Apparatus, 54(01), 164–169 (2018).
10. Jing, L., Hang, L., Longbin, Y., et al.: Influence of environmental factors on pollution accumulation characteristics of composite insulators. Science Technology and Engineering, 23(18), 7772–7778 (2023).
11. Zhicheng, H.: Research on collision adhesion model between particles and composite insulators and its applicationt. M.D. thesis, North China Electric Power University, 2023.
12. Xilin, W., Zhengyi, Z., Guoxiang, M., et al.: Research on shed crack of composite insulators used in strong wind area. Power System Technology, 37(10), 2845–2849 (2013).
13. Ying, R.: Erosion wear mechanism and evaluation of steel-structure coating in sandstorm. M.D. thesis, Inner Mongolia University, 2014
14. Yuanhong, H., Yongming, X., Shiting, Y.: Research on shed crack of composite insulators used in strong wind are. Tribology, 30(01), 26–31 (2010).

15. Heming, D., Yi, C., Wei, Z., et al.: Sand erosion simulation experiments on link hardware of transmission lines: mechanical property and micro analysis. High Voltage Engineering, 44(12), 3920–3928 (2018).
16. Zhengyi, Z., Zhidong, J., Guoxiang, M, et al.: Analysis on fracture failure mechanism of 750 kV composite insulator sheds utilized in a storm-hit region. Proceedings of the CSEE, 34(06), 947–954 (2014).
17. Junfang, H., Haiqing, L., Yuanhong, H., et al.: Finite element analysis of mechanical properties and impact resistance of polyurethane coatings on steel structures. Corrosion&Protection, 43(08), 80–85+118 (2022).
18. Bing, L., Xu, L., Kuanhai, D, et al.: Erosion mechanism of 30CrMo steel impacted by high speed solid particles. Surface Technology, 52(9), 135–145 (2023).
19. Jianqiao, M., Wenwen, W., Xiangfei, W, et al.: Analysis of sheath edge corrosion of EMU roof epoxy resin insulator in 57 m/s high speed wind tunnel. High Voltage Engineering, 46(07), 2496–2503 (2020).

Chapter 15
Numerical Simulation Study on the Surface Wear Characteristics of ± 800kV Composite Insulators Under Strong Wind and Sand Flow

JingWei Li, HuanLong Qu, ZuoMing Xu, JiangHai Geng, Wei Liu, Hao Wang, Yang Wang, and Yang Lu

Abstract Composite insulators operating in sandy desert areas are susceptible to erosion from high-speed wind and sand flow, leading to increased surface roughness of the shed skirt and edge wear, which results in reduced insulation performance and significantly shortened operational life. To address the wear issue of composite insulators in severe wind and sand environments, a numerical simulation method was employed to establish a sand erosion simulation model based on Finnie theory. This study examined the erosion-wear characteristics of the shed skirt surface under different wind speeds, sand particle sizes, and suspension methods. The results indicate that the erosion wear rate of the composite insulator surface increases with rising wind speed, exhibiting a nonlinear growth pattern. When the wind speed reaches 25 m/s, the wear rate suddenly increases. As the sand particle size increases, the wear rate of the shed surface also rises, and the maximum erosion wear rate tends to stabilize when the particle size reaches about 300 μm. Under the V-type string suspension method, the maximum wear rate of the shed skirt surface is similar to that of the I-type string; however, the insulator's suspension method primarily affects the area subjected to wear. Due to the unique flow field characteristics of inclined suspension, both the average wear rate and wear area of the V-type string's shed skirt are greater than those of the I-type string. The research findings provide a reference for the optimization design of composite insulator structures in wind-sand environments.

J. Li · H. Qu (✉) · J. Geng · H. Wang · Y. Wang · Y. Lu
Hebei Provincial Key Laboratory of Power Transmission Equipment Security Defense (North China Electric Power University), Baoding, Hebei Province, China
e-mail: long2095171127@163.com

Z. Xu
China Electric Power Research Institute, Wuhan, Hubei Province, China

W. Liu
State Grid XinJiang Co., Ltd., Electric Power Research Institute, Urumqi, Xinjiang Uygur Autonomous Region, China

Y. Li et al. (eds.), *Proceedings of the 5th Chinese Conference on High Voltage and Discharge Plasmas*, Springer Proceedings in Physics 254,
https://doi.org/10.1007/978-981-96-5711-7_15

Keywords Composite Insulator · Strong Wind and Sand Environment · Erosion Wear · Gas–solid Flow · Finite Element Calculation

15.1 Introduction

The desert, Gobi, and arid regions in western China are abundant in solar and wind energy resources. Developing large-scale clean energy bases in these areas represents a crucial direction for the future development of new energy in China [1, 2]. To effectively drive the development of large-scale energy bases and the long-distance transmission of electricity, the construction of Ultra High Voltage (UHV) transmission lines is experiencing continuous acceleration. The transmission lines that have been built or are planned to pass through regions in the sandy and arid areas, which are prone to frequent sand and dust storms, are significantly constrained by high wind speeds and dense dust, greatly hindering the development and transmission of large-scale new energy bases [3]. In the sandy and arid areas, the average annual wind speed can reach 3 m/s, and during severe sandstorms, the maximum wind speed exceeds 20 m/s. Long-term operation of composite insulators in areas with high wind and heavy dust can lead to increased surface roughness [4], reduced hydrophobicity [5, 6], thinning of the skirt, formation of holes, and even tearing [7, 8]. These issues result in a decline in insulation performance, threatening the operational safety of the transmission lines. Multiple failures have been discovered in composite insulators used in the 750 kV transmission lines in the northwest.

As of now, research on the erosion wear characteristics of insulators in strong wind and sand flows is still inadequate, with domestic scholars primarily conducting studies through numerical simulation and wind tunnel experiments. Han Yong et al. [9] carried out numerical simulation and wind tunnel experiment verification of insulator erosion on the roof in a wind and sand environment based on the k-ε; turbulence model. Their research found that the maximum wear occurs at a position where the angle between the shed edge and the incoming flow direction is 45°, and as the particle size and wind speed of the sand increase, the position of the maximum wear shifts outward. Fochi et al. [10] studied the wear patterns on the edge of shed skirts affected by sand particle diameter and wind speed. Their simulation and experiments showed that on the windward side of the insulator, the wear amount at the edge of the shed skirt increases with the increase of sand particle size and wind speed. Jianqiao et al. [11] conducted a wind tunnel erosion experiment on roof insulators with a wind speed of 57 m/s and particle size of 0.25 mm, investigating the wear conditions at different positions along the edge of the shed skirt under this condition. Their findings indicated that the wear rate of the shed skirts exhibits an 'M' shaped distribution. Changling et al. [12] used a low-speed wind tunnel to carry out simulated wind and sand experiments over different test durations, studying the applicability of composite insulator materials used on high-speed train roofs in high-altitude wind and sand environments. Long et al. [13] analyzed the collision process of high-speed train roof insulators with contaminated particles in high-speed airflows using finite element simulation methods, obtaining the wear patterns of RTV coating on train roof insulators. Their research found that at an airflow speed of 20 m/s, the side windward surface of the roof insulator experiences the most severe wear, and as the wind speed increases, both the wear severity and wear area gradually increase, with the wear region shifting towards the leeward side of the insulator.

Although domestic scholars have carried out a limited number of simulations and experimental studies on the performance changes and erosion wear of composite insulators in high-speed airflow and dust environments, the research subjects have mostly been rigid resin insulators used in locomotives, which have different material properties from silicone rubber insulators used in power systems. Additionally, the high-speed airflow and dust environment differs significantly from the practical operating environment of transmission lines. Detailed studies on the boundary conditions and process mechanisms of composite insulator wear and tearing caused by strong winds and sand in desert and Gobi regions are still lacking, and there is still a significant gap between research findings and engineering needs. This paper focuses on the wear issue of shed skirts on high-voltage transmission line composite insulators under the strong wind and sand conditions found in the northwestern regions of China. Using numerical simulation methods, it investigates the erosion wear characteristics of the shed skirt surfaces under different wind speeds, different sand particle diameters, and different suspension methods. The aim is to provide references for optimizing the design of composite insulators in wind and sand environments to improve their operational reliability.

15.2 Numerical Modeling

15.2.1 Geometric Model

The research subject is a composite insulator for ± 800 kV UHV transmission lines. The insulator features a staggered arrangement of large and small sheds, with the large shed diameter being 210 mm and the small shed diameter being 180 mm. The structure of the insulator is illustrated in Fig. 15.1.

15.2.2 Computational Domain and Mesh Generation

To avoid the influence of the wall effect on the computational results at the fluid domain boundary, the distance from the fluid domain inlet to the insulator should be greater than three times the diameter of the insulator skirt [14]. Based on this, the size of the flow field domain is set to $15 \times 20 \times 30$ m. The computational domain is illustrated in Fig. 15.2.

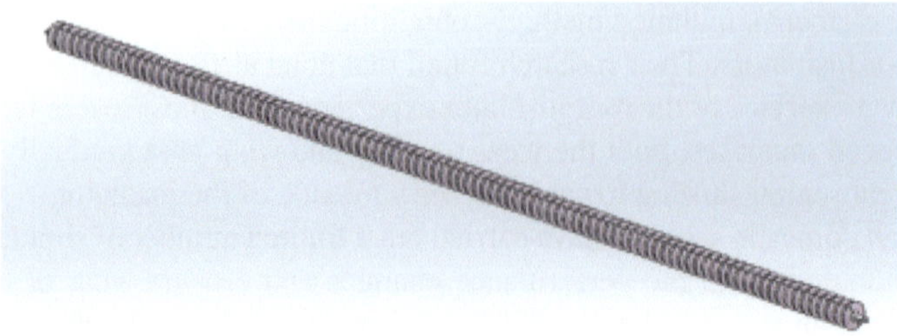

Fig. 15.1 Composite insulator for ± 800 kV UHV transmission line

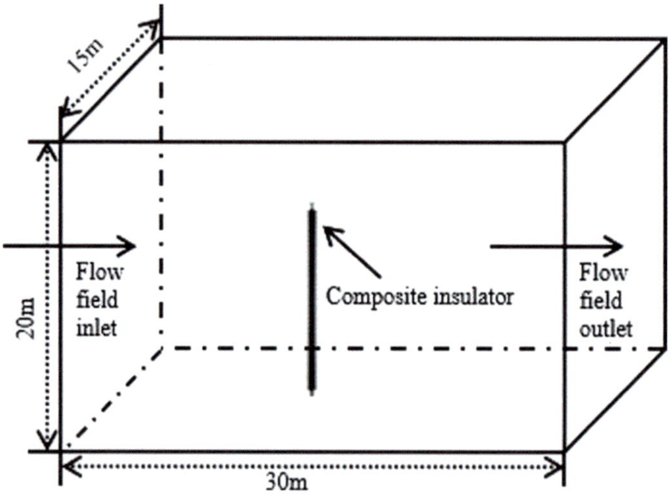

Fig. 15.2 Schematic diagram of calculation domain

Due to the complex shape of the actual insulator skirt, unstructured grids are used to mesh the fluid domain and the insulator model in order to ensure grid quality while minimizing the number of grids. To accurately calculate results such as pressure on the insulator skirt surface and erosion wear rate, the insulator is meshed with a finer grid, while the fluid domain far away from the insulator is meshed with coarser grids. Considering the number of grids, grid quality, computer memory, computation speed, and accuracy, tetrahedral grids are employed for the meshing process. The result of the grid division is shown in Fig. 15.3.

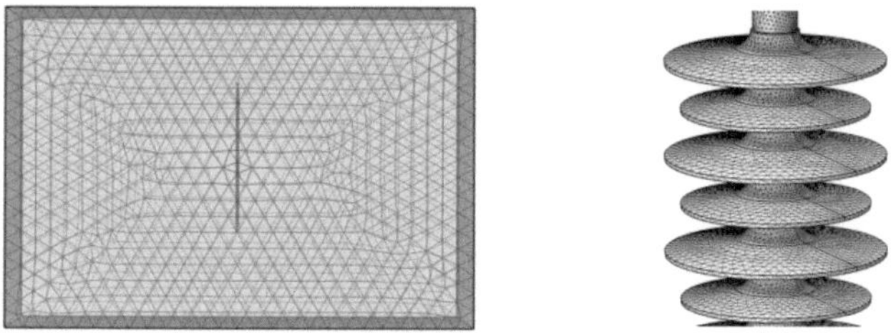

(1) Computational domain meshing (2) Insulator grid division

Fig. 15.3 Grid diagram of calculation domain

15.2.3 Mathematical Model

15.2.3.1 Airflow Control Model

The flow medium around the insulator is a mixture of air and sand particles. The air is treated as a continuous phase, while the sand particles are treated as a discrete phase. Since the volume fraction of sand particles in the air is small, the effect of sand particles on the continuous airflow can be neglected. The airflow process satisfies the continuity condition of mass conservation and the momentum conservation condition, based on which the continuity equation and momentum conservation equation are established.

The continuity equation for airflow is established as follows:

$$\frac{\partial \rho}{\partial t} + \frac{\partial \rho u}{\partial x} + \frac{\partial \rho v}{\partial y} + \frac{\partial \rho w}{\partial z} = 0 \tag{15.1}$$

where u is the velocity vector, t is time, ρ is the fluid density.

The momentum conservation equation is established as follows:

$$\frac{\delta}{\delta x_j}(\rho u_j u_i) = -\frac{\delta p}{\delta x_i} + \frac{\delta}{\delta x_j}\left(\mu \frac{\delta u_i}{\delta x_j} - \overline{\rho u_i' u_j'}\right) \tag{15.2}$$

where u is the mean velocity of the airflow, x is the displacement, μ is the dynamic viscosity coefficient of the airflow, $\overline{-\rho u_i' u_j'}$ represents the components of the Reynolds stress.

Due to the introduction of Reynolds stress components in the momentum conservation equation, it becomes difficult to obtain an exact solution. Therefore, new variables and their governing equations need to be introduced. To determine the type of flow field around the insulator, the Reynolds number of the surrounding flow field is calculated using the following formula:

$$\mathrm{Re} = \frac{\rho v L}{\mu} \tag{15.3}$$

where ρ is the fluid density, taken as 1.205 kg/m^3, μ is the dynamic viscosity of the fluid medium, taken as 1.81×10^{-5} Pa·s, v is the inlet velocity of the flow field, L is the diameter of the large umbrella skirt of the composite insulator, taken as 0.21 m.

From the calculation using Eq. (15.3), it is found that when the wind speed is 15 m/s, the Reynolds number is 2.09×10^5, indicating that the fluid clearly exhibits turbulent characteristics. Therefore, an appropriate turbulence model needs to be introduced. The standard k-ε; model has been validated by a large number of practical tests, is applicable to most turbulent flows, offers sufficient computational accuracy, and has low requirements on computing resources. Consequently, the standard k-ε; model is adopted in this paper, and its governing equations are as follows:

$$\frac{\partial(\rho k)}{\partial t} + \frac{\partial(\rho k u_i)}{\partial x_i} = \frac{\partial}{\partial x_j}\left[\left(\mu + \frac{\mu_t}{\sigma_\varepsilon}\right)\frac{\partial k}{\partial x_j}\right] + G_k + G_b - \rho\varepsilon - Y_M + S_k \tag{15.4}$$

$$\frac{\partial(\rho \varepsilon)}{\partial t} + \frac{\partial(\rho \varepsilon u_i)}{\partial x_i} = \frac{\partial}{\partial x_j}\left[\left(\mu + \frac{\mu_t}{\sigma_\varepsilon}\right)\frac{\partial \varepsilon}{\partial x_j}\right] + C_{1\varepsilon}\frac{\varepsilon}{k}(G_k + C_{3\varepsilon}G_b) - C_{2\varepsilon}\rho\frac{\varepsilon^2}{k} + S_\varepsilon \tag{15.5}$$

where G_k represents the turbulent kinetic energy term induced by the mean velocity gradient, G_b represents the turbulent kinetic energy induced by buoyancy, Y_M represents the fluctuation dilation term in compressible turbulence, $C_{1\varepsilon}$, $C_{2\varepsilon}$, and $C_{3\varepsilon}$ are constants, σ_k and σ_ε are the Prandtl numbers corresponding to turbulent kinetic energy (k) and dissipation rate (ε), respectively. The commonly used values for the model parameters are:

$$C_{1\varepsilon} = 1.44, C_{2\varepsilon} = 1.92, C_{3\varepsilon} = 0.09, \sigma_k = 1.0, \sigma_\varepsilon = 1.3$$

15.2.3.2 Sand Particle Trajectory Calculation Model

Sand particles are treated as discrete phases, and the Euler–Lagrange method is chosen to study the motion state of solid particles in the flow field. According to the basic theory of gas–solid two-phase flow, sand particles in gas–solid two-phase flow are mainly subjected to forces such as gravity, drag, and lift [15]. The calculation equation for sand particle trajectories is established in a Cartesian coordinate system as follows:

$$\frac{du_p}{dt} = f_D(u - u_p) + \frac{g_x(\rho_p - \rho)}{\rho_p} + f_x \tag{15.6}$$

where $f_D(u - u_p)$ is the drag force per unit mass on the sand particle, u is the air flow velocity, u_p is the sand particle velocity, g_x is the gravitational acceleration in the x direction, ρ_p is the density of the sand particle, ρ is the density of the air flow, and f_x is the other forces acting in the x direction.

15.2.3.3 Wear Rate Calculation Model

Erosion refers to a type of wear phenomenon where solid particles impact the material surface at a certain velocity and angle, causing material loss [16]. The Finnie model can adequately explain the phenomenon of erosion wear of rigid particles on ductile materials and establish a relationship between the erosion wear rate and the impact conditions of abrasive particles [17–21]. Based on the Finnie model, a simplified expression for calculating the erosion wear rate is:

$$M_W = \frac{Mv^2}{2} \cdot \frac{1}{P} f(\alpha) \tag{15.7}$$

$$f(\alpha) = \begin{cases} \sin 2\alpha - 3\sin^2\alpha, \alpha \le 18.5^\circ \\ \dfrac{\cos^2\alpha}{3}, \alpha \ge 18.5^\circ \end{cases} \tag{15.8}$$

$$P = \frac{k}{(1 + m_p r_p^2 / I_p)} \tag{15.9}$$

$$I_p = \frac{2}{5} m_p r_p^2 \tag{15.10}$$

$$k = \frac{F_x}{F_y} = \frac{1}{\tan\alpha} \tag{15.11}$$

where, M_W is the erosion wear mass, P is the flow stress of the target material, M is the mass of the particle bundle, v is the incident speed of the sand particles, α is the incident angle, m_p is the mass of a single sand particle, r_p is the radius of the sand particle, and I_p is the moment of inertia of the sand particle.

15.2.4 Initial Conditions and Boundary Conditions

The left side of the fluid domain is set as a velocity inlet, with wind speeds of 15, 20, 25, and 30 m/s. The right side is set as a pressure outlet with a static pressure of 0 Pa. The remaining four surfaces of the fluid domain are set as wall surfaces. Dust particles are released from the velocity inlet on the left side of the fluid domain. Based on the force analysis of the sand particles, gravity, drag force, and lift force are applied to the wind-blown sand particles. It is assumed that the shape of the sand particles is spherical, and the other components besides SiO_2 are neglected. The density is set to 2650 kg/m^3, and the particle diameters selected are 200, 300, 400, and 500 μm.

15.3 Results and Discussion

15.3.1 The Effect of Wind Speed on Canopy Erosion Wear

Figure 15.4 shows the erosion contours of the upper and lower surfaces of the insulator's large skirt for particle sizes of 300 μm at incoming wind speeds of 15, 20, 25, and 30 m/s. The results indicate that the area with the most severe erosion is located at the root of the skirt, and the main erosion zones on the skirt surface are relatively consistent across different wind speeds of the sand flow.

The effect of wind speed on the erosion wear rate of insulator surfaces is caused by various factors. Wind speed influences the flow field around the insulator, thereby altering the drag forces experienced by particles in the fluid domain and ultimately changing the trajectories of the wind-blown sand particles. Additionally, wind speed plays a crucial role in the dispersion of sand particles within the flow domain and the extent of collisions between the sand particles and the insulator surface. Under different levels of sandstorm conditions, variations in wind speed lead to certain patterns in the erosion wear characteristics of the insulator surface.

Figure 15.5 illustrates the relationship between the maximum erosion wear rate of the insulator's windward and leeward surfaces and wind speed under the impact of sand particles with a diameter of 300 μm. From Fig. 15.5, it can be observed that

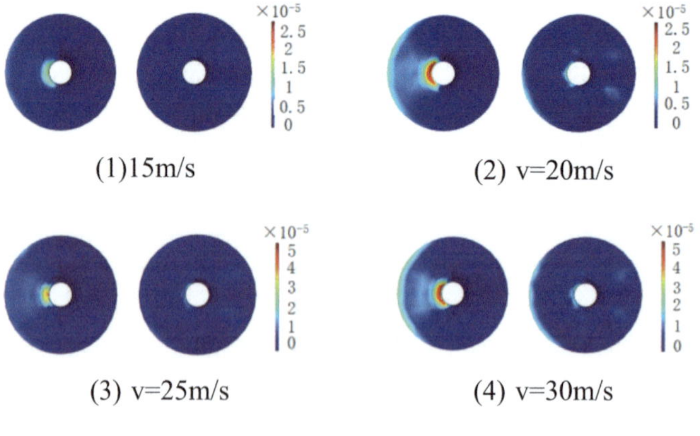

Fig. 15.4 Erosion wear rates of the upper and lower surfaces of the parachute skirt under different wind speeds

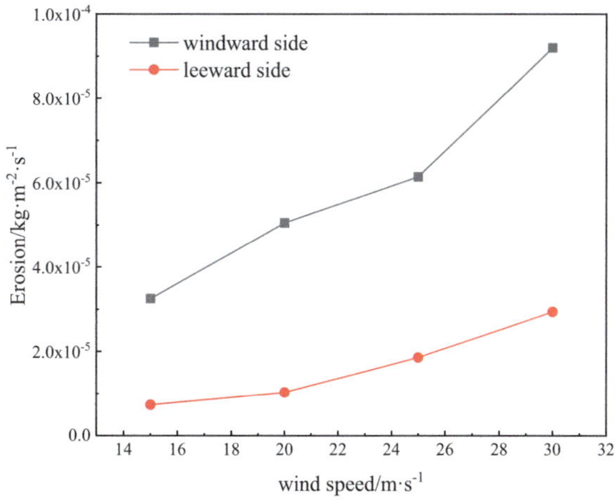

Fig. 15.5 Relation between maximum erosion wear rate and wind speed (d = 300 μm)

the erosion wear rate of the composite insulator surface increases with wind speed, demonstrating a nonlinear growth pattern. At lower wind speeds, the erosion wear rate on the windward surface of the skirt increases slowly; however, when the wind speed reaches around 25 m/s, which corresponds to a severe sandstorm level, the erosion wear rate on the windward surface increases significantly with wind speed. The growth trend of the maximum erosion wear rate on the leeward surface of the insulator is slower compared to that on the windward surface.

Figure 15.6 shows the variation of the average erosion wear rate on the insulator surface with wind speed for a particle size of 300 μm. The average erosion wear rate increases with wind speed, but the rate of increase is relatively gradual. However, when the wind speed reaches a certain range, around 25 m/s, the average growth rate of erosion we are rate suddenly increases significantly. The trend of slow initial growth followed by a sharp increase in erosion wear rate is likely caused by changes in the kinetic energy of the sand particles. Before the wind speed reaches a certain threshold, the sand particles do not possess turbulent kinetic energy and move at relatively low speeds. When the wind speed reaches approximately 25 m/s, the discrete phase (sand particles) and the continuous phase (airflow) begin to couple under the effect of turbulence, giving the sand particles turbulent kinetic energy and further increasing their speed. As a result, the intensity of collisions with the insulator surface increases, leading to a higher erosion wear rate.

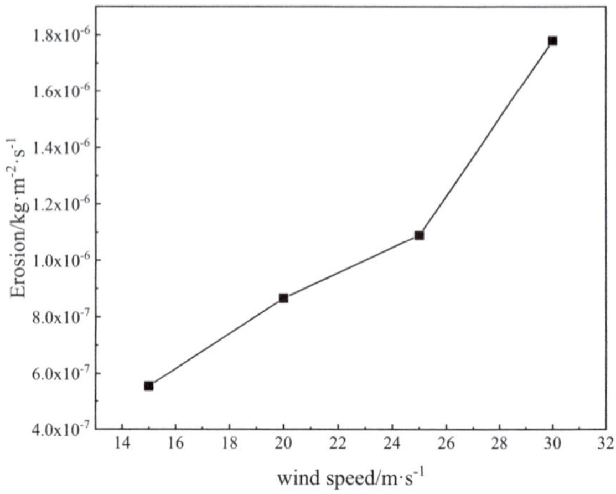

Fig. 15.6 Relation between average erosion wear rate and wind speed (d = 300 μm)

15.3.2 *The Influence of Sand Particle Size on Erosion Wear of the Skirt*

The particle size of wind-blown sand particles has a significant impact on the erosion wear rate of the insulator surface. Under the same wind speed and insulator arrangement, changing the particle size of the wind-blown sand leads to different erosion wear results on the skirt. Numerical simulations were conducted on the erosion wear rate of the insulator surface under the impact of typical particle sizes at a wind speed of 20 m/s. Figure 15.7 shows the erosion wear results of the skirt under the influence of particle sizes of 200, 300, 400, and 500 μm.

From Fig. 15.7, it can be observed that as the particle size of the wind-blown sand increases, both the maximum erosion wear rate on the insulator surface and

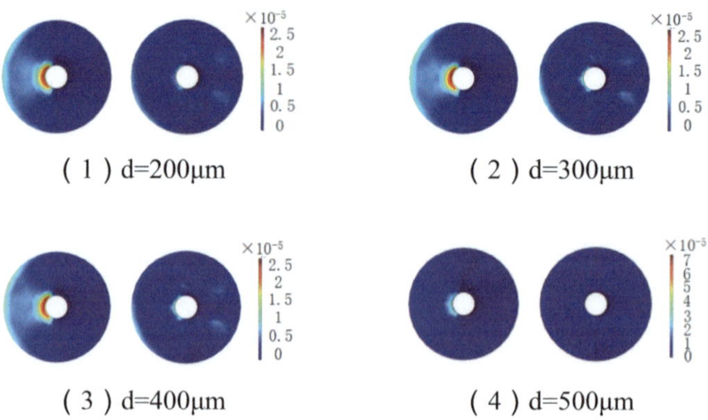

Fig. 15.7 Wear of the upper and lower surfaces of the parachute skirt under typical particle size impact

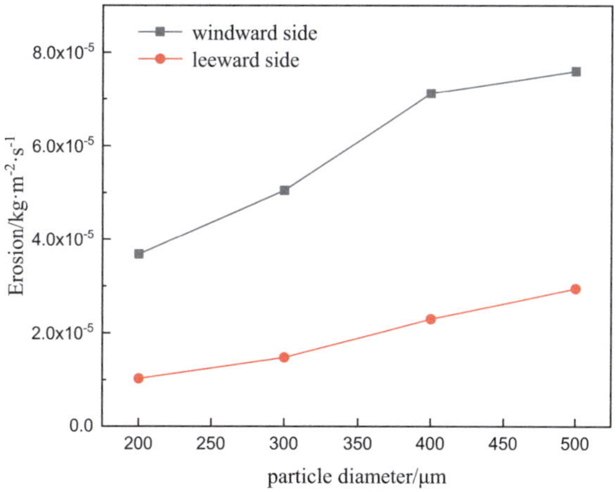

Fig. 15.8 Relationship between maximum erosion wear rate and particle size (v = 20 m/s)

the erosion wear degree of the skirt continuously increase. The more severe wear is concentrated in the ring-shaped area near the root of the skirt. As the particle size increases, the primary wear area remains largely unchanged. This is because the airflow carries the sand particles to impact the insulator surface. As the particle size increases, its kinetic energy also increases, leading to more severe erosion wear on the insulator surface.

As shown in Fig. 15.8, the curves illustrate the variation of the maximum erosion wear rate on the insulator surface under the impact of typical sand particle diameters at a wind speed of 20 m/s. From Fig. 15.8, it can be observed that the maximum erosion wear rate on the windward side increases with the growth of sand particle diameter, showing a rising trend. When the sand particle size reaches a certain range (diameter greater than 300 μm), the maximum erosion wear rate gradually stabilizes. The simulation results are consistent with the findings of Yerramareddy et al. [22]. The maximum erosion wear rate on the leeward side is much smaller than that on the windward side, and it also shows an upward trend with the increase in particle size.

As shown in Fig. 15.9, the curves illustrate the variation of the average erosion wear rate on the insulator surface under the impact of typical sand particle diameters at a wind speed of 20 m/s. The average erosion wear rate on the insulator surface fluctuates around 8.86×10^{-7} kg/(m^2·s). This is because, when the mass flow rate of the sand particles is fixed, an increase in particle diameter leads to a greater mass and kinetic energy of individual particles, intensifying the severity of collisions with the insulator. However, as the particle diameter increases, the number of sand particles decreases, resulting in a reduced surface erosion wear rate. Under the combined effect of these factors, the average erosion wear rate remains relatively constant with changes in sand particle size, indicating that the overall wear degree on the insulator is basically the same.

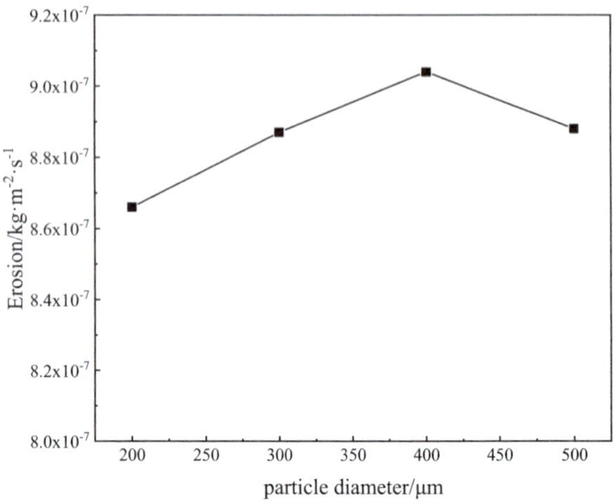

Fig. 15.9 Relationship between average erosion wear rate and particle size (v = 20 m /s)

15.3.3 *The Influence of Hanging Method on Erosion Wear of the Skirt*

The hanging methods of composite insulators on transmission towers differ, leading to variations in the direction and area exposed to the wind, subsequently affecting the erosion wear on the insulator surface. Figure 15.10 illustrates different hanging methods of composite insulators, including V-strings and I-strings. To investigate the wear characteristics under different hanging methods, changes were made to the placement of the insulators in the simulation model. For ease of description, the windward angle is defined as the angle between the axis and the wind direction. As shown in Fig. 15.11, the windward angle of the left string is 135°, and that of the right string is 45°.

Under a wind speed of 20 m/s and a sand particle size of 300 μm, the cloud maps of the erosion wear rate on the insulator surface with windward angles of 45° and 135° are shown in Fig. 15.10. When the windward angle is 45°, the wear on the underside of the umbrella skirt is mainly concentrated within the range of an angle

Fig. 15.10 Composite insulator suspension mode

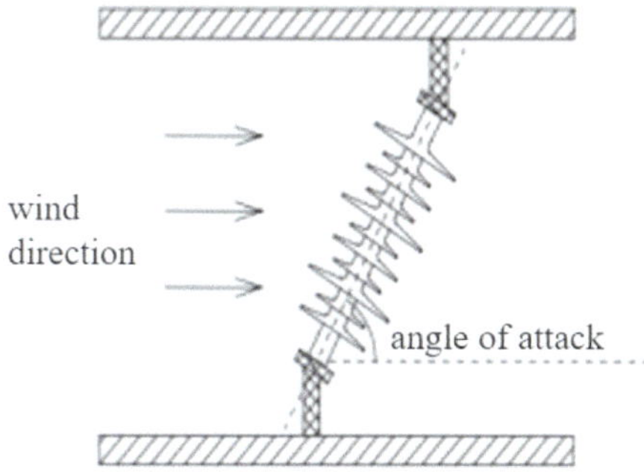

Fig. 15.11 Schematic diagram of windward angle

of −105° to 105° relative to the inflow direction. The wear on the upper surface is mainly concentrated on the sides near the root of the skirt on the crosswind side, with the wear range on the underside being much larger than that on the upper surface. The maximum erosion wear rate appears at the edge near the root on the upper surface. When the windward angle is 90°, varying degrees of erosion wear occur on the windward side of the upper surface of the umbrella skirt. When the windward angle is 135°, erosion wear appears on the upper surface within the range of an angle of −110° to 110° relative to the inflow direction. The erosion wear degree and range on the upper surface are much greater than those on the lower surface, and the maximum erosion wear rate similarly occurs near the root position on the lower surface of the umbrella disk (Fig. 15.12).

The surface wear areas of the insulators exhibit significant differences under different hanging methods. When the insulator windward angle is 135°, corresponding to the left string of the V-shaped string during actual installation, the main wear area concentrates on the upper surface of the umbrella skirt. Conversely, when the windward angle is 45°, corresponding to the right string of the V-shaped string, the main wear area focuses on the lower surface of the umbrella skirt. Due to the shielding effect of the large umbrella skirt on the small umbrella group, the erosion wear rate of the large umbrella skirt is generally higher than that of the small umbrella group. When the windward angle is 135°, wind and sand particles are more likely to collide with the upper surface of the large umbrella skirt, while the lower surface and small umbrella groups are less likely to be damaged by sand particles due to the shielding of the upper surface of the large umbrella skirt. Similarly, when the

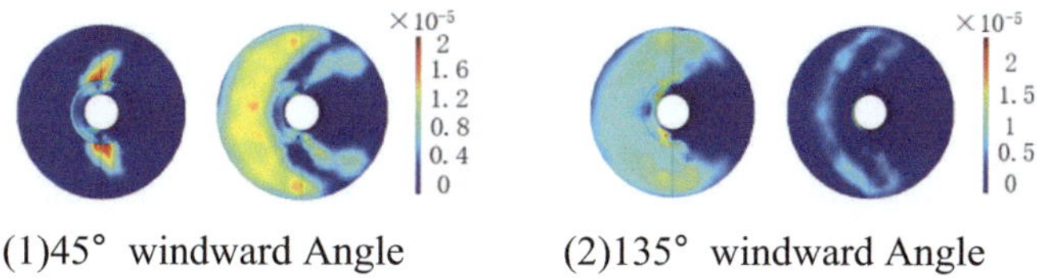

Fig. 15.12 Erosion wear rate of the upper and lower surfaces of the umbrella skirt under different suspension modes of insulators

Table 15.1 Maximum and average erosion wear rates of different suspension modes

Angle of attack	45°	90°	135°
Maximum erosion wear rate/kg m^{-2} s^{-1}	4.69×10^{-5}	3.68×10^{-5}	3.98×10^{-5}
Average erosion wear rate/kg m^{-2} s^{-1}	2.1×10^{-6}	8.66×10^{-7}	1.64×10^{-6}

windward angle is 45°, particles are more likely to collide with the lower surface of the large umbrella skirt, and the upper surface and small umbrella group are less likely to be damaged by sand particles, resulting in a lower erosion wear rate. When the windward angle is 90°, due to the downward tilt of the umbrella skirt, the upper surface is more prone to sand particle impacts, and the erosion wear rate of the upper surface of the umbrella skirt is generally higher than that of the lower surface.

Table 15.1 shows the maximum erosion wear rate of the insulator surface under different hanging methods with a wind speed of 20 m/s and a particle size of 300 μm. From Table 15.1, it can be observed that there are slight differences in the maximum erosion wear rate of the insulator surface within the same order of magnitude when the hanging method changes. This indicates that when the wind speed and sand particle size are fixed, the intensity of sand particle impacts on the insulator surface is similar, and the degree of damage in the most severely eroded areas is essentially the same. However, since the windward area of the insulator significantly increases when it is hung at an incline compared to when it is hung vertically, the area exposed to collisions with wind-blown sand particles also increases. At the same time, due to the unique flow field characteristics when the insulator is tilted, the speed of the sand particles is greater than when the insulator is vertical. Therefore, the overall erosion wear of the insulator is more severe when it is tilted.

15.4 Conclusion

This paper studies the erosion wear patterns of composite insulator umbrella skirts in strong wind-sand environments through numerical simulations, obtaining the erosion wear characteristics of the insulator surface under different wind speeds, particle diameters, and insulator hanging methods. The main research conclusions are as follows:

(a) The erosion wear rate of the surface of composite insulator umbrella skirts shows an upward trend with increasing wind speed, exhibiting nonlinear growth. The trends of the maximum and average erosion wear rates are basically the same. When the wind speed increases, the maximum surface erosion wear rate initially grows steadily, but the growth rate suddenly increases when the wind speed reaches around 25 m/s.

(b) The maximum erosion wear rate of the umbrella skirt on the windward side of the insulator shows an increasing trend with the increase in sand particle diameter. When the particle diameter grows to around 300 μm, the maximum erosion wear rate gradually stabilizes. Within the wind speed range of 15–30 m/s, the average erosion wear rate of the insulator umbrella skirt surface fluctuates around 8.86×10^{-7} kg/($m^2 \cdot s$). The average erosion wear rate remains largely unchanged with variations in sand particle diameter, indicating that the overall wear level of the insulator is similar.

(c) The hanging method of the insulator primarily affects the wear-affected areas on the umbrella skirt surface; the erosion wear rate of large skirts is generally higher than that of small ones. When the hanging method changes, the maximum erosion wear rate on the umbrella skirt surface exhibits slight differences within the same order of magnitude, and the severity of damage in the most severely eroded areas is essentially the same. Overall, the insulator experiences more severe erosion wear when arranged at an incline.

Acknowledgements The conduct of the research work in this paper was carried out under the funding of the project 'Study on the Insulation Performance of Composite Materials Considering the Effects of Strong Wind and Sand Erosion' (Project No.: 5500-202355807A-3-8-KJ). We sincerely express our gratitude for this support.

References

1. Xiaoxin. Z, Shuyong. C, Zongxiang. L.: Technology features of the new generation power system in China. Proceedings of the CSEE, 38(7): 1893–1904 (2018).
2. Yinbiao. S, Guoping. C, Jingbo. H.: Building a new electric power system based on new energy sources. Strategic Study of CAE, 23(6): 61–69 (2021).
3. Baoan. X, Mingjie. L, Jingbo. H.: Research on Security Defense System of New Power System. Proceedings of the CSEE, 43(15): 5723–5732 (2023).
4. Yitong. X, Pan. C, Yushun Z.: Overview ofweak links and optimization of roof composite insulator inhigh-speed sand environment. Journal of Electrical Engineering, 15(2): 1–10 (2020).
5. Jing. X, Bo. G, Zhenjie. S.: Study on the effect of environmental factors on hydrophobicity of composite insulator. High Voltage Apparatus, 2018(2): 103–109 (2018).
6. Yushun. Z, Yitong. X, Shanqiang. G.: Effect of surface roughness on flashover characteristics of silicone rubber. Journal of Electrostatics, 2019(99): 41–48 (2019).
7. Zhengyi. Z, Zhidong. J, Guoxiang. M.: Analysis on fracturefailure mechanism of 750 kV composite insulator sheds utilized in astorm-hit region. Proceedings of the CSEE, 36(2): 947–954 (2014).
8. Zhidong. J, Guoxiang. M, Zhengyi. Z .: Deformation and stress concentration of composite insulator in strong wind. High Voltage Engineering, 41(2): 602–607 (2015).
9. Yong. H. Study on erosion wear and insulation characteristics of EMU roof insulators under wind sand environment. M.D. thesis, North China Electric Power University, 2017.
10. Fochi. W, Yong. H, Hongyu. L.: Numerical simulation of particleimpact erosion on insulator of locomotives under windsand environment. Insulators and Surge Arresters, 06(6): 198–203 (2017).
11. Jianqiao. M, Wenwen. W, Xiangfei. W.: Analysis of sheath edge corrosion of emu roof epoxy resin insulator in 57 m/s highspeed wind tunnel. High Voltage Engineering, 46(7): 2496–2503 (2020).
12. Changling.L.: Sand Test Study of Composite Jacket Insulator Used in the Roof. Railway Locomotive & Car, 34(06): 128-130(2014).

13. Long. W, Suoli. J, Chao. F.: Study on the Abrasion Law of the Insulator Coating on the Roofand Its Influence on the Pollution Flashover Voltage. Railway Standard Design, 65(08): 134–139 (2021).
14. Boxiang. M, Hong. W.: Wind resistance performance and structural optimization of silicone rubber bird shield for high voltage transmission lines. Insulating Materials, 53(4): 64–69 (2020).
15. Ying. M, Jun. R, Yuandong. L.: Development of Research on Erosion of Materials. Journal of Lanzhou University of Technology, 31(1): 21—25 (2005).
16. Finnie. I.: Some reflections on the past and future of erosion. Wear, s186–187(part-P1):1–10 (1995).
17. Finnie.I.: Erosion of surfaces by solid particles. Wear, 3(2):87–103 (1960).
18. Finnie.I, Wolak J, Kabil Y.: Erosion of metal by solid particles. Mater, 2:682-700(1967).
19. Finnie. I.: Some observations on the erosion of ductile metals. Wear, 19(1):81–90 (1971).
20. Finnie. I, Mcfadden D H.: On the velocity dependence of the erosion of ductile metals by solid particles at low angles of incidence.Wear, 1977, 48(1):181–190 (1977).
21. Yu. Z.: Flow characteristic test of wind tunnel and erosion wear research of wind turbine blade material. M.D. thesis, Lanzhou University of technology, 2020.
22. YERRAMAREDDY.S, BAHADUR.S.: Effect of operational variables, microstructure and mechanical properties on the erosion of Ti-6A1–4V. Wear, 142:253–263 (1991).

Chapter 16
The Diffusion Law of Hcl in High-Voltage Cable Tunnel Fire

Yi Chen, Yinglong Liu, Xing Gong, Fenglang Shi, Ke Wang, Bingzhou Lu, Jie Liao, and Chuansheng Luo

Abstract In this paper, a fire model of a high-voltage cable tunnel is constructed by numerical simulation method, and the diffusion law properties of hydrogen chloride (Hcl), the characteristic gas of the fire, are analyzed under the conditions of different ignition source locations and ignition source power. The results show that the temperature in the high-voltage cable tunnel rises slowly at the beginning of the fire (25 s ago), then increases rapidly, and stabilises after 100 s. The closer the fire source is to the top of the tunnel, the more significant the temperature difference in the tunnel and the lower the diffusion rate of Hcl gas. The higher the power of the fire source, the faster the temperature rises and the more uniform the Hcl concentration distribution. When the power of the fire source reaches 6 MW, the distribution of Hcl gas in the tunnel is uniform. These findings provide an important reference for the safety design and fire protection of high-voltage cable tunnels.

Keywords High-voltage cable tunnels · Cable fires · Numerical simulation · Gas diffusion · Temperature distribution · Hcl

16.1 Introduction

16.1.1 A Subsection Sample

Gas diffusion phenomena play a crucial role in numerous engineering and scientific fields. In high-voltage cable tunnels, the occurrence of fire causes a series of complex physical and chemical processes leading to the release of large amounts of toxic gases such as hydrogen chloride (Hcl). The spread of these gases not only affects fire fighting and escape, but also poses a threat to the safe operation of equipment in the tunnel. As an important part of modern urban power transmission networks, the

Y. Chen (✉) · Y. Liu · X. Gong · F. Shi · K. Wang · B. Lu · J. Liao · C. Luo
Nanning Power Supply Bureau of Guangxi Power Grid Co., Ltd., Nanning, China
e-mail: 1144304697@qq.com

Y. Li et al. (eds.), *Proceedings of the 5th Chinese Conference on High Voltage and Discharge Plasmas*, Springer Proceedings in Physics 254,
https://doi.org/10.1007/978-981-96-5711-7_16

safety and reliability of high-voltage cable tunnels are directly related to the normal operation of cities. Once a fire occurs, the closed tunnel space will result in smoke and toxic gases that are difficult to be discharged quickly, thus exacerbating the fire hazard. Therefore, it is of great significance to study the gas diffusion law under fire conditions to improve the safety design and fire protection level of high-voltage cable tunnels [1].

Liang et al. [2] studied the development of cable compartment fire in T-type integrated pipe corridor, and analyzed the relationship between compartment roof temperature and fire source power as well as the smoke spread at the intersection of cable compartment in T-type integrated pipe corridor. Fangshun et al. [3] used numerical simulation to study the development process of cable fires in integrated tube corridors, and analyzed the influence of fire partition spacing on cable fires to obtain the fire HRR, temperature distribution, smoke spread and CO concentration distribution under different working conditions, etc. En-ze [4] and Liang [5] et al. studied the temperature distribution in different areas after the occurrence of fires in the T-type and L-type cable corridors, respectively, by means of numerical simulation. while An et al. [6] studied the temperature distribution and CO spreading caused by cable fires at different cable inclination angles in an L-type integrated corridor and found that when the fire was stabilized, the CO concentration varied in the form of a V-shaped change along the horizontal distance from the fire source. Wang et al. [7] studied the effect of curved sidewalls on the maximum temperature and longitudinal temperature distribution caused by linear fires in integrated tube corridors by carrying out five sets of experiments in which the fire source was far away from the centerline of the tunnel, taking circular tunnels as the research object. Yuhui [8] used numerical simulation methods to study the variation rules of smoke concentration, heat release rate, and longitudinal and transverse tunnel temperatures in tunnel cable fires. Chenying [9] and Mingnian [10], on the other hand, investigated the development process of cable fires under different conditions by means of experiments and numerical simulations, respectively, and obtained the temperature distribution, smoke diffusion, and carbon monoxide concentration distribution of cable fires in tube corridors. Although these studies provide an important foundation for the understanding of cable tunnel fires, most of them focus on the smoke flow and temperature distribution, and pay less attention to the diffusion law of toxic gases [11–13]. In particular, there is still a lack of systematic research on the diffusion behavior of characteristic gases in high-voltage cable tunnels under different fire source locations and fire source power conditions. Existing studies mainly focus on modeling experiments and numerical simulations, and the complexities in practical applications still need to be further explored and verified.

In this study, numerical simulations were carried out using PyroSim software, which is based on the principle of computational fluid dynamics and combines fire dynamics and heat conduction modeling to accurately simulate the gas diffusion under fire conditions. Specifically, a fire model of a high-voltage cable tunnel is constructed in this paper, and different fire source positions and power conditions are set up, and the simulation results are used to analyze the diffusion law of Hcl gas in the tunnel. The results not only provide theoretical support for the safe design of high-voltage cable tunnels, but also provide a scientific basis for the development of fire protection measures.

Modeling

16.2 Cable Compartment Modeling

16.2.1 Modeling Methodology

FDS (Fire Dynamics Simulator) is a fluid dynamics software that focuses on the simulation of fluid motion in fires. FDS is a field model-based fire simulation software developed by NIST [14]. The numerical calculations of FDS need to follow the basic governing equations such as mass and momentum conservation equations, energy conservation equations, as well as the component conservation equations and gas equation of state. The FDS numerical calculations need to follow the basic governing equations such as mass and momentum conservation equations, energy conservation equations, and component conservation equations and gas equation of state [15].

The continuity equation:

$$\frac{\partial \rho}{\partial t}+\left[\frac{\partial(\rho u_x)}{\partial x}+\frac{\partial(\rho u_y)}{\partial y}+\frac{\partial(\rho u_z)}{\partial z}\right]=0 \tag{16.1}$$

where ρ is the density (kg/m^3); t is the time (s); u_i is the velocity component (m/s).

Momentum conservation equation:

$$\frac{\partial(\rho \vec{u}\,\vec{u})}{\partial t}+\nabla\cdot(\vec{u}\,\vec{u})+\nabla p=\rho g+\vec{f}+\nabla\cdot\tau \tag{16.2}$$

where g is the acceleration of gravity (m/s^2); $\vec{f}$ is the external force vector (kg/s^2 /m); p is the pressure (Pa); t is the time (s); $\vec{u}\,\vec{u}$ is the second-order tensor of velocity; and ∇ is the Hamiltonian operator.

The energy conservation equation:

$$\frac{\partial}{\partial t}(\rho h)+\nabla\cdot(\rho h\vec{u})=\frac{\partial p}{\partial t}+\vec{u}\cdot\nabla p-\nabla\cdot\vec{q_r}$$
$$+\nabla\cdot(k\nabla T)+\sum_i\nabla\cdot(h_i\rho D_i\nabla Y_i) \tag{16.3}$$

where h is the enthalpy (J/kg); $\vec{q_r}$ is the thermal radiation flux; T is the temperature; and k is the thermal conductivity (W/(m·K)).

Component conservation equations:

$$\frac{\partial}{\partial t}(\rho Y_i) + \nabla \cdot \left(\rho Y_i \vec{u}\right) = \nabla \cdot (\rho D_i \nabla Y_i) + m_i \tag{16.4}$$

where Y_i is the mass fraction of the component i; D_i is the diffusion coefficient of the component i (m^2/s); m_i^{m} is the mass production rate of the component i ($kg/s/m^3$).

16.2.2 HV Cable Tunnel Model Construction

Based According to the 110 kV outgoing cable tunnel in a 220 kV substation, this paper establishes a high-voltage cable tunnel fire model, setting the internal dimensions of the high-voltage cable tunnel to be 8.3 m long, 1.9 m wide and 2.3 m high. This paper observes the development of cable fire in the HV cable tunnel by setting temperature measurement points and 3D slices. Based on the internal dimensions of the high-voltage cable tunnel, five groups of temperature sensors are set up (numbered A, B, C, D, E from right to left), three in each group (1, 2, 3 from top to bottom), with a horizontal spacing of 2 m between each group of sensors, a vertical spacing of 0.5 m, and a distance from the bottom of each group of 1.2 m, 1.7 m, and 2.3 m, respectively. The calculation model is shown in Figs. 16.1 and 16.2.

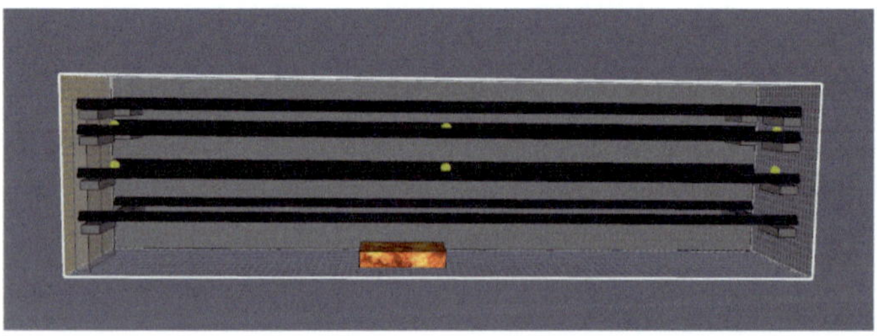

Fig. 16.1 Front view of the high voltage cable tunnel model

Fig. 16.2 Right view of the high voltage cable tunnel model

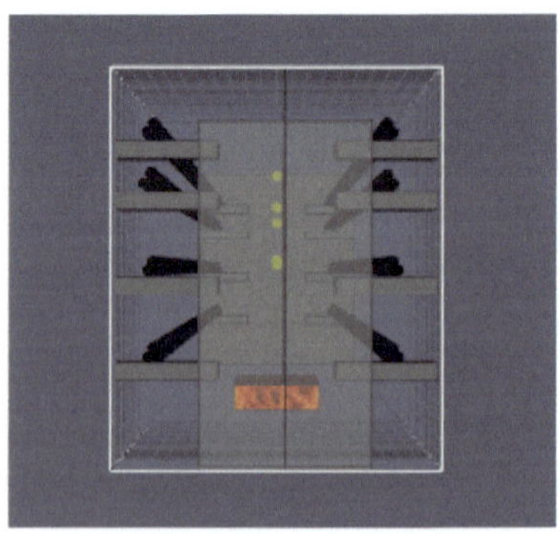

16.2.3 Boundary Condition Setting

Inside the cable compartment, the initial ambient temperature, initial speed, pressure and other factors are mainly considered, and this paper sets the simulation t = 0 s when the cable compartment: the initial temperature is 20 °C, and the initial pressure is 1.01325×10^5 Pa. The cables are symmetrically arranged on both sides of the cable compartment, with 3 groups on each side, and the upper and lower spacings of 0.5 m, and each group of 3 is arranged in zigzag shape to constitute the conductor insulation, filler and outer sheath of the power cables. The materials are combustible. The internal structure of the cable is complex, and the flammability of the insulating material is the main cause of the fire. When a fire occurs in a cable, once the insulation is ignited, the heat release rate can be very large, which greatly increases the fire risk. In order to simplify the model, the cable material is set to copper and cross-linked polyethylene (XLPE), and the relevant parameters are shown in Table 16.1.

The rate of heat release from unstable fire sources varies with time, and in order to be consistent with fire development in real situations, the t^2 model is used in the FDS numerical simulation calculations to determine the way in which the rate of heat release from the material grows in the form of equation:

$$\dot{Q} = \alpha t^2 \tag{16.5}$$

where $\dot{Q}$ is the rate of heat release from the fire source (kW) and α is the fire development factor (kW/s^2) (Table 16.2).

The heat release rate of polyethylene is 265 KW [16]. In this paper, the fire spreading level of cable protection jacket material is rapid fire, α is 0.04689, and the time to reach the maximum heat release rate is calculated to be 75 s.

Table 16.1 Cable material parameters

Materials	Density /(kg/m^3)	Specific heat / (kJ/kg·K)	Heat conductivity / (w/m·K)	Combustion heat/(kJ/kg)	Ignition temperature/°C
Copper	8940	0.38	387	/	/
XLPE	1500	1.5	0.2	2500	330

Table 16.2 Fire growth factor

Fire spread rating	Time/s to reach 1 MW	α/(kW·s^{-2})
Slow fire	584	0.0029
Medium speed fire	292	0.01127
High speed fire	146	0.04689
Ultra rapid fire	73	0.01876

Table 16.3 Cable size parameters [17]

/	Materials	Size/mm
Cable diameter	/	98
Cable core diameter	Copper	30
Insulation thickness	PVC	28
Sheath layer thick-ness	PVC	6

Table 16.4 Fire source location working conditions

Working condition number	Location of fire source
Working condition I	Bottom end of cable
Working condition II	Middle bottom of cable
Working condition III	Middle end of cable
Working condition IV	Middle middle of cable
Working condition V	Top end of cable
Working condition VI	Top middle of cable

Table 16.5 Fire source power conditions [18]

Working condition number	Fire source power
Working condition VII	1.5 MW
Working condition VIII	3 MW
Working condition IX	4.5 MW
Working condition X	6 MW

Cable joints are the high incidence of fire, so the parameters of the insulation layer at the joints set the fire source, the length of the fire source is 0.28 m, and the area of the fire source is 0.028 m^2 (Tables 16.3, 16.4 and 16.5).

16.3 Results and Discussion

16.3.1 Fire Characteristic Gas Diffusion Law

Figure 16.3 shows the change of Hcl concentration in working condition one. From the figure, it can be seen that the fire source location in the bottom end of the case, 100 s, the cable tunnel Hcl has appeared in the phenomenon of stratification, the upper layer of Hcl concentration is larger, while the Hcl has not diffused to the bottom of the tunnel, 200 s, the Hcl gas stratification is obvious, due to the role of the thermal plume, the port at the highest concentration of Hcl, 300 s when the Hcl is filled with the entire cable tunnel, the fire source location below the lowest Hcl concentration. concentration is the lowest. Figure 16.4 shows the change of Hcl concentration in

Case II. As can be seen from the figure, the fire source location in the bottom side of the middle of the case, 100 s, Hcl has appeared in the cable tunnel layering phenomenon, compared to Case 1, Hcl concentration of the two layers of the side of the more uniform, Hcl has not yet diffused to the bottom of the tunnel, 200 s, the Hcl gas layering is obvious, due to the role of the thermal plume, the Hcl concentration from the top of the tunnel downward attenuation, 300 s when the Hcl filled the whole In 300 s, Hcl fills the whole cable tunnel, and the lowest Hcl concentration is below the fire source position, compared with Case 1, the concentration is more uniform on both sides, and the temperature below the fire source at the bottom of the tunnel is lower than that at the tunnel port under the influence of the thermal plume. Comparison of Figs. 16.3, 16.4, 16.5, 16.6, 16.7, 16.8, Hcl concentration trend is similar to Case 1 and 2, the fire source location is closer to the bottom, the Reynolds number of smoke flow, turbulence phenomenon is intense, the fastest movement, the fire characteristics of the gas filled with high-voltage cable tunnel time is the shortest. The closer the location of the fire source to the top layer, Hcl concentration stratification phenomenon is more obvious, the fire characteristics of the gas diffusion rate is low, in t = 300 s, Hcl has not yet completely filled with high-voltage cable tunnel. Comparison of Figs. 16.9, 16.10, 16.11, 16.12, with the increase of fire source power, Hcl concentration is more uniformly distributed in the chamber, the concentration below the fire source is slightly lower, and the time for the upstream and downstream of the fire source to reach a higher level of Hcl concentration is shorter.

t=100s t=200s t=300s

Fig. 16.3 Variation of Hcl concentration in Case I

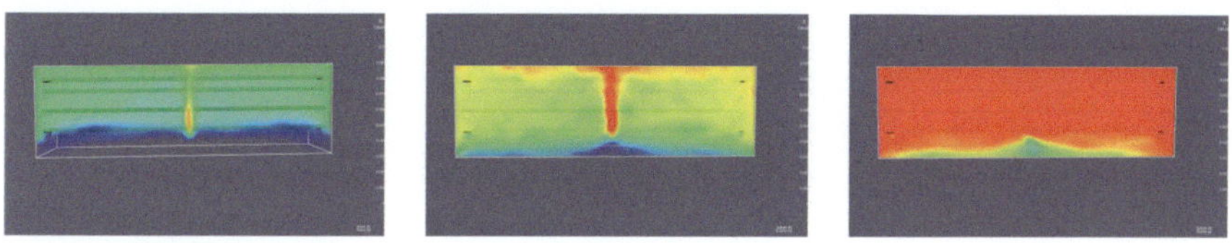

t=100s t=200s t=300s

Fig. 16.4 Variation of Hcl concentration in Case II

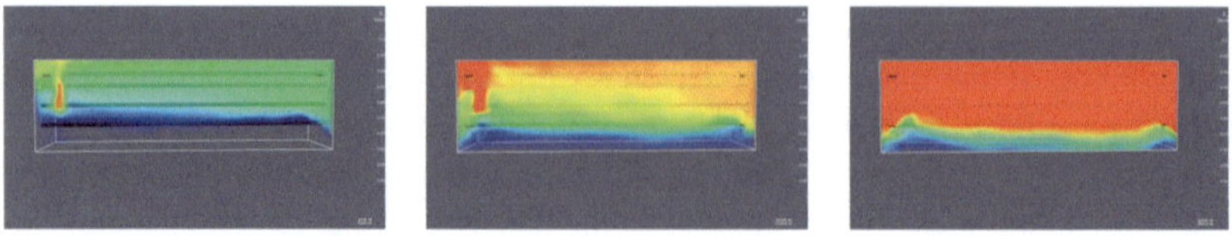

t=100s t=200s t=300s

Fig. 16.5 Variation of Hcl concentration in Case III

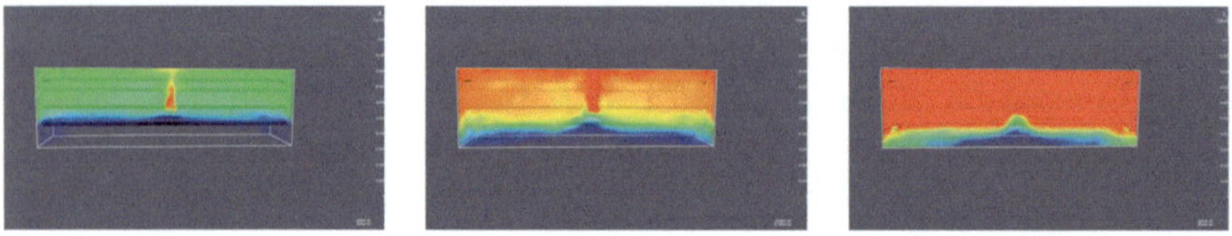

t=100s t=200s t=300s

Fig. 16.6 Variation of Hcl concentration in Case IV

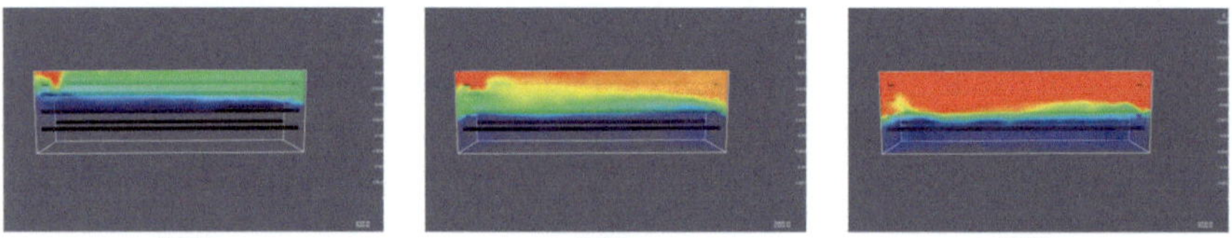

t=100s t=200s t=300s

Fig. 16.7 Variation of Hcl concentration for Case V

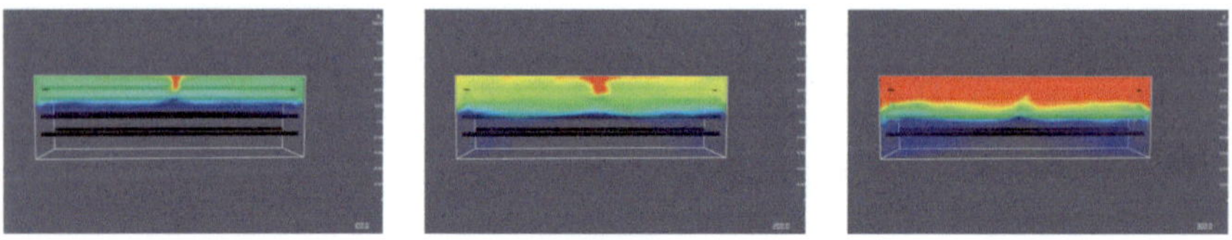

t=100s t=200s t=300s

Fig. 16.8 Variation of Hcl concentration in Case VI

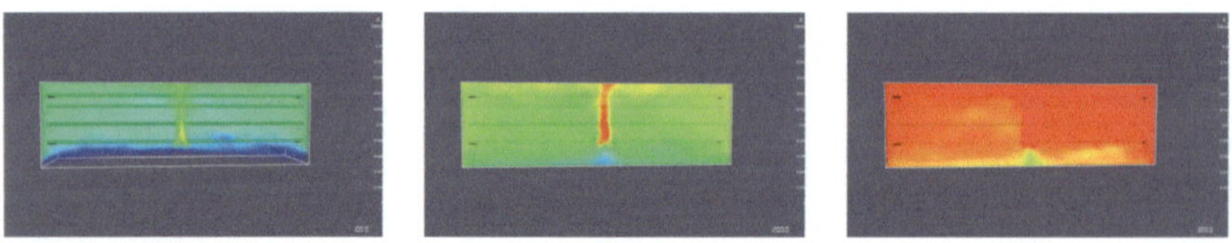

t=100s t=200s t=300s

Fig. 16.9 Variation of Hcl concentration in Case VII

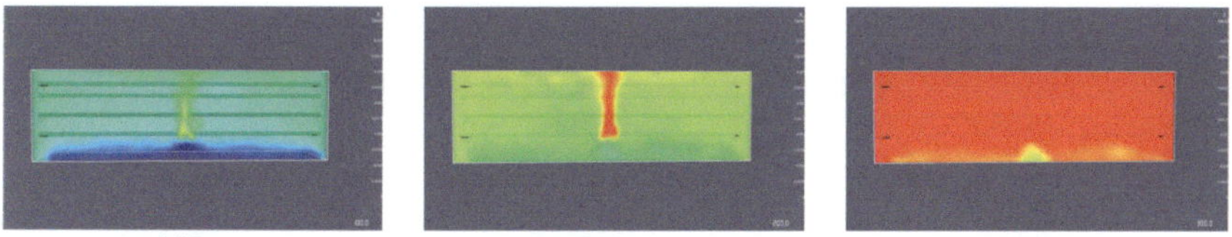

t=100s t=200s t=300s

Fig. 16.10 Variation of Hcl concentration in Case VIII

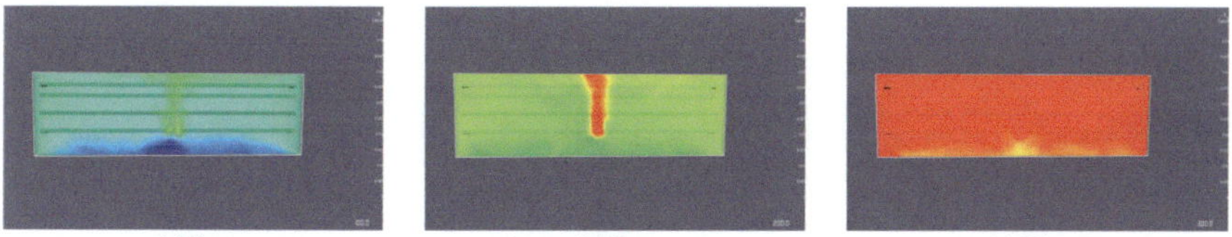

t=100s t=200s t=300s

Fig. 16.11 Variation of Hcl concentration in Case IX

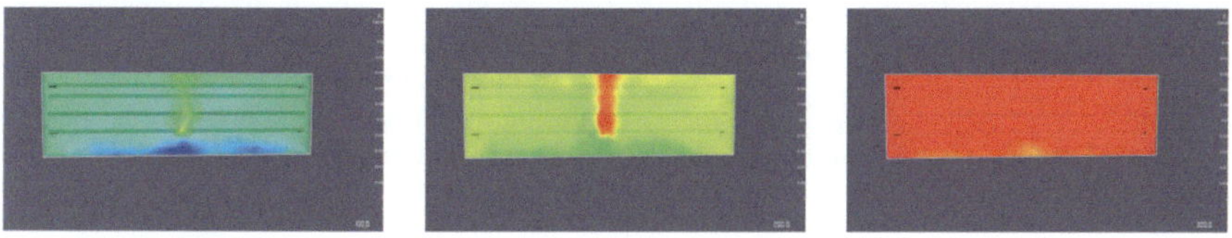

t=100s t=200s t=300s

Fig. 16.12 Variation of Hcl concentration in Case X

16.4 Conclusion

In this paper, based on the actual high-voltage cable tunnel parameters, a high-voltage cable tunnel fire model was designed. The characteristic gas combustion reaction is introduced, and the influence of the fire source parameters on the characteristic gas diffusion law of high-voltage cable tunnel fire is analyzed by changing the fire source parameters, and the following conclusions are obtained:

(1) The closer the location of the fire source is to the bottom layer, the Reynolds number of the smoke flow is large, the turbulence phenomenon is intense, the faster the smoke flow rate, the shortest time for the fire characteristic gas to fill the high-voltage cable tunnel, and the closer the location of the fire source is to the top layer, the lower the diffusion rate of the fire characteristic gas.

(2) The fire rises slowly in the early stages before 25 s, then increases rapidly and levels off after 100 s. The closer the location of the fire source is to the top of the tunnel, the more significant the temperature difference in the tunnel and the lower the gas diffusion rate. The higher the power of the fire source, the faster the

temperature rises and the more uniform the distribution of Hcl concentration. When the power of the fire source reaches 6 MW, the distribution of the fire characteristic gas in the tunnel is uniform.

(3) At t = 300 s, Hcl has not yet completely filled the high voltage cable tunnel. As the power of the fire source increases, the Hcl concentration is more evenly distributed in the chamber, with a slightly lower concentration below the fire source and a shorter time for the Hcl concentration above the fire source to reach a higher level.

References

1. Zheng X, Cai G, Guo J, et al. Combustion characteristics and thermal decomposition mechanism of the flame-retardant cable in urban utility tunnel[J]. Case Studies in Thermal Engineering, 2023, 44: 102887.
2. Liang K, Hao X, An W, et al. Study on cable fire spread and smoke temperature distribution in T-shaped utility tunnel[J]. Case Studies in Thermal Engineering, 2019, 14: 100433.
3. Wang Fangshun. Research on CFD numerical simulation of cable fire in urban comprehensive pipeline corridor[J]. Journal of Armed Police College,2017,33(08):14–18.
4. En-ze Y, Yi-ming S. Study on the Distribution of Fire Temperature Field of L-type Cable Corridor through Numerical Simulation[J]. Procedia Engineering, 2018, 211:861–870.
5. An W, Tang Y, Liang K, et al. Study on temperature distribution and CO diffusion induced by cable fire in L-shaped utility tunnel[J]. Sustainable Cities and Society, 2020, 62: 102407.
6. Mi H, Liu Y, Jiao Z, et al. A numerical study on the optimization of ventilation mode during emergency of cable fire in utility tunnel[J]. Tunnelling and Underground Space Technology, 2020.
7. Wang P, Zhu G, Pan R, et al. Effects of curved sidewall on maximum temperature and longitudinal temperature distribution induced by linear fire source in utility tunnel[J]. Case Studies in Thermal Engineering, 2020, 17:100555.03403.
8. PENG Yuhui. Numerical simulation of smoke movement law under typical cable fire conditions[J]. Marine Engineering, 2016, 45(2):5.
9. LI Chenying, CHEN Jie, LI Hongze, et al. Experimental study on fire spread and detection in underground comprehensive pipeline corridors[J]. Fire Science and Technology,2019,38(09):1258–1261.
10. WANG Mingnian, TIAN Yuan, YU Li, et al. Numerical simulation of cable fire in urban comprehensive pipeline corridor and analysis of influencing factors[J]. China Safety Production Science and Technology,2018,14(11):52–57.
11. An, Weiguang, Yanhua Tang, Kai Liang, Minglun Cai, Tao Wang, and Zhe Wang. "Study on temperature distribution and CO diffusion induced by cable fire in L- shaped utility tunnel." Sustainable Cities and Society 62 (2020): 102407.
12. Wang, Shuhan, Yongkang Zhang, Ling Wang, Qi Yuan, Chenbin Wu, Cheng Wang, and Jianbin Zang. "Fire simulation at the end of cable tunnel based on high voltage cable burning experiment." In Civil Engineering and Disaster Prevention, pp. 701–712. CRC Press, 2023.
13. ZHANG Zhikui. Research on numerical simulation and safe evacuation of building fire based on Pyrosim and Pathfinder [D]. Tianjin University of Technology, 2015.
14. Roy Gong, Guo Shou, Cheng Yang, et al. Research on the effect of fire source location on high voltage cable trench fire based on FDS[J]. Industrial Safety and Environmental Protection,2023,49(02):21–25.
15. BOWEN D, HAIMING F U. Influence of Air Curtain and Mechanical Smoke Extraction on Smoke Diffusion in Tunnel Fire[J]. Journal of Donghua University (Natural Science), 2019.

16. LIANG Y, QI X, ZhI X, et al. An experimental study on critical velocity in sloping tunnel with longitudinal ventilation under fire - ScienceDirect[J]. Tunnelling and Underground Space Technology, 2014, 43(1): 198–203.
17. Wang J , Chen B , Li Y ,et al.Multiparameter warning of lithium-ion battery overcharge-thermal runaway[J].Journal of Energy Storage, 2024.
18. Zhu X, Wu J, Bai Y, et al. Integrating FBN and FDS for quantitative risk assessment of cable fire in utility tunnel[J]. Journal of Loss Prevention in the Process Industries, 2024, 88: 105266.

Chapter 17
Simulation Analysis of the Influence of Corona Charge on Wind Turbine's Lightning Shielding Characteristics Under High-Speed Airflow

Qianhui Yang, Yeqiang Deng, Chaoying Fang, Yu Wang, Lei Lan, and Jun Xu

Abstract Lightning disaster seriously threatens the stable operation of wind turbines. However, there is still a lack of research on the influence of high-speed airflow on wind turbines' corona charge distribution and lightning shielding characteristics. Therefore, this paper quantitatively analyzes the influence of high-speed airflow on the space charge and potential of the wind turbine blade tip. Furthermore, a calculation model of wind turbine lightning shielding considering the influence of corona charge under high-speed airflow is established to analyze the influence of high-speed airflow on lightning striking distance and shielding efficiency. The results show that the corona charge will migrate toward the lightning cloud and airflow direction under the electric field and high-speed airflow combined. At the same time, high-speed airflow will lead to an increase in lightning strike distance. With the increased airflow velocity, the shielding efficiency of the lightning arrester on wind turbine tip is improved. As the size of the wind turbine becomes larger, the shielding effect of the blade tip lightning arrester is weakened, and the high-speed airflow improves the shielding efficiency of the arrester more significantly.

Keywords Corona charge · Wind turbine · Lightning · Shielding characteristics

---

Q. Yang · Y. Deng (✉) · Y. Wang · L. Lan
School of Electrical Engineering and Automation, Wuhan University, Wuhan, China
e-mail: dengyeqiang@whu.edu.cn

C. Fang · J. Xu
State Grid Fujian Electric Power Company Limited Research Institute, Fuzhou, China

Y. Li et al. (eds.), *Proceedings of the 5th Chinese Conference on High Voltage and Discharge Plasmas*, Springer Proceedings in Physics 254,
https://doi.org/10.1007/978-981-96-5711-7_17

17.1 Introduction

With the increasing height of the wind turbine, lightning stroke has gradually become one of the main factors threatening the safe and stable operation of the wind farm [1]. Because the wind turbine is often in a rotating state when struck by lightning, the physical process of the wind turbine being struck by lightning may be significantly different from the static target [2, 3]. For rotating wind turbine, there is a high-speed arc-shaped relative motion between the blade tip and the surrounding air, which can be equivalent to the static blade being subjected to high-speed arc airflow with equal speed and opposite direction. The amplitude of the airflow velocity and the positive ion migration velocity in the corona region is close to each other [4], which will change the space charge distribution in the tip region and then affect the characteristics of the upward leaders. However, due to the lack of research on the physical lightning process and lightning mechanism of rotating wind turbines, the current design of wind turbine's lightning protection system mainly relies on the lightning protection experience of static targets, which can not meet the needs of wind turbine lightning protection design.

In the current mainstream application of lightning shielding calculation models, there are mainly electrical geometric and leader development models. The electrical geometric model (EGM) [5] uses striking distance to characterize the lightning ability of the target. Many scholars have carried out long-gap discharge tests under different conditions to obtain 50% discharge voltage at different gap distances and striking distance formulas with different values [6–14]. However, the current strike distance is defined as the distance of the last strike, and the development of the upward leader is not considered, which will weaken the influence of the lightning attraction of the wind turbine. Considering that the change of space charge caused by high-speed airflow will affect the initiation and development of the upward leader, simplified models such as the electrical geometry model are no longer suitable for designing wind turbine lightning protection. Therefore, some scholars have proposed to introduce the concept of striking distance into the leader development model, use the leader development method to calculate the lightning striking distance of the towering target with the change of lightning current, and combine the electrical geometry model to calculate the complex structure target and propose a calculation method for the lightning shielding characteristics of wind turbines.

Li [15, 16] proposed a simplified leader development model based on Beccera, established a two-dimensional model of wind turbines, defined the calculation method of dynamic striking distance of wind turbine blades, and obtained the striking distance formula of wind turbine blades arranged at different angles under different lightning current amplitudes. The wind turbine blade exposure distance calculation model is proposed to calculate the lightning shielding failure probability of the wind turbine blade. Based on the above method, Li [17] studied the lightning shielding between wind turbines in large-scale wind farms, gave the relationship and criterion of lightning shielding between multiple wind turbines, and studied the influence of critical environmental parameters such as temperature, air pressure, humidity and altitude on the lightning shielding efficiency of wind turbines, and proposed guidelines for the row spacing of wind turbines in wind farms.

However, in current research, the calculation of the lightning strike distance of the wind turbine does not consider the influence of the space charge change caused by the high-speed airflow, and the commonly used strike distance formula based on the leader development model is unsuitable for studying high-speed airflow's influence on the wind turbine's lightning capacity and shielding efficiency. Therefore, this paper establishes the calculation model of corona space charge distribution under the background of a thundercloud electric field, and the model is coupled with the upward and downward leader initiation development model to establish the calculation method of lightning shielding of wind turbine considering the influence of corona upward leader's starting time, the lightning striking distance, and the shielding efficiency of the lightning arrester are analyzed, which can provide a reference for the lightning shielding design of large-size wind turbine blades.

17.2 Simulation Analysis of Space Charge Distribution

When the wind turbine blade rotates, the corona charge in the tip of wind turbine will migrate, distort the potential spatial distribution. In order to simplify the calculation, the model simulates the migration process of the induced corona charge under the combined excitation of the thundercloud background electric field and the high-speed airflow on the wind turbine tip. The simulation model is established by the method in Ref. [18].

17.2.1 Analysis of the Influence of High-Speed Airflow on Charge Distribution

The positive ion space density cloud diagram around the wind turbine tip lightning arrester is drawn when t = 20 s and the high-speed airflow velocity is $W = 0,10,20$ and 30 m/s, as shown in Fig. 17.1. Under the combined action of high-speed airflow and electric field, the positive charge tends to migrate along the upper right direction, and the greater the wind speed, the more obvious the right shift.

The average charge density on the x = 0 axis at different moments and different airflow velocities are calculated, as shown in Fig. 17.2. With the increase of high-speed airflow velocity, the charge density above the lightning arrester decreases, and the decrease in charge density shows prominent saturation characteristics.

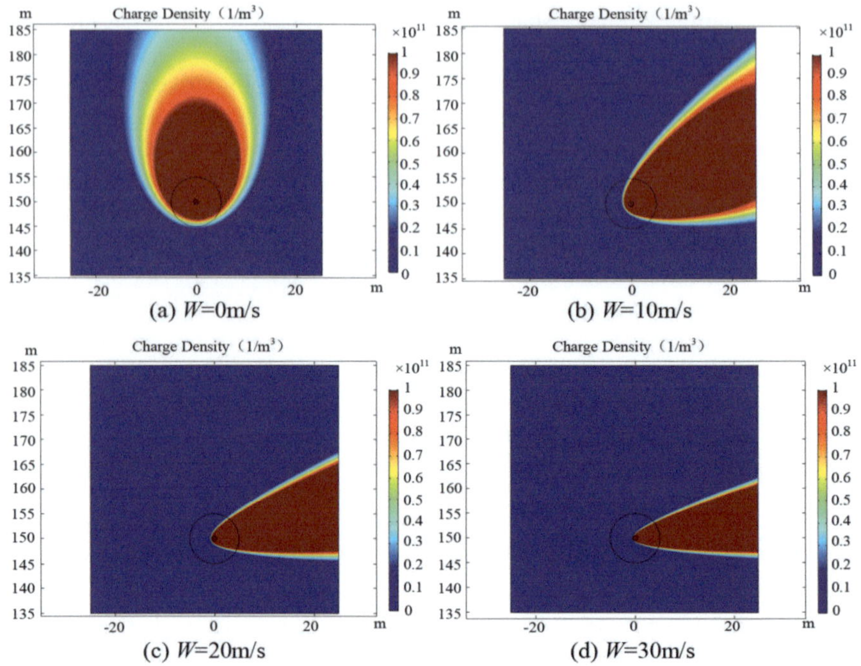

Fig. 17.1 The spatial distribution diagram of positive ion density in corona discharge of lightning arrester under different wind speeds

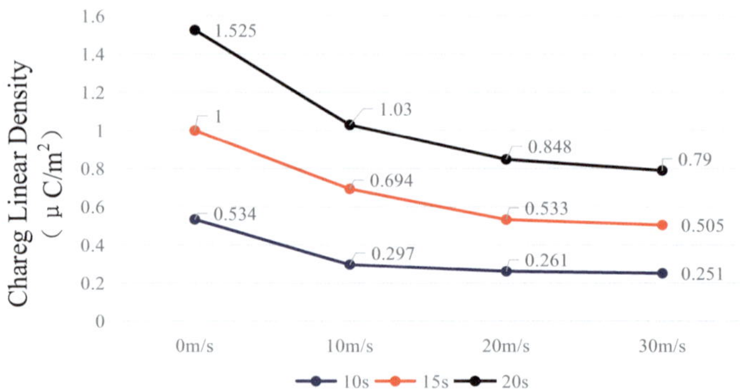

Fig. 17.2 The schematic diagram of charge density calculation results on x = 0 intercept at different times and wind speeds

17.2.2 Analysis of the Influence of High-Speed Airflow on Potential Distribution

The spatial potential distribution and electric field direction is plotted when $t = 20$ s, $W = 0$, 10, 20 and 30 m/s, as shown in Fig. 17.3. When W = 0, due to the symmetry of the spatial negative thundercloud background electric field and charge transfer, the potential distribution presents an axisymmetric distribution. The electric field direction is generally vertical upward. With the increase of airflow velocity, the charge migrates to the upper right of the lightning arrester, and the potential spatial distribution is distorted, the left potential is higher than the right potential at the same height. The direction of the maximum electric field near the lightning arrester is shifted to the left, and the electric field amplitude is also increased. With the increase of airflow velocity, this asymmetry phenomenon becomes more apparent, and the maximum electric field amplitude near the lightning arrester continues to increase.

At t = 20 s, the potential distribution within 5 m from the lightning arrester on the x = 0 axis above the lightning arrester is calculated. The results are shown in Fig. 17.4.

The results show that when the airflow velocity is 0, the potential change above the lightning arrester is the smallest. With the increase of airflow velocity, the change degree of potential above the lightning arrester is larger, and the absolute value

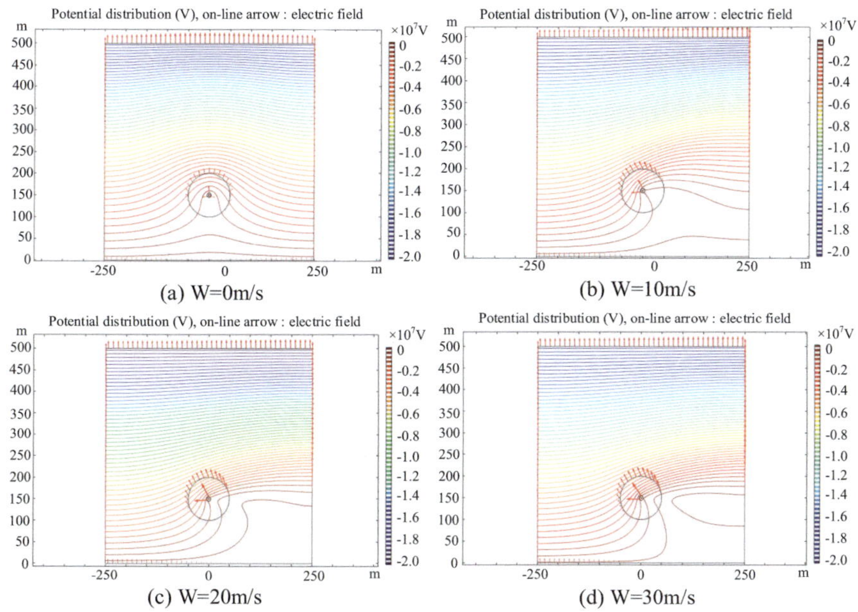

Fig. 17.3 The schematic diagram of the spatial distribution of positive ion density in corona discharge of lightning arrester under different airflow velocities at t = 20 s

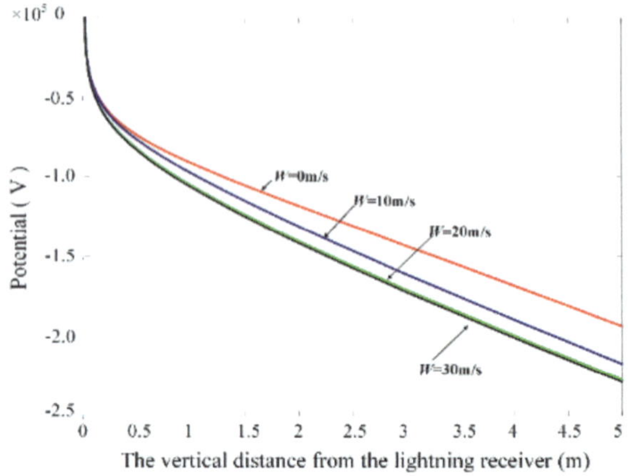

Fig. 17.4 When $t = 20$ s, the distribution curve of the line potential above the lightning arrester at different airflow velocities is plotted

of the potential at the same position is larger than that without high-speed airflow. However, with the increase of airflow velocity to a certain extent, its influence reaches a saturation trend. The analysis shows that the high-speed airflow keeps the charge away from the lightning arrester, restoring the electric field intensity at this position. When the airflow velocity reaches a certain degree and the corona discharge time reaches a specific time, most of the positive charges above the lightning arrester are far away from the tip, and the influence of the airflow velocity on the electric field reaches saturation.

17.3 Leader Development Model and Simulation Analysis Considering Tip Corona Charge

17.3.1 Negative Downward Leader Model

In this paper, the modelling of the negative leader is based on the total channel charge and charge density distribution model proposed by Cooray [19]. The downward leader adopts the vertical development model, and the downward leader development speed is selected as 2×10^5 m/s [20].

17.3.2 Upward Leader Initiation and Development Model

For the case without high-speed airflow, when considering the influence of the corona space charge, the positive charge accumulated above the target space will distort the space electric field so that the electric field in this area decreases, resulting in an increase in the potential in this area.

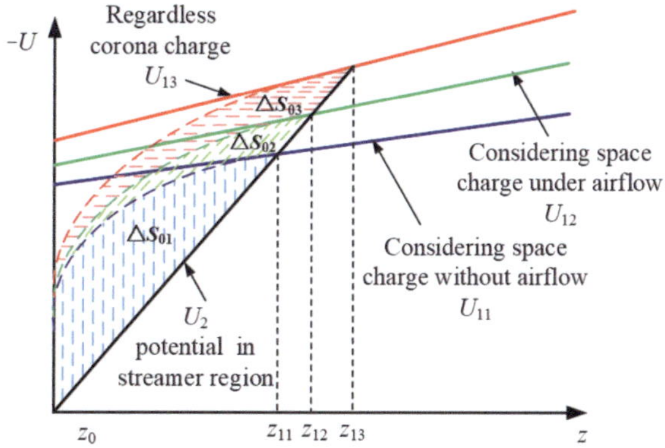

Fig. 17.5 Leader initiation model considering the influence of tip corona charge

As shown in Fig. 17.5, the potential distribution U_{13} above the target without considering the influence of space charge corresponding to the red line, and the potential at the same position is smaller than the potential distribution U_{11} considering the influence of space charge when there is no airflow corresponding to the blue line.

If the influence of space charge is considered, the high-speed airflow causes the tip corona charge to migrate, making it far away from the tip, and the space electric field recovers. The figure shows the potential spatial distribution U_{12} corresponding to the green line considering the high-speed airflow has a larger change range than U_{11}. However, due to the influence of some positive charges in space, the change range of U_{12} is still smaller than the potential distribution U_{13} without considering the influence of space charge. When considering the influence of the charge in the streamer region on the potential distribution, z_{11}, z_{12}, and z_{13} in the diagram are the length of streamer development under the influence of space charge when there is no airflow, under the influence of space charge when considering high-speed airflow and under the influence of space charge without considering space charge, respectively. At this time, the corresponding corona space charge ΔQ_{01}, ΔQ_{02}, and ΔQ_{03} are calculated as:

$$\begin{aligned} \Delta Q_{01} &= K_{\mathrm{Q}} \cdot \Delta S_{01} \\ \Delta Q_{02} &= K_{\mathrm{Q}} \cdot (\Delta S_{01} + \Delta S_{02}) \\ \Delta Q_{03} &= K_{\mathrm{Q}} \cdot (\Delta S_{01} + \Delta S_{02} + \Delta S_{03}) \end{aligned} \tag{17.1}$$

When the influence of space charge is not considered, the calculated corona charge amount ΔQ_{03} is larger than the corona charge amount ΔQ_{01} and ΔQ_{02} under the influence of space charge. Under the same calculation conditions, the time when ΔQ_{03} reaches 1 μC will be earlier than the other two, and the corresponding streamer development length z_{13} will be longer. Similarly, the space charge has the same influence on the development characteristics of the upward leader when compared with the presence or absence of high-speed airflow.

In order to reflect this effect, the treatment method in Ref. [21] is used, coupling the potential solved by the space potential equation in the leader evolution model with the initial corona space charge to the initial Laplace field potential increment. At this time, the corresponding background potential U1 can be corrected by the equation as follows:

$$U_1 = \sum_{i=1}^{N_i} U_i + \sum_{j=1}^{N_j} U_j - E_{cloud} z_1 + \Delta U \tag{17.2}$$

Among them, U_i is the potential spatial distribution of the downward lead charge in the *i* section, U_j is the potential spatial distribution of the upward lead charge in the *j* section, Ecloud z_1 is the potential distribution of the thundercloud background electric field in space, $\triangle U$ is the potential increment caused by the initial corona space charge.

17.3.3 Simulation Model

The three-dimensional model of the wind turbine is established, as shown in Fig. 17.6. The hub height is 100 m, the blade length is 50 m, and the blade material is glass fibre. A metal disc with a radius of 5 cm is arranged on the blade tip, and the metal material is copper. The down lead wire adopts a cylinder with a radius of 0.05 m, and the lightning arrester is connected to the metal hub of the wind turbine to ground it.

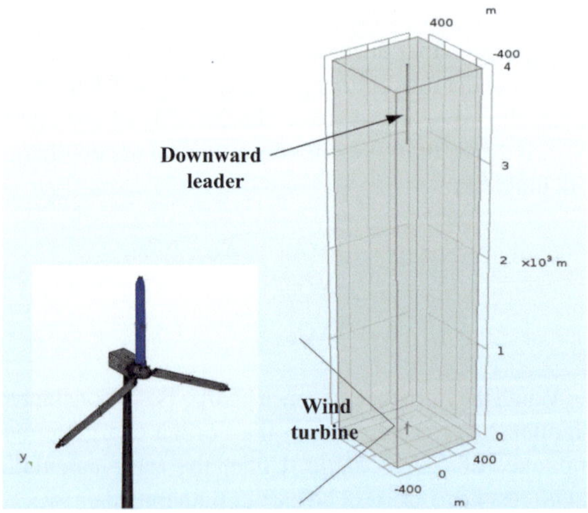

Fig. 17.6 Wind turbine model and leader development model diagram

17.3.4 Analysis of the Effect of Corona Charge on Upward Leader Initiation

Based on the upward leader model considering the influence of corona charge, the influence of tip corona charge on the initial characteristics of upward leader under different working conditions is calculated. The calculation results are shown in Table 17.1.

The calculation results show that when the corona charge is not considered, the time when the upward leader reaches the starting condition is t = 17.91 ms. The approximate distribution of the background potential and the potential distribution of the streamer area under different conditions at this time are plotted, as shown in Fig. 17.7.

At $t = 17.91$ ms, without considering the tip corona charge's influence, the background potential's approximate distribution is shown as the red line in the figure. Currently, the streamer development length z_{13} is 0.419 m, and the distance between the downward leader head and the lightning arrester is 418 m. According to the background potential red line, the black line of the potential streamer distribution, and the triangle area surrounded by the Y axis, the calculated initial corona charge is 1.006 μC greater than 1 μC, and the upward leader starts at this time.

In the absence of airflow, considering the influence of corona charge, the background potential at $t = 17.91$ ms is approximately distributed as the blue line in the diagram, and the streamer development length z_{11} is 0.214 m. Currently, the initial corona charge does not meet the leader initiation condition, and the upper leader does not start. The downward leader continues to develop. When $t = 18.36$ ms, the distance between the downward leader head and the lightning arrester is 328 m, the surface field strength is further strengthened, and the initial charge amount reaches the initial condition. In the absence of airflow, the space-accumulated charge will delay the initiation of the upward tip leader and reduce its ability to intercept the downward leader.

Table 17.1 Calculation results of discharge characteristic parameters when different working conditions reach the starting condition of the upward leader

Calculating conditions	Leader starting time (ms)	The distance between downward leader head and lightning arrester (m)	t = 17.91 ms, the streamer development length (m)
Regardless of corona charge	17.91	418	0.419
Corona charge is considered when there is no airflow	18.36	328	0.214
W = 30 m/s considering corona	18.10	380	0.320

Fig. 17.7 Potential distribution near the lightning arrester at t = 17.91 ms

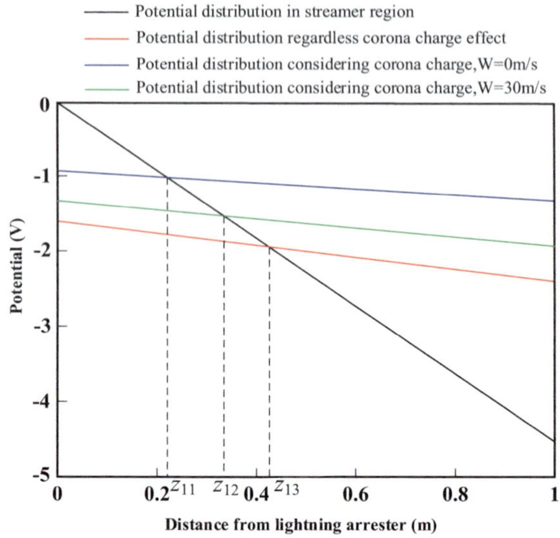

Considering the influence of the corona charge when the airflow velocity W = 30 m/s, the background potential is approximately distributed as the green line in the figure at t = 17.91 ms, and the streamer development length z_{12} is 0.320 m. Currently, the initial corona charge does not reach $1\mu C$, and the upper leader does not start. When t = 18.10 ms and the distance between the downward leader head and the lightning arrester is 380 m, the calculated initial charge reaches the initial condition of the upward leader. It can be seen that compared with the non-airflow condition, the migration of space accumulated charge caused by high-speed airflow will make the start time of tip upward leader advance by 0.26 ms.

17.3.5 Analysis of the Influence of Corona Charge on Lightning Striking Distance

Taking the blade vertical 90° as an example, the influence of tip corona space charge on lightning striking distance is analyzed, and the lightning striking distance under different conditions is calculated. The fitting curve is compared with the striking distance formula $r = 6.72I^{0.8}$ used in the classical A-W electrical geometry model [13], as shown in Fig. 17.8.

The analysis shows that the calculated lightning strike distance is the largest without considering the effect of the corona charge. In other words, if the effect of the corona charge is not considered in the simulation calculation, the lightning-attracting ability of the wind turbine lightning arrester will be exaggerated. When considering the effect of corona charge without airflow, the calculated lightning strike distance is the smallest due to the lightning shielding effect caused by the accumulation of corona

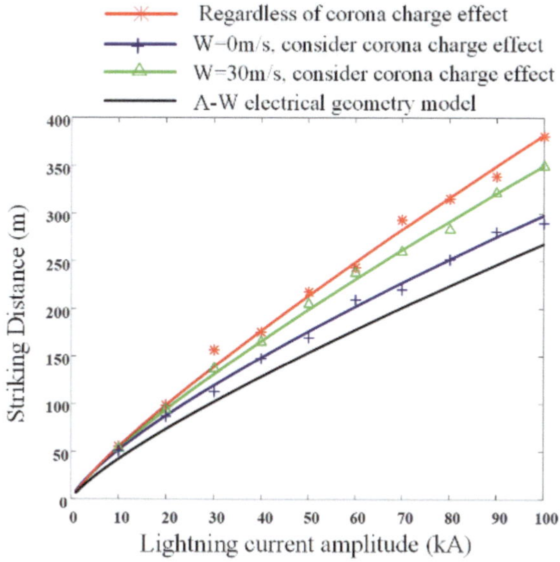

Fig. 17.8 Comparison of simulation results of lightning current amplitude lightning strike distance under different calculation methods

charge without high-speed airflow. However, wind turbine is under high-speed airflow for a long time, ignoring the influence of high-speed airflow will underestimate the lightning-attracting ability of the wind turbine lightning arrester.

17.4 Calculation Model and Simulation Analysis of Lightning Shielding Efficiency of Wind Turbine Blades

17.4.1 Calculation Model of Lightning Shielding Efficiency of Wind Turbine Blades

Based on the lightning strike distance formula of wind turbines under high-speed airflow calculated in the previous section, this section proposes a calculation method of lightning shielding efficiency considering different lightning amplitudes, lightning incident angles, and blade angles based on the method of the electrical geometry model. It quantitatively evaluates high-speed airflow's influence on the blade lightning system's shielding efficiency.

The calculation model of blade exposure and shielding distance is shown in Fig. 17.10. The dotted arc segment is the positional relationship between the exposed arc and the shielding arc of the blade under a certain lightning current amplitude. The shielding distance is defined as the horizontal distance between the farthest two points of the exposure arc at an incident angle, and the exposure distance is defined similarly. Therefore, as shown in Fig. 17.9, the maximum horizontal distance between the straight line of the vertical incident angle and each exposure arc and shielding arc is each exposure distance and shielding distance.

Under this condition, the exposure distance and shielding distance are calculated as follows:

$$W_s = |x'_B - x'_A| + \left|\frac{y'_B - y'_A}{\tan\varphi}\right|$$
$$W_e = |x'_c - x'_B| + \left|\frac{y'_C - y'_B}{\tan\varphi}\right| \quad (17.3)$$

W_s is the shielding distance of the blade, m; W_e is the exposure distance of leaves, m. The x'_A, x'_B and x'_C are the abscissa of A′, B′ and C′ points. The y'_A, y'_B and y'_C are the ordinates of A′, B′ and C′ points, and φ is the incidence angle of lightning.

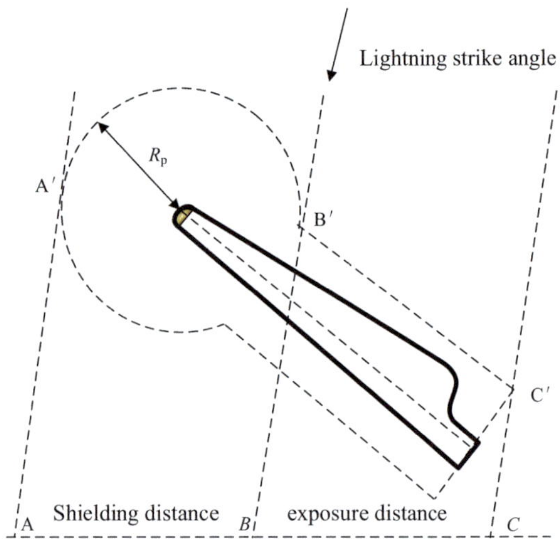

Fig. 17.9 Calculation model of blade exposure and shielding distance

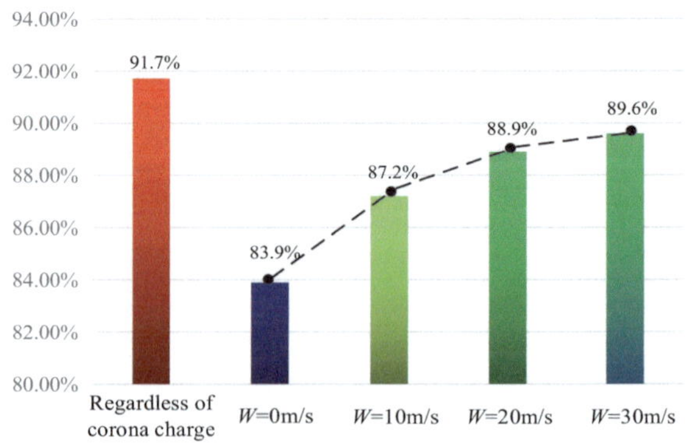

Fig. 17.10 Shielding efficiency calculation results of wind turbine blades

According to IEC 61,400-24 Technical Guidelines for Lightning Protection of Wind, Turbines [22], the protection efficiency E of the wind turbine blade flashover system is calculated by the following formula:

$$E = E_b \times E_s \tag{17.4}$$

Among them, E_b is the lightning efficiency of the wind turbine blade, and Es is the grading efficiency of the blade lighting system. Here, we generally believe the lightning system is stable and reliable enough after lightning, regardless of the lightning. After the lightning is successful, the lightning current may cause substantial damage to the blade, so the value of E_s is 1.

The blade lightning efficiency can be obtained by calculating the lightning current amplitude I, the exposure distance W_e, and the shielding distance W_s when the incident angle is ψ and the blade angle is θ. As the formula shows:

$$E_b = \frac{W_s}{W_s + W_e} \tag{17.5}$$

The calculation formula of lightning shielding efficiency of wind turbine blade lightning system is obtained by cyclic calculation of different lightning current amplitude, different lightning incident angle, and different blade lightning angles:

$$E = \frac{1}{\theta_{\max} - \theta_{\min}} \int_{I_{\min}}^{I_{\max}} \int_{\varphi_{\min}}^{\varphi_{\max}} \int_{\theta_{\min}}^{\theta_{\max}} E_b(I, \varphi, \theta) \times E_s \times P(I) \times P(\varphi) dI \cdot d\varphi \cdot d\theta \tag{17.6}$$

Among them, I_{min} is the minimum lightning current that can cause damage to the wind turbine blade. In the simulation, the damage to the wind turbine blade after lightning is not considered. Therefore, *1kA* is taken as the minimum lightning current value in the simulation. I_{max} is the maximum lightning current value. According to the observation results of lightning current, 200kA is taken as the simulation's maximum current value. *P(I)* is the probability density of the lightning current amplitude; ψ_{max} and ψ_{min} define the range of lightning incident angle. Regardless of the case of lightning negative angle from bottom to top, the range of lightning incident angle is 0–π. $P(\psi)$ is the lightning incident angle's probability density; the blade angle range is specified by θ_{max} and θ_{min}, and the blade angle is included in the simulation calculation at 0–180 °. Because the blade is simplified into an axisymmetric model, the calculation range of blade angle in the actual calculation is 0–π/2. Thus, the formula (17.8) can be expressed as follows:

$$E = \frac{2}{\pi} \int_{1}^{200} \int_{0}^{\pi} \int_{0}^{\pi/2} E_b(I, \varphi, \theta) \times E_s \times P(I) \times P(\varphi) dI \cdot d\varphi \cdot d\theta \tag{17.11}$$

17.4.2 Calculation and Comparative Analysis of the Influence of High-Speed Airflow on Lightning Shielding Efficiency of Wind Turbine Blades

The calculation results of shielding efficiency of blade tip lightning arrester under without considering the influence of corona charge, considering the influence of corona charge without airflow, and considering the influence of corona charge with different airflow velocities are compared and analyzed, as shown in Fig. 17.10.

The calculation and analysis show that the shielding efficiency of the tip lightning arrester without considering the corona charge reaches 91.7%, and the lightning probability of the tip lightning arrester considering the corona charge is only 83.9% without airflow. With the increase of airflow velocity, the shielding efficiency of the tip lightning arrester increases. Under the condition of air flow velocity of 30 m/s, the shielding efficiency of the lightning arrester is close to 6% higher than without airflow. With the increase in airflow velocity, the lifting effect shows an apparent saturation trend. It can be seen that although the high-speed airflow will enhance the lightning ability of the wind turbine, the shielding effect of the blade tip lightning arrester is also enhanced. As long as the blade lightning receiving system is safe and reliable, the lightning current is introduced into the earth, which can effectively ensure that lightning strikes do not damage the wind turbine.

Furthermore, the difference in shielding effect of high-speed airflow on wind turbine flashover under different size wind turbines is studied. The shielding efficiency of wind turbine blades with 30, 50, and 70 m blade lengths are calculated for the exact blade tip lightning arrester arrangement and the same airflow speed W = 30 m/s, as shown in Fig. 17.11.

The calculation results show that as the blade becomes longer, the shielding probability of the lightning arrester decreases with or without high-speed airflow. At this time, the second or third lightning arrester needs to be arranged to increase the

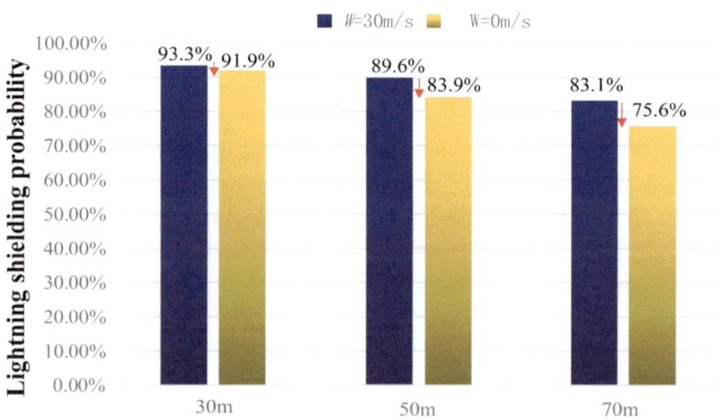

Fig. 17.11 Calculation results of the lightning probability of blade tip lightning arrester with different blade sizes and operating conditions

shielding range of the lightning arrester system. On the other hand, the longer the blade length, the more significant the difference between the shielding probability of the wind turbine under high-speed airflow and the shielding probability of the wind turbine without airflow. When the blade is 30 m, the lightning shielding efficiency of the blade tip lightning arrester is 2.4%, and the shielding efficiency of the blade tip lightning arrester is 7.5% when the blade is 70 m. The analysis shows that when the blade is short and there is no airflow, the tip lightning arrester has incorporated the wind turbine blade body into the shielding range. Although the lightning strike distance under high-speed airflow is significant, it has little effect on the result of the tip lightning probability. A single-tip lightning arrester cannot be completely shielded when the blade is long, and there is no airflow. Under high-speed airflow, the lightning strike distance becomes larger, which can increase the shielding effect of the tip lightning arrester.

Therefore, to more accurately evaluate the lightning protection effect of large-size wind turbines, it is necessary to incorporate the influence of corona charge migration under high-speed airflow into the lightning shielding design of the blade lightning receiving system.

17.5 Conclusion

(1) Based on the corona charge distribution calculation model, the influence of high-speed airflow on the space charge distribution and potential distribution of blade tip corona discharge is quantitatively analyzed. It is found that under the combined action of electric field and high-speed airflow, corona charge will migrate towards thundercloud and high-speed airflow. The space charge density on the axis above the lightning arrester decreases with the increase of airflow velocity. As the airflow velocity increases, the absolute value of the potential change on the axis directly above the lightning arrester becomes larger.

(2) The leader development model is established considering the corona charge at the wind turbine's tip. It is found that under the calculation conditions in this paper, the high-speed airflow will lead to the start time of the upward leader advance by 0.26 ms, and the lightning strike distance is the largest without considering the influence of the corona charge. Compared with no airflow condition, high-speed airflow will also lead to an increase in lightning strike distance. With the increase of current amplitude, the influence on lightning strike distance is more significant.

(3) The lightning shielding efficiency calculation model of wind turbine blades under high-speed airflow is established. The calculation of the lightning shielding efficiency of a typical MW wind turbine blade tip shows that under 30 m/s high-speed airflow, the shielding efficiency of the lightning arrester is nearly 6% larger than that of the non-airflow condition. With the increase of airflow velocity, the shielding efficiency of the blade tip lightning arrester is improved, and there is a saturation trend. As the size of the wind turbine becomes

larger, the shielding effect of the blade tip lightning arrester is weakened, and the high-speed airflow improves the shielding efficiency of the arrester more significantly.

Acknowledgements The work was supported by the Science and Technology Project of State Grid "Research on Disaster Warning and Risk Assessment Technology_for Lightning Strike in Large Wind Farms (No.5500-202221143A-1-1-ZN)".

References

1. Glushakow B. Effective Lightning Protection for Wind Turbine Generators [J]. IEEE Transactions on Energy Conversion, 2007, 22(1):214-222.
2. MONTANYÀ J, VELDE O, WILLIAMS E R. Lightning discharges produced by wind turbines[J]. Journal of Geophysical Research Atmospheres, 2014, 119(3): 1455-1462.
3. RACHIDI F, RUBINSTEIN M, MONTANYA J, et al. A review of current issues in lightning protection of new-generation wind-turbine blades[J]. IEEE Transactions on Industrial Electronics, 2008, 55(6): 2489-2496.
4. Y. Wang, Y. Deng, Y. Liu, L. Qu, X. Wen, L. Lan, J. Wang. Influence of blade rotation on the lightning stroke characteristic of a wind turbine[J]. Wind Energy, 2019, 22(8):1071-1085.
5. Golde R H. The Frequency of Occurrence and the Distribution of Lightning Flashes to Transmission Lines [J]. 1945, 64(12):902–910.
6. Wagner C F. The relation between stroke current and the velocity of the return stroke [J]. IEEE Transactions on Power Apparatus and Systems, 1963, 82(68), 609-617.
7. Golde R H. The lightning conductor [J]. Journal of the Franklin Institute, 1967, 283(6), 451-477.
8. Young F S, Clayton J M,Hileman A R. Shielding of transmission lines[J]. IEEE Transactions on Power Apparatus and Systems, 1963, PAS-82: 132–154.
9. Armstrong H R, Whitehead E R. Field and analytical studies of transmission line shielding[J]. IEEE Transactions on Power Apparatus and Systems, 1968, PAS-87(1): 270–281.
10. Brown G W, Whitehead E R. Field and analytical studies of transmission lines shielding: part II [J]. IEEE Transactions on Power Apparatus and Systems, 1969, 88(5):617-626.
11. Love R. Improvements on the lightning stroke modeling and applications to the design of EHV and UHV transmission lines [D]. Colorado, USA: University of Colorado,1973.
12. Anderson J D. Transmission line reference book:345 kV and above[M]. Palo Alto, CA: Electric Power Research Institute, 1982.
13. IEEE Working Group on Lightning Performance of Transmission Lines. A simplified method for estimating lightning performance of transmission lines [J]. IEEE Transactions on Power Apparatus and Systems, 1985, PAS-104(4): 919–932.
14. Taniguchi S, Tsuboi T, Okabe S, et al. Improved method of calculating lightning stroke rate to large-sized transmission lines based on electric geometry model [J]. IEEE Transactions on Dielectrics and Electrical Insulation, 2010, 17(1), 5412002.
15. WANG Guozheng, ZHANG Li, GUO Zixin, et al. Dynamic Striking Distance and Electrical Geometry Model of Wind Turbine Blades Based on Lightning Physics [J]. Proceedings of the CSEE,2017,37(21):6427–6436.
16. GUO Zixin, LI Qingmin, REN Hanwen, et al. Probabilistic Risk Assessment of Lightning Strike on Wind Turbine Blades Based on the Physical Mechanism of Lightning Upward Leader Inception [J]. Proceedings of the CSEE,2018,38(02):653–662+696.
17. ZHANG Li, ZHANG Yao, WANG Guozheng, et al. Lightning Shielding of Multiple Wind Turbines Based on Lightning Physics and Optimized Spatial Allocation of Wind Farm [J]. Proceedings of the CSEE,2018,38(18):5335–5342.

18. Qu L , Wang Y, Liu G , et al. Simulation Study on Positive Corona Discharge of Receptors on Rotating Wind Turbine Blade Tips under Thundercloud Electric Fields[J]. Energies, 2019, 12(24):4696.
19. Cooray V, Becerra M. Attractive radii of vertical and horizontal conductors evaluated using a self-consistent leader inception and propagation model—SLIM [J]. Atmospheric Research, 2012, 117(11), 64-70.
20. Petrov N I, Alessandro F D. Theoretical analysis of the processes involved in lightning attachment to earthed structures. J. Phys. D: Applied Physics, 2002, 35(14): 1788-1795
21. Gallimberti I. The mechanism of the long spark formation[J]. IEEE Transaction on Power Delivery, 1979, 40(C7):193-250.
22. IEC 61000–24: 2010 Wind Turbine generator systems. Part 24: Lightning protection [S]. IEC,2010.

Chapter 18
Simulation Study of Lightning Direct Effects on Aircraft with Different Material Structures

Chenghang Yu, Zhiliang Wang, Yunhua Jiao, Xiao Xu, Hongxu Liu, Honglei Sui, Xueling Yao, Jinru Sun, and Yupeng Chai

Abstract In order to study the electric field and current distribution of the aircraft with different materials and structures when it is struck by lightning in flight, the electric field distribution and current distribution of the aircraft with different materials and structures has been studied in the paper. A full-scale aircraft model with different materials is established, the distribution of lightning field is simulated by using electrostatic field technology. The influence of material characteristics on the distribution and intensity of electric field on the surface of aircraft is investigated, and the lightning attachment rule is analyzed. The simulation results show that the electrode position has an effect on the electric field distribution, the maximum electric field strength of the aircraft is approximately at the electrode position 23. Secondly, the material of aircraft has influence on its electric field strength and distribution. The electric field strength of pure metal material is the highest about 9703 V/m. The electric field strength of composite material aircraft is about 4573 V/m. Based on the above analysis, the maximum probability of the aircraft's lightning attachment partition is attached to the fairing tip area, and current conduction path. This study can provide a basis for the calculation of the electric field development process of aircraft and a theoretical basis for the protection design of aircraft in lightning environment.

Keywords Composite aircraft · Material type · Electric field distribution · Lightning attachment zone

---

C. Yu · Z. Wang · Y. Jiao · X. Xu · H. Liu · H. Sui
Beijing Institute of Electromechanical Engineering, Beijing, China

X. Yao · J. Sun · Y. Chai (✉)
State Key Laboratory of Electrical Insulation of Power Equipment, Xi'an Jiaotong University, Xi'an, China
e-mail: 2663435538@qq.com

Y. Li et al. (eds.), *Proceedings of the 5th Chinese Conference on High Voltage and Discharge Plasmas*, Springer Proceedings in Physics 254,
https://doi.org/10.1007/978-981-96-5711-7_18

18.1 Introduction

Lightning phenomenon is a kind of discharge phenomenon that occurs in nature from time to time, with the characteristics of large instantaneous current, high voltage, and strong destructive force [1], and the aircraft will inevitably approach or even traverse the thunderstorm area during flight, which is a major threat to the safe and stable flight of the aircraft [2, 3].

With the improvement of the performance requirements of the aircraft, due to the new composite materials such as corrosion resistance, high specific strength, and light specific weight have been applied to the aircraft [4, 5], and the aircraft is developed from pure metal materials to composite aircraft. However, due to the large difference in electrical conductivity and dielectric constant between insulating materials and metallic materials [6], which leads to changes in the electric field distribution of composite-type vehicles in their lightning environments [7]. In addition, compared to metal materials with good electrical conductivity, composite materials produce a large amount of Joule heat in a short time when they are struck by lightning, and the composite materials show damage such as ablation, internal delamination or even fracture [8–10]. Secondly, the damage of insulating glass fiber material is more than the lightning damage of carbon fiber composites [11], and the current in the fastener-containing composites internal conduction will be electric spark phenomenon [12, 13], easy to become a source of ignition. Finally, the mechanical properties of composites under lightning environment are degraded [14, 15], which leads to the overall performance degradation of the vehicle structure. Therefore, it is necessary to study the direct effect of lightning on aircraft, and the premise of the direct effect of lightning research is to determine the lightning zoning and current conduction paths, and analyze the environment of the aircraft lightning effect.

According to the SAE ARP5412 [16] standard, it is found that each area of the aircraft may be subjected to lightning strikes, but the degree and probability of lightning strikes are different, thus requiring different protective measures. The SAR 5414 [17] standard gives the zoning recommendations of a variety of different structures of aircraft such as airliners, rotorcraft, etc., and it is also found that the zoning of aircraft is affected by the structure of the aircraft model [18]. The new composite materials in the structure and material uniform existing standards in the aircraft type has differences, so the new structure and material aircraft lightning attachment zoning research, for its lightning protection design is particularly important.

China for the aircraft lightning zoning research started late, but its lightning environment to build and simulation research development is rapid, but for the aircraft lightning strike test laboratory environment needs high voltage lightning test equipment, its cost is high [19], the experimental process of the test methodology and other flight is also limited by the equipment [20]. Therefore, on the basis of existing experiments and standards, many scholars at home and abroad have used electrostatic field simulation and calculation methods to simulate the lightning attachment region of different structural vehicles [21–23]. Huang Liyang [21] and other scholars use comsol simulation software to calculate the "Black Hawk"helicopter and analyze its electric field distribution; Nie Ru [22] uses CST electromagnetic simulation software and the probe setting method to simulate the process of the aircraft suffered from lightning strikes. At the same time, scholars studied the influence law of different electrode position [23] factors on the electric field distribution of the aircraft, and finally used the threshold method, numerical fitting, etc. to divide the simulation results of the lightning distribution of the aircraft, to get the lightning partitioning recommendations for the required research models. The electric field distribution of the aircraft is affected by the model and material, and in the current simulation study lacks the simulation calculation of the composite aircraft combining composite and metal materials.

Based on the above research status, this paper establishes the electric field simulation model of a pure metal aircraft under 37 typical electrode positions according to the lightning strike area division experiments carried out in the SAE ARP5412 [16] standard, establishes the electric field distribution simulation model of a composite aircraft on this basis, studies the influence of the material change on the electric field strength of the aircraft, and determines the electric field distribution and lightning strike zoning proposal of a composite aircraft. The electric field distribution of a composite aircraft is determined and the lightning strike zoning is proposed to analyze the possible current conduction path of the aircraft when it is struck by lightning, so as to provide a theoretical reference basis for its protection design.

18.2 Simulation

18.2.1 Control Equations

Control Equations for Electrostatic Field Simulation

CST simulation software is used to simulate the electric field distribution of the vehicle. When the voltage acts on the vehicle, the electric field distribution is calculated using the electrostatic field simulation, and the differential form of the electrostatic field equation is shown in Eq. (18.1):

$$\begin{aligned} \nabla \times E &= 0 \\ \nabla \cdot D &= \rho \end{aligned} \tag{18.1}$$

E—electric field strength; D—electric flux density; ρ—free charge density. Where the flux density is related to the dielectric constant and the electric field as follows:

$$\begin{aligned} D &= \varepsilon_0 \varepsilon_{\mathrm{r}} E \\ E &= -\nabla U \end{aligned} \tag{18.2}$$

ε_0– vacuum dielectric constant; ε_0– relative dielectric constant; U – electrostatic field potential. Thus the Poisson equation for the electrostatic potential is given in Eq:

$$\begin{aligned} \nabla^2 U &= -\frac{\rho}{\varepsilon_0 \varepsilon_\mathrm{r}} \\ \nabla^2 U &= 0 \end{aligned} \tag{18.3}$$

For a region of space where no charge exists, Eq. 18.3 is written as:

$$\begin{aligned} \nabla^2 U &= -\frac{\rho}{\varepsilon_0 \varepsilon_\mathrm{r}} \\ \nabla^2 U &= 0 \end{aligned} \tag{18.4}$$

Based on the electrostatic field equations, simulations are carried out to calculate the electric field distribution of the vehicle.

Current Conduction Control Equation

The lightning current satisfies Ampere's law and the current conservation equation during conduction in the vehicle, as shown in Eq:

$$\begin{aligned} \nabla \times (\mu_0^{-1} \mu_\mathrm{r}^{-1} \vec{B}) &= \vec{J_\mathrm{e}} + \sigma \frac{\partial \vec{A}}{\partial t} + \sigma \vec{v} \times \vec{B} \\ \vec{J} &= \vec{J_\mathrm{e}} + (\sigma + \varepsilon_0 \varepsilon_\mathrm{r} \frac{\partial}{\partial t}) \vec{E} \\ \vec{B} &= \nabla \times \vec{A}, \vec{A} = -\nabla U \end{aligned} \tag{18.5}$$

$\vec{J_\mathrm{e}}$—applied current source current density; J—unit current density; U—electric potential; ε_0—vacuum dielectric constant; ε_r—relative dielectric constant; μ_0—vacuum permeability; μ_r—relative permeability; $\vec{A}$—magnetic vector; $\vec{v}$—velocity.

18.2.2 Flight Vehicle Model

With reference to a model of a certain type of aircraft model, combined with the actual engineering and assembly needs, using UG software segmentation of the model modeling, the model diagram is shown in Fig. 18.1.

As shown in Fig. 18.1 the length of the vehicle is 14.6 m, the length of the wing is 7 m, and the diameter of the cabin is about 0.5 m. After the assembly is completed, the simulation calculations are completed under electric and magnetic fields using the 3D electromagnetic analysis software CST.

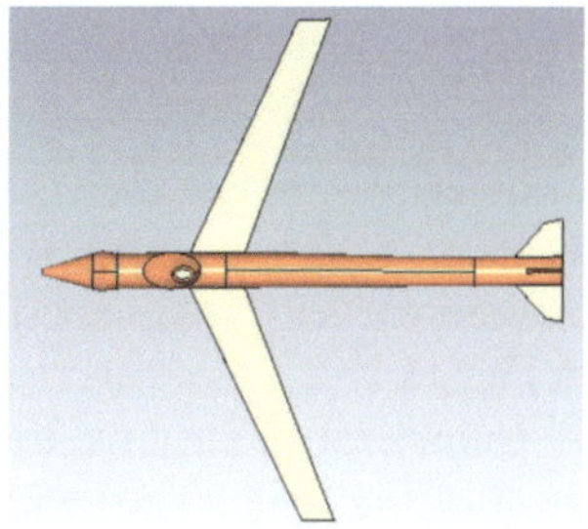

Fig. 18.1 Model of the aircraft

18.2.3 Vehicle Material Parameters

Vehicle materials are divided into two kinds, quartz (silicon dioxide) hybrid materials, titanium alloys established by the vehicle model, different materials vehicle parameters as shown in Table 18.1.

The material of each part of the hybrid material composite vehicle consists of titanium alloy and quartz (silicon dioxide) material, as shown in Fig. 18.2.

The yellow portion of the vehicle in Fig. 18.2 is a quartz blend and the orange portion is a titanium alloy.

Table 18.1 Material parameters

Material Properties	Relative dielectric constant	Conductivity (elec.)(S/m)	Conductivity (mag.)(1/Sm)	Relative magnetic permeability	Density (kg/m^3)
Sapphire	3.9	1e-15	1e-7	1	5.8e2
Titanium	30	1e6	1e-6	1	4.5e3

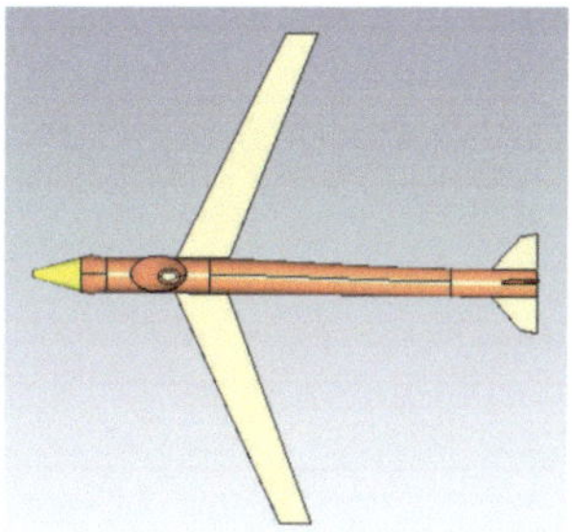

Fig. 18.2 Material layout for each part of the aircraft

18.2.4 Selection of Electrode Position

In order to simulate the electric field distribution when the vehicle is in different attitudes, high voltage electrodes are used to simulate the thundercloud effect based on the SAE ARP5412 experimental standard, and the electrode positions are shown in Fig. 18.3.

Numerical characterization of angles θ and Φ of the electrode position arrangement in Fig. 18.3. The electrode positions are varied to analyze the electric field distribution of the vehicle under different conditions for the simulation of the 37 relative positions in Fig. 18.4.

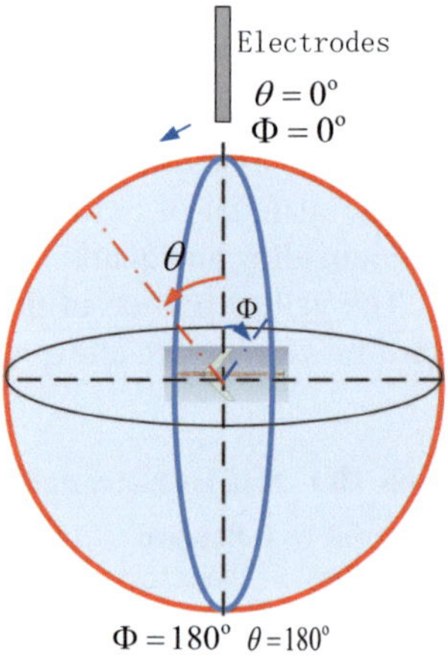

Fig. 18.3 Diagram of electrode position selection

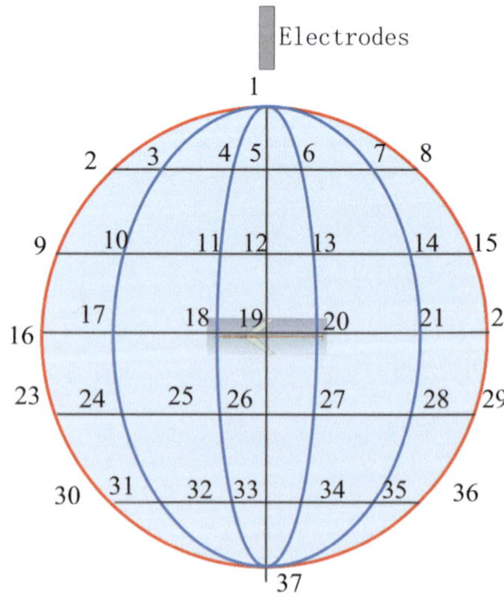

Fig. 18.4 Diagram of 37 electrode position

The different electrode positions in Fig. 18.4 simulate the flight attitude of the vehicle at different moments, where 1–15 shows the vehicle below the cloud, 16–22 shows the vehicle crossing the cloud, and 23–37 shows the vehicle above the cloud.

18.3 Simulation Results of Electric Field Distribution of the Vehicle

18.3.1 Electric Field Distribution in Metallic Vehicles

In this paper, according to the above simulation calculation model, the electric field distribution results of the vehicle are obtained, and the electric field distribution results of the vehicle under the more typical electrode positions are selected for analysis.

Electric Field Distribution When the Electrodes Are Located Directly in Front of the Vehicle

Combining the positional relationship between the electrode and the vehicle in Figs. 18.3 and 18.4, the thundercloud is located in the direction directly in front of the vehicle (position 16), the electrode voltage is 1 MV, and the result of the electric field distribution of the vehicle is shown in Fig. 18.5.

In Fig. 18.5, the tip region and electrode position have a great influence on the electric field distribution of the aircraft in the lightning environment, and the tip near the electrode position is more likely to become a charge gathering area; under the action of this lightning electrode position, the strongest electric field distribution region is located at the tip of the radome at about 8084 V/m, and the electric field intensity is also larger in the region of the top compartment section, and the tip of the tail as well as the wing is still stronger despite the fact that the electrode is relatively far away from the electrode position. The tips of the tail and the wing, although relatively far away from the electrode location, still have a strong E-field distribution in the tip part. Finally, the connection between the center section and the wing is smooth and has the smallest E-field distribution.

Based on the simulation of the electric field distribution of the vehicle, the charge density distribution of the vehicle is obtained as shown in Fig. 18.6.

It can be seen from the charge distribution of the vehicle in Fig. 18.6 that the vehicle charge distribution is similar to the electric field distribution, and the maximum charge distribution region of the vehicle is still in the front position of the radome. The charge density at the front end of the radome is 8.8e-8 C/m^2 and 2.55e-8 C/m^2 at the top compartment section.

From the analysis of electric field distribution and charge distribution results, the tip of the radome and the front area of the aircraft are most likely to become lightning attachment points; while the areas with stronger electric field strength and charge strength, such as the top section location area and the tip of the wing and tail, are also

Fig. 18.5 The electric field distribution of the aircraft

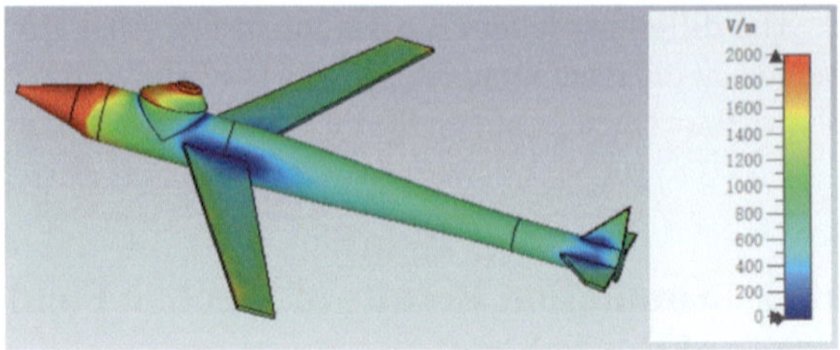

(a) Overall view

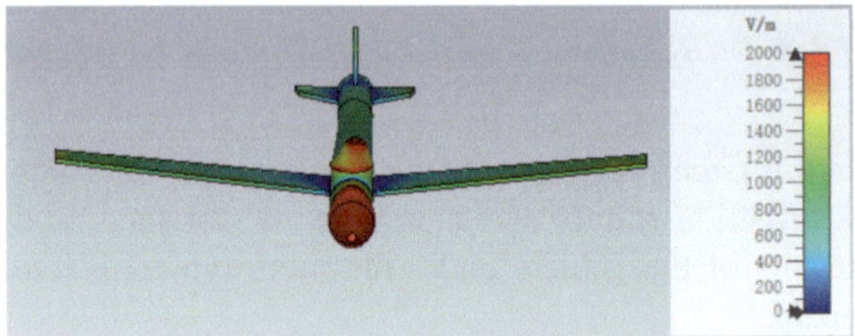

(b) Front view

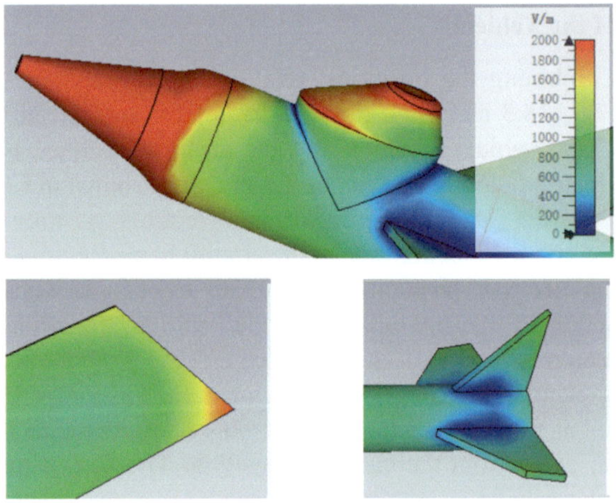

(c) Local Enlarged View

more likely to become lightning attachment points, and the intensity of the lightning environment will be slightly different.

Electric Field Distribution When the Electrode is Located Directly Above the Vehicle

Combined with the positional relationship between the electrode and the vehicle in Fig. 18.3, the thundercloud is located in the direction directly above the vehicle (position 1), the electrode voltage is 1 MV, and the electric field distribution of the vehicle is calculated as shown in Fig. 18.7.

Fig. 18.6 The charge distribution of the aircraft

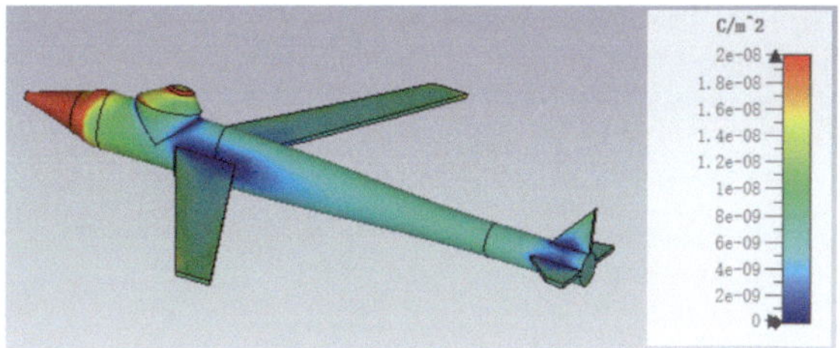

(a) Overall view

(b) Front view

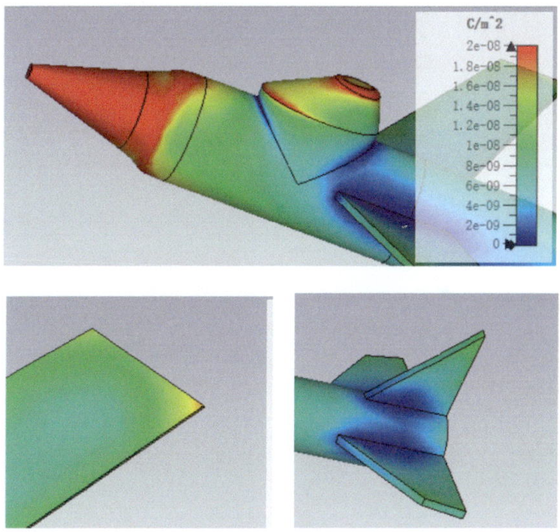

(c) Local Enlarged View

When the electrode position is located directly above the vehicle, the overall electric field strength is larger than when the electrode is located directly in front of the vehicle, the tip area and electrode position in the figure have a big influence on the electric field distribution of the vehicle in the lightning environment, and the tip of the electrode position is more likely to be the area where the electric charge is gathered; under the effect of the lightning electrode position, the electric field strength of the tip area of the radome, the area of the top section of the cabin, as well as the tip of the

Fig. 18.7 The electric field distribution of the aircraft

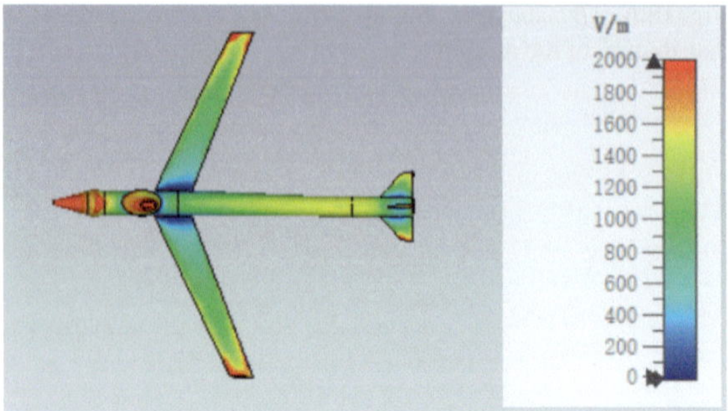

(a) View near electrode side

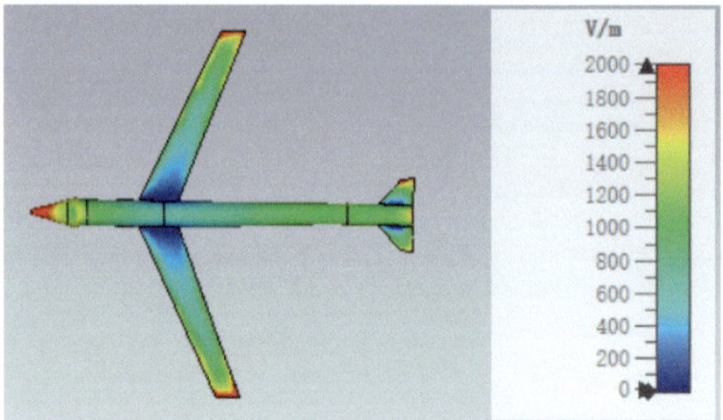

(b) View of the far electrode side

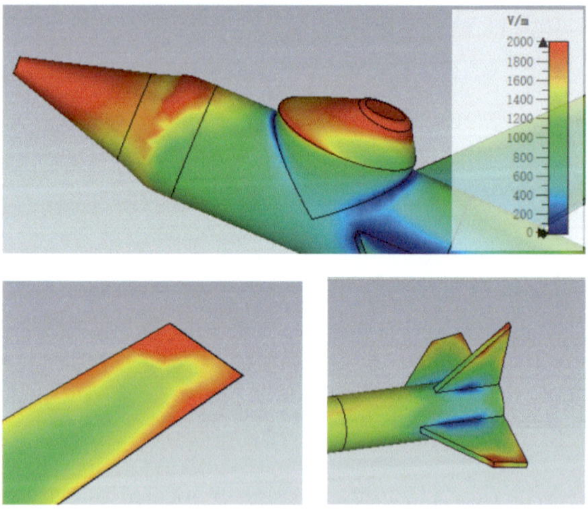

(c) Local Enlarged View

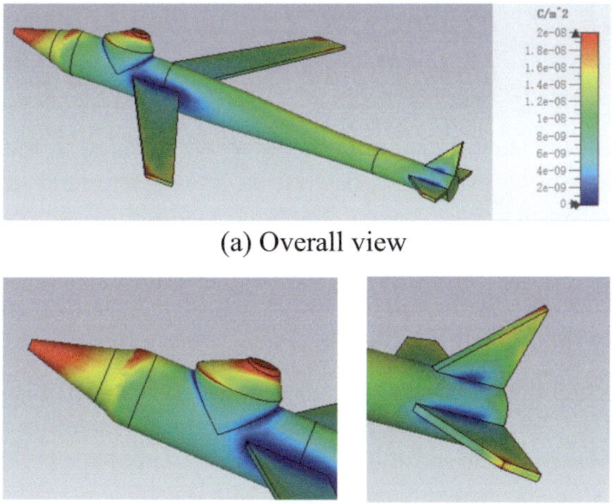

(a) Overall view

(b) Local Enlarged View

Fig. 18.8 The charge distribution of the aircraft

wings and the tail is stronger, and the strongest electric field distribution area is about 6365 V/m in the radome tip area. The strongest area of electric field distribution is at the tip of the radome, which is about 6365 V/m. Compared with the simulation results of the electric field distribution of the vehicle when the electrode position is located directly in front of the vehicle, the overall electric field strength of the vehicle decreases in this electrode position, and the area of the maximum field strength is changed, and the maximum value of the electric field strength is about 78% of that of the previous one. The simulation results of the charge density distribution of the vehicle at this electrode position are shown in Fig. 18.8.

In Fig. 18.8, it can be seen from the charge distribution of the vehicle that the charge distribution of the vehicle is similar to the electric field distribution, affected by the material of each part of the vehicle and the electrode position, the radome tip is 6.0e-8C/m^2, and the top section area is 3.5e-8C/m^2.

From the analysis of electric field distribution and charge distribution results, the aircraft radome tip location area is most likely to become a lightning attachment point; and electric field strength and charge strength of the stronger parts such as the top section location area of the electric field strength of the electrode location at the front end of the situation to become a greater probability of lightning attachment point. It can be seen that the flight attitude of the vehicle during flight has an effect on the lightning attachment partition and electric field strength.

18.3.2 *Electric Field Distribution in a Metal + Insulating Material Vehicle*

In this paper, according to the above simulation model, analyze the effect of the fairing as insulating material on the electric field distribution of the vehicle, and select the more typical electrode position of the electric field distribution of the vehicle to analyze the results.

Electric Field Distribution When the Electrodes Are Located Directly in Front of the Vehicle

Combining the positional relationship between the electrode and the vehicle in Figs. 18.3 and 18.4, the thundercloud is located in the direction directly in front of the vehicle (position 16), the electrode voltage is 1 MV, and the simulation results of the electric field distribution of the vehicle are shown in Fig. 18.9.

When the radome is an insulating material in Fig. 18.9, compared with the electric field distribution of the vehicle in Fig. 18.5, firstly, the overall electric field distribution is changed, and the region of maximum electric field intensity is no longer the tip region of the radome, but the interface between the two materials and the part of the top compartment section, which has a maximum field intensity value of about

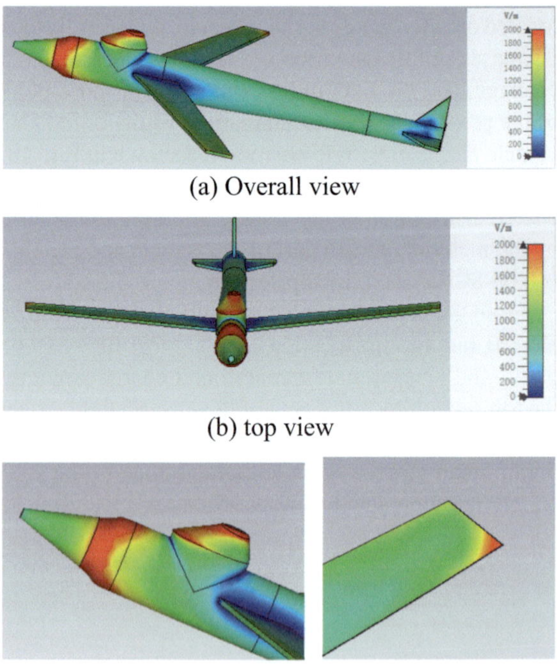

(a) Overall view

(b) top view

(c) Enlarged view of localized area

Fig. 18.9 The electric field distribution of the aircraft

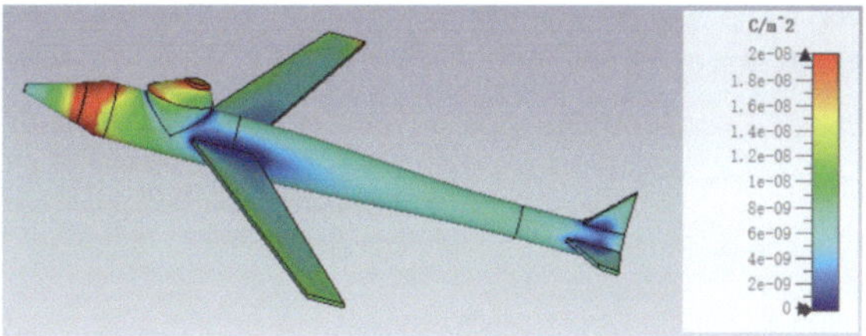

Fig. 18.10 The charge distribution of the aircraft

3238 V/m; and secondly, the strength of the electric field distribution is enhanced in the place of the combination of the top compartment section and the fuselage.

Based on the simulation calculation of the electric field distribution of the vehicle, the charge density distribution of the vehicle is obtained as shown in Fig. 18.10.

In Fig. 18.10, it can be seen from the charge distribution of the vehicle that the charge distribution of the vehicle is similar to the electric field distribution.

From the electric field distribution and charge distribution results, the electric field distribution of the metal + insulating material composite aircraft changes, and the tip of the wing surface tip portion and the location of the top section are easy to become the lightning attachment point. The joint of radome and fuselage, which is the junction of the two materials, is also prone to become lightning attachment points with stronger electric field strength and charge strength.

Electric Field Distribution When the Electrode is Located Directly Above the Vehicle

Combining the positional relationship between the electrode and the vehicle in Figs. 18.3 and 18.4, the thundercloud is located in the direction directly above the vehicle (position 1), the electrode voltage is 1 MV, and the simulation results of the electric field distribution of the vehicle are shown in Fig. 18.11.

Firstly, when the radome in Fig. 18.11 is insulated, compared with the electric field distribution of the vehicle in Fig. 18.6, it is still the tip of the wing and the top section of the vehicle that have the highest electric field distribution intensity, with a maximum field strength value of about 4143 V/m; the electric field distribution at the combination of the radome and the fuselage has changed, and the area here has become a stronger electric field area.

Secondly, compared with Fig. 18.10 where the vehicle is located directly in front (position 16), the electric field strength at the front tip of the radome decreases, while the pattern of the electric field distribution in the other regions changes less, and only the electric field strength changes.

It can be seen that the flight attitude of the composite-type vehicle vehicle during flight has a weakened influence on the lightning attachment zoning and the presence of electric field strength.

Fig. 18.11 The electric field distribution of the aircraft

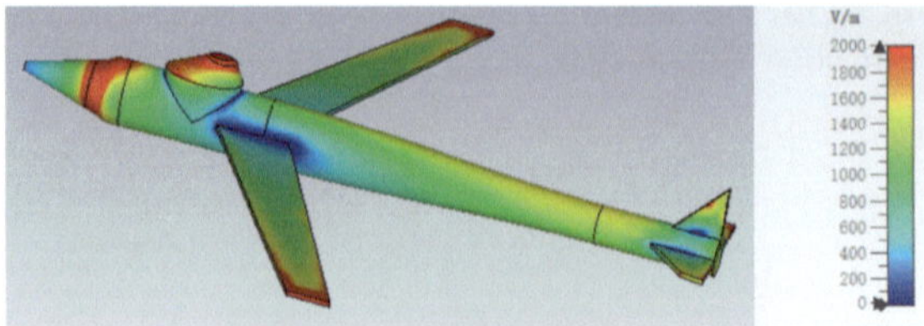

(a) Overall view

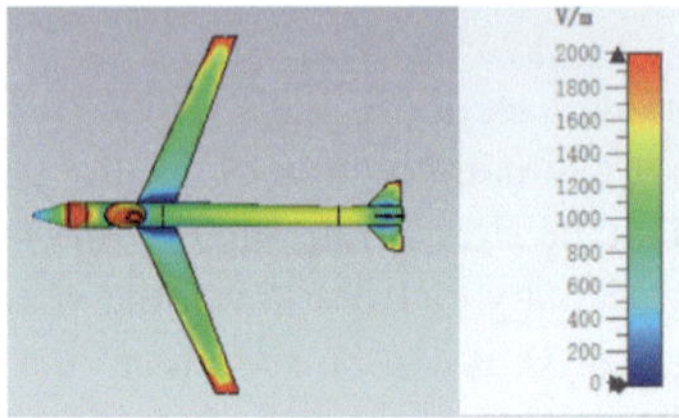

(b) Schematic of the near-electrode side

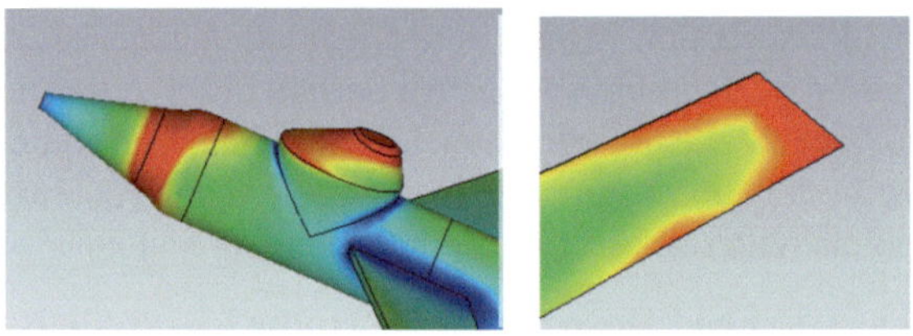

(c) Enlarged view of localized area

18.4 Analysis of Results

18.4.1 Actors Affecting the Electric Field Distribution of a Flight Vehicle

Based on the above electric field distribution research calculations of the vehicle, the resultant nodes are expanded in the model tree, and seven surface nodes are selected to analyze the electric field distribution and charge distribution under different probe positions of the vehicle, and the electric field probe position selection map is shown in Fig. 18.12.

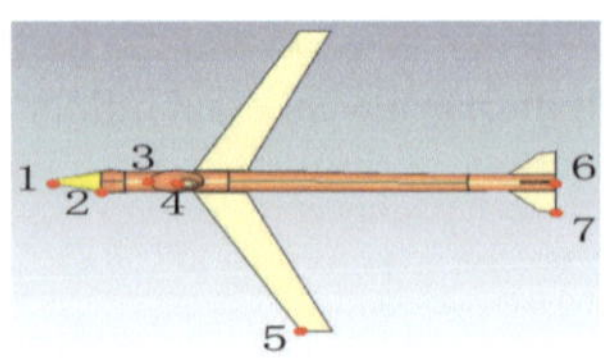

Fig. 18.12 Probe selection diagram of electric field and charge distribution for aircraft

Table 18.2 The probe position corresponds to the model region

Probe number	Parts of model
1	Front of radome
2	Radome-to-torso interface
3	Forward end of the top segment
4	Upper end of the top segment
5	Wing section tip
6	Tip of tail
7	Side of tail

Typical regions with large electric fields for the different electrode cases selected are shown in the figure, and the corresponding model positions for the probe positions are shown in Table 18.2.

Effect of Electrode Position on the Electric Field Distribution of the Vehicle

The results of the electric field distribution calculations for the metal + insulating material composite type as an example of the vehicle are shown in Fig. 18.13 for the electric field strength of the vehicle at 37 electrode positions under the same probe position.

The orange part is the electrode position with the largest electric field strength in the top bay section region, the yellow part is the electrode position with the largest electric field strength in the tail wing tip region, and the blue part is the electrode position with the largest electric field strength in the wing tip region.

The red part represents the variation of the electric field strength in the radome tip region with the electrode position, and since the radome material was replaced with insulating material, the regions with larger electric field strength are all in the wing tip region. The maximum and strongest regions of electric field strength of the vehicle with different electrode positions in Fig. 18.13 are analyzed, as shown in Table 18.3.

Influence of Materials on Electric Field Distribution in Aircraft

Based on the calculation of the electric field distribution of the metal vehicle, the electric field strength of the vehicle at 37 electrode positions under the same probe position was calculated as shown in Fig. 18.14.

The orange part in Fig. 18.14 is the electrode position with the maximum electric field strength in the front region of the radome, the yellow region is the electrode position with the maximum electric field strength in the tip region of the tail wing, and the blue region is the electrode position with the maximum electric field strength in the tip region of the wing. Comparison of electric field strengths at the same position under the two different material compositions clearly shows the effect of the change of material on the change of the lightning-strike attachment point. The maximum value and the strongest region of the electric field strength of the vehicle under different electrode positions in Fig. 18.14 are analyzed, as shown in Table 18.4.

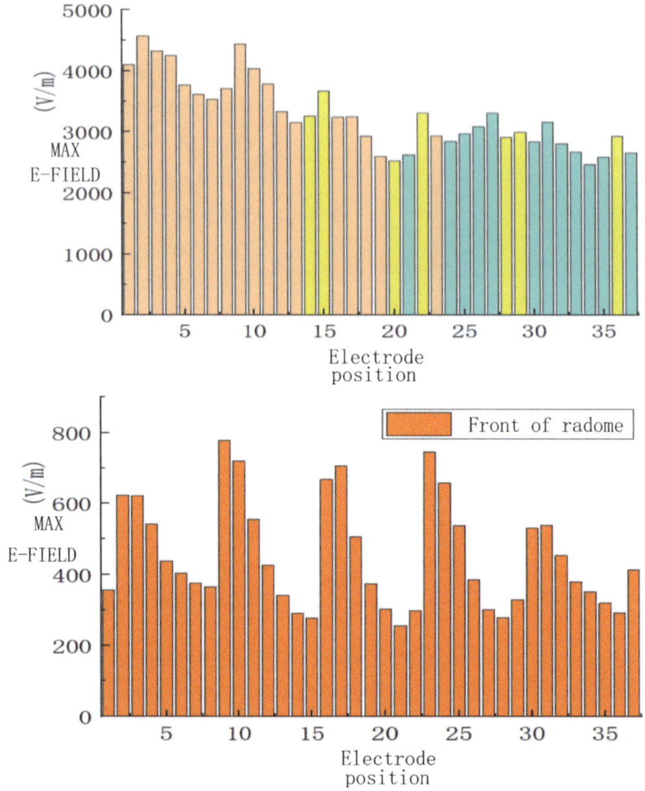

Fig. 18.13 Electric field intensity distribution of aircraft

According to the electric field distributions in Tables 18.3 and 18.4, it can be seen that because the radome is an insulating material, the electric field distributions of the two materials' vehicles change under the same electrode position. Taking electrode position 16 as an example, the strongest region of electric field strength of metal material is at the front end of the radome, while the strongest region of electric field strength of metal + insulating material is at the top compartment section region of the vehicle, and its maximum value of electric field strength is 40.05% of metal material.

Based on the distribution of the strongest electric field in the aircraft, the most likely lightning strike attachment location and the lightning current lead location of the aircraft are predicted.

Table 18.3 Electric field strength value of aircraft

Positions	E-FIELD(V/m)	Region	Positions	E-FIELD(V/m)	Region
1	4103	Top segment	20	2232	Tail tip
2	4573	Top segment	21	2211	Wing tip
3	4328	Top segment	22	2125	Tail tip
4	4253	Top segment	23	2940	Top segment
5	3771	Top segment	24	2813	Wing tip
6	3609	Top segment	25	2652	Wing tip
7	3528	Top segment	26	2341	Wing tip
8	2605	Tail tip	27	2082	Wing tip
9	4445	Top segment	28	1923	Tail tip
10	4043	Top segment	29	1920	Tail tip
11	3785	Top segment	30	2411	Wing tip
12	3331	Top segment	31	2315	Wing tip
13	3150	Top segment	32	2243	Wing tip
14	2763	Tail tip	33	2147	Wing tip
15	2724	Tail tip	34	1977	Wing tip
16	3238	Top segment	35	1946	Wing tip
17	3251	Top segment	36	1917	Tail tip
18	2931	Top segment	37	2027	Wing tip
19	2603	Tail tip			

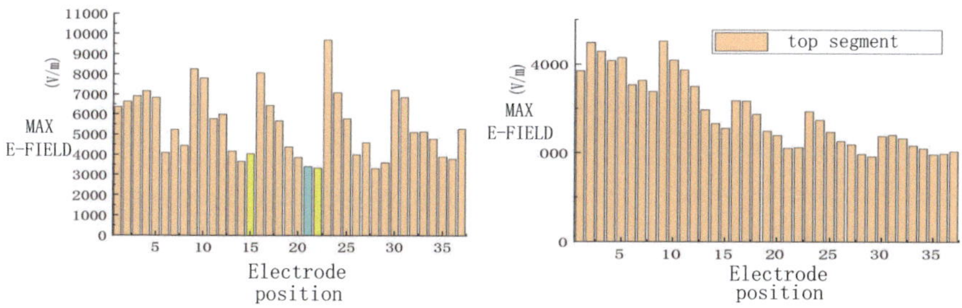

Fig. 18.14 The electric field strength of the aircraft

18.4.2 *Vehicle Lightning Attachment Point Subdivision (VLASP)*

According to the results of the electrostatic field simulation calculations of the above flight vehicle. It is found that the radome tip of the pure metal vehicle accumulates more electric charge, and the induced electric field strength is much larger than that after replacing the radome part with insulating material, and then combined with

Table 18.4 Electric field strength value of aircraft

Positions	E-FIELD (V/m)	Region	Positions	E-FIELD (V/m)	Region
1	6365	Front of radome	20	3866	Front of radome
2	6656	Front of radome	21	3415	Wing tip
3	6933	Front of radome	22	3346	Tail tip
4	7182	Front of radome	23	9703	Front of radome
5	6864	Front of radome	24	7106	Front of radome
6	4112	Front of radome	25	5793	Front of radome
7	5251	Front of radome	26	4023	Front of radome
8	4452	Front of radome	27	4601	Front of radome
9	8274	Front of radome	28	3338	Front of radome
10	7824	Front of radome	29	3615	Front of radome
11	5790	Front of radome	30	7234	Front of radome
12	6011	Front of radome	31	6868	Front of radome
13	4181	Front of radome	32	5113	Front of radome
14	3676	Front of radome	33	5133	Front of radome
15	4050	Tail tip	34	4793	Front of radome
16	8084	Front of radome	35	3916	Front of radome
17	6465	Front of radome	36	3808	Front of radome
18	5676	Front of radome	37	5280	Front of radome
19	4391	Front of radome			

the results of the literature for comprehensive consideration; 7% of the maximum electric field field strength on the surface of the vehicle serves as the discriminatory threshold, and the area exceeding this threshold is classified as the initial attachment area of lightning, area 1A.

Metallic Materials Vehicle Electric Lightning Subdivision

According to the electric field distribution of the vehicle in Fig. 18.14, the maximum value of the electric field strength of the metal material vehicle at position 16 is about 8084 V/m. Comparing with the standard partition, 7% of the maximum electric field strength on the surface of the vehicle is taken as the discrimination threshold, and the lightning attachment point partition 1A of the metal material vehicle is shown in Fig. 18.15.

Zone 1A of the vehicle in Fig. 18.15 is mainly located at the radome tip, wing tip, top section and tail tip area, while zone 1B is mostly present at the trailing edge of zone 1A. The swept area is located in the rear area of zone 1, which may cover the fuselage and can be categorized as zone 2A, while all areas other than the above are categorized as zone 3.

Fig. 18.15 Lightning attachment zone of metal aircraft

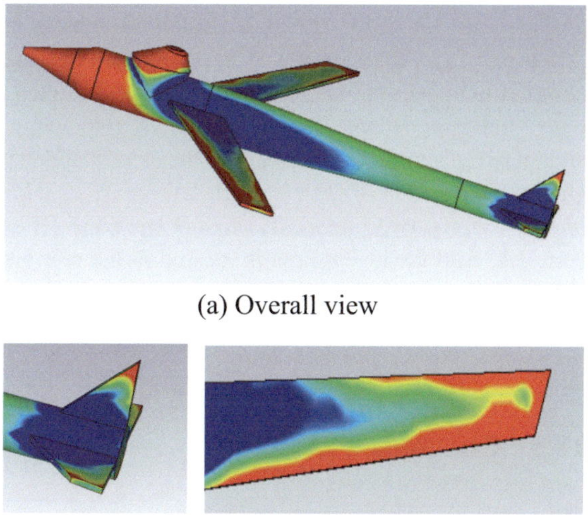

(a) Overall view

(b) Local enlargement

Metal + Insulation Vehicle Electric Lightning Subdivision

According to the electric field distribution of the vehicle in Fig. 18.13, the most probable lightning attachment position of the vehicle as well as the lightning current lead-in position are predicted, and the lightning attachment point zoning of the composite vehicle with metal + insulating material is shown in Fig. 18.16. Position in the No. 2 position case, the electrode is located directly above, the maximum electric field strength is 4573 V/m, 7% of the maximum electric field field strength on the surface of the vehicle as the discriminatory threshold, the lightning attachment point partition 1A for the metal material vehicle is shown in Fig. 18.16.

Fig. 18.16 Lightning attachment zone of metal + insulating aircraft

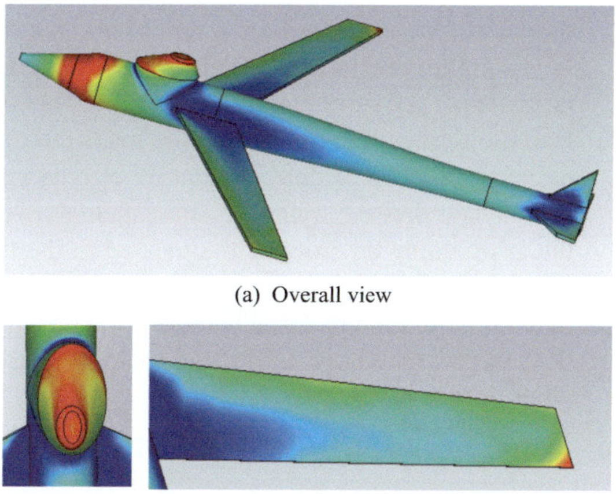

(a) Overall view

(b) Lightning Attachment Partitioning for Metal + Insulation Vehicles

The zone 1A of the metal + insulation vehicle in Fig. 18.16 is mainly located at the interface between the radome and the carcass, at the upper end of the top compartment section, and at the tip area of the wing surface. Comparison with the pure metal reveals that the area of zone 1A at the wing surface area becomes smaller, and there is a significant decrease in the electric field strength at the tip of the radome. Due to the position of the electrodes, there exists a larger zone 1A in the area of the interface of the two materials. Most of the zone 1B exists at the trailing edge of the zone 1A. And the swept area is located in the rear area of zone 1, which will cover the fuselage and can be classified as zone 2A, and all the areas other than the above are classified as zone 3.

18.5 Conclude

In this paper, the electric field distribution on the surface of the aircraft under different materials and 37 different electrode positions is simulated and calculated to analyze its lightning partition, which provides an analytical basis for the protection design and optimization of the aircraft, and the main conclusions are as follows.

(1) The simulation results found that the electrode position has an effect on the electric field distribution on the surface of the vehicle, taking the metal material vehicle as an example when the electrode position is at position 23, the electric field distribution of the vehicle is the strongest value, with a maximum value of about 9703 V/m; when the electrode position is above the vehicle, the overall high electric field distribution range of the vehicle is the largest.
(2) The material properties affect the electric field distribution and the electric field strength value of the vehicle, such as when the electrode position is directly in front of the vehicle, the maximum location of the electric field strength in the electric field distribution of the all-metal material vehicle is located at the tip of the radome at about 8,084 V/m, while the maximum region of the electric field strength in the electric field distribution of the composite material vehicle is located at the front end of the wing at about 3,238 V/m.
(3) Based on the simulation study, the maximum region of electric field strength in electric field distribution under 37 electrode positions is investigated, and it is analyzed that the lightning attachment 1A region of metal material vehicle under lightning environment is located in the tip of the radome, the tip of the wing surface, and the tail part; and the lightning attachment 1A region of the metal + insulation material vehicle is located in the interface of the radome and the carcass, the upper end of the top compartment section, and the tip of the wing surface part.

References

1. Dwyer JR, Uman MA. The physics of lightning[J]. Physics Reports, 2014, 534: 147-241.
2. Abdelal G, Murphy A. Nonlinear numerical modelling of lightning strike effect on composite panels with temperature dependent material properties[J]. Composite Structures, 2014, 109: 268-278.
3. Christian K, Fridolin H, Christian P. Protection of aircraft radomes against direct lightning strikes-An Overview[J]. Atmosphere, 2021, 12(9): 1-13.
4. Ma Linmin, ZHANG Jiazhen, et al. Application of composites in new generation of large civil aircraft [J]. Acta Materiae Compositae Sinica,2015, (02):317–322.
5. Shen J, XIE H Q. Recent progress in study and application of composite materials for aeronautical engineering[J]. Fiber Reinforced Plastics/Composites, 2006, 5: 48–54.
6. Forintos N, Czigany T. Multifunctional application of carbon fiber reinforced polymer composites: electrical properties of the reinforcing carbon fibers-a short review[J]. Composite Part B:Engineering, 2019, 162:331–343.
7. Yousefpour K, Lin WH, Wang YQ, et al. Discharge and ground electrode design considerations for the lightning strike damage tolerance assessment of CFRP matrix composite laminates[J]. Composites Part B: Engineering, 2020, 198: 1-11.
8. Chemartin L, Lalande P, Peyrou B, et al. Direct effects of lightning on aircraft structure: analysis of the thermal, electrical and mechanical constraints[J]. Aerospace Lab, 2012, 5: 1-15.
9. SUN Jinru, YAO Xueling, XU Wenjun, CHEN Jingliang, WU Yi. Lightning Damage Effects of Lightning Current Components on Carbon Fiber Reinforced Polymer Laminates. High Voltage Engineering. 2019, 45(12): 3836-3843
10. Wang B,Ming YK,et al.Fabrication of continuous carbon fiber mesh for lightning protection of large-scale wind-turbine blade by electron beam cured printing[J]. Additive Manufacturing, 2019,31(6):100967.
11. Tian XY, Yao XL, Feng RX, et al. Damage mechanism analysis for carbon fiber reinforced polymer composites with a fastener under lightning strike[J]. Composite Structures, 2022, 289: 1-11.
12. Teulet P, Billoux T, Cressault Y, et al. Energy balance and assessment of the pressure build-up around a bolt fastener due to sparking during a lightning impact[J]. The European Physical Journal Applied Physics, 2017, 77(2):20801.
13. Feraboli P, Miller M. Damage resistance and tolerance of carbon/epoxy composite coupons subjected to simulated lightning strike[J]. Composites: Part A, 2009, 40(6–7): 954–967.
14. Yin Junjie, Li Shulin, Yang Zhe, et al. Test research on damage and mechanical property degradation of composite laminate with fastener subjected to lightning strike[J].Acta Materiae Compositae Sinica,2018,35(5):1131–1138.
15. SAE ARP 5412. Aircraft Lightning Environment and Related Test Waveforms[S]. Warrendale, PA, United States: SAE International Group, 2013.
16. SAE ARP 5414. Aircraft lightning zoning[S]. Warrendale, PA, United States: SAE International Group, 2013.
17. Y.G. Kim, J.H. Jo, D.H. Kim, et al. Effects of lightning on UAM aircraft: Complex zoning and direct effects on composite prop-rotor blade [J].Aerospace Science and Technology,2022, 124: 107560.
18. LIU Henan, GUO Jun, YI Tongqiang, et al. Protection of Direct and Indirect Effects of Lightning on Aircraft[J].Astronautical Systems Engineering Technology,2019,3(4):56–62.
19. Chen Gong, et al. Analysis of standards and testmethods for aircraft and subsystem lightning effects [J]. HIGH POWER LASER AND PARTICLE BEAMS,2024,36(4).
20. Huang Liyang, Chen Xiaoning, Guo Fei, et al. Numerical simulation of initial lightning attachment points on helicopter based on stream-leader discharge theory[J]. High Power Laser and Particle Beams, 2016,28:073201.https://doi.org/10.11884/HPLPB201628.073201
21. NIE Ru.Numerical Simulation and Transient Analysis of Lightning Strike Aircraft Based on CST Simulation Software[J].Insulators and Surge Arresters,2018(1):126–131.

22. GAO Cheng SONG Shuang, GUO Yongchao, et al. Study of numerical simulation of aircraft attachment points and lightning zoning[J].CHINESE JOURNAOF RADIO SCIENCE, 2012, 27(6): 1238–1243.
23. LU Xiang,SHAN Ze-zhong,ZHAO Miao.Study of Numerical Simulation of Civil Aircrafts Attachment Points and Lightning Zoning[J].JOURNAL OF MICROWAVES, 2019,35(2):87–91.

Chapter 19
Early Warning of Lightning Activity Based on Lightning Location Data and Meteorological Data

Chaoying Fang, Jun Xu, Yeqiang Deng, Yu Wang, Haochen Zhang, and Yuzhe Chen

Abstract Due to the complexity of lightning formation and development process, the lightning proximity warning based on single data is difficult to meet the requirements in terms of prediction accuracy. In this paper, we conduct a lightning activity warning study based on lightning localization data and meteorological data. The combination of ERA5 dataset and lightning localization dataset is used as the model training and testing data, and the format unification of 1 km spatial resolution and 10 min temporal resolution is achieved with the help of cubic spline interpolation. On this basis, this paper compares the good and bad prediction performance of decision tree, support vector machine and bp neural network models for the target dataset, respectively, in which the support vector machine model performs better in terms of probability of detection (POD), while the bp neural network performs better in terms of false alarm rate (FAR).

Keywords Lightning warning · Cubic spline interpolation · Machine learning · ERA5 data

---

C. Fang · J. Xu
State Grid Fujian Electric Power Company Limited Research Institute, Fuzhou, China

Y. Deng (✉) · Y. Wang · H. Zhang · Y. Chen
School of Electrical Engineering and Automation, Wuhan University, Wuhan, China
e-mail: dengyeqiang@whu.edu.cn

Y. Wang
e-mail: yuwang@whu.edu.cn

H. Zhang
e-mail: zhanghc21@whu.edu.cn

Y. Chen
e-mail: 2019302070280@whu.edu.cn

Y. Li et al. (eds.), *Proceedings of the 5th Chinese Conference on High Voltage and Discharge Plasmas*, Springer Proceedings in Physics 254,
https://doi.org/10.1007/978-981-96-5711-7_19

19.1 Introduction

According to the study, it shows that the number of trip-out fault on transmission lines triggered by lightning strikes accounts for 40% to 70% [1, 2]. It can be seen that lightning strikes seriously threaten the safety and reliability of the power system. Therefore, the study of lightning warning is very necessary for active lightning protection. Due to the complexity of factors affecting the generation process of lightning, lightning warning based on a single type of data is difficult to make a significant breakthrough in prediction accuracy and effectiveness [3, 4]. In this paper, lightning location data and meteorological data are comprehensively considered, the POD and FAR of different modeling methods for lightning warning are compared.

There are various methods of lightning warning, and the traditional lightning warning includes threshold warning based on determining the intensity threshold of the atmospheric electric field at the time of lightning occurrence [5, 6]. These method is based on a single condition, and the threshold is set by human, the accuracy is not high. Zhang et al. [7] synthesized lightning location data, radar data and atmospheric electric field data and used Conv-LSTM neural network to train the model, which significantly reduced the prediction error. Luo et al. [8] and others synthesized historical lightning monitoring data, combined with electric field data and temperature, humidity and other meteorological data to achieve early warning of lightning activity, and it has a higher accuracy compared to single data warning. These study shows that the introduction of multi-source data can effectively improve the effectiveness of lightning warning.

Meanwhile, lightning activity and precipitation tend to be strongly correlated. Serge et al. [9] found a high spatial coherence between lightning strikes and precipitation, and also found that positive ground flashes correspond to higher precipitation compared to negative ground flashes. Battan [10] found a positive correlation between the frequency of ground flashes and precipitation. In the time dimension, Piepgrass et al. [11] found that the peak of lightning frequency is always earlier relative to the peak of precipitation intensity. Based on the study of mesoscale convective systems, Holle et al. [12] found that the ground flash frequency climbed rapidly during the storm generation phase, the proportion of positive ground flashes increased during the maturation phase, and the frequency of negative ground flashes decreased rapidly during the dissipation phase. The above studies illustrate that meteorological data and the occurrence of lightning activity are strongly correlated, and the part of meteorological parameters as a dataset for lightning warning is effective.

Combining the above factors, this paper selects lightning location data and meteorological data as the target dataset, and unifies their spatio-temporal resolution by interpolation. On this basis, the effect of different training models on the early warning of lightning activity is compared and analyzed, which is of better significance for the lightning protection of the power system.

19.2 Data Sources and Pre-processing

The meteorological data in this paper are derived from the ERA5 dataset of the European Center for Medium-Range Weather Forecasts (ECMWF), which has a resolution as accurate as 0.25° and can provide hour-by-hour data on a variety of meteorological factor parameters, including atmospheric temperature, barometric pressure, wind, rainfall, soil moisture content, and wave height. The lightning localization data is derived from the monitoring of lightning activity of a wind turbinefarm for the whole year of 2021, with a spatial accuracy of 0.001° and a temporal accuracy of microseconds.

19.2.1 Meteorological Data

Meteorological data interpolation methods are pretty much, and common interpolation methods on two-dimensional space include bilinear interpolation, spline function interpolation and Kriging interpolation. In this paper, we choose the cubic spline interpolation method to process the meteorological data, which is divided into 10-min intervals of data with a spatial resolution of 0.01°.

Cubic spline function interpolation generate smooth curves with continuous first and second order derivatives as interpolation curves through the contents of multiple data points in order to achieve an accurate fit to the actual distribution. The formula of cubic spline interpolation in one-dimensional case is shown below:

$$S_i(x) = a_i + b_i(x - x_{i-1}) + c_i(x - x_{i-1})^2 + d_i(x - x_{i-1})^3 \tag{19.1}$$

$S_i(x)$ is a part of fitting curve $S(x)$ in the subinterval $[x_{i-1}, x_i]$, $S(x)$ is a cubic polynomial function, where a_i, b_i, c_i, d_i is the coefficient to be solved and can be solved by the following equation:

$$\begin{gathered} S_i(x_i) = y_i \quad S_i(x_{i-1}) = y_{i-1} \\ S'_{i-1}(x_{i-1}) = S'_i(x_{i-1}) \quad S''_{i-1}(x_{i-1}) = S''_i(x_{i-1}) \end{gathered} \tag{19.2}$$

The above formula ensures the continuity of the first-order and second-order derivatives of the function at the sample points, while the polynomial content is determined in each sub-interval to ensure that the local features are preserved. Cubic spline function can make full use of the known data points to approximate the actual value, the advantage is high interpolation accuracy and the disadvantage is that its interpolation accuracy is affected by the boundary conditions, the use of different boundary conditions will bring errors to the interpolation results. In general, we choose the natural cubic spline function interpolation as the target function. ($S''(x_0) = 0, S''(x_N) = 0$, x_0 and x_N are the starting point and the end point of the interval. When the cubic spline interpolation is applied to two-dimensional space, it is usually used

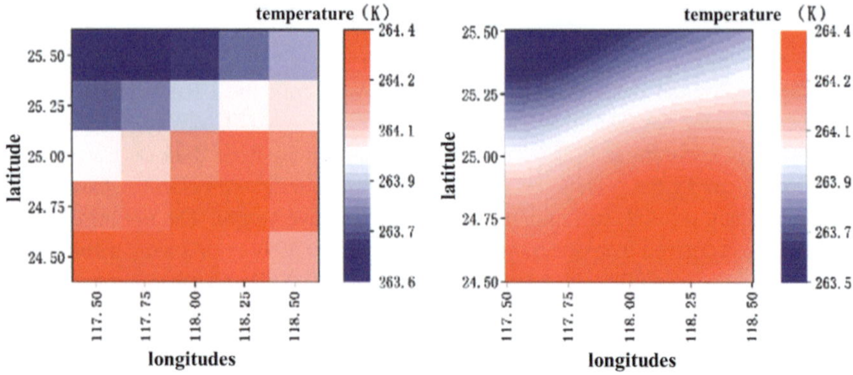

Fig. 19.1 Cubic spline interpolation results of meteorological parameters

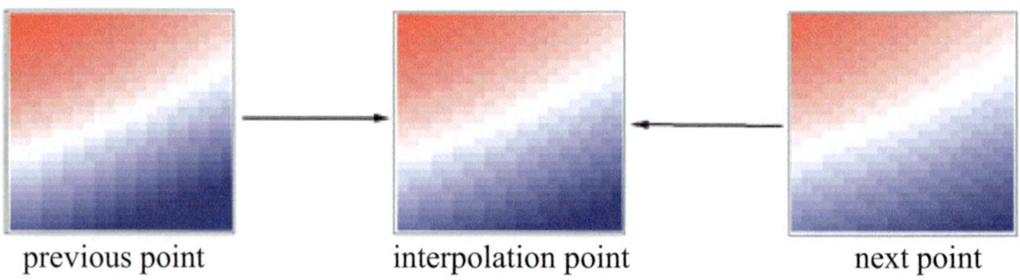

Fig. 19.2 Temporal interpolation of meteorological parameters

to generate a smooth surface with two-dimensional continuous partial derivatives by applying one-dimensional cubic spline interpolation in two dimensions, which is adopted in this paper to realize parameter interpolation in the target region, as shown in Fig. 19.1.

For the temporal interpolation processing of the input data, due to the interpolation on the time sequence axis, in order to obtain weather characteristics when lightning occurs, we take the two nearest time points as the target point for linear interpolation processing, as shown in Fig. 19.2.

19.2.2 Lightning Localization Data

The spatial resolution of the original ground-flash localization data reaches 0.001°, and the localization time is accurate to microseconds, which is much higher than the accuracy of this study. Therefore, we divide the atmospheric reanalysis data points spatial latitude and longitude within 0.005°, time before and after the 5-min range of all occurrences into one dataset. Based on the classification method, we complete the unification of lightning position data and meteorological data. At the same time, in

Fig. 19.3 Lightning localization data in the target area

order to avoid the lack of data information caused by the lack of information on the boundary of ground flash position, this design selects the internal rectangular range of ground flash positioning data as the study area, as shown in Fig. 19.3.

Meteorological data and ground flash localization data for each data point were processed as follows for differernt data points in the study area:

1. Consider points at different latitude and longitude locations or at different times with a same location as different data points;
2. All meteorological parameters at the same location at the same time are taken as input parameters;
3. Each data point is labeled differently by whether lightning occurred or not. When lightning occurs (without counting the number of occurrences), the data point is labeled as "occurrence" for a positive sample point; when no lightning occurs, the data point is labeled as "non-occurrence" for a negative sample point.

Finally, based on the full data points obtained, we produced the dataset for model training.

19.3 Lightning Warning Modeling and Evaluation

The training dataset of this study consists of 20,556 data points with a ratio of positive and negative sample points of 1:5, divided into a training set and a test set with a ratio of 9:1. The inputs are meteorological parameters (temperature, humidity, wind speed, vertical airflow intensity) under sixteen different pressure layers, and the outputs are binary classification labels of whether lightning occurs or not. Meanwhile, the input data are normalized by adopting Z-score normalization, which preserves the distribution characteristics of the original data while normalizing, as shown in the following Eq. (19.3):

$$Z = \frac{X - \mu}{\sigma} \tag{19.3}$$

Z is the normalized data, X is the original data, and μ is the original data mean, and σ is the original data standard deviation.

In order to compare the advantages and disadvantages of different models, this paper adopts two indicators of probability of detection POD and false alarm rate FAR for evaluation, which are calculated as shown in the following formula (19.4):

$$\begin{aligned} \text{POD} &= \frac{TP}{TP+FN} \\ \text{FAR} &= \frac{FP}{FP+TN} \end{aligned} \tag{19.4}$$

where TP stands for True Positives, FN stands for False Negatives, FP stands for False Positives, and TN stands for True Negatives.

19.3.1 Decision Tree Model

Decision tree is a tree-like classification structure whose composition includes nodes and branches. The input data starts from the root node, decides the branching direction based on different attribute judgment conditions at the internal nodes, and finally reaches the leaf nodes to realize the decision classification, which is a machine learning method to realize the classification through the decision path from the root node to the leaf nodes, as shown in Fig. 19.4.

In order to find the optimal number of splits for the decision tree, the model prediction results under different number of splits are evaluated in this paper, as shown in Fig. 19.5.

As can be seen from the above figure, after the maximum number of splits reaches 320, the growth rate of the model POD slows down and the change of FAR is smooth. However, after the number of splits increased from 440 to 480, the FAR suddenly increased while the POD did not change significantly, indicating that the model was overfitting. However, even if it is located in the optimal interval, the decision tree classification model predicts the probability of lightning occurrence is usually lower than 55%, while the false alarm rate is more than 30%, and its prediction effect is poor.

19.3.2 Support Vector Machine Model

Support Vector Machine (SVM) is a machine learning model specialized in solving binary classification problems. It maps the input feature variables into a high-dimensional space, and searches for the optimal hyperplane in the high-dimensional space to maximize the distinction between two different categories of different data points, so as to complete the classification of the data points, which is schematically shown in Fig. 19.6.

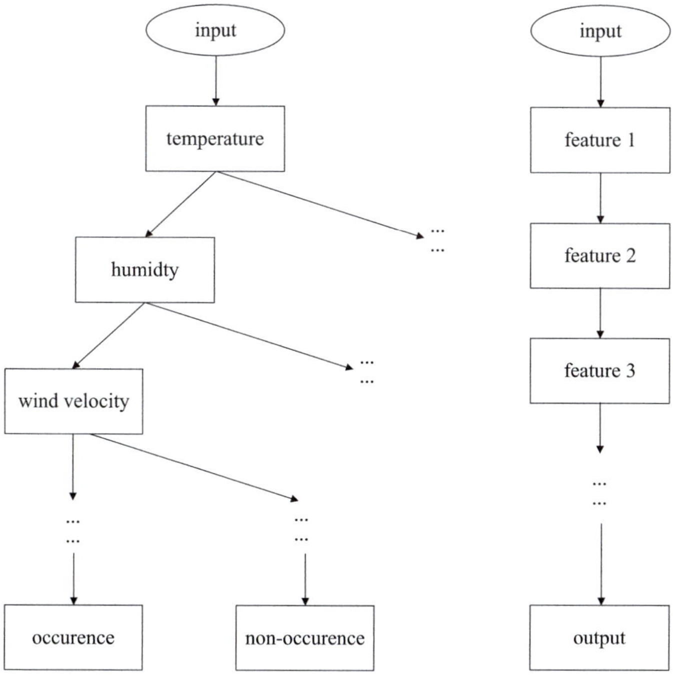

Fig. 19.4 Decision tree model

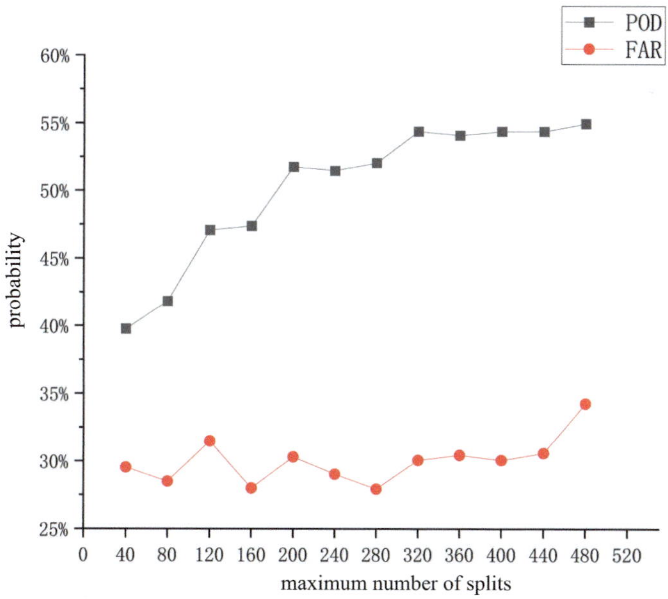

Fig. 19.5 Number of decision tree splits and prediction results

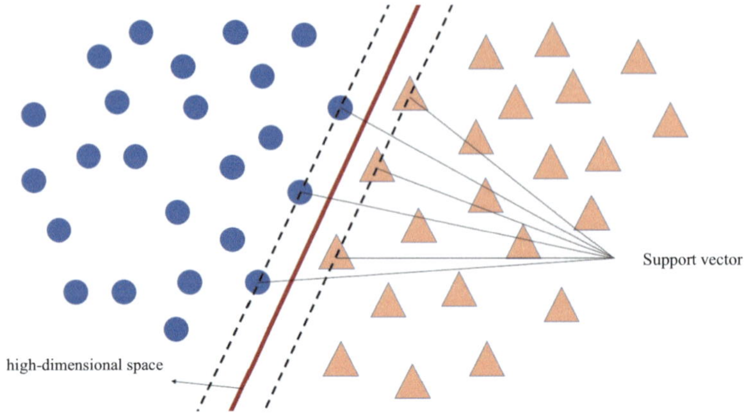

Fig. 19.6 Support vector machine classification model

For high-dimensional data or nonlinearly differentiable data, SVM maps the original data to the high-dimensional space through kernel function, and then calculates the optimal hyperplane in the high-dimensional space to realize data classification. The more common kernel functions mainly include linear kernel function, polynomial kernel function and Gaussian kernel function, etc. The formulas are shown in (5) below.

$$
\begin{aligned}
K(x, y) &= x \cdot y \\
K(x, y) &= (x \cdot y + c)^n \\
K(x, y) &= \exp\left(-\frac{\|x - y\|^2}{2\sigma^2}\right)
\end{aligned}
\tag{19.5}
$$

For the research done in this paper, the Gaussian kernel function works better, and the results of POD and FAR for the support vector machine model with different kernel scales are shown in Fig. 19.7.

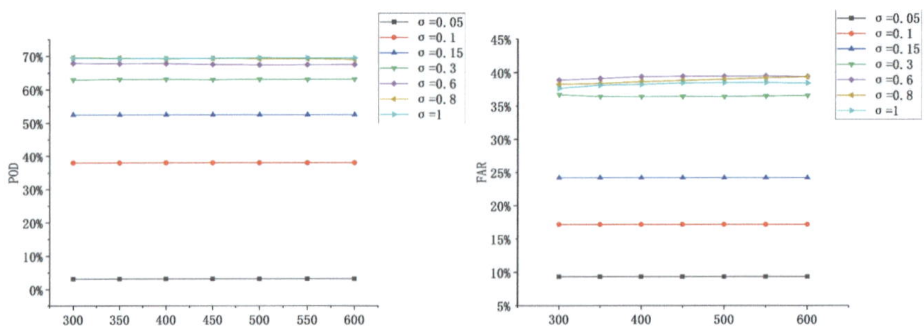

Fig. 19.7 Gaussian kernel function SVM model evaluation results

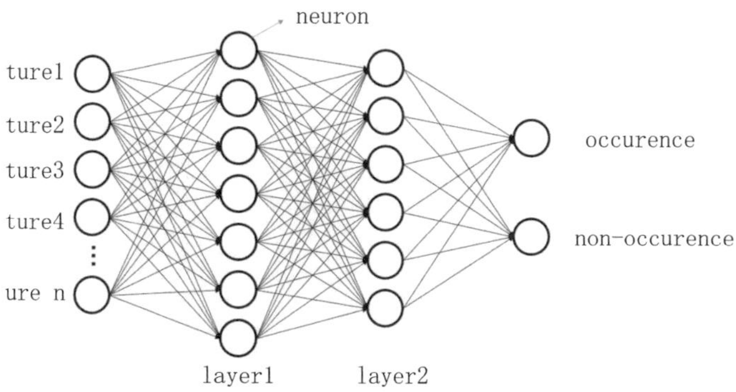

Fig. 19.8 Schematic diagram of BP neural network model

As can be seen from the figure, the size of the model POD and FAR is basically stable with the change of the box constraint level at different kernel scales, which indicates that at this kernel scale, the box constraint level is large enough to give full play to the model performance. With the increase of the nuclear scale, the model POD increases gradually and stabilizes at about 69% in the nuclear scale 0.8 ~ 1, while the value of FAR first increases with the increase of the nuclear scale, and then decreases and fluctuates when the nuclear scale is higher than 0.6.

19.3.3 BP Neural Network Model

The BP neural network extracts features by inputting data into neurons with linear weighting and bias of eigenvalues, fits complex functions using multiple hidden layers, and finally realizes result prediction by outputting data through hidden layers. At the same time, the neural network can also realize different nonlinear data transformations through different activation functions, as shown in Fig. 19.8.

In this study, three different activation functions, ReLU, Tanh and Sigmoid, were selected for training, and the results of their POD and FAR evaluations are shown in Fig. 19.9.

When the Sigmoid function is used as the activation function, the model POD is significantly improved compared to the ReLU activation function model, and is smoother compared to the Tanh activation function model; the model FAR is smaller compared to the Tanh function model, and the changes are also smoother. The final model predicts 65% POD and 32% FAR.

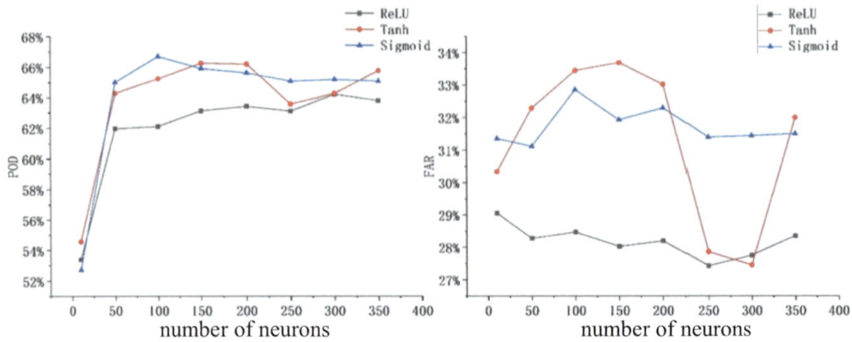

Fig. 19.9 Effect of BP neural network with different activation functions

19.4 Conclusion

In this paper, the dataset used for lightning warning is established based on the ERA5 meteorological dataset of the European Center for Medium-Range Weather Forecasts for the year 2021 and the lightning localization data of a wind turbine farm. We unify the spatial and temporal resolution between different data by the method of cubic spline interpolation. After that, we compares the advantages and disadvantages of decision tree model, support vector machine model and BP neural network model in the warning of lightning activity by using POD and FAR as the judging criteria. Above all, the decision tree model has poorer warning accuracy and false alarm rate, and the support vector machine model has higher warning accuracy while its false alarm rate is also higher. Taken together, the BP neural network based on the Sigmoid function has a higher warning accuracy while its false alarm rate is only about 30%, which is a better choice for the dataset selected in this paper.

Acknowledgements This research was funded by the Science and Technology Project of State Grid "Research on Dis-aster Warning and Risk Assessment Technology for Lightning Strike in Large Wind Farms" (No. 5500-202221143A-1-1-ZN).

References

1. GU Shanqiang; LI Jian; WU Min; ZHANG Haochen; DENG Yeqiang; WANG Yu; MA Zhiqing; Simulation analysis of lightning protection performance of dense transmission corridor with the influence of design and neighboring line erection direction and shielding [J/OL]. Grid Technology: 1–10.
2. Mei Zhen, Chen Shui-ming, Gu Qin-wei, et al. National lightning disaster statistics from 1998 to 2004[J]. High Voltage Technology,2007,33(12):173–176.
3. MONTANYA J, BERGAS J, HERMOSO B. Electric field measurements at ground level as a basis for lightning hazard warning[J]. Journal of Electrostatics, 2004, 60(2–4): 241–246.

4. Ju Zeli, Chen Lei, Pu Lu, et al. A lightning forecasting method based on latent potential forecast and radar echo characteristics[J]. Electric Porcelain Lightning Arrester, 2019(3): 6.
5. TAO H, GU S, WANG H, et al. Method of Lightning Warning Based on Atmospheric Electric Field and Lightning Location Data[C]//2016 33RD INTERNATIONAL CONFERENCE ON LIGHTNING PROTECTION (ICLP). TECNICO LISBOA; Univ Lisboa; IEEE, 2016.
6. LU Y, ZHOU Z, GU S, et al. Research on lightning warning method based on the characteristics of atmospheric electric field[J]. IEEE, 2016.
7. Zhang Zhuoyu. Research on thundercloud monitoring and lightning proximity warning method based on multi-source data [D]. Hubei: Huazhong University of Science and Technology, 2022.
8. Luo Longbo, Chen Minghui, Liu Yanping et al. Lightning early warning based on multi-weighted feature fusion[J]. Manufacturing Automation. 2023, 45(11):61–65.
9. Serge Soula, Serge Chauzy. Some aspects of the correlation between lightning and rain activities thunderstorms[J]. Atmospheric Research, 2001(56).
10. Battan L J. Some factors governing precipitation and lightning from convective clouds [J]. J AtmosSci, 1965(22).
11. Piepgrass M V, Krider E P, Moore C B. Lightning and surface rainfall during Florida thunderstorms[J]. J Geophys Res,1982, 87(13).
12. Holle R L, Watson A L, Lopez R E, et al. The life cycle of lightning and severe weather in a 3–4 June 1985 PRE - ST ORM mesoscale convective system [J]. Mon Wea Rev, 1994(122).

Chapter 20
Differential Lightning Protection Strategy and Comprehensive Evaluation of Transmission Lines

Mengfei Lei, Yong Zhang, Jian Li, Jian Wang, Hua Ren, and Liting Wang

Abstract In this paper, the differential lightning protection strategy and comprehensive evaluation of ±800 kV UHV HVDC transmission lines. The differential lightning protection configuration optimization method is described in detail, covering the calculation of lightning trip rate, analysis of main lightning protection measures, and lightning protection transformation process. By calculating the lightning protection level and trip rate of each base tower, a targeted transformation plan is formulated. Finally, the typical line lightning protection configuration scheme is given, and some base towers are taken as examples to show the change of trip rate before and after the transformation. The results show that the lightning protection transformation measures effectively reduce the line trip rate, and the average lightning trip rate of each line segment is significantly reduced, providing a reference for the differential lightning protection transformation of transmission lines.

Keyword Differentiated lightning protection · Lightning trip rate · Safety

M. Lei (✉) · J. Li · J. Wang · H. Ren
NARI Group Corporation Ltd., Nanjing, China
e-mail: 718656829@qq.com

Wuhan NARI Limited Liability Company, State Grid Electric Power Research Institute, Wuhan, China

National Energy Key Laboratory of Lightning Disaster Detection, Early Warning and Safety Protection, Wuhan, China

Hubei Key Laboratory of Power Grid Lightning Risk Prevention, Wuhan, China

Y. Zhang
State Grid Zhejiang Electric Power Co., Ltd., Hangzhou, China

L. Wang
Shaoxing Supply Company of State Grid Zhejiang Electric Power Co., Ltd, Shaoxing, China

Y. Li et al. (eds.), *Proceedings of the 5th Chinese Conference on High Voltage and Discharge Plasmas*, Springer Proceedings in Physics 254,
https://doi.org/10.1007/978-981-96-5711-7_20

20.1 Introduction[1]

Currently, scholars both domestically and internationally have conducted extensive research on differentiated lightning protection modifications [1–3]. Reference [4] establishes an accurate calculation model of lightning trip-out rates for a 35 kV collecting line in a mountain wind farm, taking into account factors such as mountain terrain, tower structure, insulation configuration, and grounding resistance. Reference [5] compares the lightning protection performance of composite materials cross-arm (CMCA) towers with steel towers for quadruple-circuit 500 kV transmission lines. Reference [6] proposes a one-circuit insulation enhanced scheme, emphasizing that the key to lightning protection for double-circuit transmission lines lies in reducing the double-circuit simultaneous trip-out rate while maintaining a reasonable total trip-out rate.

Finally, taking a ±800 kV DC transmission line as an example, the specific process of lightning protection transformation decision-making is explained, and the optimal lightning protection measures of each base tower are obtained. The lightning trip rate of each base tower before and after the transformation is compared, which can provide some reference for the differential lightning protection transformation of transmission lines.

20.2 Transmission Line Corridor Introduction

The selected ±800 kV UHV DC transmission line starts at Shuanglong Converter Station in Yibin, Sichuan and ends at Jinhua Converter Station in Zhejiang. In this study, the 3292#–3300# section of the BJ line is adopted. Since this section is a heavy icing section with fewer installed lightning protection measures and weak lightning protection ability, and there is no lightning arrester wire in the 3294#–3297# section, the risk of lightning strike is relatively high. Therefore, this section is selected for research on differentiated lightning protection transformation. The tower coordinates and line direction of the line are shown in Fig. 20.1.

[1] Project supported by the Science and Technology Project of State Grid Corporation of China (5500-202255399A-2-0-ZN).

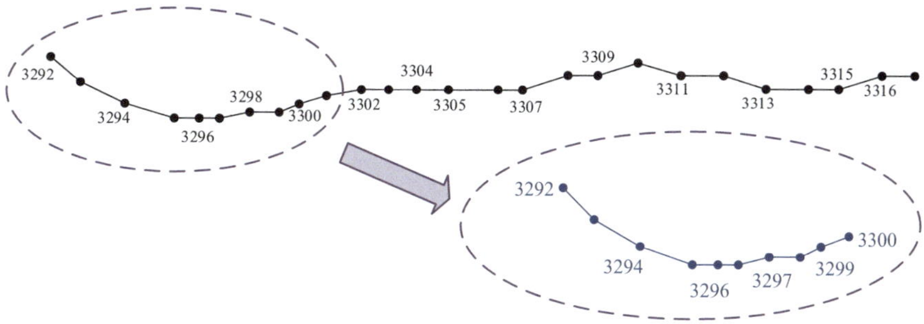

Fig. 20.1 Partial schematic diagram of the research section

20.3 Differentiated Lightning Protection Configuration Optimization Method

20.3.1 Calculation Method of Lightning Trip Rate

For 800 kV transmission lines, tripping is mainly caused by direct lightning strikes. When the thundercloud is within the width of the direct lightning strike, lightning will directly act on the line, generating direct lightning overvoltage [5, 6]. The situation of lightning acting on the line is shown in Fig. 20.2.

In Fig. 20.2, b is the length of the cross arm. y_d is the width of the direct lightning strike. y_d is closely related to the magnitude of the lightning current and the tower structure, and its value can be calculated by the improved electrical geometry model

Fig. 20.2 Schematic diagram of lightning strikes on transmission lines

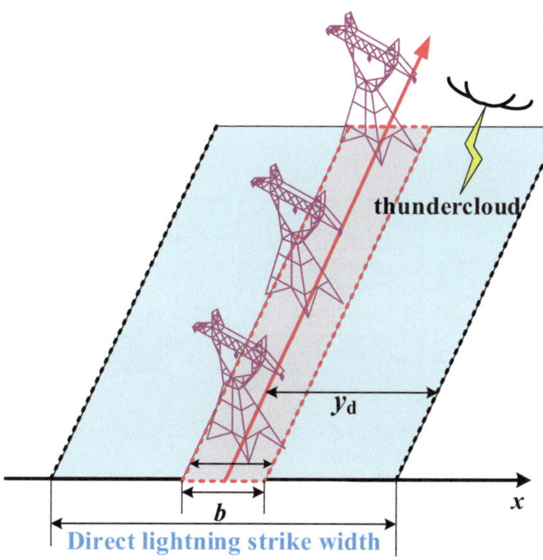

[6] and the travel distance theory. The y_d calculated by different lightning current amplitudes is different, so each lightning current amplitude corresponds to a trip rate, which can be added to obtain the total trip rate of the line. Therefore, the calculation formula of the trip rate is as follows.

$$TR_a = 0.1N_g \eta g \sum_{I_a}^{200} (2y_d + b)P(I_k) \tag{20.1}$$

$$TR_b = 0.1N_g \eta \sum_{I_b}^{200} (1 - g)(2y_d + b)P(I_k) \tag{20.2}$$

In the formula, TR_a and TR_b are the counterattack trip rate and the bypass strike trip rate, respectively. N_g is the ground flash density. η is the arc rate. g is the strike rate. $P\ (I_k)$ is the lightning current amplitude density.

20.3.2 *Differentiated Lightning Protection Transformation Process*

In order to find the best lightning protection configuration scheme as soon as possible, it is necessary to optimize the existing lightning protection configuration. The optimization direction mainly depends on the calculation results of the trip rate. When the average trip rate of the whole line does not meet the operating requirements, it is necessary to strengthen the lightning protection configuration. The lightning protection measures considered in this paper include replacing the insulator string, increasing the lightning arrester and reducing the grounding resistance. When the average trip rate of the whole line meets the operating requirements, no modification is required. The flowchart of the optimization and transformation is shown in Fig. 20.3.

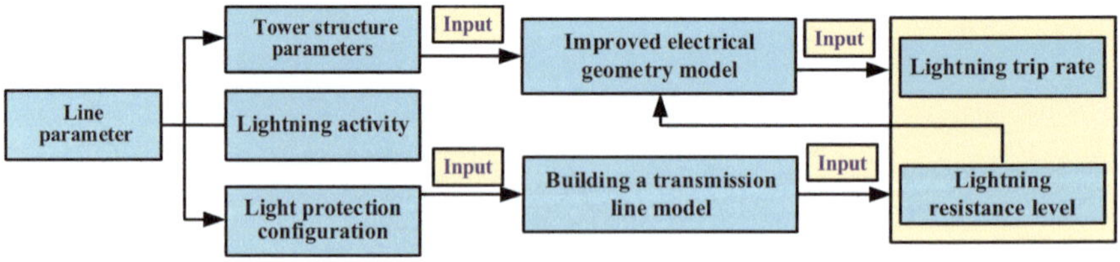

Fig. 20.3 Lightning protection transformation flow chart

Table 20.1 Calculation results of lightning resistance level

Towe number	Counterattack lightning withstand level/(kA)	Lightning shielding failure withstand level/(kA)
3292#	200	50
3293#	201	50
3294#	335	50
3295#	80	15
3296#	100	15
3297#	215	50
3298#	320	50
3299#	215	50
3300#	250	50

20.4 Typical Line Lightning Protection Configuration Optimization

20.4.1 Calculation and Analysis of Lightning Trip Rate

In this paper, the electromagnetic transient software (ATP) is used to calculate the lightning resistance level of the study section. The calculation results are shown in Table 20.1. From Table 20.1, it can be seen that the lightning resistance level amplitude of the transmission line is high, and it is not easy to cause counterattack tripping, which is consistent with the actual project.

According to the lightning resistance level, the tripping rate of each base tower is calculated as shown in Fig. 20.4.

From Fig. 20.4 and the line parameters, it can be seen that the 3295 #and 3296 #towers are not equipped with lightning protection lines, resulting in extremely high tripping rates, and the tripping rates of other towers also exceed the line operation requirements. Therefore, it is necessary to carry out lightning protection transformation in this section.

20.4.2 Typical Line Lightning Protection Optimization Configuration Scheme

It can be seen that the tripping rates of towers 3293 #, 3297 #and 3299 #are lower, and the transformation measures of replacing insulators to strengthen insulation are used. The other six base towers have a higher tripping rate. Among them, towers 3295 #and 3296 #are affected by no lightning shield line, and the tripping rate is the highest. They are all retrofitted by installing lightning arresters. The tripping rate of

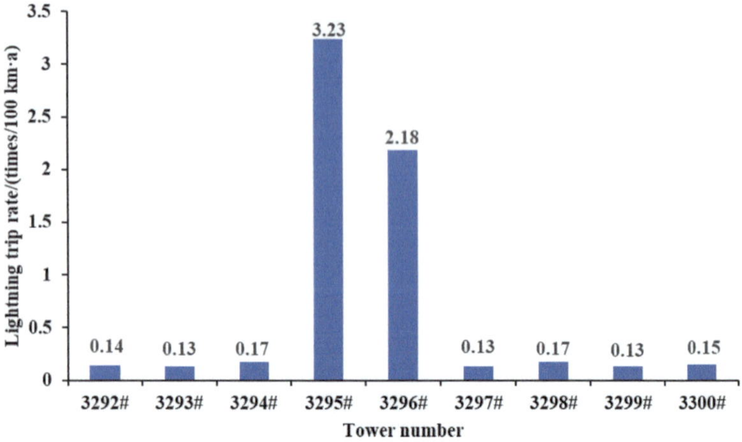

Fig. 20.4 Trip rate before modification

each base tower before and after the renovation is shown in Fig. 20.5. The risk level before and after the reformation is shown in Fig. 20.6.

It can be clearly seen from Fig. 20.5 that after the lightning protection modification measures, the trip rate of the line was significantly reduced and stabilized, and the average lightning trip rate of each line section decreased by 0.162/time/100 (km·a). As can be seen from Fig. 20.6, in the section without lightning arresters, the risk level of the line is reduced from level 4 to level 2 by installing lightning arresters, and the rest of the towers are reduced to level A by strengthening insulation.

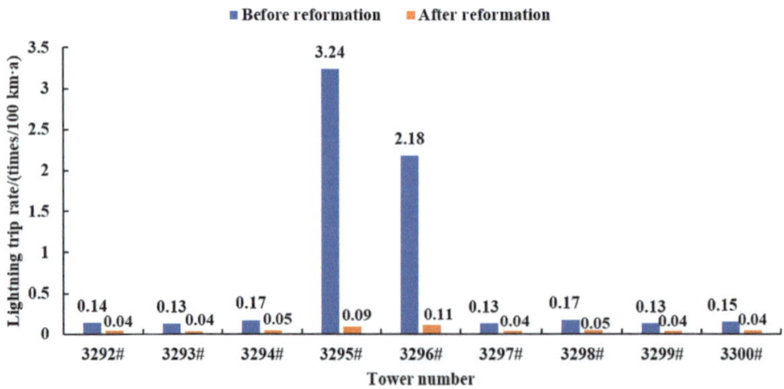

Fig. 20.5 Comparison of trip rate before and after reformation

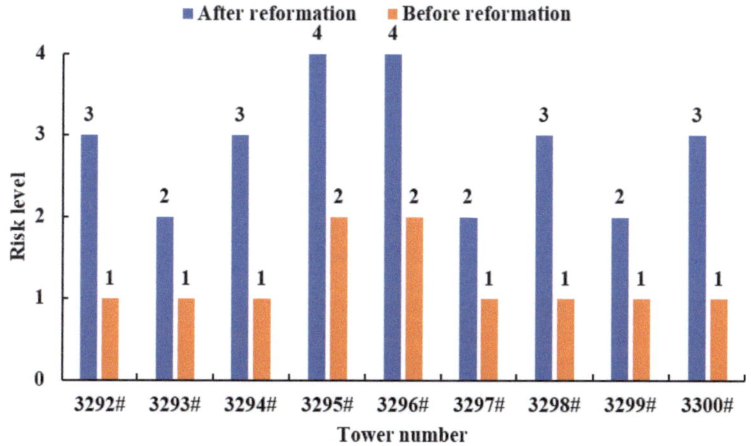

Fig. 20.6 Comparison of risk levels before and after reformation

20.5 Conclusion

(1) This paper proposes a differentiated lightning protection strategy and comprehensive evaluation method for ±800 kV·UHV HVDC transmission lines. This method uses protective measures such as installing lightning arresters, strengthening insulation and reducing resistance to differentiate lightning protection for ice-covered reformed line sections, which significantly reduces the risk of lightning strikes in this line section.

(2) 3294 #-3297 #is a severely icing section, and lightning arresters and other protective measures are not installed, resulting in a lightning trip rate far exceeding the safe value. In this paper, by installing arresters in this section, the average trip rate is reduced from 2.71 times/100 (km·a) to 0.1 times/100 (km·a), and the risk of lightning strikes on the line is significantly reduced.

(3) The lightning risk level of the 3292 #- 3300 #line section is higher, among which the 3-base tower is the 2-level risk, the 4-base is the 3-level, and the 2-base is the 4-level. After the differentiated lightning protection transformation, the risk of the 7-base tower is reduced to the 1-level, the 2-base is reduced to the 2-level, and the risk level is greatly reduced, which proves the effectiveness of the proposed differentiated lightning protection transformation method.

References

1. Yang, Y, Wang, S, Wang, SF, et al.: Lightning trip-out risk assessment and differential lightning protection of 35 kV transmission lines in mountain wind farm. Energy Reports, 2022, 8, 581–588. Presented at: 8th International Conference on Power and Energy Systems Engineering (CPESE), Fukuoka, JAPAN, SEP 10–12, 2021.
2. Li, HC, Xiang, NW, Bao, H, et al.: Lightning protection performance of quadruple-circuit 500 kV transmission lines on the same tower with composite cross arm. Energy Reports, 2022, 8, 520–526. Presented at: 8th International Conference on Power and Energy Systems Engineering (CPESE), Fukuoka, JAPAN, SEP 10–12, 2021.
3. Huang, WG.: Lightning performance of 500-kV double-circuit line schemes for the three-gorge project. Ieee Transactions on Power Delivery, 2006, 21(2), 736–743.
4. Matsuura, Susumu; Noda, Taku; Asakawa, Akira; Yokoyama, Shigeru.: A distribution line model for lightning overvoltage studies. Electrical Engineering in Japan, 2010, 173(1), 11–23.
5. Florkowski, Marek; Furgal, Jakub; Kuniewski, Maciej.: Lightning impulse overvoltage propagation in hvdc meshed grid. Energies, 2021, 14(11), 3047.
6. Brown, G. W.; Whitehead, E. R.: Field and analytical studies of transmission line shielding: part. Ieee Transactions on Power Apparatus and Systems, 1969, 5, 617–626.

Chapter 21
Research on Lightning Strike Risk and Differentiated Lightning Protection Modification of Transmission Lines Based on Multi-Factor Quantitative Classification

Junbo Liu, Jingyao Luan, Changlong Gao, Yufeng Tai, Peng Jin, Huafei Zhang, Chunshen Wang, Yuzhe Chen, Yeqiang Deng, and Yu Wang

Abstract Lightning strike tripping of transmission lines poses a serious hazard to the safe and stable operation of power grids, and this paper carries out a research on lightning strike risk assessment and differentiated lightning protection transformation for a 110 kV transmission line. We obtained line tower parameters, ground flash data, terrain elevation, etc., using the improved EGM and ATP electromagnetic transient simulation to calculate the winding lightning with-stand levels, rebound lightning withstand levels and trip rates, fine assessment of the lightning risk of 57 towers, put forward based on multi-factor quantitative grading of tower high-risk cause finding and lightning protection and renovation measures refinement of the intelligent recommendation method process, effectively improving the de-cision-making efficiency, taking into account the technical economy of various types of measures. The results show that all 27 high-risk towers on the line are reduced to low risk after differentiated lightning protection. This paper can provide a reference for the lightning protection of electric power system.

Keywords Transmission lines · Multi-factor staging · Lightning resistance levels · Risk assessment · Differentiated lightning protection

---

J. Liu (✉) · J. Luan
Jilin Electric Power Scientific Research Institute Co. Ltd, Changchun, China
e-mail: 13894819779@139.com

C. Gao · Y. Tai · H. Zhang · C. Wang
Electric Power Research Institute of State Grid Jilin Electric Power Co., Ltd, Changchun, China

P. Jin
State Grid Jilin Electric Power Co., Ltd, Changchun, China

Y. Chen · Y. Deng · Y. Wang
School of Electricity and Automation, Wuhan University, Wuhan, China

Y. Li et al. (eds.), *Proceedings of the 5th Chinese Conference on High Voltage and Discharge Plasmas*, Springer Proceedings in Physics 254,
https://doi.org/10.1007/978-981-96-5711-7_21

21.1 Introduction

As an important component of the power grid connection, the safe operation of overhead transmission lines is of great significance to the safety and economy of the power grid. However, its operation is affected by extreme weather conditions, including lightning strikes and other numerous hazardous factors. Research has shown that lightning strikes are the primary factor causing faults and tripping of overhead transmission lines in China [1, 2].

With the rapid development of the power grid and the increase of severe convective weather, lightning damage faults occur frequently; There are many trips caused by direct lightning strikes in high-voltage transmission lines in China. Therefore, in order to improve the reliability of power system operation, it is necessary to conduct in-depth research on lightning damage risk assessment of transmission lines. In recent years, many scholars have conducted research in this area, often calculating lightning resistance levels based on regulatory methods or electromagnetic transient simulation methods, and then calculating trip rates based on electrical geometric models [3] and other methods. However, for the lightning protection renovation of high-risk towers, decisions such as increasing the length of insulator strings and installing lightning arresters are generally made directly based on calculation results [4–6], which may not be quantitative, differentiated, and intelligent enough, and lack systematic theoretical support considering multiple factors.

Based on existing lightning risk assessment methods, this paper proposes a method for recommending differentiated lightning protection measures for transmission lines that uses multi-factor quantitative classification. The lightning protection calculation and retrofit measures were integrated into a single process, considering factors like lightning resistance levels, tripping rates, and damage risk levels, considering tower differences and technical and economic feasibility. Subsequently, a study was conducted on a certain 110 kV transmission line, resulting in a good lightning protection effect.

21.2 Transmission Line Models and Multi Factor Quantitative Classification

21.2.1 Basic Parameters and Model Construction of Transmission Lines

The research object of this article is a 110 kV AC transmission line, which includes 58 towers. The tower structure, grounding resistance, lightning parameters, ground inclination angle and other parameters are comprehensively collected, and all towers are straight towers.

The transmission line model was constructed using ATP electromagnetic transient analysis software; the main structure is shown in Fig. 21.1, where the lightning current uses a double oblique wave and the transmission line uses the J. Marti model. Most tower types are cat head towers, while some are drum towers with single circuit wiring, as Fig. 21.2 illustrates.

Furthermore, Fig. 21.3 plots the ground flash density of the line corridor using a 2*2 km grid size, as suggested by standard Q GDW 11,452. The line is evidently located in the C1 and C2 lightning zones, necessitating the completion of the lightning risk assessment.

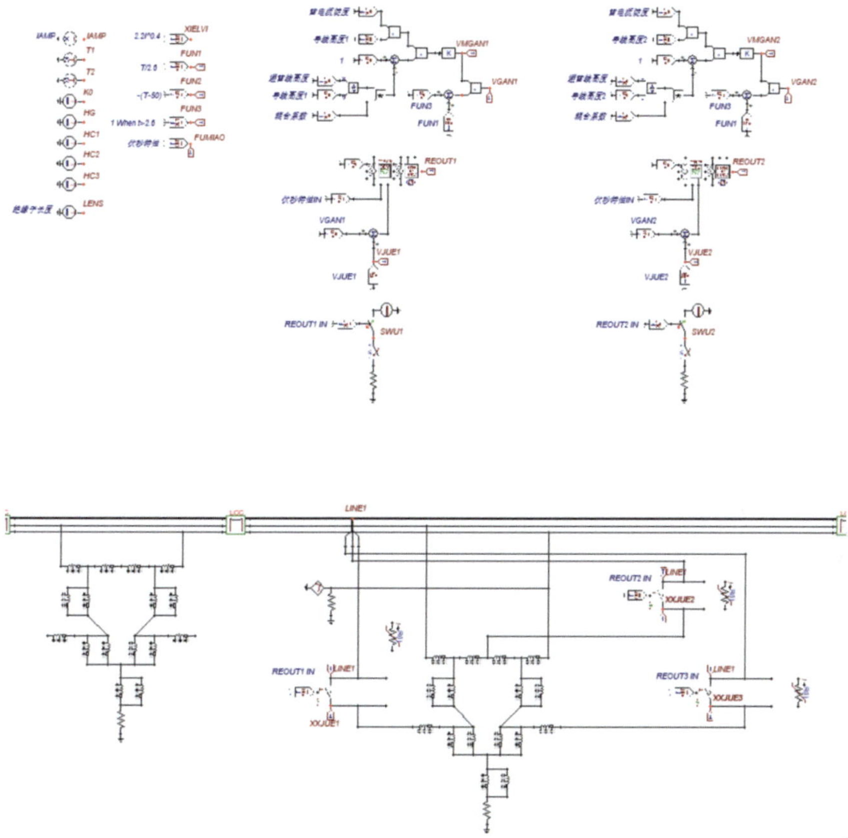

Fig. 21.1 Computational model local graph

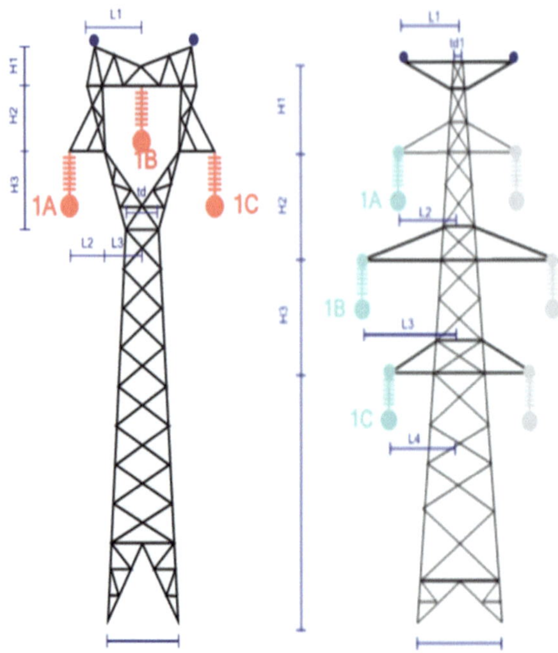

Fig. 21.2 Cat's head and drum tower

21.2.2 Multi-Factor Quantitative Classification

According to the actual operation experience of the line as well as reviewing the literature [7, 8], it is known that the main factors affecting the risk of lightning strikes on transmission lines include: insulator string length is too short, grounding resistance is too large, the height of the tower is too high, the angle of inclination of the arbitrary ground on both sides is too large, and the angle of protection of the lightning conductor is too large. Therefore, multi-factor quantitative classification is adopted for the above factors. Considering the benchmark parameter values of 110 kV pole towers specified in the standard GB/T 50,064–2014, this paper gives the specific grading values, as shown in Table 21.1.

Figure 21.4 displays the tower specifications in three dimensions at each level for the 110 kV transmission line used in this paper. The towers' latitude and longitude are represented by the x- and y-axes, and their level of parameters is indicated by the z-axis. It is evident that several towers have parameters in the more severe level 1 and 2.

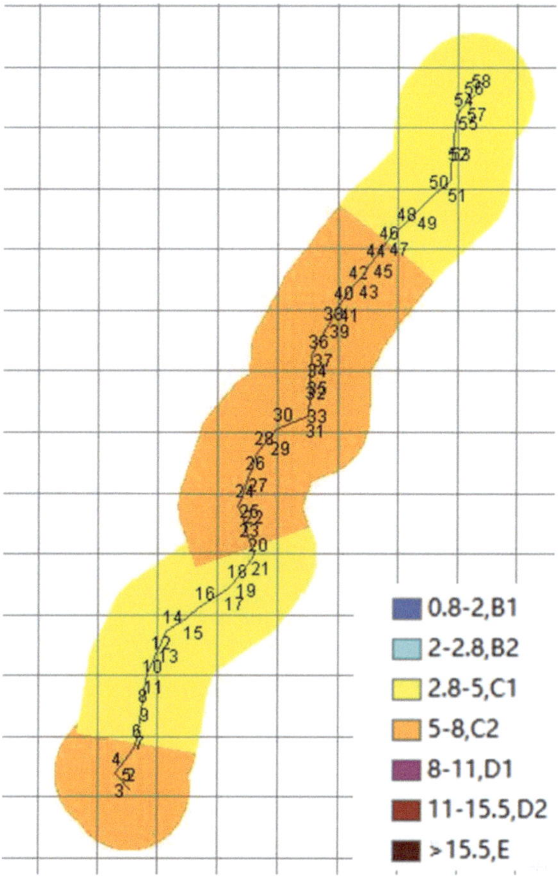

Fig. 21.3 Line corridor ground-flash density map (Unit:times/km^2/year)

Table 21.1 Comparison table for grading of influencing factors

		Level 1	Level 2	Level 3	Level 4	Level 5
Insulator string length	÷ Standard	<1.0	1.0~1.1	1.1~1.2	1.2~1.3	>1.3
	Concrete value /mm	<1314	1380	1511	1642	2200
Grounding resistance		>20Ω	15−20Ω	10−15Ω	5−10Ω	≤5Ω
Tower height	÷ Standard	>1.3	1.2~1.3	1.1~1.2	1.0~1.1	<1.0
	Drum tower	>36.4	33.6–36.4	30.8–33.6	28–30.8	<28.0
	Else	>26.78	24.72–26.78	22.66–24.72	20.6–22.66	<20.6
Ground inclination		>15°	10–15°	5–10°	0–5°	<0°
Protection angle		>5°	0–5°	(−5)−(0)	(−10)~(−5)	<−10°

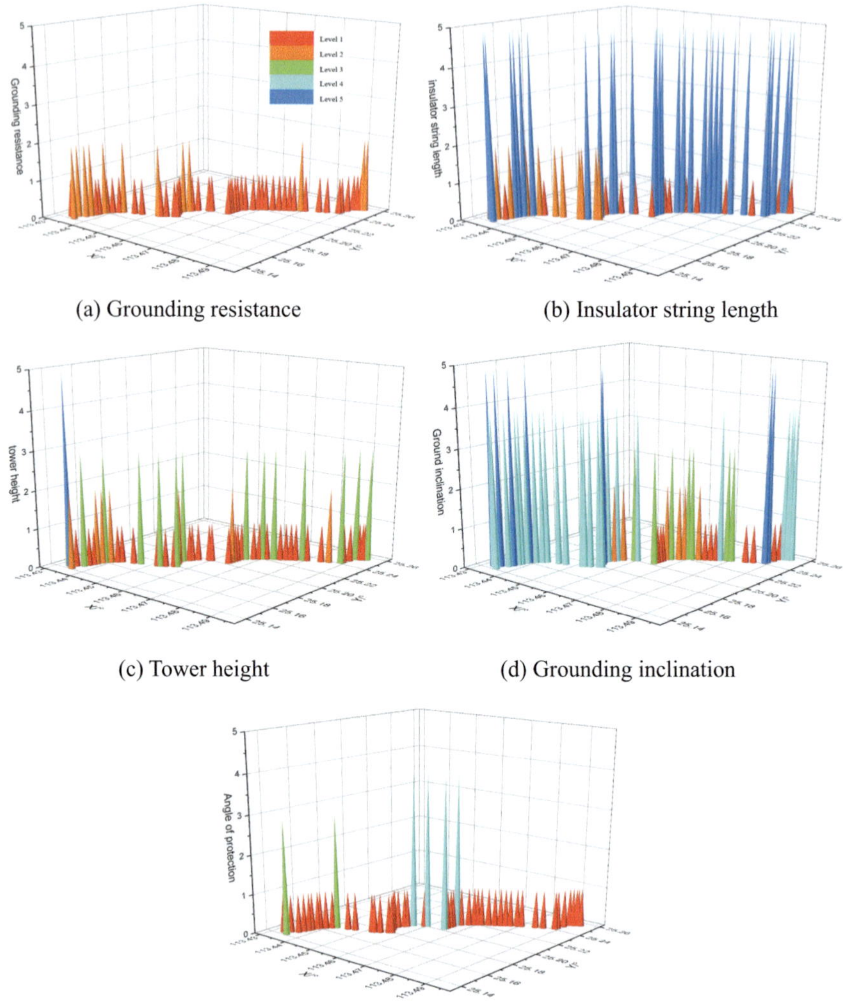

(a) Grounding resistance

(b) Insulator string length

(c) Tower height

(d) Grounding inclination

(e) Angle of protection

Fig. 21.4 The factor level of each tower

21.3 Lightning Risk Assessment

21.3.1 Calculation of Lightning Resistance Levels and Trip Rates

In order to simulate the bypass and counterstrike, respectively, the lightning current (current source) model is applied to the transmission line and the tower top lightning line using the ATP-EMTP electromagnetic transient analysis software. In the simulation software, first set the initial values of the lightning current Imin and Imax so that the insulator does not flashover, and then use the dichotomy method to gradually reduce the range of lightning amplitude, the final lightning current Ix, which is the size of the level of lightning resistance, and its error is within the control range. Among them, whether the insulator flashover is the use of the "intersection method" to determine. In order to make the calculation and solution process faster, this paper adopts the Java program based on semantic recognition call to realize the automated calculation of dichotomous method, which is not repeated here.

After calculating the lightning resistance level, the electrical geometric modeling method (EGM) is used and improved to some extent according to the actual situation to calculate the winding strike trip rate of the tower. The counterstrike trip rate is calculated according to the standard GB T 50,064–2014. The ATP software is the kernel in the whole calculation process, combined with the java calling program to form a set of lightning resistance level and trip rate batch automatic calculation system, which improves the calculation efficiency.

21.3.2 Criteria and Results of Lightning Risk Assessment

The risk assessment rules for the lightning strike trip rate of the line are determined on the basis of the Technical Guidelines for Lightning Protection of Overhead Transmission Lines issued by the Southern Power Grid Corporation, with reference to the actual operating experience of the line and other factors, The lightning strike trip rate for overhead transmission lines of 110 kV voltage level is converted to the baseline reference value *S' = 0*.525 times/(100 km-a) at 40 thunderstorm days per year (i.e., ground flash density of 2.78 times/(km2-a)). The control value control reference value S for each tower is calculated by the following formula.

$$S = S'N_{gav}/2.78 \quad (21.1)$$

where N_{gav} is the actual ground flash density of the line corridor in times/(km^2·a). The transmission line lightning strike risk classification rules are shown in the following equation, where: R_{i} is the tower lightning strike trip rate.

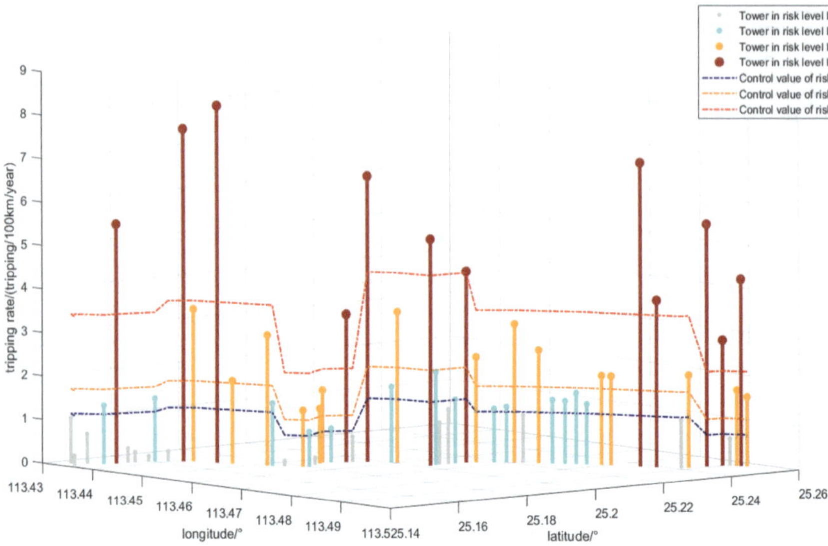

Fig. 21.5 Lightning strike risk rating chart

$$\begin{cases} R_i < 1.0S & (I,\ lowest\ risk) \\ 1.0S \le R_i < 1.5S & (II,\ low\ risk) \\ 1.5S \le R_i < 3.0S & (III,\ high\ risk) \\ R_i > 3.0S & (IV,\ very\ high\ risk) \end{cases} \tag{21.2}$$

Based on the tripping rate obtained from the previous calculation, the lightning strike risk level of the line is plotted as shown in Fig. 21.5. Where the x-axis and y-axis are the latitude and longitude of the towers, the bars of different colors on the z-axis indicate the trip rate $_{\mathrm{Ri}}$ of each tower, and the dashed lines of different colors indicate 1. *0S*, 1. *5S* and 3. *0S* in Eq. (21.2), respectively. The lightning risk of each tower can be visualized, of which 47.37% of the towers are at a high risk of 3 or 4, urgently need to carry out differentiated lightning protection modification.

21.4 Transmission Line Differentiated Lightning Protection Modification

21.4.1 Finding the Causes of High Risk in the Tower

For the higher risk towers, the possible causes are: excessive grounding resistance, insufficient insulator string length, no lightning arrester installed, etc. Figure 21.7 uses a flowchart to develop a specific cause analysis for the high risk towers. The

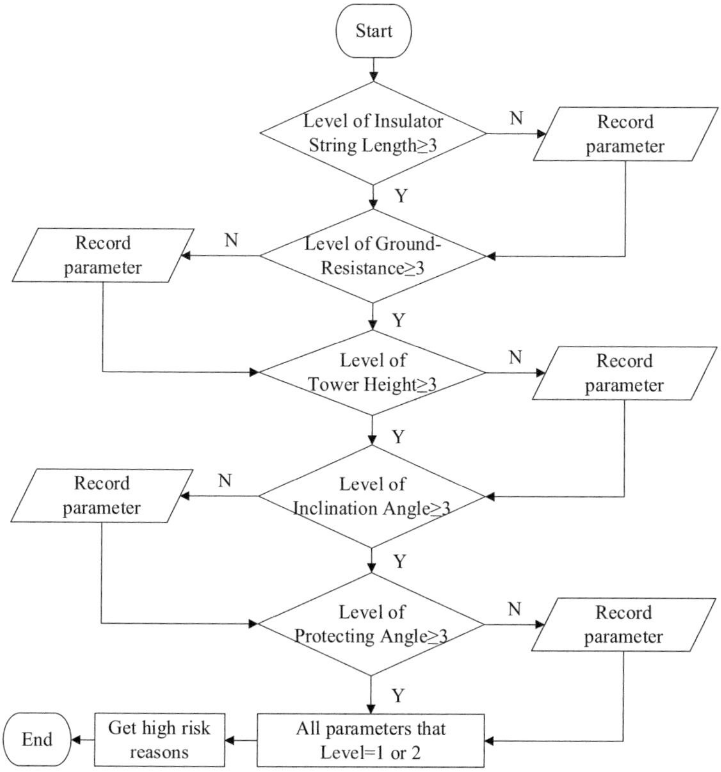

Fig. 21.6 Flowchart for analyzing the causes of high-risk towers

causes of high-risk towers on this line vary according to the respective conditions of each tower, as shown in Fig. 21.4.

21.4.2 Intelligent Proposals for Improving Lightning Protection Retrofit

According to the actual operation of the line, increasing the length of the insulator string can effectively reduce the tripping rate of the shielding failure, while lowering the grounding resistance can improve the tower's ability to drain the lightning current and thus reduce the generation of back flashover overvoltage. These two methods are relatively low technical difficulty, high economic lightning protection measures, and thus preferred. The tower's height and ground inclination angle are too high and large, making it more susceptible to lightning strikes, but these parameters are predetermined and difficult to alter. The EGM explains that the angle of protection

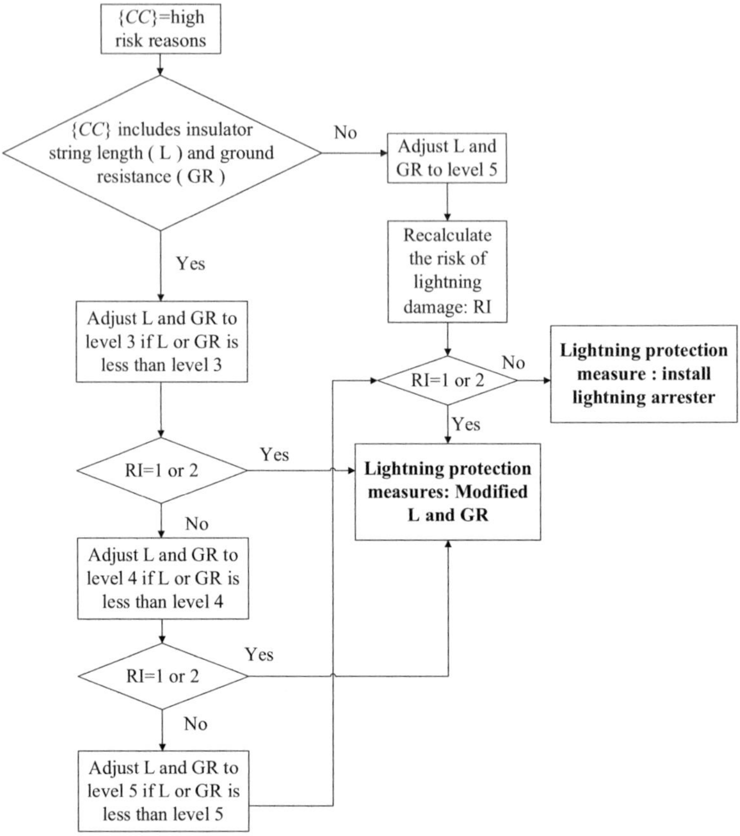

Fig. 21.7 Recommended flowchart for differentiated lightning retrofit programs

of the lightning line significantly impacts the shielding effect of lightning shielding failure, which is influenced by the tower head structure and the method of hanging insulators in the string, which is already designed based on construction drawings and other factors. In this calculation example, the tower line has been built rather than in the design stage, so it is not appropriate to make substantial changes. In addition, the installation of lightning arresters can effectively improve the level of lightning resistance of the transmission lines to lighting shielding failure. In theory, it can be assumed that there will be no more shielding failure tripping faults after installing lightning arresters, while the lightning arrester cost is high, so considering the economy, it will be the last recommended lightning protection measure as a last resort.

In summary, considering the technical difficulty and economy of each measure, this paper proposes a differentiated lightning protection retrofit process for transmission lines, as shown in Fig. 21.7, based on the aforementioned idea of multi-factor quantitative classification and the process of analyzing the causes of high-risk towers. Starting from the results of risk level calculation, different lightning protection measures are recommended according to the different conditions of the tower parameters, which realizes the differentiation and improves the efficiency of decision-making.

Figure 21.7 depicts the process of modifying the insulator string length and grounding resistance in stages; if the risk of lightning strike cannot be reduced to a low risk under the level 3 parameter, the parameter will be adjusted to level 4 or even 5, and if the risk of lightning strike remains high, a lightning arrester will be installed.

Figure 21.8 shows the effect of each stage of lightning protection modification. After the retrofit measures are adjusted to level 3, among the remaining high-risk towers, the shielding failure trip rate is the dominating factor producing the risk. Figure 21.9 depicts the statistical changes in the shielding failure and back flashover trip rates before and after the parameters are modified to level 4. The main cause of its high shielding failure trip rate is that the pole tower's protection angle is excessively large, ranging from 11.8 to 24.9°, making the shielding effect of the lightning conductor on the conductor negligible. However, as previously stated, the lightning conductor's protection angle should not be significantly changed; thus, given the technical nature of differential lightning protection, we continue to prioritize the modification of grounding resistance and insulator length. For the last parameter update to level 5, which is still in the high risk category of the eight towers, it is recommended to install a lightning arrester, which can significantly minimize the danger of shielding failure. Furthermore, the price of a lightning arrester for a 110 kV AC line is not prohibitively expensive, and the installation has a more mature operation procedure, so the differentiated lightning protection notion provided in this research is both applicable and sensible.

The magnitude of parameter changes of quantitatively differentiated lightning protection modification is analyzed separately according to the risk level sorted from high to low, as shown in Fig. 21.10: The trend of the magnitude of tripping rate reduction has an obvious correlation with the trend of grounding resistance changes, while the correlation with the magnitude of insulator string changes is weaker, which indicates that the reduction of the grounding resistance is more effective in reducing the tripping rate in the present line. As analyzed above, this is because the conventional lightning protection measures have no obvious effect on the reduction of the shielding failure trip-out rate under the condition of large protection angle, and the grounding resistance mainly affects the counterattack performance.

As shown in Fig. 21.11, the entire line has been reduced to a low risk (Class I and Class II) through the implementation of differentiated lightning protection measures.

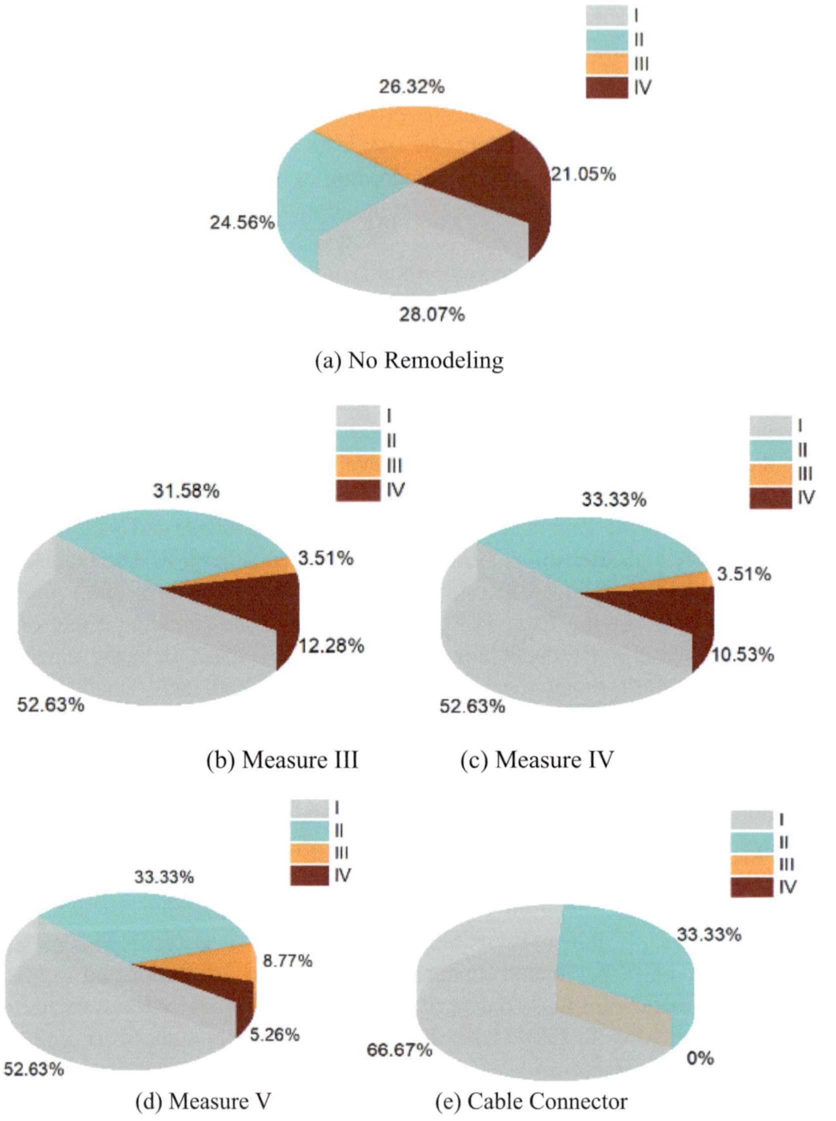

Fig. 21.8 Lightning protection retrofit effect of each stage

21.5 Conclusion

This study uses a 110 kV transmission line as the research object, first gathering the essential information about the line towers required for lightning protection computation, and then screening out all of the parameters influencing the probability of lightning damage and conducting quantitative classification. The upgraded EGM

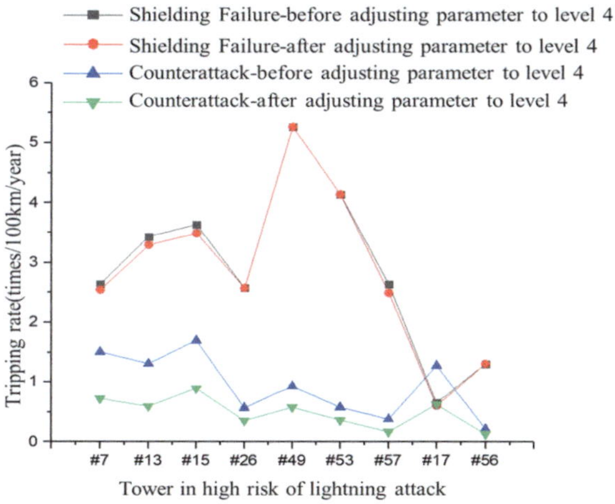

Fig. 21.9 Shielding failure and counterattack trip rates before and after parameter adjusted to level 4

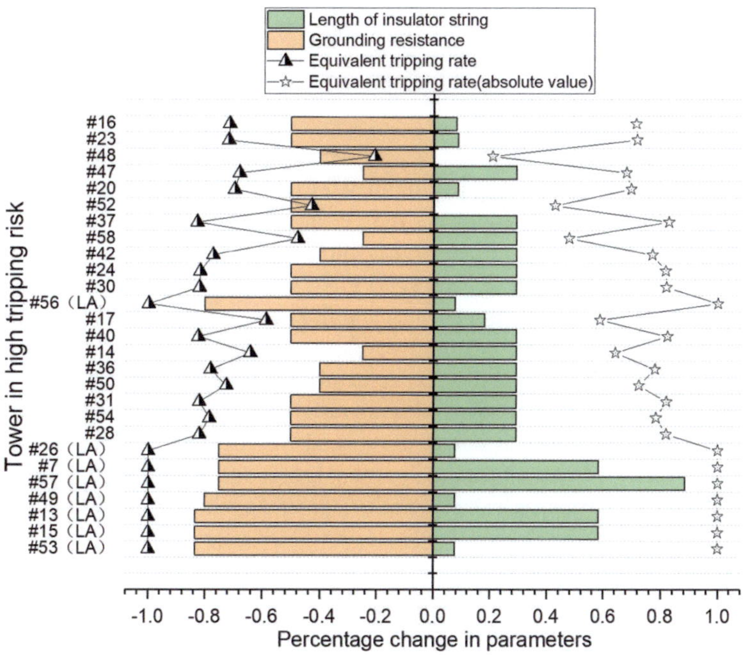

Fig. 21.10 Amplitude of parameter changes and tripping rate changes for lightning protection modification

and ATP algorithms were then used to calculate line shielding failure, back flashover trip rate, and finely analyze the lightning risk of 57 towers, giving the conclusion that roughly half of the towers were at high risk. After quantitatively examining the

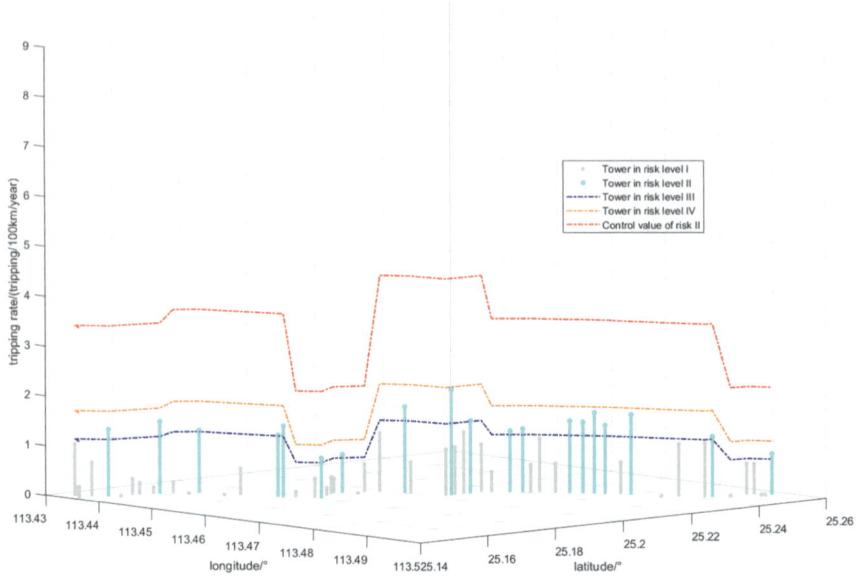

Fig. 21.11 Lightning risk level map after lightning protection retrofit

reasons for the high-risk towers, Differentiated lightning protection renovation plan in stages was carry out, and all towers were eventually rated as low risk.

The method proposed in this paper takes into account the technical economy of all kinds of lightning protection measures, prioritizes the reduction of grounding resistance and the growth of insulator strings, and finally considers the addition of lightning arrester, which effectively improves the decision-making efficiency. Furthermore, this paper discovers that the change in grounding resistance in the calculation example has a greater impact on the tripping rate, whereas the insulator string length has a smaller impact, which is due to the special case of the tower protection angle being too large, so the recommended lightning protection measures proposed in this paper have certain limitations, and a certain degree of optimization is required in the future.

The research in this paper can provide an idea and reference value for the effective differentiated lightning protection of transmission lines.

Project funding The project funded in this paper is the research on lightning risk assessment and precise lightning protection transformation technology of transmission lines based on multi-source data, No.2024–47.In this paper, the formulation of technical scheme and simulation calculation and analysis work are completed with the strong support of the staff of State Grid Jilin Electric Power Co., Ltd., and sincere thanks are expressed to him (her).

References

1. Jiang T, He B, Liao S, et al.: Research on key technologies and multi-scenario application of quantitative prediction of short-term meteorological disaster risk. Meteorological, Hydrological and Marine Instruments 41(05), 105–108 (2024).
2. Silveira F H , Almeida F S , Visacro S.: Assessing the influence of peak current distributions of first return strokes on the lightning performance of transmission lines: Instrumented tower distributions versus standard distributions. Electric Power Systems Research 214, 1–8 (2023).
3. Jahangiri T , Wang Q , Silva F F D ,et al.: Lightning Shielding Performance of Fully Composite Pylon (2020).
4. Zeng Cheng, Zhou Yusheng, Zhou You, et al.: Research on lightning withstand performance of hybrid voltage transmission lines. High voltage electrical apparatus 57(09), 131–137 (2021).
5. Luo J.: Analysis of Lightning Protection Status of 35~110kV Transmission Lines and Countermeasures. Journal of Physics Conference Series 2033 (2021).
6. Tao Z , Dai L U , Hong-Bin W ,et al.: Analysis of Lightning Stroke Fault of 110kV Transmission Line and Research on Lightning Protection Reform. Telecom Power Technology (2019).
7. Trotsenko Y, Peretyatko J, et al.: Overview of factors affecting the estimation of lightning shielding performance of overhead transmission lines. (2022).
8. Chen H, Chen C.: Research on lightning control method of 10kV distribution line. Electric porcelain lightning arrester (02), 8–13 (2022).

Chapter 22
Self-Healing and Degradable Epoxy Resin Insulating Material with Boronic Ester Bonds Synergized with Permanent Cross-Linking Cites

Zhaoxin Teng and Mingxiao Zhu

Abstract The epoxy resin insulating components are influenced by electric fields during service, leading to the formation of dendritic electrical damage channels within the material, which can result in insulation breakdown. Furthermore, the insoluble and infusible characteristics of traditional epoxy resins pose challenges for the recycling and disposal of decommissioned equipment. In this study, based on the theory of dynamic chemical bonds exchange reactions, we introduced dynamic boronic ester bonds into a conventional epoxy resin system. This resulted in an epoxy resin insulation material that is cross-linked by both boronic ester bonds and traditional anhydride curing agents. We employed a needle-plate electrode model to investigate the insulation properties and self-repair capabilities of this material. The results indicate that the material exhibits excellent insulation performance as well as self-healing abilities; after thermal healing of dendritic defects caused by electrical stress, its breakdown strength was restored to 82.22% of its initial value. Additionally, we conducted degradation recycling experiments to explore the dynamic exchange and degradation characteristics of the resin. Our findings reveal that cross-linking with boronic ester bonds endows the resin network with excellent dynamic properties, enabling environmentally friendly degradation of resin materials. This research provides new insights and options for self-repairing internal dendritic defects in epoxy resin insulating materials as well as for harmless recycling processes for decommissioned equipment.

Keywords Boronic ester bonds · Electrical tree · Self-healing · Degradation and recovery

Z. Teng · M. Zhu (✉)
College of New Energy, China University of Petroleum (East China), Qingdao, China
e-mail: zhumx@upc.edu.cn

Y. Li et al. (eds.), *Proceedings of the 5th Chinese Conference on High Voltage and Discharge Plasmas*, Springer Proceedings in Physics 254,
https://doi.org/10.1007/978-981-96-5711-7_22

22.1 Introduction

Epoxy resins are widely used in the power industry for the manufacture of insulating components such as insulating bushings, support insulators, contact boxes, insulating cylinders and poles due to their excellent insulating properties, mechanical properties, good workability and simple curing [1]. However, during operation of these devices, polymeric materials are subjected to long-term stresses from high electric fields, mechanical vibrations, and extreme temperatures, and physical and electrical damage can occur internally [2, 3]. Among them, electrical tree triggered by injected charges and localized discharges is a common type of microscale dendritic tubule electrical damage [4]. Unlike gas and liquid insulating media recoverable insulation, electrical damage in solids is often considered to be a permanent damage defect, and this tiny electrical tree will continue to expand, resulting in a gradual decline in the insulating properties of the insulating parts and even triggering the insulation breakdown, leaving a serious risk to the safe and stable operation of the electrical equipment [5]. Existing defect detection methods, such as visible/infrared/ultraviolet (V/IR/UV) imaging, surface potential measurements, and partial discharge detection [6–11], are difficult to effectively identify and solve the problem of internal electrical trees in insulating parts. In addition, the stable cross-linking structure brings excellent performance to the material while making it difficult to dissociate, which brings challenges to the decommissioning recycling treatment of power equipment and the recovery of internal materials. Therefore, it is of great significance to develop epoxy resin insulating materials with excellent insulation properties, while also enabling the resin to have self-healing properties and the ability to decompose and recycle.

Inspired by the biological self-healing phenomenon, some scholars focus on the insulating material itself, and utilize the means of introducing microcapsules, microvessels or reversible chemical bonding into the material to make the material have the ability to self-repair the electrical trees, and maintain the insulating property of the material. Microcapsule-type self-repairing materials refer to the filling of microcapsules or microvessels containing repairing agents inside the material, and when the material suffers internal damage, the microcapsules realize the repairing of the damage by releasing the repairing agents inside them and initiating the polymerization by contacting with the catalyst in the material matrix [12–15]. Subsequent paragraphs, however, are indented. Lesaint et al. [16] explored the healing of electrical trees in epoxy resin matrix using a polyurea-formaldehyde encapsulated DCPD microcapsule/Grubbs' catalyst system, and the experimental results revealed that the presence of the microcapsules not only allowed for the repair of the discharge channel through the curing agent released from the microcapsules, but also inhibited the development of electrical trees. However, after self-healing of microcapsule-based self-healing materials, the residual capsule shell of the matrix introduces new defects into the material, which may result in a degradation of the insulating properties of the material. In addition, the number of repairs is limited by the complexity of the preparation process for microcapsule-based materials and the fact that a low concentration of microcapsules is usually used in order to maintain the insulating properties of the composite material. Currently, the main reversible chemical bonds that can be used to construct self-healing materials are disulfide bonds [17–19], Diels–Alder bonds [20–22], dynamic boronic ester bonds [23–25], ionic bonds [26], hydrogen bonds [27], and Π—Π stacking [28]. Jiaye Xie et al. [29] used 3-furanic acid and maleic anhydride to prepare a dynamic cross-linking agent for epoxy resins, and introduced D-A bonds into the resin system, and the results showed that the material could realize self-healing through the transformation from a fully cross-linked state to a partially cross-linked state, and after 10 transformations, the mechanical and insulating properties of the material could still be maintained. Disulfide bonds have high self-healing efficiency, but the repaired structures become unstable in the presence of light, heat, or mechanical forces [30, 31]. Diels–Alder bonds are thermally initiated covalent bonds, and thus self-healing processes based on the Diels–Alder reaction usually require harsh external conditions, such as high temperatures and long heating times [32]. Hydrogen bonding, for example, is capable of achieving significant repair efficiency under milder conditions. However, due to the low bonding energy of non-covalent bonds, inherent properties such as mechanical, thermal and insulating properties are degraded while repairing the damage. In particular, the valence reaction of dynamic bonding narrows the band gap of polymers, which reduces the breakdown strength and makes it difficult to be applied to insulating materials [17].

However, the self-healing capability often contradicts the intrinsic properties including mechanical strength and insulating properties due to the fact that the self-healing process tends to require high molecular chain mobility, which in turn reduces the mechanical and insulating properties. Therefore, it is a major challenge to harmonize the contradiction between the inherent properties of insulating materials and their self-healing capabilities. In addition, repairing electrical damage under mild conditions is also an urgent problem. The introduction of boronic ester bonds into epoxy resin systems is a promising strategy for efficient self-healing of organics. The bond exchange reaction can be triggered at room temperature and the reaction occurs within seconds, much faster than most types of dynamic covalent bonds [33].

In this paper, dynamic boronic ester bonds are introduced into conventional epoxy resin, and epoxy resin insulating materials co-crosslinked with boronic ester bonds and conventional anhydride curing agents are prepared, which endow their self-healing properties and ensure their mechanical strength at the same time. Needle plate electrode structure was used to study its electrical trees initiation voltage and breakdown voltage under industrial frequency electric field conditions to characterize the insulating properties of the material. The self-constructed electrical trees observation platform is used to observe the morphology of electrical trees before and after self-healing to visualize the self-healing performance of the material, and the breakdown voltage and leakage current are tested before and after degradation and self-healing to characterize the self-healing performance of the material. Finally, based on the excellent dynamic properties of the constructed cross-linking network, the degradation recycling and hot-pressing remodeling experiments were carried out. The experiments demonstrate that the material can realize self-healing of internal electrical trees channels, efficient degradation and hot-pressing recycling.

22.2 Experimental Methods

22.2.1 Experimental Materials

The experimental materials and manufacturers used in this experiment are shown below, and all reagents were used directly without purification.

Bisphenol A-type epoxy resin (DGEBA, Changzhou Runxiang Chemical Co., Ltd.), 3-mercapto-1,2-propanediol (MP, Myracle), trimethyl borate (TMB, Aladdin), 1,8-diazabicyclo[5,4,0]undec-7-ene (DBU, Aladdin), methyl tetrahydrophthalic anhydride (MTHPA, Changzhou Runxiang Chemical Co., Ltd.), 2,4,6 Ltd.), 2,4,6-tris(dimethylaminomethyl)phenol (DMP-30, Changzhou Runxiang Chemical Co., Ltd.). All reagents were used directly without purification.

22.2.2 Sample Preparation

The self-healing epoxy insulators co-crosslinked with boronic ester bonds and acid anhydride curing agent were named EP(B-O&MTHPA), and the reaction principle for their preparation is shown in Fig. 22.1a and b: on the one hand, the epoxy groups reacted with and crosslinked with MTHPA under the promotion of DMP-30 [34]. On the other hand, in an alkaline environment, the epoxy group reacts with the sulfhydryl group in the MP molecule to produce a molecular fragment containing β-dihydroxy end groups [35–38]. Further, these β-dihydroxy end-groups react with trimethyl borate [39–42] under the alkaline catalytic action of DBU to remove methanol and cross-link to form a polymer network. The preparation method was as follows: weighing a certain amount of TMB, MP, DGEBA, MTHPA, and DMP-30 (the mass ratio of DGEBA, MTHPA, and DMP-30 for curing by anhydride only was 100:80:1; the molar ratio of sulfhydryl groups to epoxy groups for cross-linking by boronic ester bonds only was 1:1, and the molar ratio of trimethyl borate to 3-mercapto-1,2-propanediol was 1:2. The exact amount is determined by the ratio of the B-O bond to the permanent cross-linking point in the reaction system, and the ratio chosen herein is 1:1). The materials were homogenized in a thermostatically heated magnetic stirrer at high speed for 1 h at 60 °C, followed by vacuum debubbling for 0.5 h in a vacuum drying oven at a temperature of 60 °C. In order to prevent the generation of air bubbles in the samples, a casting method in which the needle electrodes were pre-implanted was used in this study. Specifically, the mixed solution after vacuum defoaming was poured into a mold in which the needle electrode had been placed, and at this time, both the mold and the needle electrode had been preheated to 60 °C. Subsequently, the molds containing the mixed solutions were placed into a vacuum drying oven and cured at 90 °C for 8 h. After the curing was completed, the molds were naturally cooled to ambient temperature in the vacuum drying oven and then removed from the molds for subsequent testing. The crosslinked network formed by the reaction is shown in Fig. 22.1c, and the whole reaction flow is shown in Fig. 22.2.

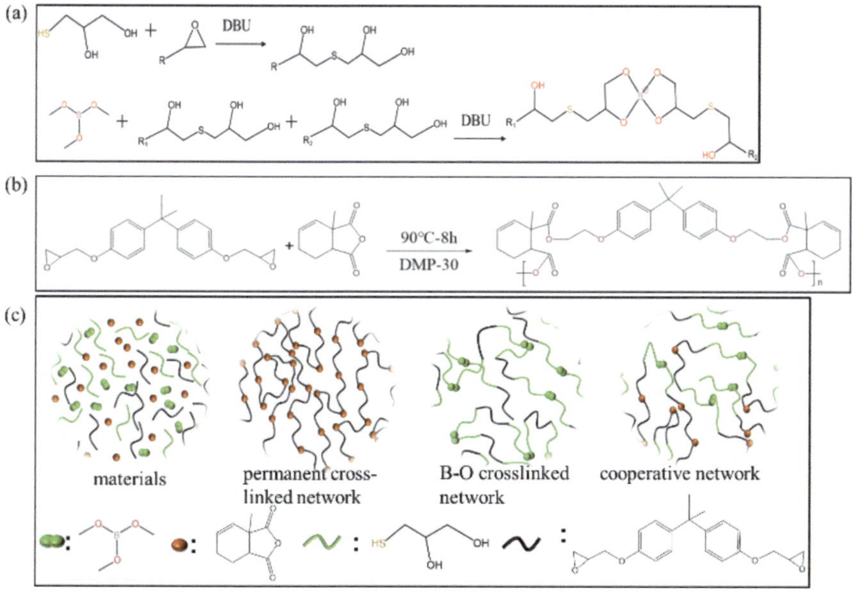

Fig. 22.1 Reaction principle and crosslinking network

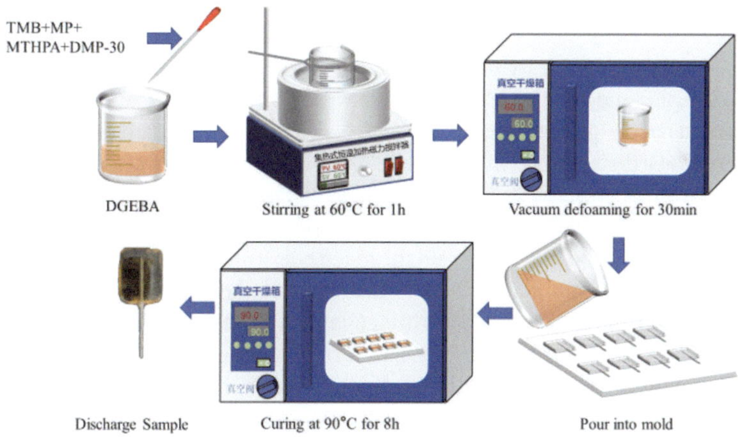

Fig. 22.2 Preparation process

22.2.3 Testing and Characterization

FTIR

The resin system was analyzed for the introduction of boronic ester bonds using Fourier Transform Infrared Spectroscopy (FTIR) spectroscopy (Thermo Scientific iN10, USA). The raw materials and final products were tested, all in ATR test mode measuring wave numbers in the range of 400 to 4000 cm^{-1}.

DSC

The glass transition process of the two resin systems was studied using a differential scanning calorimeter (Mettler Toledo –DSC3, Switzerland). The samples were sealed in a crucible. During the test, the samples were heated from 25 to 120 °C and first cooled from 120 to 25 °C to eliminate the thermal history. Then, it was heated from 25 to 120 °C to study the glass transition process. The rates of temperature change were all 10 °C/min and the test atmosphere was N2.

Tensile Strength Test

The tensile strength and Young's modulus of EP and EP (B-O & MTHPA) were tested using an electronic universal testing machine (INSTRON 5982, USA).

DMA

Dynamic mechanical analysis (DMA) method (TA-Q800, USA) was used to determine the dynamic mechanical properties of EP (B-O&MTHPA) under single cantilever fixture. The sample size was about 30 × 10 × 5 mm. The test temperature range was from 30 to 100 °C, and the heating rate was 3 °C/min. The amplitude was 10 μm and the frequency was 1 Hz.

Work-frequency Electrical Trees Initiation, Growth and Observation

The accelerated aging test by needle-plate electrode produces electrical trees inside the epoxy resin, and the test platform is shown in Fig. 22.3a. The experiment was conducted in needle-plate discharge mode, where a needle was inserted into the epoxy resin sample at a distance of 2 mm from the plate. The tail of the needle was connected to the high voltage to serve as the high voltage electrode. The epoxy resin plate was pressed with conductive silicone rubber and copper plate, and the copper plate was grounded as the ground electrode, and the electrode structure is shown in Fig. 22.3b. Using a needle tip to simulate defects in the material, the externally applied voltage was stepped up from 3 kV, and each voltage level was held for 5 min until electrical trees began to form inside the sample. The corresponding voltage value was defined as the starting tree voltage when the electrical trees in the specimen grew from the tip to 10 μm in the laboratory environment. This parameter is a key indicator of the occurrence of electric dendrites and is a critical factor in assessing the insulating properties of materials [43]. After recording the voltage readings when the length of the electrical trees reached 10 μm, the voltage was set to 15 kV in the experiment and the aging process was continued. After the formation of electrical trees defects of different lengths, the experiment was stopped to obtain the aged samples. The aged samples were placed in a vacuum drying oven under heated conditions for self-healing and the self-healing effect was observed under a microscope. In order to prevent flashover along the surface of the specimen during the pressurization process and to ensure the transparency of the observation system, the experimental samples were submerged in dimethylsiloxane oil throughout the experimental process. In

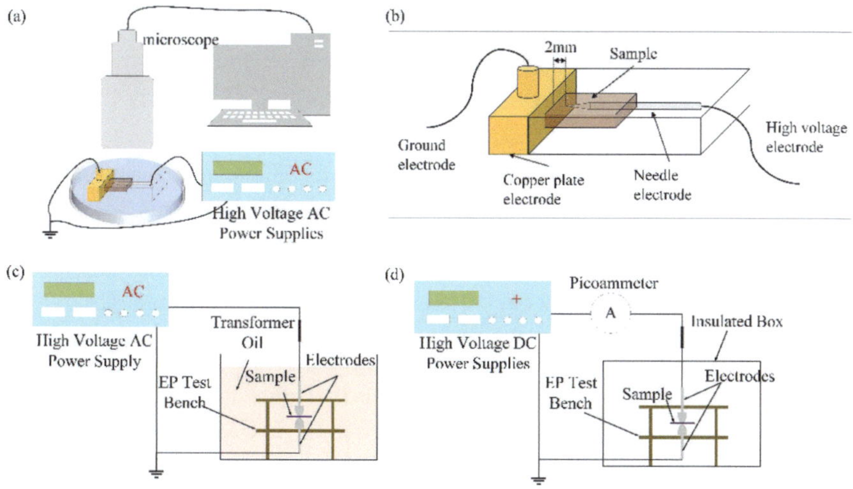

Fig. 22.3 Experimental equipment

order to prevent flashing along the surface of the specimen during the pressurization process and to ensure the transparency of the observation system, the specimen was always surrounded by dimethylsiloxane oil during the experiment.

Breakdown Test

The breakdown performance test was carried out using an artificially controlled AC breakdown test device. The high-voltage power supply was produced by Dongwen High Voltage Power Supply (Tianjin) Co., Ltd. with a rated voltage of 10 kV and an output current of 2 mA, and the schematic diagram of the AC breakdown test system is shown in Fig. 22.3c. The test electrodes are copper hemispheres with a diameter of 3 cm, the AC breakdown test environment is insulating transformer oil at room temperature, and the test bench is made of epoxy resin (EP) insulating material, and the test is conducted with a linear voltage increase at a boost rate of 1 kV/s, with a set boost maximum value of 40 kV.

Leakage Current Test

The sample is placed in an insulated box, and a DC power supply with positive polarity is connected to a resistor connected to the positive terminal of the sample. A series Picoammeter is connected to observe the leakage current, and the test circuit is shown in Fig. 22.3d. A voltage of 10 kV was applied to the DC power supply. After applying voltage for five minutes, when the leakage current value is stabilized, read the number on the Picoammeter, read ten values and take the average value as the leakage current value of the sample.

22.3 Experiments and Analysis of Results

22.3.1 *Thermal and Mechanical Properties of Materials*

Figure 22.4a shows the FTIR spectra of the reaction raw materials and products, from which it can be seen that the asymmetric stretching vibrational absorption peak of −SH at around 2557 cm^{-1} [44], and there is no peak located at 2557 cm^{-1} in EP(B-O&MTHPA), and at the same time, the stretching vibrational peak of the epoxide group located at 1229 cm^{-1} [45] and the bending vibrational peak of the epoxide group located at 827 cm^{-1} were also significantly reduced in intensity, proving that the sulfhydryl group reacted with the epoxy group. TMB had a strong B-O stretching vibration peak at 1342 cm^{-1}, while the stretching vibration absorption peak of B-O (newly formed boronic ester bonds) in EP(B-O&MTHPA) was shifted to 1322 cm^{-1}; compared to the infrared spectral pattern of B-O of TMB, EP(B-O&MTHPA) formed new B -O with different peak profiles, proving that the reaction successfully introduced boronic ester bonds in the epoxy resin system [46]. The DSC test results of EP(B-O&MTHPA) are shown in Fig. 22.4b, and its glass transition temperature (Tg) is 85.72 °C. The DMA test results of EP(B-O&MTHPA) are shown in Fig. 22.4c, and its storage modulus is maximum at about 30 °C, which is about 587 MPa, and its tanδ reaches its maximum at 89.61 °C, indicating that the glass transition temperature of EP(B-O&MTHPA) is 89.61 °C. It is generally believed that DMA is more sensitive than DSC in the measurement of Tg [47]. The tensile strength test results of EP and EP(B-O&MTHPA) are shown in Fig. 22.4d and e. The tensile strengths of EP and EP(B-O&MTHPA) are 57.13 MPa and 47.92 MPa, respectively, and the two materials exhibit good tensile strengths. EP and EP(B-O&MTHPA) have Young's modulus of 2.893 GPa and 3.992 GPa, respectively, and both materials exhibit transparent hard glassy body at room temperature. The maximum strains of EP and EP(B-O&MTHPA) before tensile breakage are 0.393% and 0.195%, respectively, and both materials have good rigidity.

22.3.2 *Characterization of Insulation Properties*

The experimental data of electrical trees initiation voltage of the specimens were statistically analyzed using a two-parameter Weibull distribution model, and the voltage value corresponding to 63.2% probability of tree initiation was defined as the electrical trees initiation voltage of the resin system. As shown in Fig. 22.5a, the electrical trees initiation voltages of EP and EP(B-O&MTHPA) were 18.47 kV and 17.53 kV, respectively, under industrial frequency voltage. In addition, the breakdown voltage tests were conducted in accordance with the IEC 60,243–1–2013 standard, and the Weibull distributions of the breakdown field strengths of EP and EP(B-O&MTHPA) are shown in Fig. 22.5b. From Fig. 22.5b, it can be observed that the breakdown field strengths of EP and EP(B-O&MTHPA) are 16.73 kV/mm and

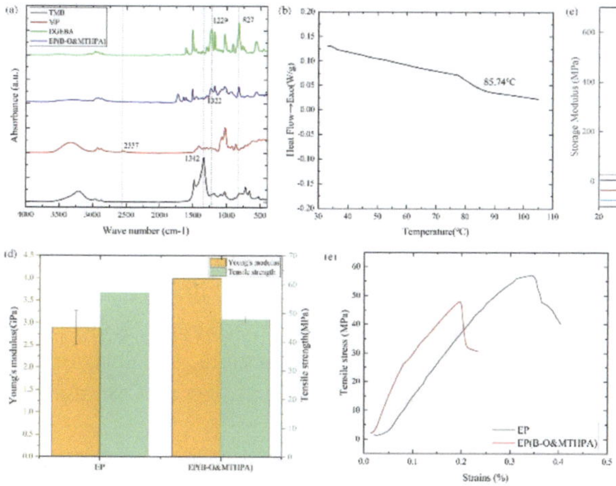

Fig. 22.4 Thermal and mechanical properties of materials

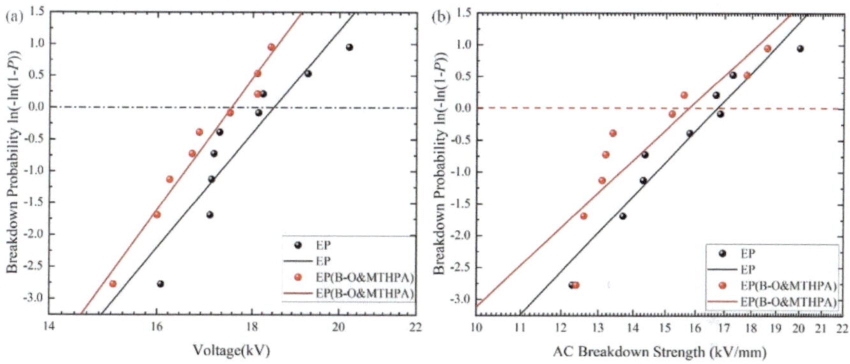

Fig. 22.5 Insulation properties of EP, EP(B-O&MTHPA)

15.75 kV/mm, respectively, under the action of the industrial frequency voltage. The experimental data show that the epoxy resin system is still able to maintain high dielectric strengths even after the introduction of the dynamic chemical bonding.

22.3.3 Electrical Trees Morphology and Self-Healing Properties

In this experiment, the distance from the tip of the needle to the farthest point on the growing end of the electrical trees was defined as the growth length of the electrical trees at that moment [48]. A constant voltage of 15 kV was maintained after the triggering of the electrical trees, and the morphology of the electrical trees was observed using a self-built electrical tree observation platform. Figure 22.6 shows the morphology of electrical trees after the initiation of different EP (B-O&MTHPA) samples and the morphology of branches after self-healing at 80 °C for 24 h. The morphology of electrical trees after the initiation of different EP (B-O&MTHPA) samples is shown in Fig. 22.6, which shows the morphology of electrical trees after the initiation of different EP samples. It can be seen from the figure that, with the increase of pressurization time, the electrical trees in the resin matrix extended to the ground electrode, and the electrical tree branches expanded in the resin matrix, which resulted in the continuous degradation of the resin matrix, and further led to the breakdown of the insulation. The samples after self-healing, the deterioration channel of the electrical trees is obviously shallow, which proves that the material has good self-healing performance.

In order to further verify the self-healing performance of the samples, the insulation performance of the samples before and after deterioration and after self-healing were also tested in this experiment. Figure 22.7a below shows the Weibull distribution of the AC breakdown strength of EP(B-O&MTHPA) samples in the initial state, after aging and after self-healing. The initial breakdown strength of EP(B-O&MTHPA) is 15.75 kV/mm, and the breakdown strength of the sample after deterioration is destroyed, and the breakdown strength drops to 10.15 kV/mm, which is only 64.44% of the initial one, thus the insulation performance is greatly reduced. It

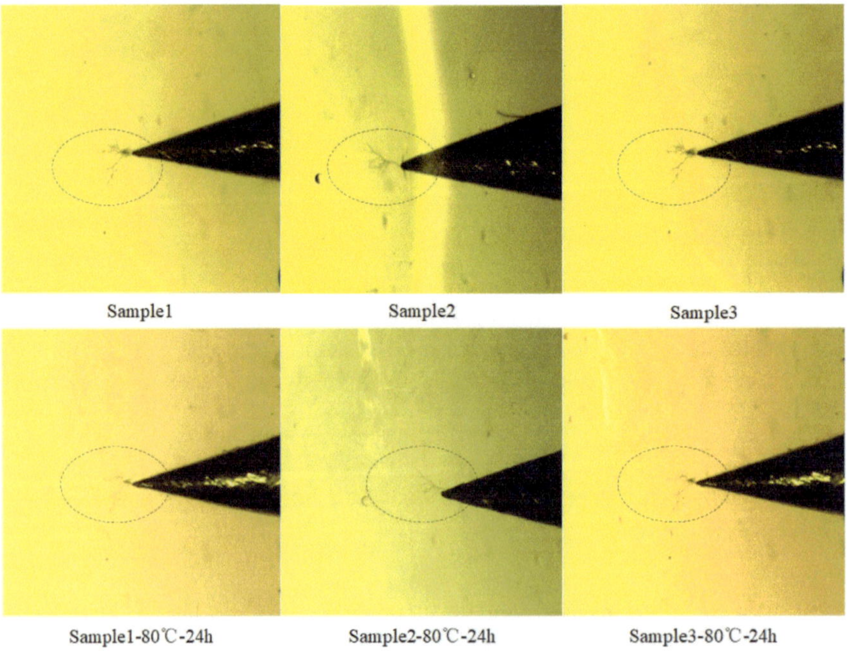

Fig. 22.6 Dendritic morphology of different samples after deterioration and self-healing

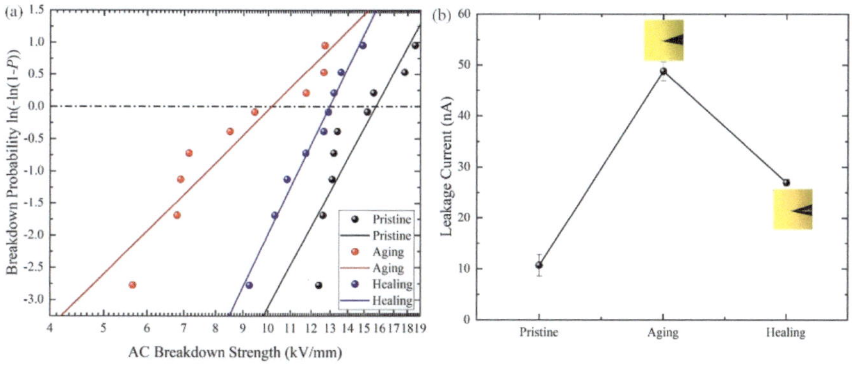

Fig. 22.7 Characterization of self-healing properties of materials

is only 64.44% of the initial one, and the insulation performance is greatly reduced. After self-healing, the breakdown strength of the sample is restored to 12.95 kV/mm, the breakdown strength is restored to 82.22% of the initial, and the insulation performance is restored. The conductive current of a dielectric is formed by the resistive current formed by the movement of ions, also known as leakage current, which is an important basis for evaluating the insulating properties of a material. In this experiment, the leakage current of EP(B-O&MTHPA) before and after aging and after self-healing was tested using a high-voltage DC power supply as shown in Fig. 22.3d, and the results are shown in Fig. 22.7b. The initial leakage current of EP(B-O&MTHPA) was 10.67 nA, and after aging of the sample, the leakage current increased to 48.79 nA, and the insulating property was destroyed, and the leakage current of the sample recovered to 26.94 nA after self-healing. The breakdown strength and leakage current tests proved that the resin system has good self-healing performance after the introduction of dynamic chemical bonding.

After the introduction of boronic ester bonds in the epoxy resin system, its self-healing ability mainly comes from the dynamic reversible properties of boronic ester bonds. The boronic ester bonds can be re-formed after fracture, thus realizing the reconstruction of the polymer network and the self-healing of the damage, and its self-healing mechanism is shown in Fig. 22.8. Under thermal stimulation or other suitable conditions, the exchange rate of boronic ester bonds will be greatly accelerated, which helps to change the topological network structure of the polymer, realizing the reconstruction of the network and self-healing of the damage. In addition, the free hydroxyl groups in the reaction system can promote the dynamic exchange process between borate ester molecules, and at the same time, the hydroxyl groups, as free binding sites, can be re-bonded with the dissociated borate ester molecules, thus increasing the rate of dynamic fracture-reorganization of borate ester molecules [49]. Compared with EP, the dynamic reversible cross-linking property possessed by EP (B-O&MTHPA) endows it with damage repairing properties, which can substantially restore the insulation strength, delay the development of electrical trees, and prolong

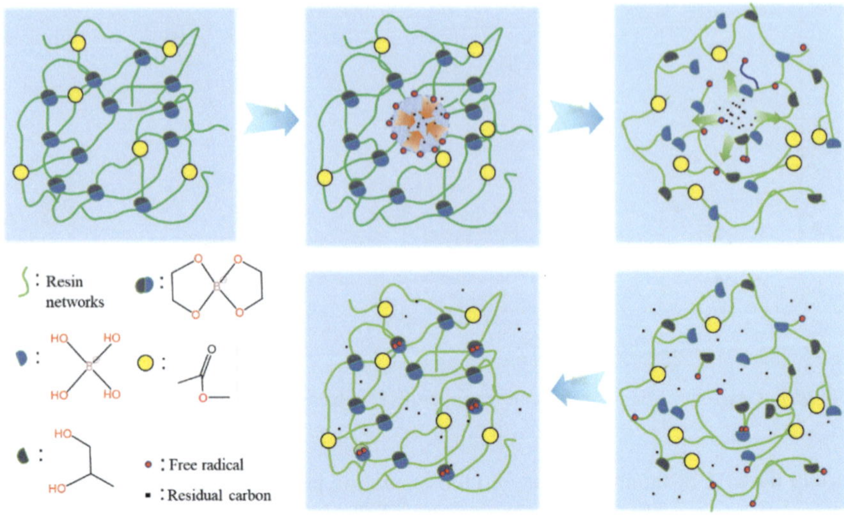

Fig. 22.8 Self-healing mechanism of materials

the operating life of the equipment. However, the characteristics of carbonized particles and electrolytic gasification produced by electrical damage do not achieve 100% repair effect like physical damage [50]. In the early stage of the growth of electrical trees, make full use of the dynamic cross-linking characteristics of the resin cross-linking network, which can effectively prevent the non-conductive channels from being transformed into conductive channels, and realize the self-healing of electrical trees.

22.3.4 Degradation and Recovery

The chemical inertness and dense three-dimensional network structure of epoxy resins give them excellent insulating properties and mechanical strength, but also increase the difficulty of their recycling. Due to the limited biodegradability of epoxy resins, they are prone to release harmful gases during the incineration process, so landfills have become the main disposal method. This disposal strategy not only consumes a large amount of land resources, but also poses a potential long-term pollution risk to the environment. The epoxy resin system crosslinked with boronic ester bonds has excellent dynamic properties, and the boronic ester bonds can be esterified by hydrolysis or alcoholysis, which can realize efficient degradation of the resin and nearly lossless recovery of internal materials at room temperature.

Degradation experiments were carried out using 75% aqueous ethanol solution. Both EP and EP(B-O&MTHPA) were placed in 75% aqueous ethanol solution, and after standing for 12 h, EP was unchanged, while EP(B-O&MTHPA) was completely degraded in the aqueous ethanol solution, and the degradation process was shown in Fig. 22.9a. In addition, the degraded oligomer was poured into the mold again after evaporating the aqueous ethanol solution and cured at 90 °C for 8 h. The oligomer could be re-cured and recycled, and the recycling process was shown in Fig. 22.9b. Compared with the insoluble and immiscible properties of traditional resins, EP(B-O&MTHPA) can realize the harmless and green treatment of decommissioned equipment as well as the convenient and non-destructive recycling of high-value materials such as copper windings and carbon fibers cast inside the epoxy insulating parts. In the degradation solvent, in addition to the exchange of dynamic boronic ester bonds, the boronic ester bonds can undergo exchange reactions with ethanol and water molecules, leading to the dissociation of the polymer network. The degradation mechanism is shown in Fig. 22.9c. Alcohol small molecules can effectively diffuse into the resin crosslinked network at room temperature, and the reactive hydroxyl groups in the alcohol molecules can be fully transesterified with the boronic ester bonds in the epoxy resin crosslinked network through the embedded DBU catalyst, and the polymer crosslinked network inside the resin is gradually cut off, and becomes epoxy oligomers dissolved and dispersed in the solvent [51]. The boronic ester bonds endow EP(B-O&MTHPA) with excellent dynamic exchange properties, which enable the material to be degraded in the solvent to form soluble or dispersible small molecules or oligomers for degradation and recycling.

Epoxy resin is a thermosetting material, and once cured and molded, its molecular structure is difficult to reprocess under heated conditions, which makes the recycling and reuse of epoxy resins more complicated than thermoplastic materials. However, boronic ester bonds can undergo reversible dynamic reactions; therefore EP(B-O&MTHPA) can be recycled after destruction. The recycling process was as follows, the prepared samples were crushed and poured uniformly into the molds, the hot press was adjusted to 100°C and 5 MPa molding process, and the molds were demolded after cooling down to room temperature to reacquire the intact samples. The physical objects of the resin system before and after physical hot press recycling are shown in Fig. 22.10a, which shows that the resin powder after hot press treatment is re-crosslinked and molded again, and the recycling of resin can be realized. The

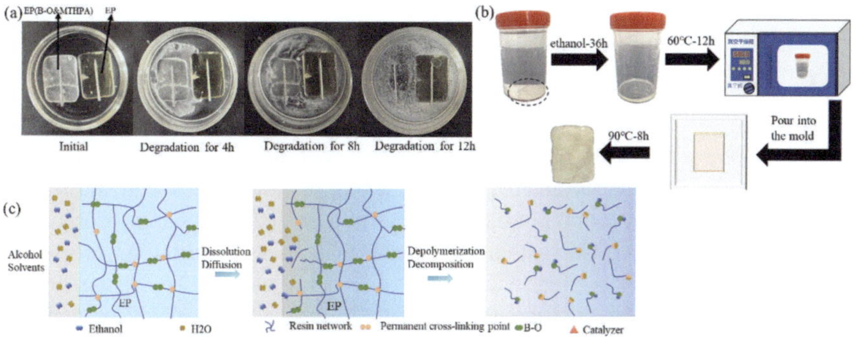

Fig. 22.9 Degradation and recycling processes and degradation mechanisms of materials

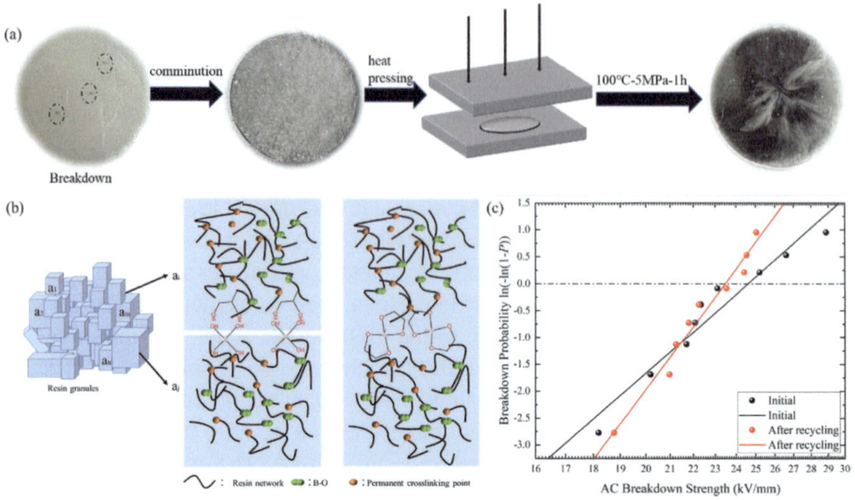

Fig. 22.10 Recycling process, mechanism and breakdown strength after material pulverization

re-crosslinking process and principle of resin particles during hot pressing is shown in Fig. 22.10b. Boronic ester bonds can be broken and reorganized under certain conditions, and this dynamic reversibility is the key to achieving material recycling. The application of heat and pressure promotes the exchange of boronic ester bonds, allowing the movement and rearrangement of chain segments in the polymer network. This dynamic reversibility allows the material to re-melt and flow under certain conditions, thus enabling remodeling. Hot pressing can provide the necessary energy to allow the rearrangement and cross-linking of molecular chain segments in the material to form new network structures [52]. The breakdown strength of the samples before crushing and after recycling are shown in Fig. 22.10c, before crushing, the breakdown strength of the samples is 24.52 kV/mm, after crushing and recycling, the breakdown strength of the samples is 23.37 kV/mm, which is 95.31% of the initial one, and after recycling, the material still has excellent insulating properties.

22.4 Conclusions

In conclusion, we report an insulating material with self-healing capability and capable of green degradation and recycling. The boronic ester bonds were introduced into the epoxy resin by a one-pot method to form a reversible dynamic crosslinked network in the epoxy resin system. The material possesses insulating properties comparable to those of conventional epoxy resins, and can realize self-healing of mechanical damage and electrical trees by reconfiguring the crosslinking network through dynamic chemical bonding. After repair, the breakdown strength of EP(B-O&MTHPA) was restored to 82.22% of the initial one, respectively. In addition, the materials were completely degraded in ethanol aqueous solution and recovered by hot pressing after being destroyed, and the recovered breakdown strength was 95.31% of the initial one. It is proposed to change the ratio of boronic ester bonds to permanent cross-linking sites in the material in order to find insulating materials with both high insulating properties and good self-healing properties. This work provides a new idea for the research of electrical materials with self-healing function and green recycling.

References

1. CAO Wenbin. Application of epoxy resin insulation in electric equipment[J]. Electrical Manufacturing, 2009, (12): 2.
2. WANG Jian, ZHANG Xueyue, SHAN Weiwei. Effect of the mechanical destruction liability risk on the aging pyrolysis performance of the cross-linked polyethylene insulation material[J]. Journal of Safety and Environment, 2018, 18(5): 1811–1816.
3. GAO Yu, WANG Xiaofang, LI Nan, *et al.* Advances in characterization of carrier traps in polymer insulating materials and the effect of traps on insulation breakdown[J]. High Voltage Technology, 2019, 45(7): 12.
4. DU Boxue, ZHANG Ying, KONG Xiaoxiao, *et al.* Progress on the degradation of epoxy resin insulating electric dendrites[J]. Journal of Electrotechnology, 2022, 37(5): 1128–35.
5. Sun W, Xu J, Song J, et al. Self-healing of electrical damage in insulating robust epoxy containing dynamic fluorine-substituted carbamate bonds for green dielectrics[J]. Materials Horizons, 2023, 10(7): 2542–2553.
6. Zha H, Guo Y, Liu Y, et al. The characteristic analysis of ESDD and NSDD detection of composite insulators based on hyperspectral technology[J]. IEEE Trans Instrum Meas, 2023, 72: 1–8.
7. WANG Hongmei, DING Dengwei, JIA Zhijie, et al. Characterization of long intermittent discharges from surface defects of GIS basin insulators under constant voltage[J]. Chinese Journal of Electrical Engineering, 2018, 38(20): 6188–94.
8. WANG Jian, SHU Shengwen, JIN Ming, et al. Multi–source characterization and pattern recognition of composite insulator defects on ±1100kV transmission lines[J]. Electric Porcelain Surge Arrester, 2022, (02): 183–8.
9. WANG Kesheng, ZHAO Yanping, YUAN Shuai, et al. Research on crack defect detection method of 220kV GIS basin insulator based on electric field calculation and modal analysis[J]. Grid and Clean Energy, 2021, 37(08): 32–8+47.
10. XU Yuan, LIU Weidong, CHEN Weijiang, et al. Partial discharge characteristics of micrometallic particles on the surface of GIS insulators based on high-sensitivity measurements[J]. High Voltage Technology, 2019, 45(09): 2707–14.
11. WU Xutao, ZHAO Jinfei, MA Yunlong, et al. Diagnosis method of GIS mechanical defects based on vibration response under multi-frequency excitation[J]. Power Capacitor and Reactive Power Compensation, 2022, 43(04): 108–15.
12. Jones A R, Blaiszik B J, White S R, et al. Full recovery of fiber/matrix interfacial bond strength using a microencapsulated solvent-based healing system[J]. Composites Science & Technology, 2013, 79(apr.): 1–7.
13. Han J, Liu T, Hao C, et al. A Catalyst-Free Epoxy Vitrimer System Based on Multifunctional Hyperbranched Polymer (vol 51, pg 6789, 2018)[J]. Macromolecules, 2018, (20): 51.
14. Pansare A V, Khairkar S R, SHEDGE A A, et al. In Situ Nanoparticle Embedding for Authentication of Epoxy Composites[J]. Advanced Materials, 2018, 30(33): 1801523.
15. Zou Z, Zhu C, Li Y, et al. Rehealable, fully recyclable, and malleable electronic skin enabled by dynamic covalent thermoset nanocomposite[J]. Science Advances, 2018, 4(2): eaaq0508.

16. Lesaint C, Risinggård V, Hølto J, et al. Self-healing high voltage electrical insulation materials[C]//2014 IEEE Electrical Insulation Conference (EIC). IEEE, 2014: 241–244.
17. Guo H, Han Y, Zhao W, et al. Universally autonomous self-healing elastomer with high stretchability[J]. Nature communications, 2020, 11(1): 2037.
18. Wang Z, Liu Y, Zhang D, et al. Mussel-inspired self-healing PDMS/AgNPs conductive elastomer with tunable mechanical properties and efficient antibacterial performances for wearable sensor[J]. Composites Part B: Engineering, 2021, 224: 109213.
19. Wang Z, Liu Y, Zhang D, et al. Tough, stretchable and self-healing C-MXenes/PDMS conductive composites as sensitive strain sensors[J]. Composites Science and Technology, 2021, 216(4): 109042.
20. Li L, Qin X, Mei H, et al. Reprocessed and shape memory networks involving poly(hydroxyl ether ester) and polydimethylsiloxane through Diels-Alder reaction[J]. European Polymer Journal, 2021, (160-): 160.
21. Zhou M, Yan Q, Fu Q, et al. Self-healable ZnO@ multiwalled carbon nanotubes (MWCNTs) /DA-PDMS nanocomposite via Diels-Alder chemistry as microwave absorber: A novel multifunctional material-ScienceDirect[J]. Carbon, 2020, 169: 235–47.
22. Yilmaz D, Lansade D, Lewandowski S. Combination of permanent hydrosilylation and reversible Diels-Alder reactions for self-healing poly(dimethylsiloxane) materials with enhanced ageing properties[J]. Materials Today Chemistry, 2022, (24-): 24.
23. Wu Q, Xiong H, Peng Y, et al. Highly stretchable and self-healing "solid–liquid" elastomer with strain-rate sensing capability[J]. ACS applied materials & interfaces, 2019, 11(21): 19534–40.
24. Chen Y, Pu X, Liu M, et al. Shape-Adaptive, Self-Healable Triboelectric Nanogenerator with Enhanced Performances by Soft Solid-Solid Contact Electrification[J]. ACS Nano, 2019, 13(8).
25. Huang H, Wang X, Yu J, et al. Liquid-Behaviors-Assisted Fabrication of Multidimensional Birefringent Materials from Dynamic Hybrid Hydrogels[J]. ACS nano, 2019, 13(4): 3867–74.
26. Wu J, Cai L H, Weitz D A. Tough self-healing elastomers by molecular enforced integration of covalent and reversible networks[J]. Advanced Materials, 2017: 1702616.
27. Mo J, Chen X, Fu Y, et al. A solvent-free, transparent, self-healing polysiloxanes elastomer based on unsaturated carboxyl-amino ionic hydrogen bonds[J]. Polymer, 2021, 228: 123903.
28. Burattini S, Colquhoun H M, Fox J D, et al. A self-repairing, supramolecular polymer system: healability as a consequence of donor-acceptor pi-pi stacking interactions[J]. Chemical Communications, 2009, 44(44): 6717.
29. Xie J, Yang M, Liang J, et al. Self-healing of internal damage in mechanically robust polymers utilizing a reversibly convertible molecular network[J]. Journal of Materials Chemistry A, 2021, 9(29): 15975–15984.
30. Cai Y, Zou H, Zhou S, et al. Room-temperature self-healing ablative composites via dynamic covalent bonds for high-performance applications[J]. ACS Applied Polymer Materials, 2020, 2(9): 3977–87.
31. Black S P, Sanders J K, Stefankiewicz A R. Disulfide exchange: exposing supramolecular reactivity through dynamic covalent chemistry[J]. Chemical Society Reviews, 2014, 43(6): 1861–72.
32. Utrera-barrios S, Verdejo R, López-manchado M A, et al. Evolution of self-healing elastomers, from extrinsic to combined intrinsic mechanisms: A review[J]. Materials Horizons, 2020, 7(11): 2882–902.
33. 34. Peters J A. Interactions between boric acid derivatives and saccharides in aqueous media: Structures and stabilities of resulting esters[J]. Coordination chemistry reviews, 2014, 268: 1–22.
34. CHEN Ping, CHEN Hui, Jian Xigao, et al. Curing reaction of epoxy resin systems and dielectric properties of composites[J]. Polymer Bulletin, 2003, (4): 1–9.
35. 36. Zhang Q, Anastasaki A, Li G Z, et al. Multiblock sequence-controlled glycopolymers via Cu (0)-LRP following efficient thiol–halogen, thiol–epoxy and CuAAC reactions[J]. Polymer Chemistry, 2014, 5(12): 3876–83.
36. 37. Yoshimura Y, Yamazaki Y, Saito Y, et al. Synthesis of 1–(5, 6-dihydro-2H-thiopyran-2-yl) uracil by a Pummerer-type thioglycosylation reaction: the regioselectivity of allylic substitution[J]. Tetrahedron, 2009, 65(45): 9091–102.

37. 38. Brändle A, Khan A. Thiol–epoxy 'click'polymerization: efficient construction of reactive and functional polymers[J]. Polymer Chemistry, 2012, 3(12): 3224–7.
38. 39. Binder S, Gadwal I, Bielmann A, et al. Thiol-epoxy polymerization via an AB monomer: Synthetic access to high molecular weight poly (β-hydroxythio-ether) s[J]. Journal of Polymer Science Part A: Polymer Chemistry, 2014, 52(14): 2040–6.
39. 40. Brooks W L, Sumerlin B S. Synthesis and applications of boronic acid-containing polymers: from materials to medicine[J]. Chemical reviews, 2016, 116(3): 1375–97.
40. Duncan T T, Berrie B H, WEISS R G. Soft, peelable organogels from partially hydrolyzed poly (vinyl acetate) and benzene-1, 4-diboronic acid: applications to clean works of art[J]. ACS applied materials & interfaces, 2017, 9(33): 28069–78.
41. 42. Angelova L, Leskes M, Berrie B, et al. Selective formation of organo, organo-aqueous, and hydro gel-like materials from partially hydrolysed poly (vinyl acetate) s based on different boron-containing crosslinkers[J]. Soft Matter, 2015, 11(25): 5060–6.
42. 43. Yesilyurt V, Webber M J, Appel E A, et al. Injectable self-healing glucose-responsive hydrogels with ph-regulated mechanical properties[J]. Advanced Materials, 2016, 28(1): 86–91.
43. ZHOU Yuanxiang, CHEN Ming, ZHANG Yunxiao, et al. Influence of temperature on the onset characteristics of silicone rubber electric dendrites under DC field[J]. High Voltage Technology, 2018, 44(12): 3784–90.
44. 45. Li Z, Liu R, Zhang Y, et al. Designing Schiff-Based Hyperbranched Polysiloxane for Simultaneously Enhancing Epoxy Resin with Mechanical Properties, Thermal Stability, and Recyclability[J]. ACS Applied Polymer Materials, 2024, 6(6): 3115–27.
45. 46. Hu J, Feng H, Rong Y, et al. Recyclable bio-based epoxy resins containing hybrid cross-linking networks[J]. Polymers for Advanced Technologies, 2023, 34(5): 1599–607.
46. 47. JI Fei. Preparation and performance study of self-healing polymer materials based on borate ester bonds[D]. Beijing: China University of Petroleum, 2020: 7–9.
47. 48. Startsev O, Vapirov Y M, Lebedev M, et al. Comparison of glass-transition temperatures for epoxy polymers obtained by methods of thermal analysis[J]. Mechanics of Composite Materials, 2020, 56(2): 227–40.
48. 49. LIU Hechen, GUO Zhanpeng, LI Yan, et al. Comparative study on the properties of itaconic acid-based epoxy resin and bisphenol A epoxy resin[J]. Journal of Electrotechnology, 2022, 37(9): 2366–76.
49. 50. Zhao Z H, Wang D P, Zuo J L, et al. A tough and self-healing polymer enabled by promoting bond exchange in boronic esters with neighboring hydroxyl groups[J]. ACS Materials Letters, 2021, 3(9): 1328–38.
50. Burnette M L, Martinez R, Faubel J L, et al. Poly (dicyclopentadiene) Coatings for the Mitigation of Surface Flashover in Gas Switch Housings[C]//2022 IEEE Conference on Electrical Insulation and Dielectric Phenomena (CEIDP). State of Colorado: IEEE, 2022: 483–486.
51. 52. Mu Q, An L, Hu Z, et al. Fast and sustainable recycling of epoxy and composites using mixed solvents[J]. Polymer Degradation and Stability, 2022, 199: 109895.
52. LIU Hechen, SUN Zhanglin, LIU Yunpeng, et al. Preparation and properties of recyclable glassy epoxy resin based on ester exchange[J]. Journal of Electrotechnology, 2023, 38(15): 4019–29.Author, F.: Article title. Journal 2(5), 99–110 (2016).

Chapter 23
Study of the Effect of Micro-Water on the Decomposition Characteristics of SF_6/N_2 Gas Mixture

Chuansheng Luo, Dibo Wang, Jian Qin, Fusheng Zhou, Chunlei Li, Yao Zheng, Linli Cui, and Jiahe Yu

Abstract The SF_6/N_2 gas mixture, as an alternative to SF_6 in gas-insulated systems within the power industry, has gained significant attention and application. However, the effect of trace amounts of moisture on its decomposition characteristics remains unclear. To further investigate the feasibility of replacing SF_6 with SF_6/N_2, this study employs molecular dynamics simulations to explore the decomposition mechanism of a 30% SF_6/70% N_2 mixture in the presence of trace moisture at the microscopic level. H_2O molecules introduce hydrogen radicals, which attract F radicals, thus influencing the decomposition process of the SF_6/N_2 mixture. As the moisture content increases, the number of hydrogen radicals rises, leading to an increase in SF_6 decomposition. In contrast, N_2 decomposition fluctuates but is eventually suppressed as the amount of H_2O molecules grows. The decomposition products of the SF_6/N_2 mixture tend to form stable structures such as SF_2, SF_4, and NF_3, with SF_2 and SF_4 easily reacting further to generate SOF_2. The formation of SOF_2 in the equipment can reduce insulation performance, and the simultaneous production of HF, a highly corrosive substance, degrades the performance of gas-insulated devices. Therefore, measures should be taken to monitor its concentration. This research provides theoretical support and engineering guidance for understanding the decomposition mechanism of the SF_6/N_2 mixture under trace moisture conditions.

Keywords SF_6/N_2 gas mixture · Micro-water · Decomposition properties · ReaxxFF

C. Luo · J. Qin · C. Li · L. Cui
Nanning Power Supply Bureau, Guangxi Power Grid Co., Ltd, Guangxi, China

D. Wang (✉) · F. Zhou · Y. Zheng · J. Yu
China Southern Power Grid Research Institute Co., Ltd, Guangdong, China
e-mail: 2641684212@qq.com

Joint Laboratory of Advanced Electrical Materials and Equipment, China Southern Power Grid Co., Ltd., Guangdong, China

Y. Li et al. (eds.), *Proceedings of the 5th Chinese Conference on High Voltage and Discharge Plasmas*, Springer Proceedings in Physics 254,
https://doi.org/10.1007/978-981-96-5711-7_23

23.1 Introduction

Sulfur hexafluoride (SF_6) gas is widely used in gas-insulated electrical equipment, such as circuit breakers, gas-insulated transmission lines (GIL), gas-insulated switchgear (GIS), and switch cabinets, due to its excellent insulation, arc-quenching properties, and stability [1–5]. However, SF_6 is an extremely potent greenhouse gas, with a global warming potential (GWP) 23,500 times that of CO_2 and a lifespan in the atmosphere of up to 3,200 years. It is one of the six gases whose emissions are restricted under the Kyoto Protocol [6, 7]. In response to national initiatives on energy conservation and emission reduction, and to support the goal of achieving "carbon peak" and "carbon neutrality," reducing the use of SF_6 and other greenhouse gases is imperative.

Currently, research on alternative insulating gases includes conventional gases (air, N_2, CO_2), SF_6 gas mixtures, and strongly electronegative gases and their mixtures [8]. Nitrogen (N_2), with its advantages of being non-toxic, non-polluting, having low greenhouse effects, and a low liquefaction temperature, is often added to SF_6 to form a gas mixture that reduces SF_6's sensitivity to local electric field distortions. This mixture effectively mitigates the greenhouse effect and high liquefaction temperature issues of pure SF_6 [9, 10]. Presently, gas mixtures with low SF_6 concentrations are already widely used in GIS electrical equipment [11].

Research on the insulating, electrical, and thermal decomposition characteristics of SF_6/N_2 gas mixtures has made considerable progress. For example, Fuping Zeng from Wuhan University and Yansong Ji from the China Electric Power Research Institute studied the decomposition characteristics of SF_6/N_2 mixtures under partial discharge and overheating conditions, identifying key decomposition products such as SOF_2, SO_2F_2, SO_2, and H_2S [12]. Yanliang He from Xi'an Jiaotong University compared the negative corona characteristics of pure SF_6 and SF_6/N_2 mixtures under DC voltage, concluding that the lower limit of positive ion flux during the normal glow stage in SF_6/N_2 mixtures is 6 to 7 orders of magnitude higher than the upper limit during the Townsend stage [13]. Youping Tu developed a $\pm$ 100 kV high-voltage DC SF_6/N_2 GIL and successfully passed type tests [14].

In summary, SF_6/N_2 gas mixtures show potential for replacing pure SF_6 in gas-insulated equipment [15–17]. During installation or maintenance, gas-insulated equipment may come into contact with moisture and other impurities, leading to partial decomposition of the insulating medium and trace amounts of water, which can generate free radicals and trigger complex chemical reactions that form various decomposition products. Currently, the effects of different moisture levels on the decomposition characteristics of SF_6/N_2 mixtures remain unclear, necessitating further research to provide valuable data for future operations and monitoring.

To explore the impact of trace moisture on the decomposition process of SF_6/N_2, this study utilizes molecular dynamics simulations to investigate the influence of H_2O molecules on the decomposition mechanism of SF_6/N_2 mixtures. Based on the analysis under varying moisture conditions, engineering recommendations are provided for the application of SF_6/N_2 gas mixtures.

23.2 Theoretical Research and Methodology

ReaxFF molecular dynamics (ReaxFF-MD) simulates the microscopic behavior of materials by calculating the interactions and movements of molecules. ReaxFF is a molecular dynamics method specifically designed for simulating chemical reactions. Unlike traditional force fields, ReaxFF can handle the formation and breaking of chemical bonds, thereby simulating the dynamic process of chemical reactions. In the ReaxFF-MD reactive force field, the energy distribution of the system is described as follows [18–23]:

$$E_{system} = E_{bond} + E_{over} + E_{angle} + E_{torsion} + E_{vdWals} + E_{coulomb} + E_{specific} \quad (23.1)$$

where: E_{system} is the total energy of the system; E_{bond} represents the bond energy between atoms; E_{angle} and $E_{torsion}$ are the three-body angle correction term and the four-body torsion term, respectively; E_{over} is an over-coordination correction term; E_{vdwals} and $E_{coulomb}$ describe the van der Waals and Coulomb forces, respectively, accounting for electrostatic and dispersion effects as non-bonded terms; $E_{specific}$ is a system-specific energy term, used for interactions such as lone pairs, conjugated electrons, hydrogen bonds, and C2 corrections.

This study investigates the mechanism of trace moisture's influence on the reaction process of SF_6/N_2 using reactive molecular dynamics (ReaxFF-MD) and density functional theory (DFT). First, a gas mixture system containing 900 SF_6 molecules and 2,100 N_2 molecules is constructed using the Amorphous Cell module in the Materials Studio software. The system box size is set to 472.7 × 472.7 × 472.7 Å, resulting in an initial density of 0.003 g/cm^3 at room temperature, which closely matches the actual density of the gas mixture. Subsequently, trace amounts of water molecules are introduced into the gas mixture system. Considering that the number of water molecules based on the actual moisture concentration would be too small to yield clear results in the simulation, the number of H_2O molecules is set to 6, 12, 18, and 24. The system is then relaxed and geometrically optimized to achieve the lowest energy and most stable structure. Figure 23.1 illustrates the SF_6/N_2 gas mixture with trace moisture model.

The molecular dynamics simulations in this study are conducted using the ReaxFF module in the AMS platform, which supports multi-scale computational chemistry. To simulate chemical reactions within a reasonable simulation time, the reaction temperature is increased [24, 25]. Considering various factors, the gas mixture system is simulated at 3000 K under a constant number of atoms, volume, and temperature (NVT) ensemble. The Berendsen thermostat is applied for temperature control with a damping constant set to 75 fs. The total simulation time is set to 500 ps with a timestep of 1 fs.

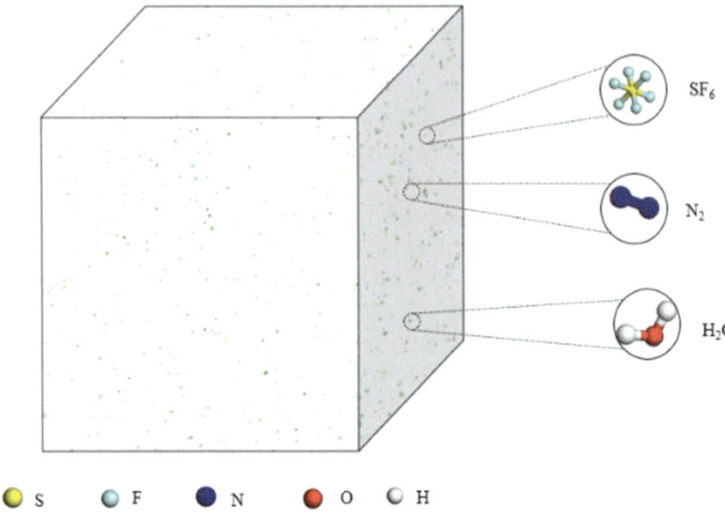

Fig. 23.1 SF_6/N_2 gas mixture micro-water modeling2. Simulation results and analysis

23.3 Simulation Results and Analysis

Figure 23.2 illustrates the effect of moisture on the decomposition of the SF_6/N_2 gas mixture. It is evident that as the number of H_2O molecules increases, the degree of SF_6 decomposition also rises, indicating that the presence of trace moisture facilitates the breakdown of SF_6. In contrast, the decomposition of N_2 first increases and then decreases with the addition of H_2O; specifically, when the number of H_2O molecules reaches 12, N_2 decomposition is promoted. However, overall, the presence of H_2O tends to inhibit N_2 decomposition. This may be attributed to H_2O molecules capturing free radicals generated during the decomposition, altering the reaction pathway.

During the decomposition of SF_6, the breaking of S-F bonds leads to the formation of various low fluorinated sulfur compounds, such as SF_5, SF_4, SF_3, and SF_2. These low fluorinated compounds can further react to yield final decomposition products. Figure 23.3 shows the variation in the generation of these low fluorinated compounds with different moisture levels. As the number of H_2O molecules increases, the conversion of SF_5 to SF_4 significantly rises when the count reaches 12; however, this promoting effect diminishes at 18 H_2O molecules and encourages the conversion of SF_3 to SF_2 when reaching 24. This is likely due to the more stable symmetrical structures of SF_4 and SF_2, indicating that the presence of H_2O facilitates the formation of more stable decomposition products from SF_6. In actual reactions, SF_5 readily reacts with H_2O to form SO_2F_2, while the generated SF_4 and SF_2 are more likely to react with H_2O to form SOF_2.

To clarify the influence of moisture on N_2 decomposition, Fig. 23.4 presents the variations in the generation of N free radicals under different moisture conditions. The production of NF is notably reduced when the number of H_2O molecules reaches

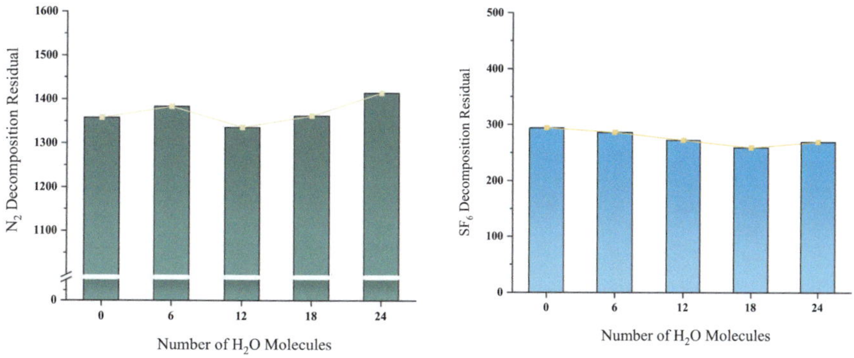

(a) (b)

Fig. 23.2 Effect of H_2O on the residual of SF_6/N_2 gas mixture decomposition

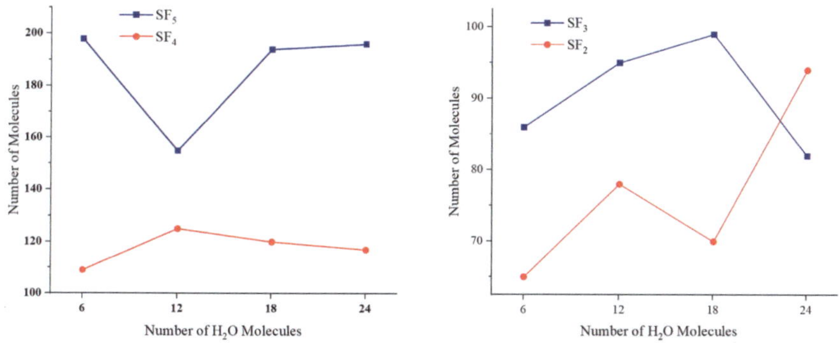

(a) SF_5、SF_4 (b) SF_3、SF_2

Fig. 23.3 Variation of low fluorine sulfide production at different micro-water contents

18 and 24, suggesting that NF tends to further react to form NF_2 or NF_3 in the presence of H_2O. The quantity of NF_3 generally increases with the number of H_2O molecules, peaking at 18 and then slightly decreasing, likely due to the hydrolysis of NF_3 by the additional H_2O molecules.

The presence of H_2O molecules leads to the generation of various H and O free radicals, impacting the decomposition processes of SF_6 and N_2. Based on the ChemTraYzer 2 module within AMS software, the main reaction pathways and quantities of H_2O involvement are summarized in Fig. 23.5. It is clear that H_2O contributes H, O, and OH free radicals that primarily interact with F free radicals, affecting the decomposition of the SF_6/N_2 gas mixture. F free radicals are highly oxidative and can enhance the decomposition reactions. As the number of molecules increases, reactions such as $F + H \rightarrow HF$, $F + HF \rightarrow HF_2$, and $F + HF_2 \rightarrow HF_3$ increase, with the combination of H and F free radicals promoting SF_6 decomposition and producing more F free radicals. Consequently, the addition of 12 H_2O molecules accelerates

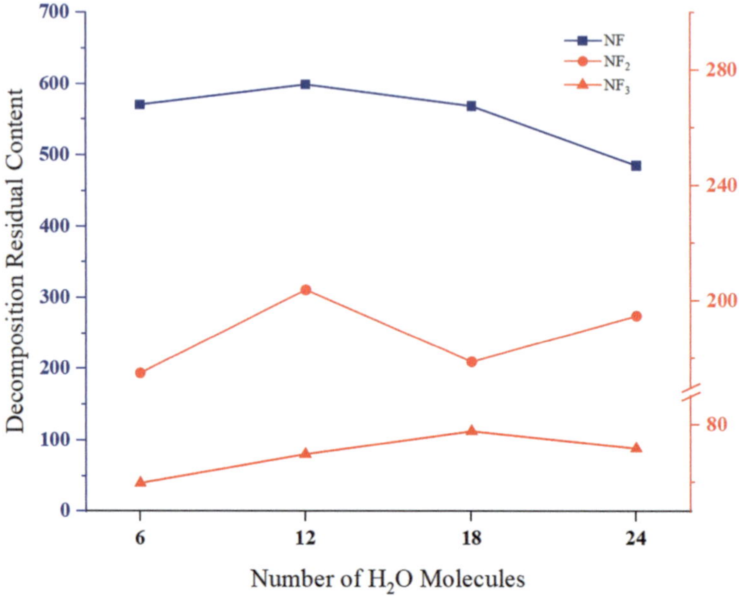

Fig. 23.4 Variation of NFx radical generation at different micro-water contents

N_2 decomposition; however, with further increases in H_2O molecules, the continued combination of H and F free radicals again inhibits F's ability to decompose N_2. The decomposition pathways include $NF2 + HF \rightarrow NF_3 + H$, $NF2 + HF_2 \rightarrow NF_3 + HF$, $HF_4N \rightarrow NF_3 + HF$, and $NF_3O \rightarrow NF_3 + O$, facilitating the formation of stable NF3 structures from N_2 decomposition.

In summary, the H_2O molecules in the system attract F free radicals to form a series of new free radicals. A certain concentration of H_2O promotes the decomposition of

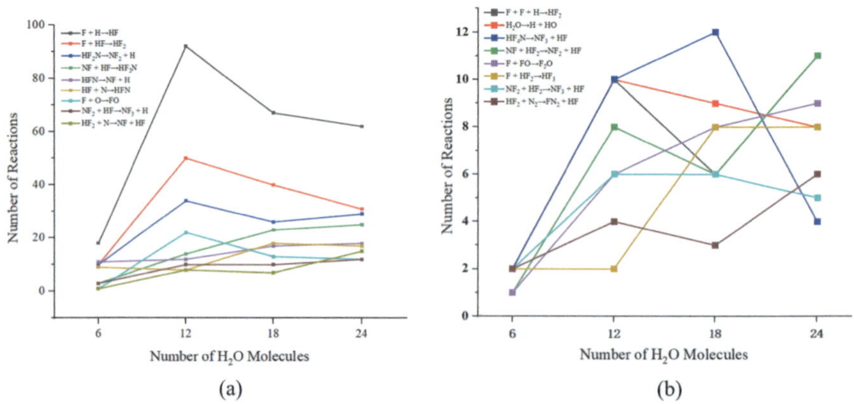

Fig. 23.5 Number of H and O radical-containing reactions at varia-tion micro-water contents

SF_6 while inhibiting that of N_2, resulting in decomposition products that are more likely to form stable structures.

23.4 Conclusion

This study investigates the impact of trace moisture on the decomposition characteristics of the SF_6/N_2 gas mixture, with the following conclusions:

(1) The presence of H_2O molecules attracts F free radicals, reducing their concentration and consequently affecting the decomposition pathway of the system. A certain concentration of H_2O promotes the decomposition of SF_6 while inhibiting that of N_2.

(2) A specific amount of H_2O molecules facilitates the formation of stable structural products such as SF_2, SF_4, and NF_3. In the decomposition pathway, SF_2 and SF_4 can produce SOF_2, which may lead to a decrease in insulation performance. Additionally, the generated HF is highly corrosive, which can further degrade the performance of gas-insulated equipment; therefore, monitoring its concentration is essential.

References

1. L. G. Christophorou, J. K. Olthoff, and R. J. Van Brunt, Sulfur Hexafluoride and the Electric Power Industry[J] IEEE Electrical Insulation Magazine 1997(13): 20–24.
2. Maiss M, Brenninkmeijer C M. Atmospheric SF_6: Trends, sources, and prospects [J]. Environmental Science & Technology, 1998, 32(20):3077–3086.
3. Chun F Y. SF_6 decomposition in gas-insulated equipment [J]. IEEE Trans on Electrical Insulation, 1986, 21(5):593–275.
4. Hermann Koch. Gas-Insulated Transmission Lines (GIL)[M]. IEEE Press, 2012.
5. ZHANG Xiaoxing, YAO Yao, TANG Torch et al. Current status and development of SF_6 discharged decomposition gas component analysis[J]. High Voltage Technology, 2008.4, 34(4):664–669.
6. WU Bin, SHU Lichun, SIMA Wenxia et al. SF_6 gas insulation and the earth's greenhouse effect[J]. High Voltage Electrical Apparatus, 2000, 36(6):23–26.
7. Kieffel Yannick, Biquez Francois, Ponchon Philippe et al. SF_6 alternative development for high voltage switch-gears [C]. IEEE Power and Energy Society General Meeting 2015.
8. ZHANG Xiaoxing, TIAN Shuangshuang, XIAO Song et al. A review of SF_6 alternative gases[J]. Chinese Journal of Electrical Engineering, 2018(12):2883–2893.
9. W. Boeck, T. R. Blackburn, A. H. Cookson, et al. N_2/SF_6 mixtures for gas insulated systems [J]. Cigre Technical Brochure, 2004(10):260.
10. KUFFFEL E. Comparison of SF_6/N_2 and SF6/CO_2 gas mixtures as alternative to SF_6 gas [J]. IEEE Trans. on DEI, 1999, 6(6):892–895.
11. XIAO Dengming. Development prospects of environmentally friendly insulating gases [J]. High Voltage Technology, 2016, 42(4):1035–1046.
12. JI Yansong; ZHANG Min; WANG Chengyu et al. Detection method and experimental study of SF_6/N_2 gas mixture decomposition products[J]. High Voltage Electrical Apparatus,2020,56(12):97–102.

13. HE Yanliang, DING Wei, SUN Anbang, et al. Influence of electric field inhomogeneity coefficient on the negative DC corona current pulse characteristics of SF_6/N_2 gas mixture[J]. Journalof Electrotechnology,2021,36(15):3124–3134.
14. Tu Y, Chen G, Li C et al. ±100-kV HVDC SF_6/N_2 Gas-Insulated Transmission Line[J]. IEEE transactions on power delivery,2020,35(02):735–744
15. ZHAO Yanjie;CHEN Lian;YUAN Chuanzhen.Characterization of local discharge of SF_6/N_2 gas mixture for 252kV GIS[J]. High Voltage Electrical Apparatus,2023,59(09):169–175+183.
16. ZHANG Min,ZHANG Ran,XIAO Peiwei,et al.Development and testing of current switching test circuit for small-capacity busbar conversion of SF_6/N_2 mixed gas GIS disconnectors[J]. High Voltage Electrical Apparatus,2022,58(02):82–88.
17. LI Zhibing, XU Xiaodong, ZHANG Ran, et al. Development of DC 80 kV SF_6/N_2 plasma jet triggering gap for ±800 kV hybrid DC transmission project[J]. High Voltage Technology,2023,49(03):937–945.
18. F. Castro-Marcano, A.M. Kamat, M.F. Russo, A.C.T. van Duin, J.P. Mathews, Combustion of an Illinois No. 6 coal char simulated using an atomistic char representation and the ReaxFF reactive force field, Combust. Flame (2012).
19. Nomura K. I, Kalia R. K, Nakano A, et al. "A scalable parallel algorithm for large-scale reactive force-field molecular dynamics simulations". Computer Physics Communications. Vol:178, no:2, pp: 73–87,2008.
20. Song L, Zhao F, Xu S, et al. ReaxFF study on combustion mechanism of ethanol/ nitromethane[J]. Fuel, 2021, 303: 121221.
21. Lindgren E B, Monteiro J G S, dos Santos A R, et al. Initiation mechanisms and kinetics of the combustion of cyclopentane and cyclopentene from ReaxFF molecular dynamics[J]. Fuel, 2021, 303: 121205
22. A, Bharti, T. Banerjee, Reactive force field simulation studies on the combustion behavior of n-octanol, Fuel Process. Technol.(2016).
23. S. Monti, M. Pastore, C, Li, F, De Angelis, V. Carravetta, Theoretical investigation ofadsorption, dynamics, self-aggregation, and spectroscopic properties of the D102indoline dye on an anatase (1 0 1)substrate, J. Phys. Chem.C (2016).
24. Ashraf C, Shabnam S, Jain A, et al. Pyrolysis of BinaryFuel Mixtures at Supercritical Conditions: A Reaxff Molecular Dynamics Study [J]. Fuel, 2018, 235(JAN.1):194–207.
25. Hong D, Cao Z, Guo X. Effect of calcium on the secondary reactions of tar from Zhundong coal pyrolysis: a molecular dynamics simulation using ReaxFF[J]. Journal of Analytical and Applied Pyrolysis, 2019, 137: 246–252.

Chapter 24
Design of High Voltage Pulse Power Supplies Based on Inductively Isolated Marx Circuits

Yifan Liu, Xiaolong Wang, Huaifeng Sun, Yuchao Liu, and Ying Sun

Abstract The instantaneous high-voltage characteristic of the pulsed power supply can effectively generate high-density plasma, increase the rate of plasma chemical reaction, and reduce the energy loss. For the application requirements of high-voltage pulsed discharge and low-temperature plasma, this paper designs a high-voltage pulsed power supply based on an inductance-isolated Marx circuit. The main circuit of this power supply adopts the classical Marx principle of capacitor charging in parallel and discharging in series, uses diodes instead of isolation resistors, and adds a boost current-limiting inductor in front of the first-stage Marx circuit, and uses insulated gate bipolar transistors (IGBTs) as the switching device, with each stage consisting of IGBTs, high-voltage diodes, and inductorless capacitors. The feasibility of the two driving schemes of optocoupler floating drive and transformer drive are designed and verified respectively, and the experiments show that the transformer drive scheme is more feasible. Accordingly, a 20-stage Marx high-voltage pulse power supply was constructed, which can realize 12 kV pulse output, pulse width of 4–12 us, pulse frequency of 1–5 kHz, and is successfully applied to argon medium blocking jet discharge.

Keywords High voltage pulsed power supply · Marx generator · Synchronous drive · Plasma

---

Y. Liu · X. Wang (✉) · Y. Sun
School of Electrical Engineering, Shandong University, Jinan, China
e-mail: wangxiaolong@sdu.edu.cn

H. Sun · Y. Liu
School of Civil Engineering, Shandong University, Jinan, China

Y. Li et al. (eds.), *Proceedings of the 5th Chinese Conference on High Voltage and Discharge Plasmas*, Springer Proceedings in Physics 254,
https://doi.org/10.1007/978-981-96-5711-7_24

24.1 Introduction

Pulsed power supply is a kind of electric device that can release high energy in a short period of time, and is widely used in the fields of plasma physics, particle acceleration, laser drive and environmental protection. Compared with conventional continuous power supplies, pulsed power supplies are characterized by high instantaneous power, low average power, and highly controllable pulse width and frequency. These characteristics make them particularly important in applications where high peak power is required but low total energy consumption is required [1–3]. In the field of plasma technology, the use of pulsed power supplies is particularly critical. Pulsed power supplies are capable of generating plasma through short-duration high-voltage discharges, which are widely used in environmental protection, biomedicine, and materials processing. For example, pulsed discharge plasma can be used for pollutant degradation in water treatment, air purification, and sterilization of medical devices [3, 4]. In addition, pulsed plasma has gradually become a new research hotspot due to the instantaneous high-voltage characteristics of the pulsed power supply, which can effectively generate high-density plasma, which helps to increase the rate of plasma chemical reaction and reduce the energy loss [5].

Although pulsed power supplies have a wide range of applications in many fields, the design and development of pulsed power supplies face many challenges, especially under the conditions of high voltage, high frequency, and fast response, and their performance stability and energy transfer efficiency are important factors affecting their practical applications. Therefore, the in-depth study of pulsed power supply technology not only has important theoretical significance, but also can provide a solid technical foundation for many industrial and scientific fields.

In this paper, based on the inductively isolated Marx circuit, through the design and verification of optocoupler floating drive and transformer drive the feasibility of the two drive schemes, successfully developed the output of up to 12 kV, pulse width of 4–12 us, frequency of 1–5 kHz high-voltage pulsed power supply, and successfully realized argon dielectric blocking jet discharge.

24.2 Principle and Design

24.2.1 System Framework

The overall framework of the system is shown in Fig. 24.1, where the multi-stage Marx circuit is charged by a DC power supply, and the signal generating circuit controls the structural changes of the multi-stage Marx circuit through a driver circuit to realize the pulse output. Insulated Gate Bipolar Transistor (IGBT) is used as the internal switching device of Marx circuit. The human–computer interaction module controls the output state of the circuit through the buttons and the display. The

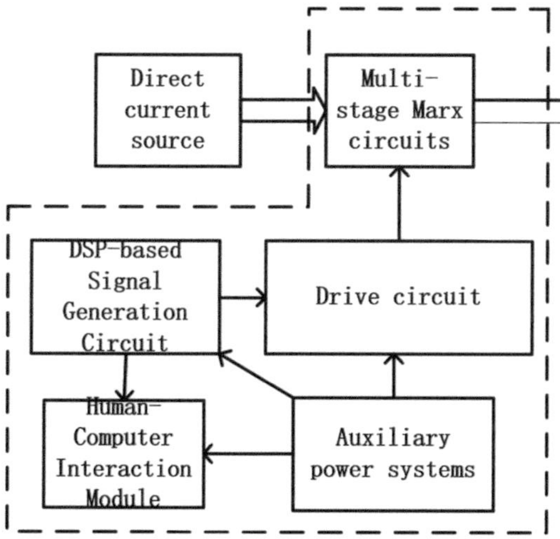

Fig. 24.1 Overall system framework

auxiliary power supply system is responsible for powering the HCI module, the DSP-based signal generation circuit, and the driver circuit.

24.2.2 Multi-stage Marx Circuit Analysis

The traditional Marx circuit generally consists of n switches, capacitors, and isolation resistors [6, 7]. In this design, diodes are used instead of isolation resistors, and a boost current limiting inductor is added before the first stage charging circuit; compared with the traditional circuit this design not only reduces the power loss, but also suppresses the transient changes of the circuit current, and reduces the requirement of the DC source. The circuit structure is shown in Fig. 24.2a, L1 is a boost current limiting inductor, and each stage consists of two high-voltage diodes, an IGBT and a capacitor.

The charging circuit is shown in Fig. 24.2b, when all the IGBTs are in the off state, the diodes are all on, and the input power supply constitutes a closed loop through the inductor, diode and capacitor, and at this time the capacitor is charged in parallel. The voltage at ends of the load is the sum of the diode D_2, D_4, ……D_{2n} conduction voltage drop in series.

Discharge circuit shown in Fig. 24.2c, when the IGBT all conduction, diode D_2, D_3, D_4, ……D_{2n} withstand reverse voltage shutdown, all levels of capacitance through the switching tube discharge in series, to provide the load with negative high-voltage pulse. In addition, diode D_1 is subjected to positive voltage, D_1 conduction, the input power supply through the inductor, diode D_1, IGBT and capacitors C_2–C_n constitute another discharge circuit.

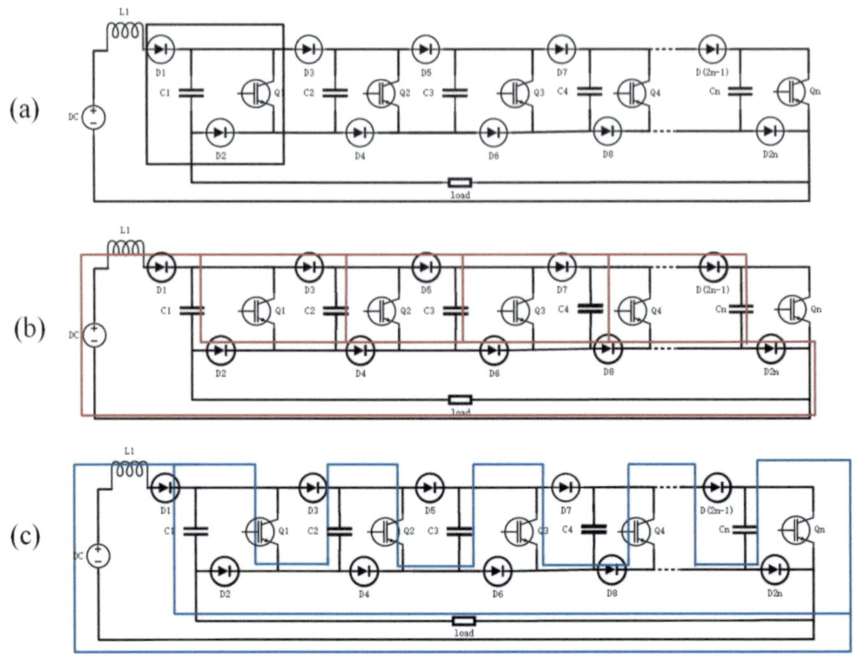

Fig. 24.2 Multi-stage Marx circuit diagram (**a** Circuit structure diagram, **b** Charge circuit diagram, **c** Discharge circuit diagram)

The inductor is charged all the time during the discharge period, and during the charging period the DC constant voltage source and the inductor are connected in series to charge the capacitors at all levels, which produces a boosting effect with respect to a BOOST circuit. When the inductor current is continuous, the expression (24.1) for the output voltage U_O and the inductor continuity condition (24.2) can be derived based on the volt-second equilibrium principle:

$$U_0 = -\frac{nU_I}{1 - nD} \tag{24.1}$$

$$L \geq \frac{(1 - D)(1 - nD)T_S R}{2n} \tag{24.2}$$

where U_I is the input voltage, n is the number of circuit stages, L is the inductive reactance of the inductor, D is the duty cycle, T_S is the pulse period and R is the output load of the circuit.

When the inductor current is intermittent, each capacitor voltage is not equal and satisfies Eq. (24.3):

$$U_c = \frac{T_S D^2 U_O^2}{2LI_O} + U_O \tag{24.3}$$

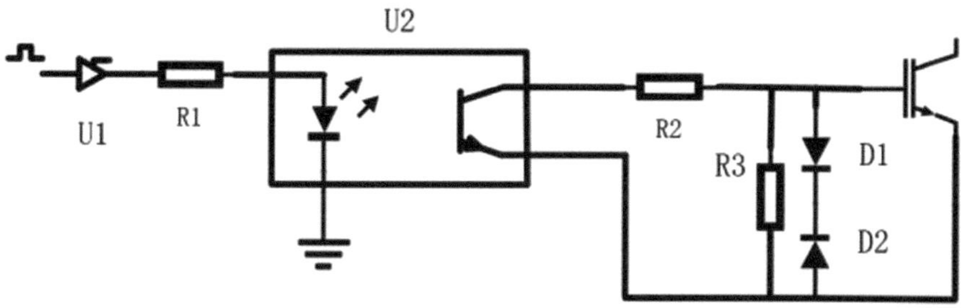

Fig. 24.3 Optocoupler based floating drive circuit structure

where I_O is the average output current.

In the case of continuous inductor voltage, the boosting effect is negligible due to the low duty cycle in the practical application of pulsed power supply; in the case of intermittent inductor current, the envisioned boosting effect can be achieved by appropriately selecting the inductor parameters.

24.2.3 Driver Circuit Design

In order to drive the IGBT to realize the structural changes in the circuit, respectively, design based on optocoupler floating drive circuit and transformer isolated drive circuit for driving verification.

24.2.3.1 Based on Optocoupler Floating Drive Circuit

Pulse signal through the buffer 74HC245D and optocoupler isolation FOD3120 to get the isolation and power to strengthen the effect, after the resistance value of 20 ohms drive resistor to drive the IGBT. D_1, D_2 for the voltage regulator diode, R_3 for the pull-down resistor, the circuit structure is shown in Fig. 24.3, the output waveforms are shown in Fig. 24.4, which can achieve a reliable shutdown and start of the IGBT.

24.2.3.2 Transformer Driver Circuit

Transformer drive circuit shown in Fig. 24.5, the pulse signal through a single-channel gate driver chip 1EDB7275 pulse signal power amplification, through the control resistor R_1, R_2 control MOS tube Q_1. MOS tube Q_1 open and close to realize the circuit pulse signals from the transformer primary side coil through electromagnetic induction will be transmitted to the secondary side coil pulse signal. The signal

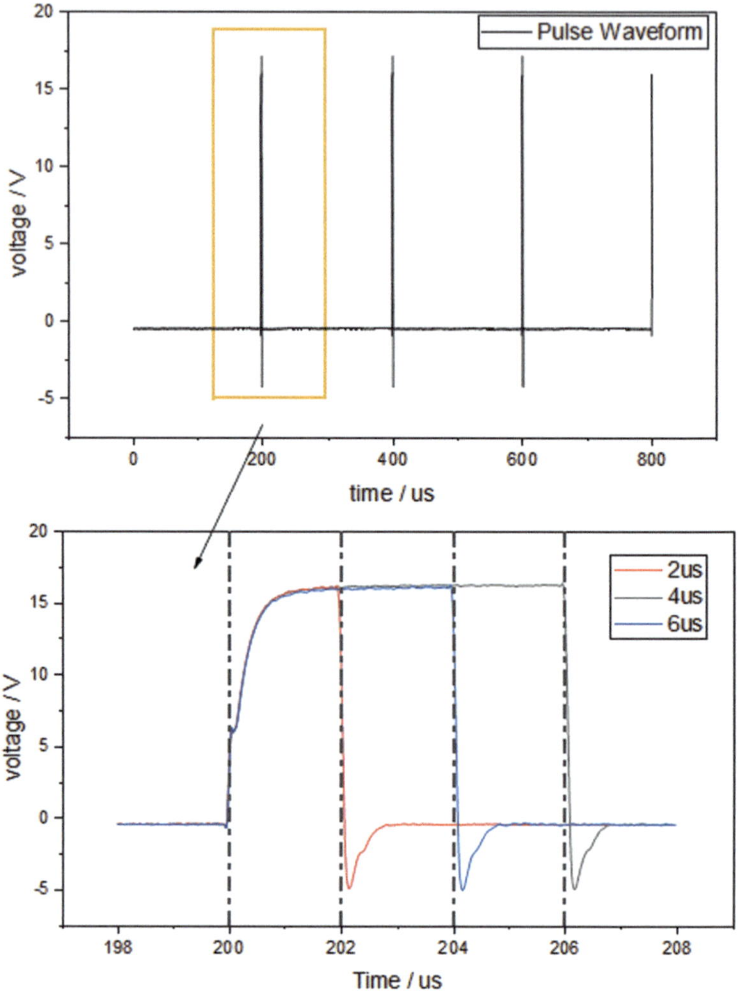

Fig. 24.4 Optocoupler float drive circuit output waveforms

is transmitted to the secondary coil through the signal conditioning circuit for rectification to realize the high-voltage side drive. Among them, R_1, R_2, R_4 is a 20 Ω control resistor, R_3, R_5 is a matching resistor with a resistance value of 200 ohms and 100 ohms, respectively, D_1, D_2 is a renewable diode, D_3, D_4 is a current limiting diode, D_5, D_6 is a voltage stabilizing diode, R6 is a 20 Ω bias resistor, and Q_3 is a triode.

When the high level of the pulse signal passes through the primary side circuit, the MOS tube Q_1 conducts, and the transformer primary coil generates a positive pulse; when the low level passes through the MOS tube Q_1 is turned off, the diodes D_1 and D_2 provide current continuation, and the transformer primary coil generates a

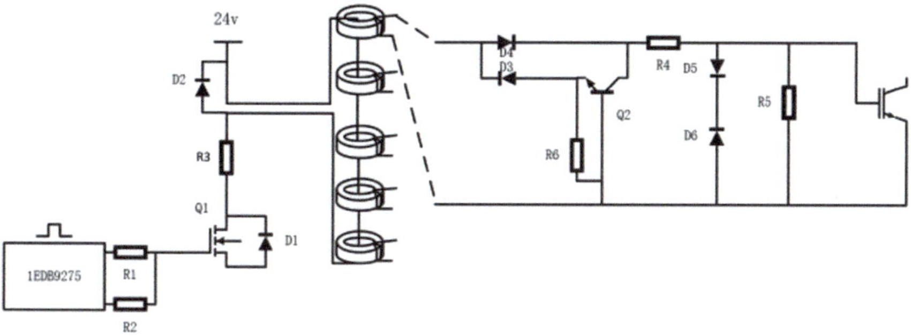

Fig. 24.5 Transformer drive circuit structure diagram

negative voltage. As a result, the pulse signal is transmitted to the secondary side via the primary side. The secondary coil produces positive voltage current through the current limiting diode D_4, resistance R_4, when the voltage at the ends of R_5 is higher than 18 V D_6 is broken, play the role of clamping; produce negative voltage, diode D_3 conduction, and in the role of resistor R_6, transistor Q_2 conduction, to accelerate the output point of the potential drop, to ensure that the speed of the IGBT shutdown, when the voltage is lower than 5.1 V, D_5 is broken, to play the role of clamping. role. Drive signal shown in Fig. 24.6.

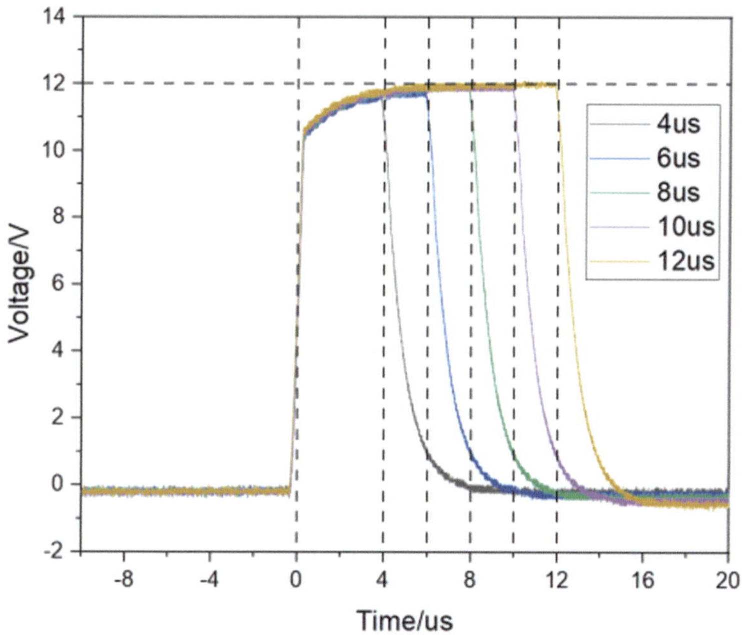

Fig. 24.6 Transformer drive circuit output waveform

24.2.4 Parameter Design

The operating frequency of this power supply is 1–5 kHz, which requires the switching tube to have high voltage and current resistance, and Infineon's IHW30N160R5 (1.6 kV, 60 A) IGBTs are selected as the switching tubes for all levels. The diode adopts 10 A, 20 kV high voltage diode. Self-wound inductors (10 mH, 20 kV) are used for the inductors. The energy storage capacitance mainly depends on the pulse width and the pulse voltage landing allowed by the circuit, the capacitance of this design is selected 1 uf.

Transformer core is the focus of transformer design, in order to meet the high-frequency characteristics, to prevent core saturation, need to meet the following formula:

$$B_{\max} = \frac{VT_{on}}{NA_e} \leq B_s \tag{24.5}$$

where B_{max} is the maximum magnetic flux density, B_s is the saturation magnetic induction intensity, V is the supply voltage, Ton is the maximum conduction time, N is the primary coil, and Ae is the core cross-sectional area.

Common magnetic cores on the market are selected for comparison, and finally an iron-based nanocrystalline core is chosen. This core is widely used in circuits with EMC common-mode inductors, EMI filters, high-frequency transformers, current transformers, magnetic amplifiers, and drive transformers. The size of the core is 50 × 32 × 20 (mm), primary coil 4 turns, secondary coil 20 turns.

24.3 Results

The optocoupler floating driver circuit is used to drive the Marx circuit. Since this solution uses an isolated power supply to drive the optocoupler module, it is affected by the voltage withstand level of the isolated power supply, and can only be applied to the 8-stage Marx circuit to realize 6 kV output (see Fig. 24.7).

The use of pulse transformer drive circuit has a good effect, can be applied to the 20-stage Marx circuit to achieve 12 kV output, and compared with the optocoupler output, the drive in the actual application of the process of IGBT shutdown faster, wider range of applications. And can realize argon gas injection (see Fig. 24.8).

24.4 Conclusions

This design uses diodes instead of isolation resistors on the basis of the traditional Marx circuit, and adds a boost current limiting inductor before the first stage charging circuit, which is able to reduce the power loss and the requirement of charging

Fig. 24.7 Optocoupler float drive circuit driving Marx circuit dielectric barrier atmospheric discharge test at 15 kPa

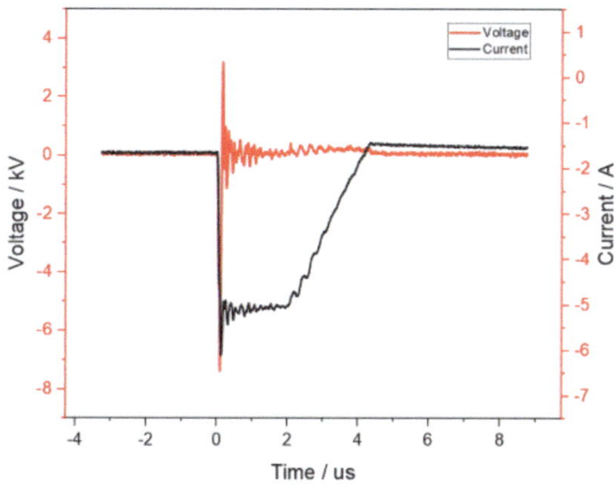

(a) Voltage and current waveforms

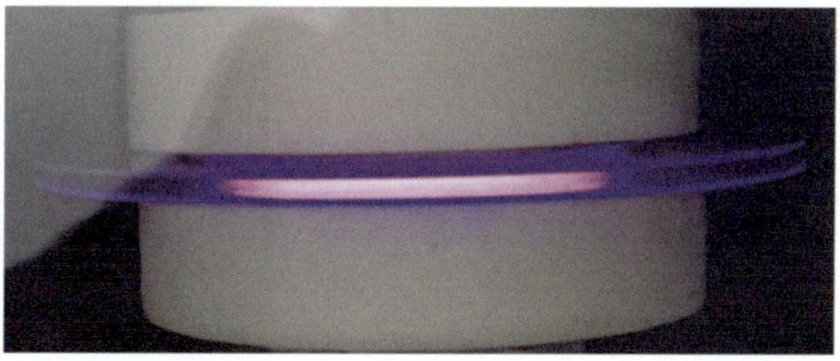

(b) Discharge image

power supply. The design verifies the feasibility of two driving schemes: optocoupler floating drive and transformer drive. Experiments show that the transformer-driven program is more feasible, and this program is based on the assembly of 20-stage Marx circuit, developed to achieve 12 kV pulse output, pulse width of 4–12 us, repeat frequency of 1–5 kHz high-voltage pulse power supply, and successfully drive the argon jet discharge.

Fig. 24.8 Transformer driver circuit drives Marx circuit to realize argon jets

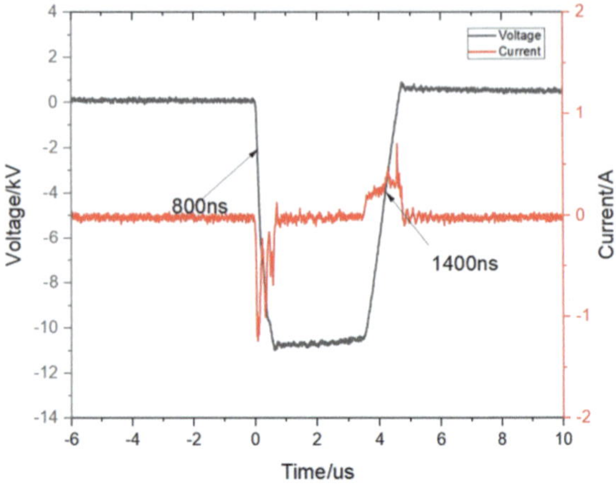

(a) Voltage and current waveforms

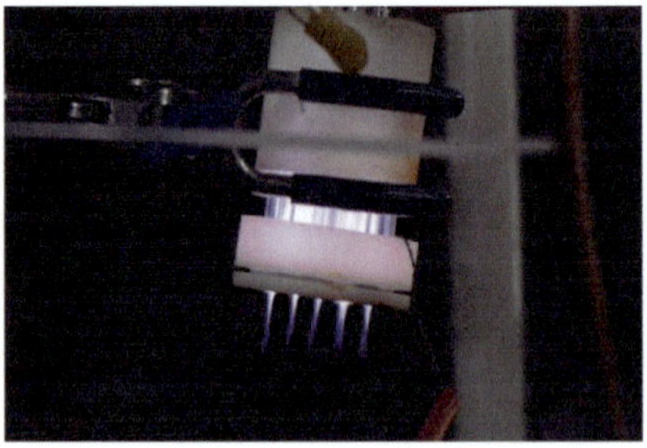

(b) Discharge image

References

1. ZHANG L, YANG D Z, WANG W C, et al. Needle-array to Plate DBD Plasma Using Sine AC and Nanosecond Pulse Excitations for Purpose of Improving Indoor Air Quality [J]. Scientific Reports, 2016, 6.
2. MORABIT Y, HASAN M I, WHALLEY R D, et al. A review of the gas and liquid phase interactions in low-temperature plasma jets used for biomedical applications [J]. European Physical Journal D, 2021, 75(1).
3. WANG D. Environmental and Biological Applications for Pulsed Discharge Plasma [M]. Springer Series in Plasma Science and Technology. 2023: 675–710.
4. GAO L H, SUN L, WAN S G, et al. Degradation kinetics and mechanism of emerging contaminants in water by dielectric barrier discharge non-thermal plasma: The case of 17β-Estradiol [J]. Chemical Engineering Journal, 2013, 228: 790–8.
5. ZHUANG J, SHI F, GUO J. Biomedical Applications of Pulsed Discharge and Pulsed Electric Field [M]. Springer Series in Plasma Science and Technology. 2023: 737–60.
6. PINGUET S, DUPéROUX J P, DELMOTE P, et al. Short-Pulse Marx Generator for High-Power Microwave Applications [J]. Ieee Transactions on Plasma Science, 2013, 41(10): 2754–7.
7. LI H, WANG C, WANG L, et al. 500 kV all-solid-state Marx generator [J]. High Power Laser Part Beams (China), 2012, 24(4): 917–20.

Chapter 25
Molecular Dynamics Simulation Study of the Degradation of Organic Contaminants by Energetic Reactive Species in Plasma

Zixu Wang, Yuhai Li, Mingzhi Zhu, Baoxu Wang, Qingshun Bai, and Peng Zhang

Abstract Current plasma cleaning technologies have been proven effective in removing organic contaminants from the surfaces of optical components; however, they still suffer from low cleaning efficiency and an uncontrollable cleaning process. The fundamental reason for these shortcomings lies in the lack of a profound understanding of the microscopic mechanisms by which plasma removes organic contaminants. Therefore, this study proposed using reactive molecular dynamics simulations to elucidate the microscopic physicochemical mechanisms through which energetic reactive species in plasma degrade organic contaminants. This research aims to investigate the microscopic reaction processes and the factors influencing plasma's cleaning efficiency in removing organic contaminants. Firstly, a physicochemical interaction model was constructed based on reactive force fields to describe the interactions between typical reactive oxygen species in plasma and representative organic contaminants in a vacuum system. The molecular-scale dynamics of bond breaking and reconfiguration, reaction products, and the mechanism of layer-by-layer removal during the plasma degradation of organic contaminants are described over picosecond timescales. Secondly, the study systematically discusses the effects of key cleaning parameters, such as plasma energy, plasma flux, and environmental temperature, on the removal efficiency of organic contaminants during plasma cleaning. The findings provide a theoretical foundation and technical support for enhancing the efficiency

Z. Wang · M. Zhu (✉) · B. Wang
Institute of Systems Engineering, China Academy of Engineering Physics, Mianyang, China
e-mail: zhumz@caep.cn

Y. Li (✉)
Research Center of Laser Fusion, China Academy of Engineering Physics, Mianyang, China
e-mail: 15546027805@163.com

Q. Bai · P. Zhang (✉)
School of Mechatronics Engineering, Harbin Institute of Technology, Harbin, China
e-mail: zp@hit.edu.cn

Y. Li et al. (eds.), *Proceedings of the 5th Chinese Conference on High Voltage and Discharge Plasmas*, Springer Proceedings in Physics 254,
https://doi.org/10.1007/978-981-96-5711-7_25

of plasma cleaning for organic contaminants on the surfaces of optical components and for controlling the cleaning process.

Keywords Plasma cleaning · Organic contaminant · Reactive molecular dynamics · Energetic active particle

25.1 Introduction

Inertial confinement fusion (ICF) achieved through large-scale laser systems is considered one of the most promising methods for humanity to obtain clean and safe energy, thereby addressing sustainable development challenges [1]. Large aperture optical components are the core components to ensure efficient and stable operation of intense laser systems [2, 3]. However, in a vacuum environment, a significant amount of organic contaminants gradually accumulate on the surfaces of large optical components as system operating time increases. Under the influence of intense lasers, this contamination not only leads to enhanced electric fields on the component surfaces, causing localized ablation, but can also result in melting or fracturing (producing particulate debris and liquid jets, among other effects). Consequently, the laser-induced damage threshold (LIDT) of optical components can decrease by more than four times, and the resulting debris and liquid jets can contaminate other optical elements in the vacuum system, causing irreversible damage [4, 5]. Due to these contamination issues, many optical components must undergo offline cleaning after being used in laser systems for a certain period. However, disassembling, sealing, transporting, and precisely reinstalling optical components during offline cleaning can damage the components and consume substantial time and resources, significantly reducing the operational efficiency of laser systems [6, 7]. Therefore, developing in-situ, non-destructive cleaning methods for large-aperture optical components has become an urgent issue.

Plasma cleaning technology has been widely confirmed as effective in removing organic contaminants adhered to the surfaces of optical components [8–10]. Maffini et al. [11] utilized argon radio-frequency plasma to clean a 500 nm thick rhodium (Rh) film deposited on a one-inch mirror substrate, and scanning electron microscopy (SEM) and X-ray photoelectron spectroscopy (XPS) analyses confirmed that no contaminants remained post-cleaning, with the mirror's reflectivity fully restored. Devadevan et al. [12] cleaned substrates contaminated by fingerprints, photoresist, and pump oil with a radio-frequency plasma cleaner, reducing the water contact angle of pump oil-contaminated substrates from approximately 70° to around 25°. Li et al. [13] employed low-pressure plasma cleaning technology, resulting in a 7% increase in light transmission and a 22.37% enhancement in LIDT for optical components. However, plasma cleaning technology still suffers drawbacks, such as low cleaning efficiency and uncontrollable cleaning processes [14, 15]. In their experiments with radio-frequency argon plasma cleaning, Maffini et al. [11] observed a deterioration in the surface quality of the mirrors after plasma treatment, which could evolve into severe mirror damage after repeated cleanings. Adelodun [16] summarized the technical characteristics of using non-thermal plasma (NTP) for the treatment of volatile organic compounds (VOC) and highlighted the issue of inefficiency associated with this treatment method. Patricia et al. [17] investigated the cleaning of fluorine-doped tin oxide (FTO) with oxygen plasma and noted that the cleaned FTO became deactivated, primarily due to the injection of oxygen anions into oxygen vacancy sites during the plasma cleaning process, which reduced the oxygen vacancy concentration in the samples. Li et al. [18] discovered that after helium ion irradiation, cracks and pores appeared on the surface of the sol–gel silica film, resulting in reduced anti-contamination performance and a subsequent decline in laser damage resistance. These studies demonstrate that plasma cleaning technology still has many limitations, fundamentally arising from a lack of a deep understanding of the microscopic mechanisms by which plasma removes organic contaminants.

The plasma removal of organic contaminants involves the intricate physical and chemical interactions between energetic active particles excited by an electric field and organic contaminants on the surfaces of optical components. Sanito et al. [19] employed a microwave plasma reactor at atmospheric pressure to treat polystyrene foam (PSF), utilizing scanning electron microscopy (SEM) and Fourier transform infrared spectroscopy (FTIR) to analyze the surface morphology and the reactive functional groups formed during the reaction process. Liang et al. [20] investigated the degradation process of benzene in non-thermal plasma (NTP) through synchrotron vacuum ultraviolet photoionization mass spectrometry (SVUV-PIMS), analyzing the main products to gain insights into the details of the degradation reaction mechanism. Li et al. [21] utilized finite element methods to analyze the cleaning mechanism of atmospheric pressure plasma in removing organic contaminants from aluminum alloy surfaces, validating their simulation results through experimental data. Kim et al. [22] simulated hydrogen plasma's collision, adsorption, permeation, diffusion, and absorption on the sample surface using Reactive Molecular Dynamics (RMD). In summary, while experimental methods can reveal the reaction products resulting from plasma interactions with organic contaminants under current experimental conditions, they do not explain the underlying physicochemical reaction mechanisms or the layer-by-layer removal of organic contaminants. Reactive molecular dynamics simulations can address these technical limitations and bridge the gap in researching relevant mechanisms.

This paper investigates the micro-reaction processes involved in the plasma removal of organic contaminants through reactive molecular dynamics simulations. It aims to elucidate the microscopic mechanisms of organic contaminant degradation and the effects of key cleaning characteristic parameters during the plasma cleaning process on the removal efficiency of organic contaminants. Firstly, a physicochemical interaction model for the cleaning process is established based on the reactive force field (ReaxFF), selecting typical energetic reactive particles from the plasma and representative organic contaminants from the vacuum systems of intense laser devices. The model is analyzed at the molecular level with a time step of femtoseconds to study the chemical bond reorganization processes, the generation of reactants, and the stepwise removal of contaminants. Secondly, variations in plasma energy, plasma flux, and environmental temperature are introduced to analyze the cleaning results, exploring the influence and mechanisms of the key cleaning characteristic parameters on the efficacy of organic contaminant removal. The insights gained from understanding the microscopic mechanisms of plasma cleaning of organic contaminants, as well as the influence of key parameters on cleaning efficacy, can provide theoretical foundations and guidance for enhancing the cleanliness of large-aperture optical components, improving energy output in intense laser devices, and advancing the research progress in inertial confinement fusion.

25.2 Simulation Methods and Details

25.2.1 Construction of Reactive Molecular Dynamics Model of Organic Contaminants' Degraded by Plasma

ReaxFF is a bond-order-based molecular force field that combines quantum chemical calculations with classical force field simulations. In the reactive force field, non-bonded interactions exist between all atoms, and to more accurately describe the complexity of molecules, perturbations such as bond lengths, bond angles, and dihedrals are also included, similar to classical non-reactive force fields. The ReaxFF force field generally divides the system's energy into the following components [23, 24]:

$$E_{sys} = E_{bond} + E_{over} + E_{under} + E_{val} + E_{pen} + E_{tors} + E_{conj} + E_{vadal} + E_{coulb} \tag{25.1}$$

In Eq. (25.1), E_{sys} represents the system energy, E_{bond} denotes the bond energy, E_{over} indicates the over-saturated bond energy, E_{under} refers to the under-saturated bond energy, E_{val} signifies the bond angle energy, E_{pen} represents the covalent bond correction energy, E_{tors} denotes the torsional angle energy, E_{conj} refers to the conjugation energy, E_{vadal} indicates the van der Waals interaction energy, and E_{coulb} represents the electrostatic interaction energy. The ReaxFF force field does not require the connectivity between atoms within a molecule to be fixed; during the simulation, chemical bonds between atoms can freely break and form, thus allowing for the modeling of chemical reaction processes.

Dibutyl phthalate (DBP), which contains a benzene ring, ester groups, and carbon chains, has been demonstrated in multiple studies to be a typical organic contaminant in the vacuum operating environment of intense laser systems [10]. Therefore, this study selects DBP as the organic contaminant in the model. At room temperature, DBP is in a liquid state; thus, DBP molecules are randomly generated, and their intermolecular distances are adjusted to obtain a liquid DBP model with a density of 1.13 g/cm^3. The model is then subjected to geometric optimization, followed by canonical ensemble (NVT) relaxation at 300 K to prevent excessive internal energy during the simulation calculations, which could lead to non-convergence.

The constructed model is shown in Fig. 25.1. The atoms at the bottom of the sample are fixed to prevent overall movement. At the same time, positions in the X and Y directions are randomly generated beyond the cutoff radius of the sample surface in the Z-direction. The Berendsen thermostat method is employed to simulate heat exchange between the system and the environment, thereby maintaining the system's temperature. Other parameters are presented in Table 25.1.

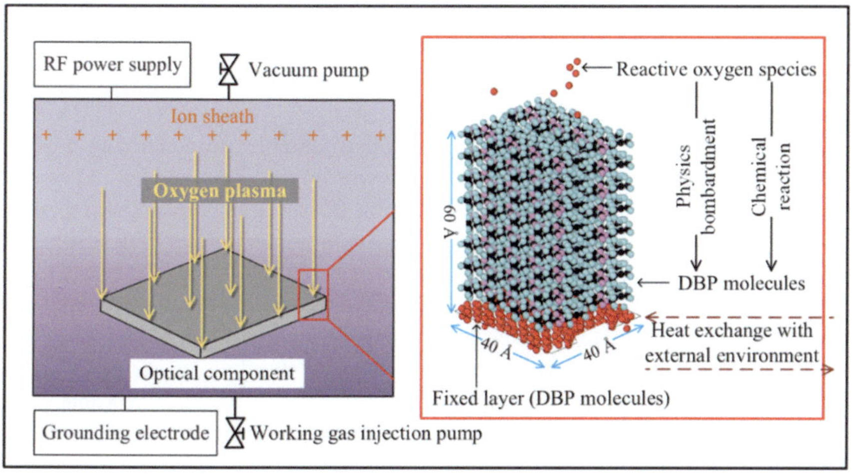

Fig. 25.1 Reactive molecular dynamics model of the reaction between reactive oxygen species in plasma and DBP as representative organic contamination in a vacuum system

Table 25.1 Simulation parameters for the plasma cleaning process

Model parameters	Numerical values
Size	$40 \times 40 \times 60$ (Å^3)
Number of molecules in contaminant sample	3456
Number of reactive oxygen species	400
Ensemble	NVT
Boundary conditions	PPF
Force field	ReaxFF
Time step	0.25 fs
Total time	100 ps
Initial temperature	300 K
Range of plasma energy	2–50 eV
Range of plasma flux	1–42 (m^{-2} s^{-1})
Range of environmental temperature variation	250–350 K

25.2.2 Details of Simulation with Different Cleaning Parameters

In the process of plasma cleaning of organic contaminants, the energy of the plasma is typically determined by the power of its power supply and the type of working gas. Plasma flux refers to the flow of particles (such as ions, electrons, and excited state gases) within the plasma. During the cleaning process, these two parameters interact with the environmental temperature as key characteristic parameters of the plasma, collectively determining the effectiveness of organic contaminant removal. Therefore, optimizing them to ensure the sufficient and efficient removal of organic contaminants is essential.

This study employs the method of controlled variables to adjust and analyze the effects of these three characteristic parameters on plasma-mediated removal of organic contaminants under different conditions. Based on the molecular dynamics modeling of plasma-assisted removal of organic contaminants described in Sect. 2.1, the following distinct simulation conditions are established to investigate the influence of plasma energy, plasma flux, and environmental temperature on cleaning efficacy. First, the simulation conditions are set with a plasma flux of 10.17 $m^{-2}\ s^{-1}$ and an environmental temperature of 300 K, with bombarding speeds of 110, 156, 135, 220, and 245 Å/ps, corresponding to plasma energies of 10, 20, 30, 40, and 50 eV, respectively. Oxygen atoms are bombarded perpendicularly onto the organic contaminants to examine the impact of plasma energy on the cleaning process. Secondly, with the plasma bombardment energy set at 10 eV and the environmental temperature at 300 K, the ion flux is varied by altering the spacing of the inserted bombarding particles. The plasma flux values are set at 1.13, 4.52, 10.17, 18.65, 28.8, and 41.81 $m^{-2}\ s^{-1}$, with oxygen atoms bombarding the surface of the samples perpendicularly to investigate the effect of plasma flux on the cleaning process. Finally, since DBP is in a liquid state at room temperature, temperature significantly influences its properties. It has a larger intermolecular distance and weaker attractive forces, resulting in more vigorous molecular motion. DBP will vaporize at 337 °C, with a flash point at 117 °C. As the temperature rises above 100 °C, diffusion phenomena begin to manifest in the model, and the thermal motion of the molecules intensifies. At this point, the dissociative action of energetic reactive particles on DBP no longer becomes dominant. Therefore, the temperature range selected for the simulation process is 250–500 K. During the simulation, with an ion flux of 10.17 $m^{-2}\ s^{-1}$ and an ion bombardment energy of 10 eV, simulations are conducted at temperatures of 300, 310, 320, 330, 340, and 350 K to examine the effect of environmental temperature on the cleaning process.

25.3 Results and Discussion

25.3.1 Process of Organic Contaminants Degraded by Reactive Oxygen Species in Plasma

During cleaning, the interaction between plasma and organic contaminants primarily consists of physical bombardment and chemical reactions. Physical bombardment involves striking surfaces with high-energy ions, spilling atoms from the material. Chemical reactions refer to the process in which reactive atoms and groups react with the material to remove layers of surface atoms [25]. Therefore, analyzing the reaction process between reactive oxygen species, as the main reactants, and DBP at the molecular scale within femtosecond time frames is of significant importance for elucidating the microscopic mechanisms through which plasma removes organic contaminants.

The reaction process between reactive oxygen species and DBP is simulated based on the conditions established in the molecular dynamics modeling and simulation outlined in Chapter Two. For this study, we select a plasma flux of 10.17 $m^{-2}{\cdot}s^{-1}$, an environmental temperature of 300 K, and a bombardment velocity of 135 Å/ps (corresponding to a plasma bombardment energy of 30 eV) to investigate the chemical bond cleavage and reconstruction of DBP under the bombardment of reactive oxygen species, resulting in the formation of decomposition products at a given moment. During the bombardment of DBP by energetic plasma particles, the branched chains attached to the phenyl ring in the DBP molecular structure are the most susceptible to degradation, dissociating into smaller hydrocarbon molecules. Conversely, the phenyl ring is the most challenging to dissociate and remove. Due to its relatively large molecular mass, the non-cyclic phenyl ring will continue to adhere to the surface of optical components, preventing the effective removal of organic contaminants. Consequently, the dissociation process of the phenyl ring warrants significant attention. The cleaning efficacy can thus be characterized by statistically analyzing the number of phenyl rings and the total number of atoms present.

Figure 25.2 illustrates the types of reaction products and the various forms generated during the degradation process of DBP, including the dissociation of the branched chains and the phenyl ring. Initially, the chemical bonds of the branched chains of the phenyl ring are bombarded by reactive oxygen species, combining with different quantities of carbon (C) or hydrogen (H) atoms to decompose into small molecules such as H_2O, CO_2, CH_2O, and CH_4. However, the phenyl ring, due to the stability of its π bonds, has not yet undergone ring opening. As the degradation reaction progresses, the phenyl ring eventually opens into long-chain carbon structures, which decompose into shorter-chain carbon structures. Ultimately, these are similarly decomposed into small molecules such as H_2O, CO_2, CH_2O, and CH_4.

Figure 25.3 illustrates the dissociation process of the benzene ring. In Fig. 25.3a, the branched structure of the DBP molecule has dissociated, leaving only the benzene ring, which continues to be bombarded by reactive oxygen species. In Fig. 25.3b, the bombardment of the benzene ring by reactive oxygen species results in the ring's

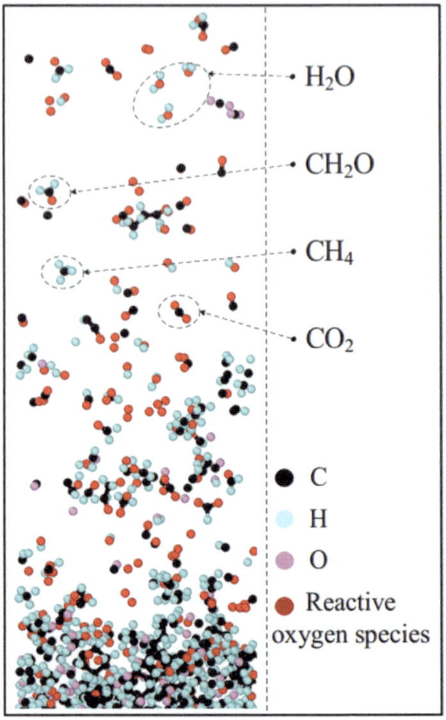

Fig. 25.2 Decomposition products of DBP during degradation

opening, leading to the formation of long carbon chains, as shown in Fig. 25.3c. In Fig. 25.3d, these long carbon chains are further dissociated into smaller molecular hydrocarbons and oxygenated compounds under the continued bombardment of reactive oxygen species, as depicted in Fig. 25.3e and f.

Figure 25.4 presents the surface morphology of organic contaminants after being bombarded by reactive oxygen species for 400 cycles, illustrating the mechanism of layer-by-layer removal of organic contaminants under plasma bombardment. The figure shows that the surface of DBP is relatively loose, with dissociation of surface DBP molecules into small hydrocarbon compounds occurring even at a low bombardment energy of 10 eV. Unlike the surface of solid film layers, the DBP's thickness variation is not significantly affected by energy changes. This phenomenon is attributed to the liquid state of DBP at room temperature, which, unlike solids, possesses a larger intermolecular distance and weaker intermolecular attractions, resulting in more vigorous molecular movement without a fixed shape. As the removal process progresses, deeper-layer DBP molecules may detach from the optical component surface either through physical bombardment by the plasma or by undergoing chemical reactions with reactive oxygen species, leading to their decomposition into small hydrocarbon compounds. In Fig. 25.4f, at an energy level of 50 eV, clumped DBP contaminants are observed detaching from the oil layer. This occurrence is attributed to the prevention of bombardment particle penetration phenomena, where

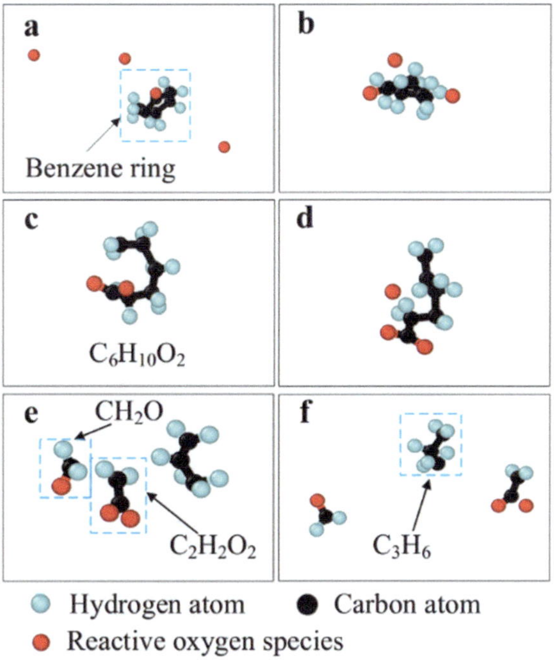

Fig. 25.3 The process of a chemical bond broken and reconstruction of DBP under the bombardment of reactive oxygen species in the plasma. **a** the undegraded benzene ring; **b** the benzene ring initially oxidized; **c** cleavage of the benzene ring; **d** long-chain carbon molecules released from the benzene ring; **e** short-chain carbon molecules generated from further cleavage; **f** small molecular hydrocarbons generated ultimately

the lowest layer of the model is configured as a reflective wall. As a result, when bombardment particles penetrate the lowest layer, their energy remains unchanged, with only the direction of their velocity altered. Therefore, when the energy of the bombardment particles increases to a certain extent, the rebounding particles can carry DBP molecules away in clusters from the oil layer. The plasma cleaning of organic contaminants on the surface of optical components occurs within a vacuum chamber, where the decomposed small hydrocarbon compounds are removed from the vacuum chamber through the injection of working gas and the extraction by vacuum pumps, thereby achieving the goal of decomposing organic contaminants and removing them from the surface of the optical components.

25.3.2 The Influence Patterns of Crucial Cleaning Process Parameters

Based on the analysis presented above, it is evident that the removal of organic contaminants by plasma occurs primarily through physical bombardment and chemical reaction mechanisms. The reaction rate is influenced by factors such as plasma energy, plasma flux, and environmental temperature. Therefore, it is crucial to investigate further the effects of key cleaning characteristic parameters on the cleaning

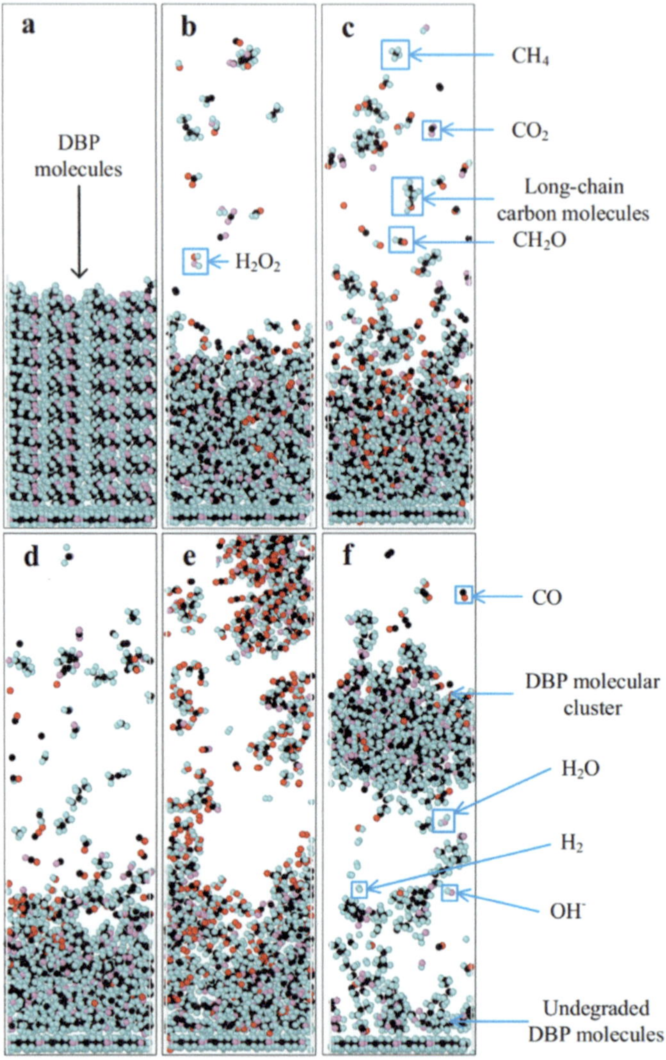

Fig. 25.4 Surface morphology image of organic contaminants after being bombarded by reactive oxygen species 400 times. **a** unimpacted DBP molecules; **b** the decomposition of surface DBP molecules at plasma energy of 10 eV; **c** the DBP molecules are decomposed as they are physically bombarded and ejected from the surface at plasma energy of 20 eV; **d** the benzene ring is dissociated at plasma energy of 30 eV; **e** Large amounts of DBP molecules are sputtered from the surface at plasma energy of 40 eV; **f** DBP molecules are removed from the surface in clusters at a plasma energy of 50 eV

efficacy and the underlying mechanisms in plasma cleaning, thereby providing theoretical guidance for enhancing plasma cleaning efficiency and process control. As discussed in Sect. 3.1 regarding the decomposition of organic contaminants, it is noted that the benzene ring is the most difficult structure to decompose and tends to accumulate on the surfaces of optical components. Additionally, due to the limitations imposed by the simulation conditions, under higher plasma energy, the rebound of energized active particles can carry DBP molecules away from the oil layer, thus hindering the characterization of cleaning efficacy based on morphology and thickness. To assess the cleaning efficacy of plasma in removing organic contaminants, this study quantifies the total number of remaining atoms and the count of benzene rings in the simulated box after 400 bombardments under various conditions. Under the simulation conditions outlined in Sect. 2.2, data regarding the remaining number of atoms and the quantity of benzene rings was obtained, with results illustrated in Figs. 25.5 and 25.6. It can be observed from Figs. 25.5 and 25.6 that the influence of plasma energy, plasma flux, and environmental temperature on cleaning efficacy exhibits similar trends. As plasma energy, plasma flux, and environmental temperature increase, the remaining number of atoms and the quantity of benzene rings show a consistent linear decrease (notably, when the plasma energy reaches 50 eV, the rebound phenomenon, constrained by simulation dimensions, leads to a sharp decline in the remaining number of atoms and benzene rings).

The following discusses the mechanisms by which different parameters of plasma cleaning affect the cleaning efficacy. First, as the plasma energy increases, the velocity of reactive oxygen species also rises, resulting in a more vigorous physical bombardment of organic contaminants. This intensified action causes deeper DBP molecules to be dislodged from the surface of optical components and dispersed within the vacuum chamber, thereby facilitating decomposition reactions. Additionally, the increase in plasma energy implies a higher activation energy for the oxygen species, leading to a more vigorous chemical decomposition of DBP, with a corresponding

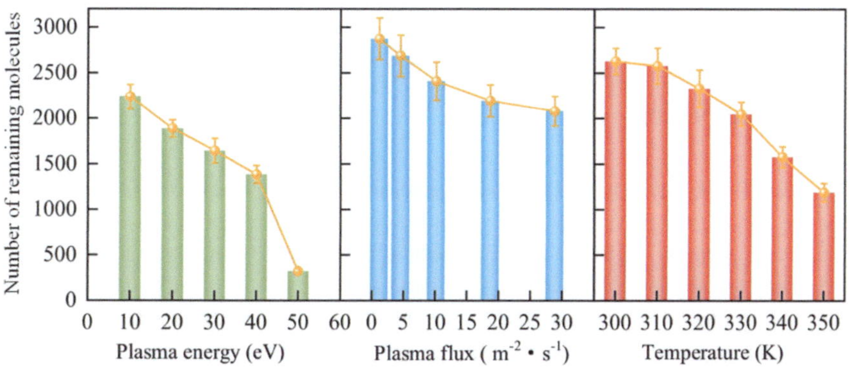

Fig. 25.5 The variety of the total particle numbers of the environment with different initial conditions after cleaning

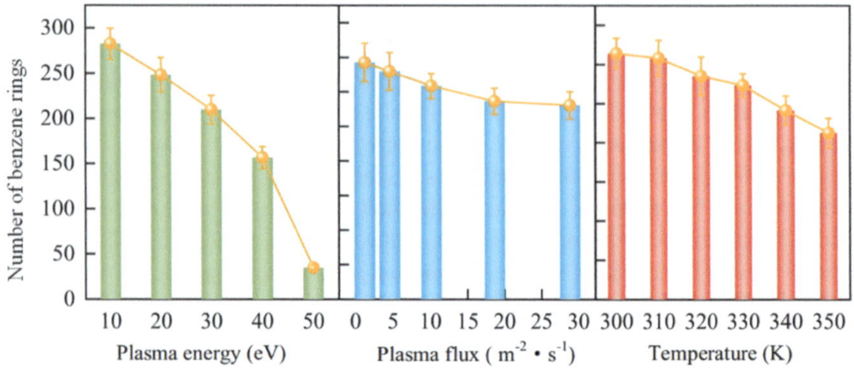

Fig. 25.6 The variety of the number of benzene rings under different conditions

increase in the number of DBP molecules undergoing decomposition. This indicates that higher plasma energy correlates with more intense DBP decomposition, enhancing plasma cleaning effectiveness and improving cleaning efficiency. Second, an increase in plasma flux results in more reactive oxygen species participating in the reactions within the chamber, allowing the DBP decomposition reactions to proceed faster. This further substantiates that increased plasma flux correlates with more vigorous DBP decomposition, thus enhancing plasma cleaning effectiveness and efficiency. Finally, an increase in ambient temperature leads to a gradual reduction in the number of remaining atoms and benzene rings within the simulation chamber. This phenomenon can be attributed to temperature being a macroscopic descriptor of molecular states; on a microscopic level, it reflects the intensity of thermal motion among the molecules. Higher temperatures correspond to more vigorous molecular motion, making it relatively difficult for the molecules to maintain their original structures. Consequently, as the temperature rises, the dissociation of DBP molecules becomes easier, resulting in improved cleaning efficiency.

25.4 Conclusion

The present study investigated the physicochemical reaction mechanisms of energetic reactive particles in plasma interacting with organic contaminants on the surfaces of optical components, utilizing reactive molecular dynamics simulations. Firstly, by integrating a reactive force field, we simulated and elucidated the physical bombardment and chemical reaction processes of reactive oxygen species on DBP during the plasma cleaning of organic contaminants on a picosecond timescale. DBP is physically ejected from the surface of the optical components due to the bombardment, while simultaneously undergoing chemical reactions with reactive oxygen species. The short-branched carbon chains of DBP dissociated from the benzene ring and were subsequently decomposed, leading to the gradual breakdown of the more chemically stable benzene ring: depolymerisation—long-chain carbon—short-chain carbon—hydrocarbon molecules. The main decomposition products of DBP were small molecules such as H_2O, CO_2, CH_2O, and CH_4. Under the dual influence of physical bombardment and chemical reactions, the organic contaminants on the surfaces of optical components were removed layer by layer. Secondly, through controlled variable experiments, we found that an increase in key cleaning parameters such as plasma energy, plasma flux, and environmental temperature enhances the cleaning effect and efficiency of the organic contaminants. The underlying principle is that the increased energy and density of energetic reactive particles in the plasma intensify DBP's decomposition and removal reactions, resulting in a greater quantity of organic contaminants being eliminated within the same treatment time. In summary, the findings of this study comprehensively reveal the microscopic mechanisms underlying the removal of organic contaminants by plasma, providing a theoretical basis and technical guidance for a deeper understanding of plasma cleaning technology principles, enhancing cleaning efficiency, and process control. This research is significant for ensuring stable transmission in intense laser system and improving the operational efficiency of such devices.

Acknowledgements This work was supported by the National Natural Science Foundation of China (Grant No. 62405296, U2030109), the Laser Fusion Research Center Funds for Young Talents (Grant No. RCFCZ7-2024-7) and the Postdoctoral Fellowship Program of CPSF under Grant Number (Grant No. GZC20233520).

Declaration of Competing Interest The authors report no conflicts of interest. The authors alone are responsible for the content and writing of this article.

References

1. Saha, Sourabh K.: Additively manufactured nanoporous foam targets for economically viable inertial fusion energy. Societal Impacts 3.C (2024).
2. Yang D, Zhao L, Cheng J, et al.: Unveiling sub-bandgap energy-level structures on machined optical surfaces based on weak photo-luminescence. Nanoscale 15.45 (2023): 18250–18264.
3. Yang Z, Zhao L, Chen M, et al.: Evolution of intrinsic defects and ring structures on the surface of fused silica optics after CO_2 laser conditioning. Optics Letters 48.21 (2023): 5727–5730.
4. Hoffman, B. N., N. Savidis, and S. G. Demos.: Monitoring and characterization of particle contamination in the pulse compression chamber of the OMEGA EP laser system. High Power Laser Science and Engineering 11 (2023): e39.
5. Li Y, Zhu B, Li G, et al.: Analysis of grating surface organic contamination in chirped pulse amplification system. Fifth International Symposium on High Power Laser Science and Engineering (HPLSE 2023). Vol. 12974. SPIE, 2023.
6. Wang S, Ye Z, Pu G, et al.: In-situ non-destructive removal of tin particles by low-energy plasma for imitation of EUV optical mirrors self-cleaning. Vacuum 212 (2023): 111963.
7. Ström, Petter, and Daniel Primetzhofer.: Ion beam tools for nondestructive in-situ and in-operando composition analysis and modification of materials at the Tandem Laboratory in Uppsala. Journal of Instrumentation 17.04 (2022): P04011.
8. Sanito, Raynard Christianson, Sheng-Jie You, and Ya-Fen Wang.: Degradation of contaminants in plasma technology: An overview. Journal of Hazardous Materials 424 (2022): 127390.

9. Golovanov, A. V., N. V. Luparev, and B. P. Sorokin.: Modification of diamond surface during photolithography, plasma-chemical cleaning, and magnetron deposition. ChemChemTech 63.11 (2020): 49–56.
10. Li Y, Bai Q, Guan Y, et al.: In situ plasma cleaning of large-aperture optical components in ICF. Nuclear Fusion 62.7 (2022): 076023.
11. Maffini A, Moser L, Marot L, et al.: In situ cleaning of diagnostic first mirrors: an experimental comparison between plasma and laser cleaning in ITER-relevant conditions. Nuclear Fusion 57.4 (2017): 046014.
12. Devadevan M M, Krishnanand J, Shilpa K B, et al.: A Compact, Fast, and Efficient Plasma Cleaner. IEEE Transactions on Plasma Science 49.11 (2021): 3576–3581.
13. Li Y, Bai Q, Shen R, et al.: In-situ cleaning organic contaminants on SiO2 sol-gel antireflection film by low-pressure air plasma. Laser-Induced Damage in Optical Materials 2021. Vol. 11910. SPIE, 2021.
14. Huang Y, Yu Q, Li M, et al.: An overview of low-temperature plasma surface modification of carbon materials for removal of pollutants from liquid and gas phases. Plasma Processes and Polymers 18.3 (2021): 2000171.
15. Zhuang B, Li S, Li S, et al.: Ways to eliminate PMMA residues on graphene—superclean graphene. Carbon 173 (2021): 609–636.
16. Adelodun, Adedeji A.: Influence of operation conditions on the performance of non-thermal plasma technology for VOC pollution control. Journal of Industrial and Engineering Chemistry 92 (2020): 41–55.
17. Díaz P, Riveros G, Faúndez M, et al.: Is the oxygen plasma cleaning technique indicated for any electrochemical purpose?: the case of FTO electrodes. Electrochimica Acta (2024): 145149.
18. Li B, Guo J, Yuan Z, et al.: Microstructure evolution and performance degradation of sol-gel silica films induced by helium ion irradiation. Optical Materials 147 (2024): 114689.
19. Sanito R C, Chen Y W, You S J, et al.: Hydrogen and methane production from Styrofoam Waste using an atmospheric-pressure microwave plasma reactor. Aerosol and Air Quality Research 20.10 (2020): 2226–2238.
20. Liang Y, Li J, Xue Y, et al.: Benzene decomposition by non-thermal plasma: A detailed mechanism study by synchrotron radiation photoionization mass spectrometry and theoretical calculations. Journal of Hazardous Materials 420 (2021): 126584.
21. Li Y, Bai Q, Yang D, et al.: Mechanism and Verification of Plasma Cleaning of Organic Contaminant on Aluminum Alloy Surface. China Surface Engineering 33.6 (2020): 58–67.
22. Kim B, Kim M, Yoo S, et al.: Atomistic insights on hydrogen plasma treatment for stabilizing High-k/Si interface. Applied Surface Science 593 (2022): 153297.
23. Xin L, Liu C, Liu Y, et al.: Thermal decomposition mechanism of some hydrocarbons by ReaxFF-based molecular dynamics and density functional theory study. Fuel 275 (2020): 117885.
24. Guo W, Fan K, Guo G, et al.: Atomic-scale insight into thermal decomposition behavior of polypropylene: A ReaxFF method. Polymer Degradation and Stability 202 (2022): 110038.
25. Zhu T, He Y, Tan T, et al.: Insight into reactive oxygen plasma characteristics and reaction mechanism on SRF accelerator plasma cleaning. Plasma Sources Science and Technology 32.11 (2023): 115002.

Chapter 26
Application of DBD Dielectric Barrier Discharge Plasma in UAV Wing De-icing

Xuejing Yang, Xincheng Wang, Yu Wang, Qipeng Li, and Zhengchao Yang

Abstract This paper explores the application method of a Plasma Synthetic Jet Actuator (PSJA) for ice removal on unmanned aerial vehicle (UAV) wings. Traditional wing anti-icing and de-icing technologies, such as engine bleed air and electric heating, suffer from issues like high power consumption and ineffective ice removal, rendering them unsuitable for small-scale UAVs. Dielectric Barrier Discharge (DBD) plasma technology offers a novel solution, ensuring ice removal on the leading edge of the **wing** while preventing overflow ice formation on the middle and rear sections. Compared to traditional electric heating methods for wing anti-icing and de-icing, this solution consumes less energy, requires simpler equipment, and is easier to control. Experimental studies have demonstrated the significant effectiveness of DBD plasma technology in ice prevention. By optimizing the layout and parameter settings of the plasma actuator, de-icing efficiency can be further enhanced. Such applications can provide solutions to the rapidly evolving problem of anti-icing and de-icing for UAVs, enhancing the ice protection performance of UAV wings while ensuring their operational endurance.

Keywords Unmanned aerial vehicle · Dielectric barrier discharge (DBD) · Wing anti-icing and de-icing · Overflow ice · Energy consumption

---

X. Yang · X. Wang · Y. Wang (✉) · Q. Li · Z. Yang
College of Urban Construction, Nanjing Tech University, Nanjing, China
e-mail: yu-wang@njtech.edu.cn

X. Wang · Y. Wang
College of Civil Aviation Safety Engineering, Civil Aviation Flight University of China, Guanghan, China

Sichuan Province All-Electric Navigation Aircraft Key Technology Engineering Research Center, Guanghan, China

Y. Li et al. (eds.), *Proceedings of the 5th Chinese Conference on High Voltage and Discharge Plasmas*, Springer Proceedings in Physics 254,
https://doi.org/10.1007/978-981-96-5711-7_26

26.1 Introduction

Aircraft icing has always been one of the major factors threatening flight safety. High-impact air weather events such as aircraft icing have a great impact on flight safety. Ice freezing can significantly affect the stability and control stability of the aircraft, and affect the flight safety and performance. This phenomenon primarily occurs when supercooled droplets (water droplets below the freezing point) in the atmosphere collide with and accumulate on the aircraft. When aircraft fly at high altitudes, icing can occur when the droplets are at approximately 5 °C. Icing disrupts the original aerodynamic shape of the aircraft, and the flow after icing exhibits strong unsteadiness, nonlinearity, and randomness [1]. Therefore, aircraft icing severely impacts flight safety by disrupting the aerodynamic shape of the aircraft surface, increasing flight resistance, and reducing the lift coefficient of the aircraft [1–3]. According to statistics and cause investigation reports on flight accidents by NASA, accidents caused by icing account for approximately 9% of all flight accidents [4]. The issue of aircraft icing is increasingly serious.

The technology for anti-icing and de-icing on the outer surface of aircraft mainly includes methods for anti-icing and de-icing, detection of icing on the outer surface, and improvements to the aircraft's outer surface structure itself. By selecting appropriate materials and designing the structure, it becomes difficult for supercooled droplets in the atmosphere to adhere to the aircraft surface, achieving the purpose of anti-icing. De-icing refers to allowing a small amount of ice to form on certain aircraft components, but then promptly removing it using corresponding equipment [5]. Traditional anti-icing and de-icing methods require a significant amount of external energy, affecting the aerodynamic shape of the aircraft or increasing its structural weight and reducing the effective payload. The technology for improving the outer surface structure of aircraft started late, and the current breakthrough in this technology lies in the materials used on the outer surface of aircraft [6]. With the development of technology, unmanned aerial vehicles (UAVs) are increasingly being used in both military and civilian sectors, with longer flight durations and higher flight altitudes. Inevitably, they must perform missions in rain and snow weather around the clock, and wing icing has become a significant hidden danger for UAV flight safety. However, research on wing anti-icing and de-icing technologies for UAVs is urgently needed.

Existing anti-icing and de-icing methods include mechanical de-icing technology, liquid anti-icing technology, and thermal anti-icing technology. Mechanical de-icing technology can be further divided into aerodynamic band de-icing, ultrasonic de-icing, and electro-impulse de-icing technologies. Thermal anti-icing technology is divided into electrothermal anti-icing and pneumatic thermal anti-icing technologies [7]. Mechanical anti-icing systems consume high energy. For example, the main disadvantage of engine bleed air systems is that excessive bleed air can often cause insufficient power in the aircraft engine, leading to serious flight accidents. If a continuous electrothermal anti-icing system is used on aircraft, it will consume a significant amount of electrical energy, increasing the load on the aircraft's electrical and energy storage systems. Currently, some new technologies are being developed, such as loop heat pipe anti-icing technology, dielectric barrier discharge (DBD) plasma anti-icing technology, bio-inspired surfaces for anti-icing, superhydrophobic passive anti-icing technology, infrared technology for anti-icing and de-icing systems, etc. A schematic diagram of various aircraft anti-icing and de-icing systems is shown in Figs. 26.1, 26.2 and 26.3 [8–11].

When a high AC voltage is applied to the electrodes in Dielectric Barrier Discharge (DBD), the air on the encapsulated electrodes is ionized and generates a plasma stream. During this process, the ambient air on the encapsulated electrodes is also heated by the plasma. Therefore, the DBD anti-icing and de-icing system can directly heat the gas layer on the wing surface to achieve anti-icing and de-icing. Liu et al. [12] conducted a comparative study using the DBD plasma de-icing method and the electrothermal anti-icing method, finding that under the same power input, the effect of using the DBD plasma method for wing anti-icing and de-icing is comparable to traditional electrothermal anti-icing under extreme conditions. Furthermore, Liu et al. [13] discovered that, at the same power input, compared to the continuous drive mode, the plasma drive in the duty cycle mode has a higher instantaneous voltage during operation, with the total power consumption of the plasma actuator

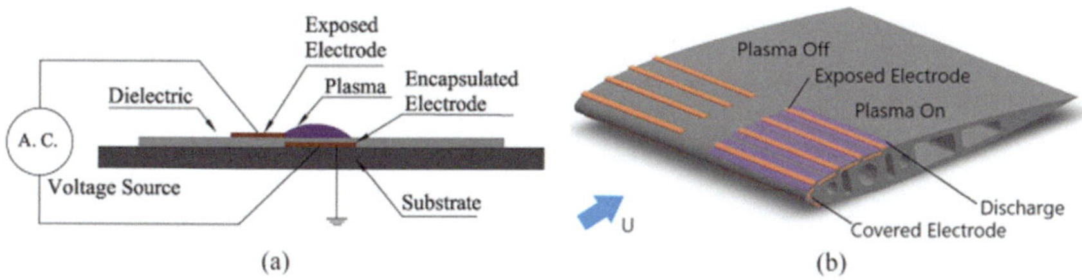

Fig. 26.1 Schematic illustration of airfoil model and plasma actuators. **a** Single SDBD plasma actuator. **b** Actuators mounted on airfoil model [8, 9]

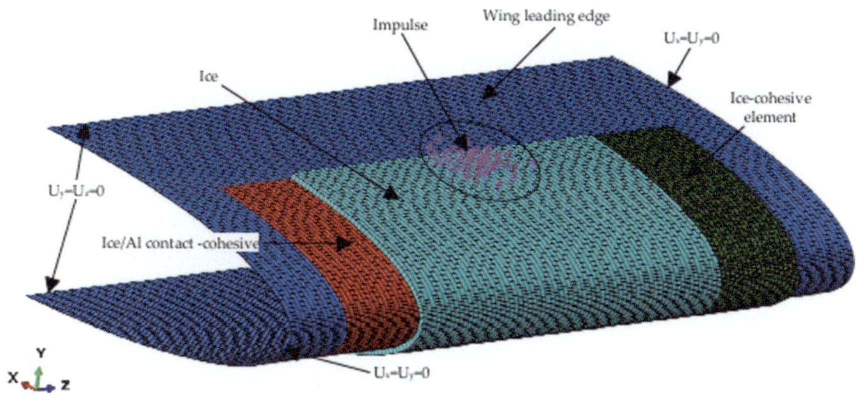

Fig. 26.2 Finite element modeling of wing icing [10]

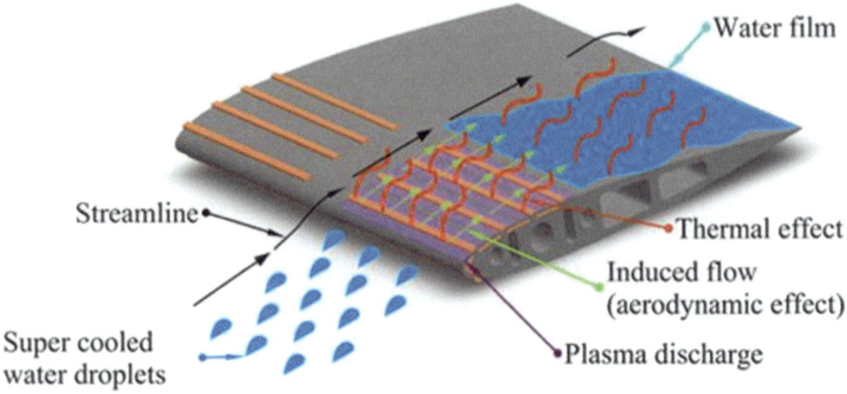

Fig. 26.3 Mechanism illustration of plasma icing control [11]

remaining unchanged over time. This results in a stronger thermal effect (twice the instantaneous input power density of traditional electrothermal film anti-icing), thereby improving anti-icing and de-icing performance. Pulse Electro Thermal De-icing (PETD) removes ice from an evaporator by introducing a very short-duration, high-power electric pulse to a very thin metal foil heater located at the interface of the ice and the evaporator's surface [14]. Compared to electro-impulse de-icing technology, which can reduce the fatigue life of the wing and pose issues such as the arrangement and orientation of excitation devices, DBD plasma has great potential in anti-icing and de-icing applications In this study [15], we fabricated a high substrate binding and mechanical durability anti/de-icing superhydrophobic coatings (SL-AP/SPS-SiO_2) with self-healing performance by spray-coating a photothermal self-healing organic layers (AP/SPS-SiO_2) plasma-sprayed porous Al_2O_3–13%TiO_2 underlayer (SL), the resulting soft-hard interlocking structure significantly increased the adhesion strength of the coating to 16.7 $\pm$ 0.36 MPa while reducing the ice adhesion strength to 46.5 $\pm$ 3.0 kPa. The influence of the detailed voltage profile on the heating effects of a Surface Dielectric Barrier Dis-charge driven by Nanosecond Pulses (NS-SDBD) is investigated, and a comparison of the anti-icing performance among different configurations of streamwise plasma heat knife is made, a three-level configuration of the plasma heat knife is proposed, and its anti-icing performance is found to be much better than that of the original configuration [16].

Although DBD plasma actuators have certain effects in anti-icing and de-icing, there are still some problems to be solved. The main requirements are: the narrow discharge gap leads to high processing requirements, and the high-performance DBD device has high requirements for the dielectric constant of the dielectric layer material; At the same time, numerical simulation methods and experimental observation techniques involving multi-physics coupling and multi-time scales are also key breakthroughs in solving related scientific problems. These problems limit the further application and development of DBD plasma actuators in anti-icing and de-icing. So, further research is needed to optimize DBD plasma actuators and further

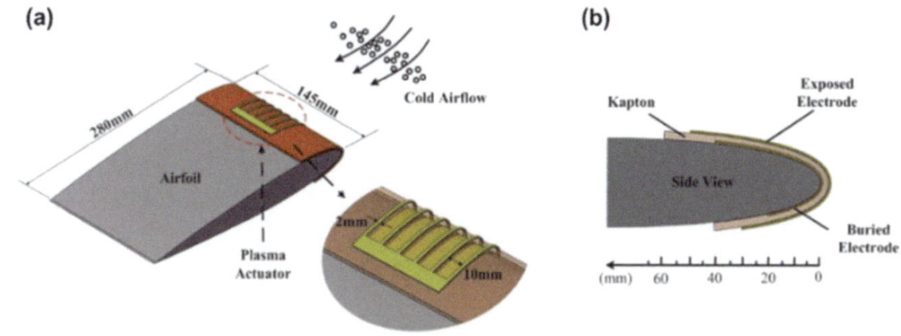

Fig. 26.4 Mechanism diagram of plasma ice control [17]

enhance anti-icing and de-icing effectiveness. The mechanism diagram of the plasma anti-icing and de-icing device is shown in Fig. 26.4 [17].

26.2 Design of Plasma-Based De-icing System for Unmanned Aerial Vehicles (UAVs)

To ensure that the anti-icing and de-icing system for UAVs operates with low power consumption and high efficiency, the system is divided into two parts:

First, the Plasma Excitation Device: The DBD plasma anti-icing and de-icing system operates by integrating a pulsed power supply unit to energize the plasma excitation device. The role of the plasma excitation device is to remove ice accumulation on the leading edge of the wing and heat the air on the wing surface while the aircraft is in flight.

Second, the Loop Heat Pipe Energy Storage System: This system transports waste heat from the UAV's hydraulic devices to a phase change energy storage device for use during wing icing. The function of the loop heat pipe energy storage system is to activate the plasma anti-icing and de-icing system before takeoff to store heat in the phase change energy storage device. After takeoff, the loop heat pipe collects waste heat from the UAV's hydraulics and transports it to the phase change energy storage device for energy recovery and reuse. Additionally, highly thermally conductive metals are used in the middle and rear sections of the wing to efficiently transfer heat and heat up overflow droplets, thereby eliminating overflow ice. The system design diagram is shown in Fig. 26.5.

The DBD plasma excitation device is shown in Fig. 26.6. This DBD plasma excitation device consists of an integrated pulsed power supply unit, an upper electrode, a cover electrode, dielectric material, and positive and negative wires. A thin film of dielectric material is bonded between the upper electrode and the cover electrode. The cover electrode is embedded into the surface of the wing skin, flush with its

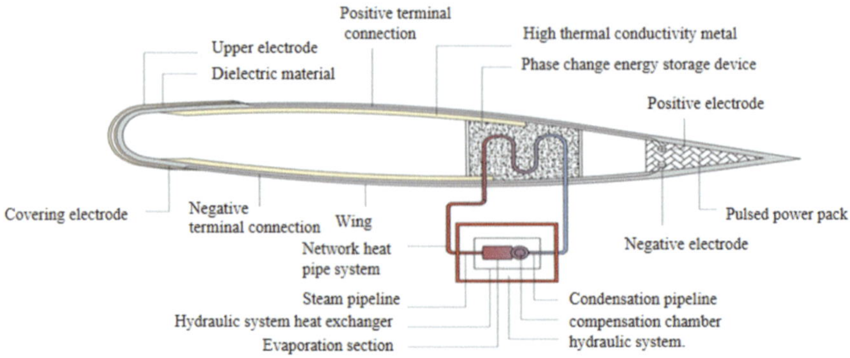

Fig. 26.5 Design diagram of the anti-icing and de-icing system

surface, and connected to the positive and negative terminals of the integrated pulsed power supply unit through wires.

During flight, the wing surface of the UAV encounters supercooled droplets that impact it head-on, causing ice accumulation on the leading edge of the wing. When the DBD plasma excitation device is powered on, it generates a significant amount of heat through the dielectric loss angle of the dielectric film. This heat melts the ice accumulated on the leading edge of the wing through thermal conduction, forming droplets that flow towards the middle and rear sections of the wing. At the same time, the plasma excitation device ionizes the dispersed air in the ice-melting area to form a plasma, which further heats the air on the wing surface. The heated air on the wing surface then exchanges heat with the droplets flowing towards the middle and rear sections of the wing, preventing the droplets from forming overflow ice.

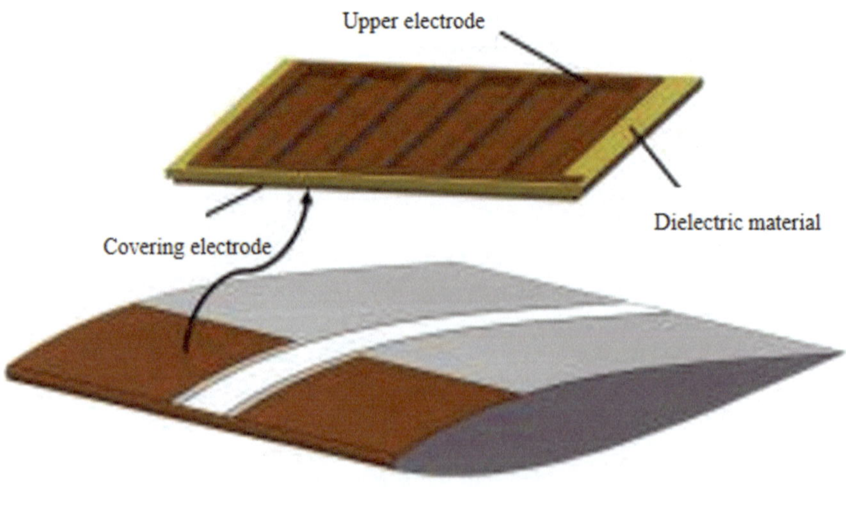

Fig. 26.6 Diagram of the DBD plasma excitation device

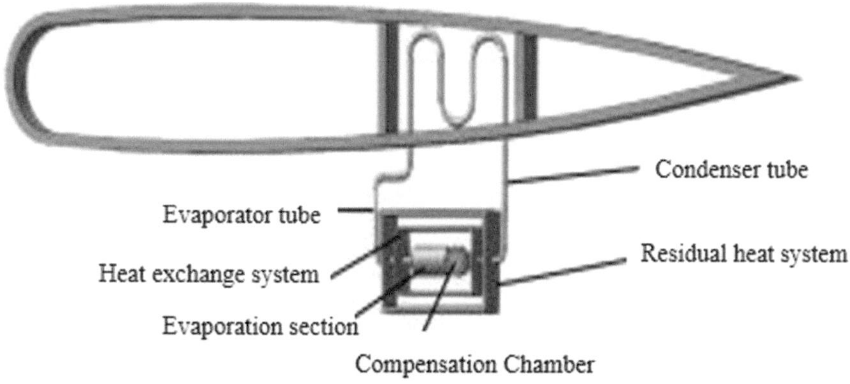

Fig. 26.7 Diagram of the energy storage device for the loop heat pipe system

The energy storage device of the loop heat pipe system is shown in Fig. 26.7. This device comprises a loop heat pipe, a phase change energy storage unit, a hydraulic system heat exchanger, and high thermally conductive metals. The phase change energy storage unit of the loop heat pipe system collects waste heat from the aircraft's hydraulic system and stores it in phase change materials through the loop heat pipe. The phase change materials continuously heat the wing through thermal conduction by high thermally conductive metals, preventing ice accumulation on the wing surface.

The loop heat pipe system consists of an evaporator line, a condenser line, and an evaporator section. A compensation chamber is designed on the right side of the evaporator section of the loop heat pipe to maintain stable heat transfer under high G-loads, preventing the evaporator section from drying out due to such loads. The loop heat pipe system transfers waste heat from the hydraulic system to the phase change energy storage unit, which then heats the skin surface through high thermally conductive metals. Under the dual thermal action of the wing surface and its air, the temperature of the droplets after ice melting is maintained above 0 °C, preventing the formation of overflow ice. The specific parameters of the system are shown in Table 26.1.

Table 26.1 Table of designed system performance parameters

Parameter category	Parametric description	Parametric description
Plasma excitation device size	Electrodes, cover electrodes, and dielectric material dimensions	Diameter of top electrode/cover electrode: 50 mm; Dickness of dielectric material: 0.2 mm
Plasma intensity	Peak discharge current and peak voltage	Nanosecond pulse power supply with a peak voltage of 8–10 kV
Pumping frequency; pump-frequency; driving frequency; excitation frequency	The discharge frequency of the plasma exciter	25 kHz
Heat pipe diameter of the loop heat pipe system	Diameter of the loop heat pipe	5–10 mm

26.3 Results and Analysis

26.3.1 Establishment of Flow Field Model and Grid Independence Verification

A flow field model was established using COMSOL software, as shown in Fig. 26.8. The wing model adopts the NACA 0012 standard airfoil with a chord length C = 0.5334 m. After meshing, the model was imported into the DROP3D and ICE-3D plugins of the icing calculation software FENSAP-ICE to calculate the water droplet collection efficiency and ice shape characteristic parameters, respectively. After meshing, a comparison was made with experimental data from literature using the following parameters:

Wing chord length: 0.5334 m

Flow field velocity: v = 100 m/s

Angle of attack (AOA): 0°

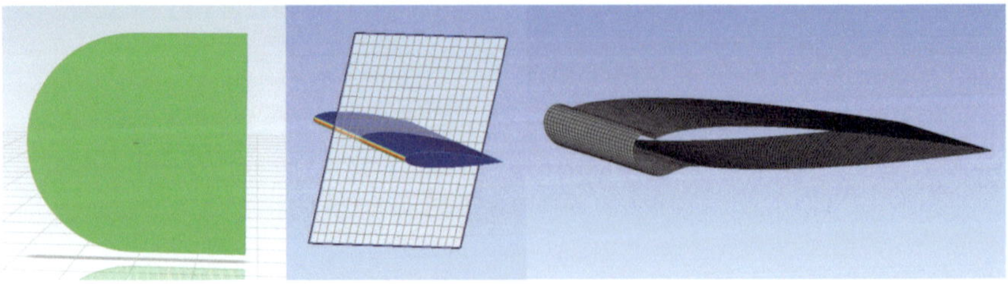

Fig. 26.8 Flow field model diagram

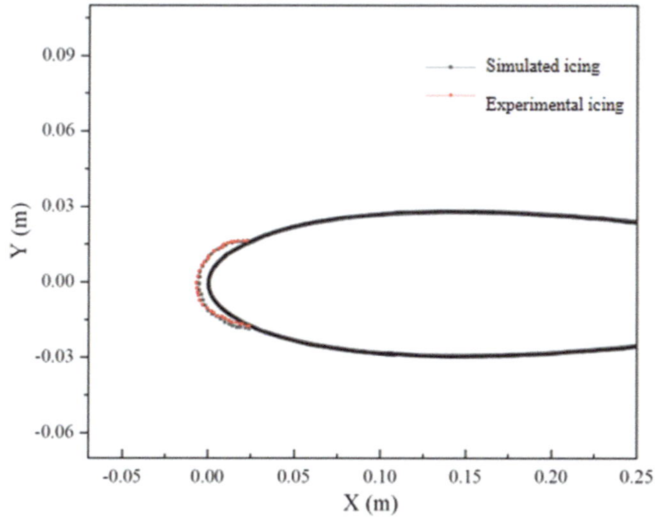

Fig. 26.9 Icing model diagram for independence verification

Atmospheric pressure: 61,640 Pa

Liquid water content (LWC): 0.3 g/m^3

Median volume diameter (MVD) of droplets: 25 μm

Ambient temperature: 248.15 K

Simulated icing time: S = 600 s.

Upon verification, the simulation results deviated by 9.31% from the experimental results from literature, indicating good grid independence, as shown in Fig. 26.9.

26.3.2 Numerical Simulation of Icing

Using the FENSAP-ICE software, simulations were conducted on an aircraft wing modeled with the NACA 0012 standard airfoil. The icing conditions were based on those outlined in Appendix O of the Federal Aviation Administration (FAA)'s icing regulations, with a flight speed of 0.5 Mach, a liquid water content (LWC) of 0.3 g/m^3, and an air temperature of 263.15 K. The simulation incorporated calculations for supercooled large droplets (SLD) based on the Pilch & Erdman model for droplet deformation and breakup in gas–liquid systems, as well as the Mundo model to capture water droplet splashing and rebound characteristics. The results of this simulation, which modeled the chordwise water droplet collection efficiency and the geometric characteristics of ice formation on the wing under freezing drizzle conditions, are shown in Figs. 26.10 and 26.11.

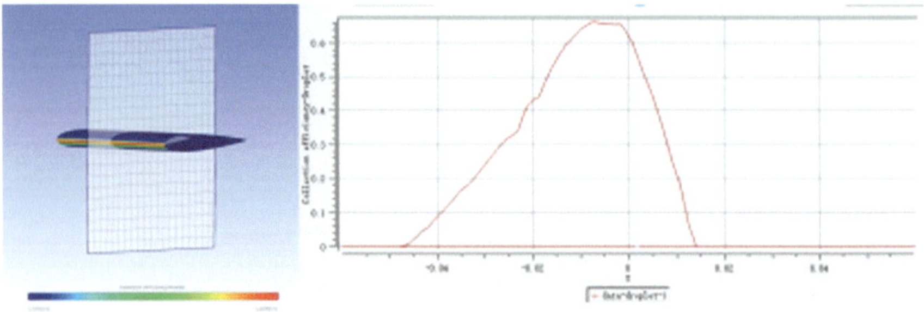

Fig. 26.10 Chordwise water droplet collection efficiency

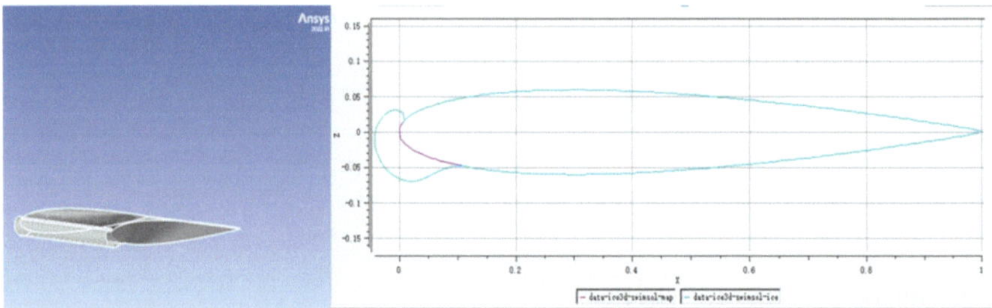

Fig. 26.11 Geometric characteristics of ice formation

After meshing, the DROP-3D plugin within the FENSAP-ICE software was utilized to calculate the impact characteristics of supercooled droplets. By setting the distribution and size of supercooled droplets, the simulation replicated the gas–liquid flow field during flight and the impact of differently sized and distributed supercooled droplets on the ice shape formed on the wing.

After performing the droplet impact calculations, using the conditions from those calculations, the ICE-3D plugin was utilized to simulate ice formation on the wing following droplet impact. The resulting geometric characteristics of ice formation revealed that the ice accumulated on the wing in a wrapping manner, covering the leading edge. Due to the aircraft's angle of attack being set at 5°, the aircraft was in an upward climbing state (i.e., ascending through clouds into the stratosphere after takeoff), resulting in a larger ice accumulation area on the underside. Therefore, according to the geometric characteristics of ice formation, the leading edge of the wing is a key protection area.

26.3.3 Simulation of DBD Plasma De-icing

The COMSOL software was used to simulate the plasma actuation device, establishing heat transfer models for electrical current, solid heat transfer, fluid heat

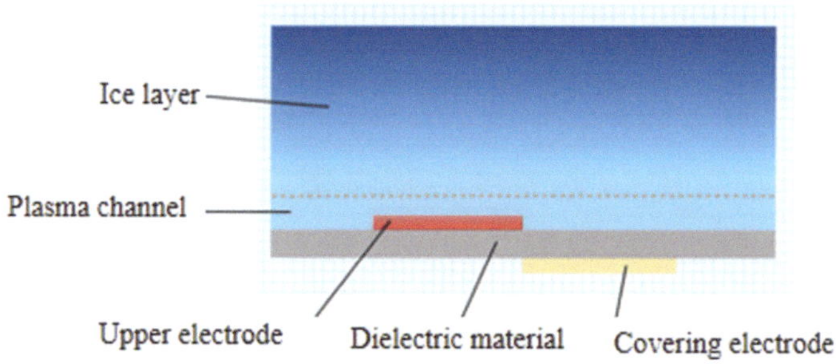

Fig. 26.12 Modeling of the plasma actuation device

transfer, and plasma. The modeling is shown in Fig. 26.12. The ice layer thickness was 8 mm, with a dielectric material of polyimide film with a thickness of 0.2 mm. The upper electrode and the covering electrode were made of copper foil with a thickness of 0.03 mm. A nanosecond pulsed power supply was used with voltages of 8 kV and 10 kV, respectively.

Comparison of Ice Melting Effects between AC-DBD and NS-DBD

This device employs NS-DBD as its power supply. The following Fig. 26.13 compares the ice melting effects of NS-DBD and AC-DBD. It can be observed that under the same voltage input conditions, the ice prevention and removal range of NS-DBD is approximately 42% larger. This is because the discharge filaments of NS-DBD extend longer and form plasma channels, resulting in a wider discharge range. The temperature of both AC-DBD and NS-DBD gradually decreases from the electrode-ice interface in both horizontal and vertical directions. AC-DBD's heat release is mainly concentrated near the upper electrode, and its primary ice removal mechanism is heating the dielectric layer. Figure 26.14 shows the isotherm map of AC-DBD. The temperature on the right side of the upper electrode in NS-DBD expands in a "filament-like" pattern. This is because NS-DBD first removes some of the accumulated ice by heating the dielectric film. The heat source for this heating is the heat generated by the dielectric loss angle. As the actuator begins to discharge, it forms plasma to heat the surrounding air. The plasma channel further expands, accelerating the ice removal rate. Additionally, NS-DBD-based ice removal can rapidly and instantaneously heat the gas to 1100 K [18] within a shorter time scale. NS-DBD exhibits excellent thermal performance, efficiently removing ice from critical areas of the wing to enhance flight safety margins.

Ice Layer Temperature Distribution on the Surface of NS-DBD Actuator at Different Voltages

Simulations were conducted with simulated voltages of 8 and 10 kV, an NS-DBD excitation frequency of 25 kHz, and an ice layer thickness of 8 mm. After 150 s of NS-DBD actuator discharge, the ice melting thicknesses at 8 kV and 10 kV were 5 mm and 6.3 mm, respectively. Along the X-axis, the ice melting length of the

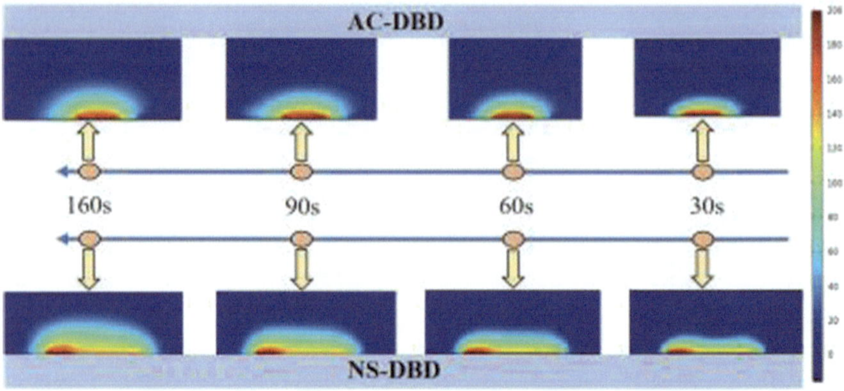

Fig. 26.13 Comparison of ice melting effects between AC-DBD and NS-DBD

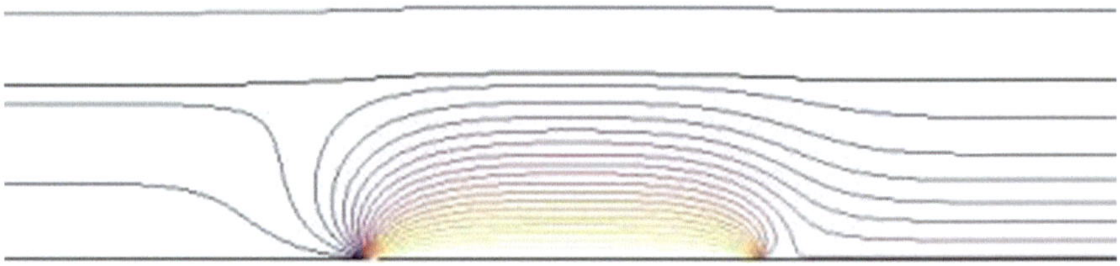

Fig. 26.14 Isotherm map of heat release from AC-DBD actuator

NS-DBD actuator at 10 kV was 35.7% longer than that at 8 kV. Along the Y-axis, the ice melting length at 10 kV was 33.6% longer than at 8 kV, resulting in an overall 11.99% larger ice melting area in cross-section. The overall contour plot is shown in Fig. 26.15. It can be seen that as the excitation voltage increases, the discharge range expands, the filamentary discharge intensifies, and the length of the ion channel also increases.

Comparison of NS-DBD Experimental and Simulation Results

For ice removal at voltages of 8 and 10 kV, an experimental setup was built to validate the simulations. As shown in Fig. 26.16, in both cases, the NS-DBD excitation frequency was 25 kHz, the ice layer thickness was 8 mm, and the ice melting thickness was measured using a 0–20 mm laser thickness gauge.

Table 26.2 presents a comparison between simulated and experimental data on ice melting length for the NS-DBD actuator under 8/10 kV voltage at various time points. Analysis of the data in the table reveals that there is a discrepancy between experimental and simulated data, with error values within 20%. The discrepancy in experimental and simulated data may be attributed to the higher ambient temperature during experiments compared to the set temperature in numerical simulations, leading to faster ice melting speeds and greater ice melting distances in experimental

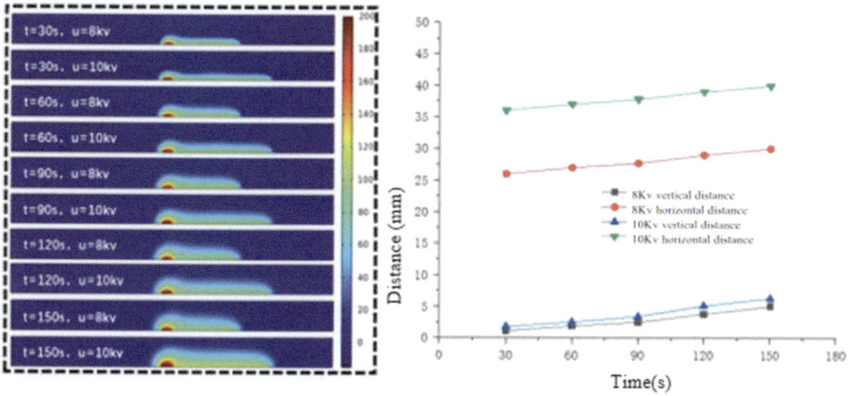

Fig. 26.15 Ice melting length of NS-DBD actuator at 8/10 kV over time

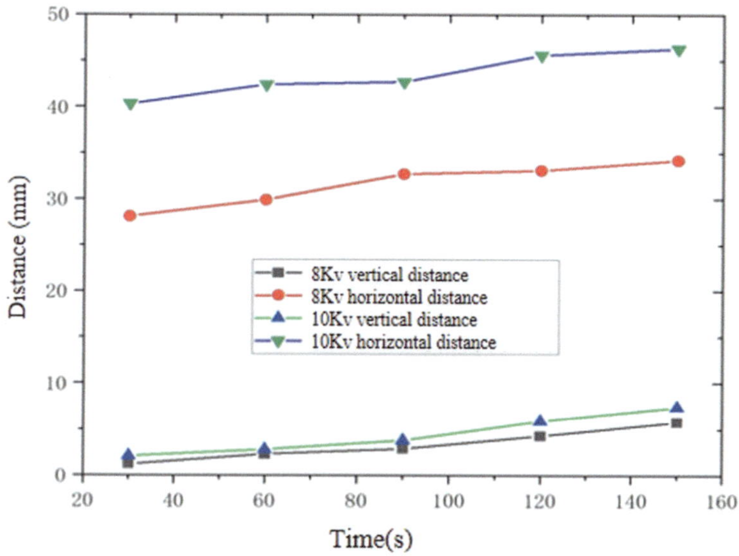

Fig. 26.16 Graph of ice melting length at different times for the NS-DBD actuator under 8/10 kV voltage in experimental conditions

results. Based on this analysis, the accuracy of both experimental and simulated data can be verified.

Table 26.2 Simulation and experimental data for ice melting results

Vertical distance at 8 km v (mm)	Simulation	1.10	1.90	2.50	3.80	5.00
	Test	1.20	2.30	2.90	4.30	5.80
Percentage difference%		9.09	21.05	16.00	13.16	16.00
Horizontal distance at 8 km v (mm)	Simulation	26.00	27.00	27.70	29.00	30.00
	Test	28.10	29.90	32.70	33.10	34.20
Percentage difference%		8.08	10.74	18.05	14. 14	14.00
Vertical distance at 10 kV (mm)	Simulation	1.80	2.50	3.40	5.10	6.30
	Test	2.10	2.80	3.80	5.90	7.40
Percentage difference%		16.67	12.00	11.76	15.69	17.46
10 kV horizontal straight distance (mm)	Simulation	36.00	37.00	37.80	39.00	40.00
	Test	40.30	42.40	42.70	45.60	46.30
Percentage difference%		11.94	14.59	12.96	16.92	15.75

26.4 Conclusion

This paper presents the design of a miniaturized and lightweight anti-icing and de-icing system for unmanned aerial vehicle (UAV) wings based on DBD plasma technology. The system employs an active wet anti-icing and de-icing method (DBD plasma de-icing technology) to significantly reduce UAV energy consumption and maintenance costs. By utilizing a nanosecond pulsed power supply (NS-DBD), instantaneous thermal excitation is achieved, reducing energy consumption and enhancing UAV flight range and lightweight capabilities. Moreover, the NS-DBD plasma actuator leverages dual heating from air and dielectric materials to quickly remove ice from the wing surface while recovering waste heat from the hydraulic system to further enhance de-icing effectiveness.

The conclusions are as follows:

(1) In simulations, SLD supercooled large droplets were modeled, with the Pilch & Erdman model used to analyze droplet deformation and breakup, and the Mundo model employed to study droplet sputtering and bouncing, as well as the water droplet collection coefficient and icing geometric characteristics of the wing under freezing drizzle conditions.. It was found that the leading edge of the wing is the primary icing region, and under a 5° angle of attack during climb, the icing area on the lower side of the wing is larger.

(2) Simulations of the plasma excitation device were performed using COMSOL software. A comparison of AC-DBD and NS-DBD ice melting effects was conducted, revealing that the anti-icing range of NS-DBD is approximately 42% larger under the same voltage.

(3) The “filamentary” structure of NS-DBD enhances ice melting efficiency. In simulations of NS-DBD actuators under different voltages, the ice melting thicknesses at 8 kV and 10 kV were 5 mm and 6.3 mm, respectively. The ice melting

length at 10 kV was significantly greater than at 8 kV, with an overall increase in ice melting area by 11.99%.

(4) Experimental validation was conducted for de-icing scenarios at 8 and 10 kV, using a laser thickness gauge to measure ice melting thickness, verifying the accuracy of simulation results.

Acknowledgements This paper is funded by Key Laboratory of Icing and Anti/De-icing of CARDC (Grant No. IADL20230410), Sichuan Province All-electric Navigation Aircraft Key Technology Engineering Research Center (Grant No. CAFUC202401KF009) and Natural science foundation of Postgraduate Research and Practice Innovation Program of Jiangsu Province (SJCX240506).

References

1. Rabiei, B. A., Kim, Y., Myong, S.R., Adaptive robust servo LQR control for aircraft under a wide range of Icing conditions[J]. International Journal of Aeronautical and Space Sciences,2024,(prepublish):1–23.
2. Liu, H., Peng, S., Fang, R., et al. Analysis of the meteorological conditions and atmospheric numerical simulation of an aircraft icing accident[J]. Atmosphere,2024,15(10):1222–1222.
3. Cui , Y., Chen, W., Dai N., et al. Integrated technologies for anti-de-icing functions and structures of aircraft: current status and development Trends[J]. Aerospace,2024,11(10):821–821.
4. Belcastro, C. M., Foster, J. V., Aircraft loss-of-control accident analysis[R]. AIAA-2010-8004, 2010.
5. Liu, J., Huang, Q., Han, Y., et al. Bi-level optimization for de-Icing position allocation and unmanned de-Icing vehicle fleet routing problem[J]. Biomimetics,2024,9(1):
6. He, H., Guo, Z.G., Superhydrophobic materials used for anti-icing theory, application, and development[J]. iScience, 2021, 24(11), 103357.
7. Du, Q., Wang, P., Li, D., Aircraft de-icing based on large vibration of wings.[J]. PloS one,2024,19(9):e0308358.
8. Tong, Y.C., Hang, W., Ling, L., et al. Mechanism of biomimetic surface protection / de-icing and its application in aircraft[J]. Transactions of Nanjing University of Aeronautics and Astronautics,2022,39(05):541–560.
9. Long, Y., Li, H., Meng, X., et al. Structure optimization of the AC-SDBD plasma actuator under duty-cycle mode[J]. Modern Physics Letters B,2018,32(26):
10. Meng, X. S., Hu, H. Y., Li, C., et al. Mechanism study of coupled aerodynamic and thermal effects using plasma actuation for anti-icing[J]. Physics of Fluids, 2019, 31(3): 037103.
11. Phillips, A., Wert, K, Loop heat pipe anti-icing system development program summary[C]// 30th International Conference on Environmental Systems, Toulouse: SAE International, 2000: 2493.
12. Liu, Y., Kolbakir, C., Hu, H.Y., A comparison study on the thermal effects in DBD plasma actuation and electrical heating for aircraft[J]. Journal of Heat and Mass Transfer, 2018, 124: 319–330.
13. Liu Y, Kolbakir C, StarikovskiyA, et al, An experimental study on the thermal characteristics of NS-DBD plasma actuation and application for aircraft icing mitigation[J]. Plasma Sources Science and Technology, 2019, 28(1): 14001.
14. JOHN TOMCZYK. Pulse Electro Thermal De-Icing[J]. Air conditioning heating & refrigeration news,2005,226(10):9–0.

15. Yang X, Liu Y, Zhong Y , et al. Anti/de-icing superhydrophobic coating with durability and self-healing by infiltrating photothermal self-stratifying organic layers into plasma-sprayed porous Al2O3–13%TiO2 underlayer[J]. Surfaces and Interfaces,2024,54105305–105305.
16. ZHI SU, HUA LIANG, HAOHUA ZONG, et al. Geometrical and electrical optimization of NS-SDBD streamwise plasma heat knife for aircraft anti-icing[J]. Chinese Aviation Journal (English version),2023,36(2):87–99.
17. Frederico, R., Mohammadmahdi, A., João N., et al. Recent Developments on Dielectric Barrier Discharge (DBD) Plasma Actuators for Icing Mitigation[J]. Actuators,2022,12(1):5–5.
18. Zhu, Y., Wu, Y., Cui, W., et al. Modelling of plasma aerodynamic actuation driven by nanosecond SDBD discharge[J]. Journal of Physics D: Applied Physics, 2013, 46(35): 355205.

Chapter 27
Numerical Study on Mechanism of Oxytetracycline Degradation Induced by Cold Atmospheric Plasmas

Hong-yan Zhang, Fang Ma, Zhao-nan Chai, and Yuan-tao Zhang

Abstract In recent years, a variety of antibiotics in water, such as oxytetracycline (OTC), tetracycline (TC) and norfloxacin, have received increasing attention because they cannot be effectively degraded due to their toxic and recalcitrant performance. Experimental results reveal that Cold Atmospheric Plasma (CAP) show great potential in the field of environmental remediation. However, the degradation pathways and products should be further understood. In this paper, the reaction processes of OTC with four Reactive Oxygen Species (ROS), such as hydroxyl radical, oxygen atom, hydrogen peroxide and ozone, were numerically studied by means of reactive Molecular Dynamic (rMD) simulations coupled with ReaxFF field. In the simulation, the breaking of important chemical bonds (such as C–N, C–O, C–C bond), destruction of functional groups (dimethyl amine, amide, hydroxyl), ring-opening reaction, and formation of low molecules (such as formic acid, formaldehyde, carbon dioxide, and nitric acid) are observed. The generation of non-toxic products that cannot combine with bacterial ribosomes means elimination of biological toxicity of oxytetracycline degradation.

Keyword Cold atmospheric plasma · Antibiotic degradation · Reactive oxygen species · rMD simulation

Faced with the increasingly serious antibiotic environmental pollution crisis, numerous experts and scholars have conducted extensive research and proposed various solutions [1–8]. It is proved that discharge plasma can effectively degrade

The National Basic Research Program (973 Program) (2006CB2003-03, 2006CB2003056).

H. Zhang
School of Medical Devices, Shenyang Pharmaceutical University, Shenyang, China
e-mail: 253856426@qq.com

F. Ma · Z.-n. Chai · Y. Zhang (✉)
School of Electrical Engineering, Shandong University, Jinan, China
e-mail: ytzhang@sdu.edu.cn

Y. Li et al. (eds.), *Proceedings of the 5th Chinese Conference on High Voltage and Discharge Plasmas*, Springer Proceedings in Physics 254,
https://doi.org/10.1007/978-981-96-5711-7_27

various antibiotics, including carbamazepine, diclofenac, ciprofloxacin, and tetracycline [9–12]. However, current research mostly focuses on degradation results and efficiency, and the specific process and mechanism of the interaction between plasma and antibiotics still need to be further explored [13–15].

Based on the method of reactive molecular dynamics, this article conducted molecular dynamics simulation experiments on representative tetracycline molecules and reactive oxygen species particles in the ReaxFF force field. The specific reaction process and products of key functional groups (such as dimethylamino, amide, hydroxyl, etc.) in tetracycline molecules under the action of reactive particles were given, and the results were in good agreement with relevant experimental results, theoretically indicating the intrinsic mechanism of plasma degradation of antibiotics.

27.1 Model Description

27.1.1 Production of Active Particles

Gas discharge under atmospheric pressure generates a large number of charged or neutral particles in ground or excited states due to frequent collisions, some of which have strong oxidizing properties. The types of active particles in atmospheric pressure plasma include reactive oxygen species and reactive nitrogen species. These diverse and high concentration active particles are the main carriers for the interaction between plasma and tetracycline molecules.

27.1.2 Molecular Structure of Tetracycline Antibiotics

As shown in Fig. 27.1, there are many active functional groups in the molecular structure of tetracycline. The antibacterial activity of tetracycline antibiotics is mainly reflected in these active functional groups, and the destruction of these functional groups is the most effective way to degrade tetracycline antibiotics.

Figure 27.1a shows the molecular structure of oxytetracycline ($C_{22}H_{24}N_2O_9$). The black numbers represent carbon atom numbers, and the red numbers represent oxygen atom numbers. From the figure, it can be seen that its skeleton consists of four membered fused rings numbered A, B, C, and D from right to left, where the D ring is a stable benzene ring and A, B, and C are three hexagonal carbon rings.

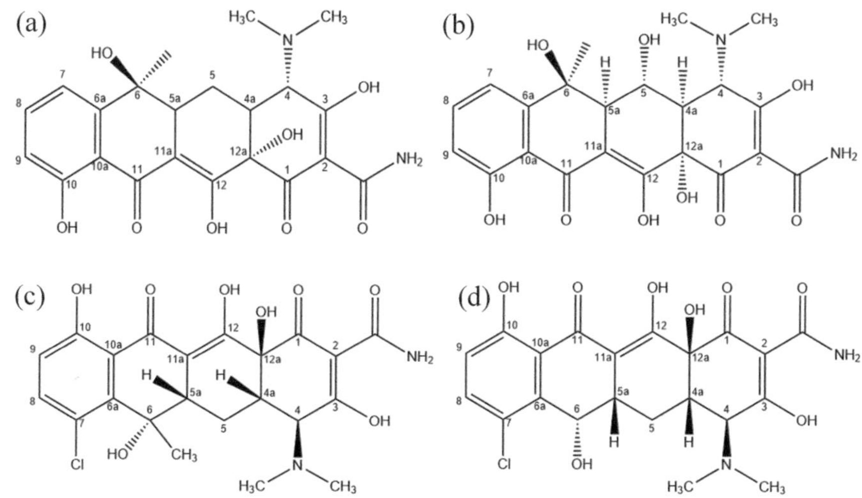

Fig. 27.1 Schematic representation of tetracyclines molecule structure

27.1.3 Reaction Molecular Dynamics Simulation

This article explores the reaction process and products of hydroxyl radicals, oxygen atoms, ozone, hydrogen peroxide, and other four reactive oxygen species particles with organic matter at a time scale of 300 hundred picoseconds to ensure that all reactions can be completed, using the method of reactive molecular dynamics, coupled with ReaxFF reaction force field [13].

Figure 27.2 shows a simulation box of 30 hydroxyl radicals and two oxytetracycline molecules, where the active particles and oxytetracycline molecules are randomly placed inside the box and subjected to periodic boundary conditions. This article aims to change the concentration of active particles by varying the number of intracellular active particles (10, 30, 50, 100) of the same volume. Molecular dynamics simulations were conducted on the reaction process between single active particles of different concentrations and oxytetracycline. The detailed parameters used in the simulation are given in Table 27.1.

27.2 Degradation Process and Products of Tetracycline by Discharge Plasma

This section selectively focuses on analyzing the reaction mechanisms and ring opening effects between four types of active particles and the amide group on the C_2 site, dimethylamino group on the C_4 site, hydrogen atom on the C_{5a} site, and

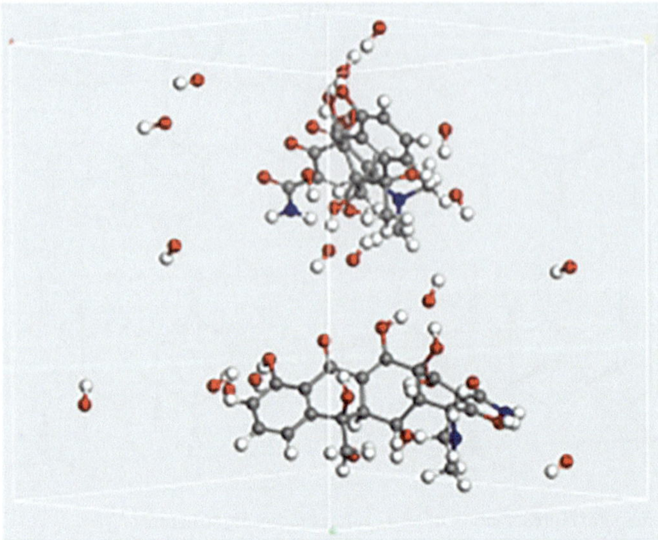

Fig. 27.2 A reactive MD simulation box with 2 OTC molecules and 30 OH radicals

Table 27.1 Reactive MD simulation parameter settings

Force field	ReaxFF 6.0
Ensemble	NVT
Temperature	300K
Relaxation time	50ps
Simulation time	300ps
Step	0.1fs

hydroxyl group on the C_{12a} site of oxytetracycline molecules, in order to reveal the substantive process of discharge plasma degradation of oxytetracycline.

27.2.1 Elimination of Amide Group at C_2 Site

Figure 27.3 illustrates the mechanism by which ozone molecules react with amide groups to form carboxyl or carbonyl groups. When ozone molecules first capture the hydrogen atom on the amino group (a), they generate hydroxyl radicals and oxygen atoms. Then nitrogen atoms combine with hydroxyl radicals, but this product is not stable. The hydrogen atoms on nitrogen and in hydroxyl groups can still be taken away by ozone molecules, generating N=O and detaching from amide groups. This leads to unsaturated chemical bonds of C_{2a} atoms. Then, the free hydroxyl radicals combine with C_{2a} atoms to form carboxylic acids, resulting in the product (b). On the

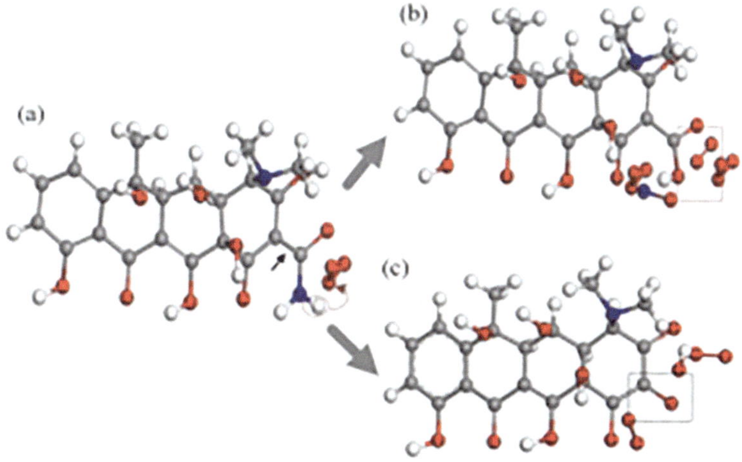

Fig. 27.3 The reaction processes of O_3 with acylamino

other hand, ozone molecules can also attack C_2–C_{2a}, directly leading to the separation of amide groups from the tetracyclic structure, resulting in the product (c) shown in Fig. 27.3.

27.2.2 Loss of Dimethylamino Group at C_4 Site

As shown in Fig. 27.4a, hydroxyl radicals first take the hydrogen atom on the methyl group in dimethylamino, forming a molecule of water. Then sp3 hybridization is converted into sp2 hybridization (Fig. 27.4b). When there are free hydroxyl radicals around, methylene radicals are prone to undergo addition reactions with hydroxyl radicals (Fig. 27.4c), forming hydroxymethyl groups. Hydroxymethyl is highly susceptible to oxidation in spaces filled with active particles. The hydrogen atoms on it and on the hydroxyl group (Fig. 27.4d) can be taken away by hydroxyl radicals, forming two molecules of water. This hydroxymethyl group is then converted into an aldehyde group (Fig. 27.4e). In order to maintain the stability of the nitrogen atom, another methyl group on it falls off and becomes a free methyl group (Fig. 27.4f). This free methyl group can be further oxidized by hydroxyl radicals to carbon dioxide. Under the attack of oxygen atoms or ozone molecules, the C–N bond between the nitrogen atom and the C4 atom is broken, causing the dimethylamino group to completely separate from the host structure (Fig. 27.4g). The active particles react with dimethylamino through hydrogen abstraction reaction, and through a chain reaction, ultimately lead to the separation of dimethylamino from the main structure, achieving the effect of reducing the activity of oxytetracycline molecules.

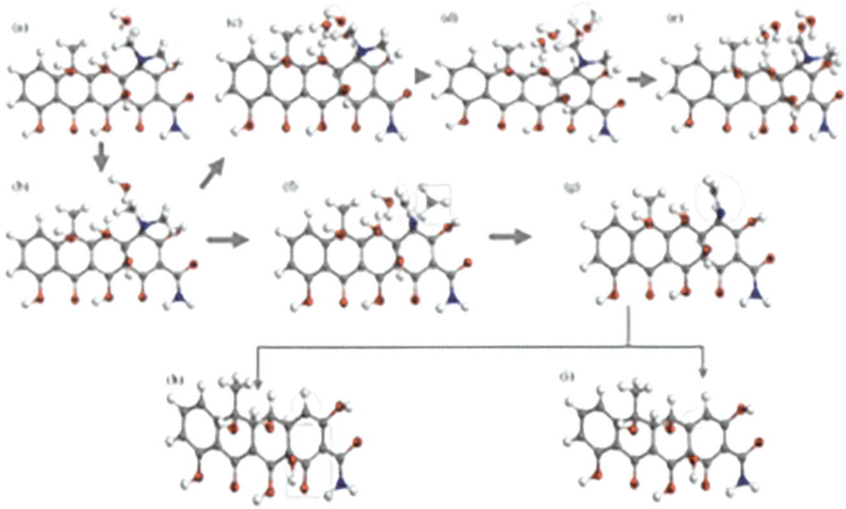

Fig. 27.4 The reaction processes of OH with dimethylamino

27.2.3 *Hydrogen Atom Drops to C=C Bond at C_{5a} Site*

After the hydroxyl group at the C_{12} site is deprived of hydrogen (a), the oxygen atom on this hydroxyl group has unpaired valence electrons, making it easy to form a $C_{12}=O_8$, which triggers the breaking of the $C_{12}=C_{11a}$ double bond into a single bond. To maintain stability, C_{11a} atoms tend to form double bonds with surrounding atoms. When the concentration of active particles is high, it leads to a higher probability of hydrogen atoms at the C_{5a} site being taken away, resulting in the formation of more stable double bonds between C_{5a} atoms and C_{11a} atoms (see Fig. 27.5).

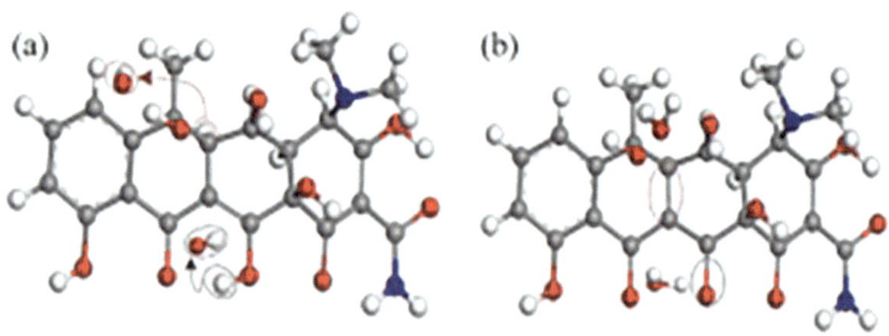

Fig. 27.5 The reaction processes of C=C bond formation upon the impact of OH

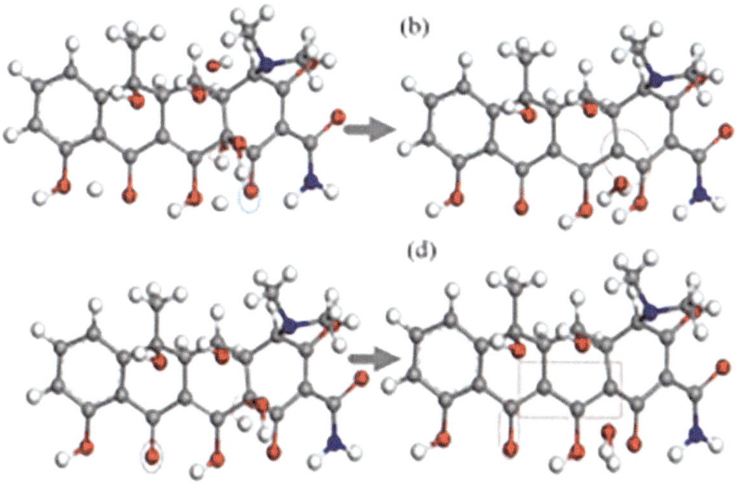

Fig. 27.6 The reaction of dehydroxylation in H_2O_2 impacts

27.2.4 *Dehydration Reaction Occurs at C_{12a} Site*

Simulations have found that dehydroxylation reactions can occur at low simulated concentrations of hydrogen peroxide. Because low concentrations of hydrogen peroxide are easily decomposed into H and HOO, and the hydroxyl group at the C_{12a} site has strong activity and is easy to remove. Therefore, the hydrogen atom binds to the hydroxyl group, causing the hydroxyl group at the C_{12a} site to detach from the tetracyclic skeleton and form a molecule of water. When the carbonyl group on the C_1 site (blue circle in (a)) is reduced to a hydroxyl group by hydrogen peroxide decomposition, the carbon oxygen double bond on the C_1 atom is released, and then forms a carbon carbon double bond with the C_{12a} atom to obtain the product (b) in Fig. 27.6; When the carbonyl group on the C_{11} atom (black circle in (c)) is reduced to a hydroxyl group by a hydrogen atom, $C_{11}–C_{11a}$ is converted to $C_{11}=C_{11a}$, $C_{11a}=C_{12}$ is converted to $C_{11a}–C_{12}$, and $C_{12}–C_{12a}$ is converted to $C_{12}=C_{12a}$ through intramolecular electron rearrangement, resulting in the product shown in Fig. 27.6d.

27.2.5 *Open Loop Effect*

As shown in Fig. 27.7, the hexagonal carbon ring has three ring opening mechanisms under the action of ozone molecules.

When ozone molecules approach the benzene ring (dashed line in Fig. 27.7a), they can break the carbon bond between C_8 and C_9, forming $C_8–O$ and $C_9–O$, forming a unstable carbon oxygen seven membered heterocyclic ring. Both C_8 and C_9 atoms

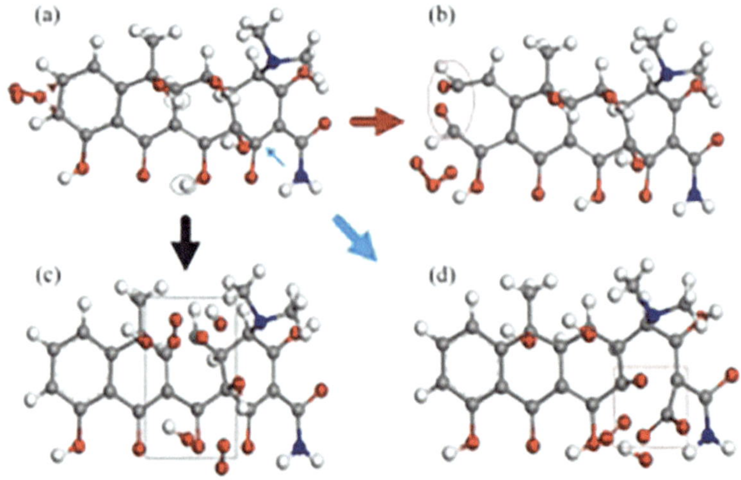

Fig. 27.7 The reaction processes of ring-opening in the case of O_3

can undergo addition reactions with free ozone molecules. From then on, the benzene ring opens and forms two stable carbonyl structures (Fig. 27.7b).

As shown by the black dashed line in Fig. 27.7a, when the hydrogen atoms in the hydroxyl groups at the C_5 and C_{12} sites are taken away by ozone molecules in a short period of time, C_5=O_4 and C_{12}=O_8 are formed at the two sites, respectively, converting the original C_{11a}=C_{12} to C_{11a}–C_{12} and C_{5a}–C_{11a} to C_{5a}=C_{11a}. At this point, the chemical bonds on C_{5a} and C_5 atoms are both in a supersaturated state. To maintain their stability, the C–C bond between the two is broken, resulting in the opening of the hexagonal ring, as shown in the black box in Fig. 27.7c.

As shown by the blue arrow in Fig. 27.7a, ozone undergoes an addition reaction with C_1 atom to generate oxygen. The chemical bond of C_1 atom is supersaturated, resulting in the disconnection of C_1–C_{12a}, and the C_1 site changes from carbonyl to carboxyl. At this time, the C_{12a} atom is in an unsaturated state. Due to the hydrogen atom in the hydroxyl group at the C_{12} site being easily taken away by free ozone molecules, hydroxyl radicals, etc., there are unpaired electrons on O_9. Therefore, O_9 is prone to form a carbon oxygen double bond with the C_{12a} atom, and the hydroxyl group at the C_{12} site changes to carbonyl, resulting in the product (d) in Fig. 27.7.

27.3 Conclusion

This article uses reactive molecular dynamics to study the reaction processes and products of the interaction between four types of reactive oxygen species particles and tetracycline in plasma. Simulation results show that reactive oxygen species particles can cause:

(1) Destruction of key functional groups of tetracycline. Such as the amide group on the C_2 site, dimethylamino group on the C_4 site, hydrogen atom on the C_{5a} site, hydroxyl group on the C_{12a} site, etc. The destruction of these functional groups indicates that tetracycline has truly undergone degradation.
(2) The opening of the quaternary fatty ring. The integrity of the linear arrangement of the tetraphenylene nucleus structure is a prerequisite for the antibacterial activity of oxytetracycline molecules. Therefore, the open-loop reaction can have a fatal impact on the antibacterial activity of tetracycline molecules.
(3) The breaking of important chemical bonds in tetracycline, such as N–H bonds, O–H bonds, C–N bonds, C–C bonds, etc. The breaking of chemical bonds can disrupt the balance between zwitterionic and non ionized states, leading to the conversion of oxytetracycline molecules into inactive homologs, thereby reducing or even eliminating the antibacterial activity of oxytetracycline.

Thanks

This work was supported by the Natural Science Foundation of China (No. 12375201).

References

1. Wang J, Wang S. Preparation, modification and environmental application of biochar: a review[J]. Journal of Cleaner Production, 2019, 227: 1002–1022.
2. 韩爽, 肖鹏飞. 过硫酸盐活化技术在四环素类抗生素降解中的应用进展[J]. 环境化学, 2021, 40(9): 2873–2883.
3. HAN Shuang, XIAO Pengfei. Application progress of persulfate activation technology in the degradation of tetracycline antibiotics[J]. Environmental chemistry, 2021, 40(9): 2873–2883.
4. Prado N, Ochoa J, Amrane A. Biodegradation and biosorption of tetracycline and tylosin antibiotics in activated sludge system[J]. Process Biochemistry, 2009, 44(11): 1302–1306.
5. Zhuang S, Liu Y, Wang J. Covalent organic frameworks as efficient adsorbent for sulfamerazine removal from aqueous solution[J]. Journal of Hazardous Materials, 2020, 383: 121126.
6. Zhuang S, Cheng R, Wang J. Adsorption of diclofenac from aqueous solution using UiO-66-type metal-organic frameworks[J]. Chemical Engineering Journal, 2019, 359: 354–362.
7. 陈勇, 胡鹭, 谭旎等. Al0-Gr-Fe0活化O2类Fenton氧化降解土霉素的研究[J]. 中国环境科学, 2021, 41(10): 4645–4653.
8. CHEN Yong, HU Lu, TAN Yi, et al. Study on Al0-Gr-Fe0 activated O2 based Fenton oxidation degradation of oxytetracycline[J]. China environmental science, 2021, 41(10): 4645–4653.
9. Zhuang S, Liu Y, Wang J. Mechanistic insight into the adsorption of diclofenac by MIL-100: Experiments and theoretical calculations[J]. Environmental Pollution, 2019, 253: 616–624.
10. Wang J, Zhuang S. Removal of various pollutants from water and wastewater by modified chitosan adsorbents[J]. Critical Reviews in Environmental Science and Technology, 2017, 47(23): 2331–2386.
11. Krause H, Schweiger B, Prinz E, et al. Degradation of persistent pharmaceuticals in aqueous solutions by a positive dielectric barrier discharge treatment[J]. Journal of Electrostatics, 2011, 69(4): 333–338.
12. Kim K S, Yang C S, Mok Y S. Degradation of veterinary antibiotics by dielectric barrier discharge plasma[J]. Chemical Engineering Journal, 2013, 219: 19–27.

13. Tang S, Yuan D, Rao Y, et al. Persulfate activation in gas phase surface discharge plasma for synergetic removal of antibiotic in water[J]. Chemical Engineering Journal, 2018, 337: 446–454.
14. 李阳, 陈永铎, 赵红杰等. 窄前沿高压脉冲放电等离子体降解水中苯胺[J]. 环境工程学报, 2014, 8(12): 5361–5366.
15. LI Yang, CHEN Yongduo, ZHAO Hongjie, et al. Degradation of Aniline in Water by Narrow Front High Voltage Pulse Discharge Plasma[J]. Journal of Environmental Engineering, 2014, 8(12): 5361–5366.
16. Bogaerts A, Yusupov M, Van der Paal J, et al. Reactive molecular dynamics simulations for a better insight in plasma medicine[J]. Plasma Processes and Polymers, 2014, 11(12): 1156–1168.
17. Van Duin A C T, Dasgupta S, Lorant F, et al. ReaxFF: a reactive force field for hydrocarbons[J]. The Journal of Physical Chemistry A, 2001, 105(41): 9396–9409.
18. Tian S Q, Wang X L, Zhang Y T. Numerical study on interactions of atmospheric plasmas and vegetable oils by reactive molecular dynamic simulations[J]. Plasma Processes and Polymers, 2021: e2100124.

Chapter 28
Development of Nanosecond Pulse Source for VFTO Research and Assessment Testing

Fangfang Wu, Hao Wu, Yibo Gao, Xuyang Zhao, Xue Lin, and Shengyi Xie

Abstract The widespread use of high-voltage equipment such as Gas Insulated Switchgear (GIS) in power systems has led to the emergence of VFTO (Very Fast Transient Overvoltages, VFTO) during switching operations. VFTO poses a significant risk to the insulation integrity of power equipment. Traditional lightning impulse voltage generators, with their relatively long front times, can no longer meet the insulation testing requirements for VFTO pulses in modern power equipment, resulting in a general lack of experimental research capabilities. This paper designs and develops a fast pulse source system centered on a fast Marx generator, utilizing a compact Marx pulse source with a rise time in the tens of nanoseconds and a peak voltage reaching 1 MV. The system can simulate VFTO pulses to assess the insulation performance of power equipment such as high-voltage cables. It effectively prevents potential hazards caused by VFTO, including flashover failures in disc insulators, false fault reports, or malfunctions of relay protection devices, thereby ensuring the operational safety of transmission and distribution networks.

Keywords Power equipment · VFTO · Nanosecond pulse source · Testing · Assessment

28.1 Introduction

Currently, in order to address various types of overvoltages such as lightning, electrical equipment in transmission and distribution networks widely undergoes lightning impulse tests under microsecond front pulse voltages [1]. After decades of research, the associated risks are now basically controllable, effectively ensuring the safety and stability of the power system.

F. Wu (✉) · H. Wu · Y. Gao · X. Zhao · X. Lin · S. Xie
Zhejiang Huadian Equipment Testing and Research Institute Co., Ltd, Hangzhou, Zhejiang, China
e-mail: 13819198545@163.com

Y. Li et al. (eds.), *Proceedings of the 5th Chinese Conference on High Voltage and Discharge Plasmas*, Springer Proceedings in Physics 254,
https://doi.org/10.1007/978-981-96-5711-7_28

In recent years, with the widespread use of Gas Insulated Switchgear (GIS) in substations, and the gradual increase in voltage levels in transmission networks, Very Fast Transient Overvoltage (VFTO) pulses with rise times in the tens of nanoseconds are frequently generated during disconnector operations, posing a threat to the safety of grid [2]. The VFTO front time is generally on the order of tens of nanoseconds, with pulse amplitudes reaching several hundred kilovolts to several megavolts, and pulse steepness of approximately tens of kV/ns [3]. The primary threat posed by VFTO is its impact on the insulation of primary equipment (e.g., causing flashover faults in disc insulators). Additionally, due to the compact layout of intelligent substation equipment, where primary and secondary equipment are in close proximity, secondary equipment is also prone to electromagnetic interference, leading to malfunctions and false alarms (e.g., false fault reports from relay protection devices) or, in more severe cases, complete failure (e.g., damage to density relays)[4–6]. However, the lack of a testing platform capable of simulating fast transient overvoltage pulse waveforms severely restricts research into failure mechanisms and testing and evaluation trials for typical power equipment, posing a significant risk to grid.

Currently, equipment testing under pulse voltage conditions in power systems typically employs lightning impulse voltage generators. Existing lightning impulse generators have front times ranging from approximately 1.2 μs to 250 μs, while VFTO pulses in power systems generally have front times in the tens of nanoseconds, making it difficult for these generators to meet the testing requirements for VFTO [11]. Moreover, existing lightning impulse voltage generators have several drawbacks, such as an inefficient layout, large size, and exposed high voltage components, making it difficult to achieve nanosecond pulse discharge through technological modifications.

Given these demands, it is necessary to conduct preliminary research on key technologies for nanosecond pulse devices aimed at testing power equipment and to develop prototypes. This will provide a platform to support the testing and evaluation of power equipment under VFTO and other nanosecond pulse conditions.

28.2 Basic Principles of Nanosecond Pulse Source

The nanosecond pulse source adopts the technical approach of generating high-power pulses using a fast Marx generator. The principle, as illustrated in Fig. 28.1, is based on the operation of a Marx generator: multiple stages of capacitors are charged in parallel, and after charging is completed, the trigger of the primary Marx switch causes it to break down and discharge. The resulting overvoltage from the breakdown of the primary switch induces the subsequent switches to break down sequentially. This process results in series discharges, producing a high-voltage output at the end of the Marx generator. Compared to lightning impulse voltage generators, the fast Marx approach employs low-inductance design in its discharge units, which makes the internal structure more compact, reducing the overall size of the device. The high-voltage components are enclosed within a sealed metallic chamber, with no external exposure, thereby significantly enhancing safety [7–10]. Additionally, the number

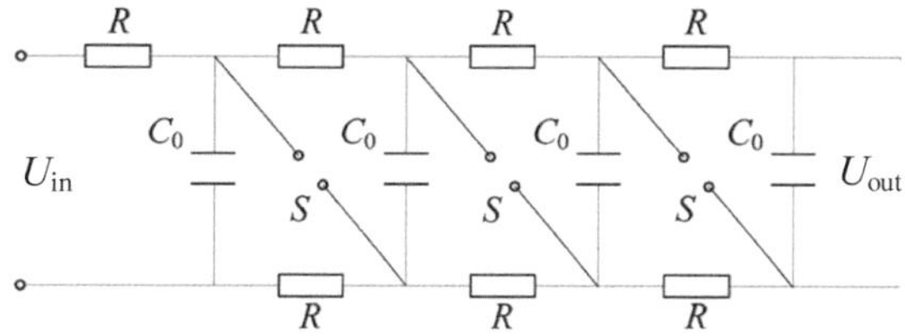

Fig. 28.1 Schematic diagram of the fast Marx generator

of gas switches is reduced and only the first few switches need to be triggered, while the remaining switches operate in a self-breakdown mode. This reduces technical complexity and improves the reliability of the device.

The MV-level fast Marx generator circuit was simulated. The simulation model is shown in Fig. 28.2, and the voltage waveform at the generator output obtained from the simulation is shown in Fig. 28.3. The simulation results indicate that by using a seven-stage fast discharge branch in series and connecting the fast Marx generator to different load circuits, it is possible to adjust the output pulse parameters. In the direct load connection mode, the fast Marx generator can produce pulses with a peak voltage exceeding 800 kV, a rise time of approximately 100 ns, and a pulse width of around 400 ns. In the output mode with peak-shaping capacitors and sharpening switches, the fast Marx generator can generate pulses with a peak voltage exceeding 1 MV, a rise time on the order of tens of nanoseconds, and a pulse width of approximately 100 ns.

This section explains the fundamental principles of the nanosecond pulse source and verifies the effectiveness of its operating principles through simulation.

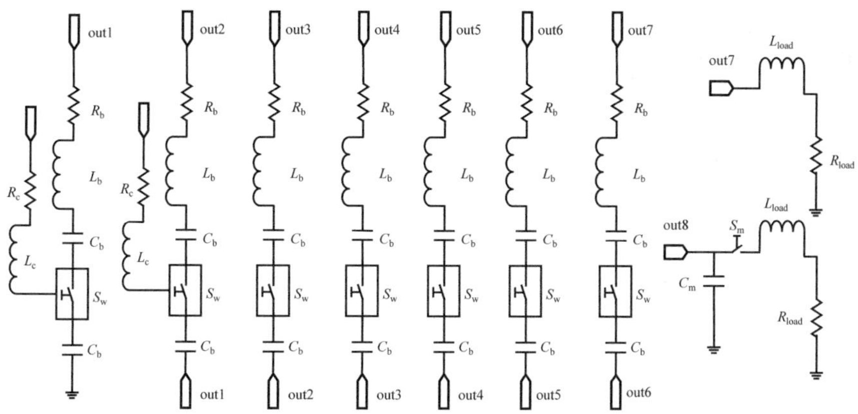

Fig. 28.2 Nanosecond pulse source circuit simulation model

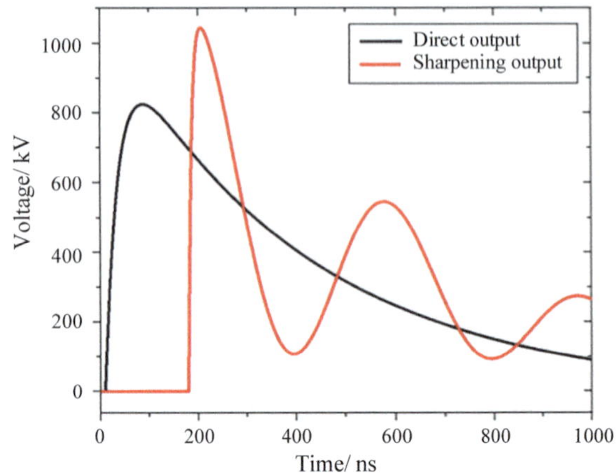

Fig. 28.3 Nanosecond pulse source output simulation results

28.3 Nanosecond Pulse Source Structure Design

To meet the requirements for testing and evaluation of VFTO fast pulses with rise times in the tens of nanoseconds, a nanosecond pulse source based on a fast Marx generator was designed and developed, building on circuit simulations. This pulse source is capable of producing output pulse voltages ranging from several hundred kilovolts to megavolts, with rise times in the tens of nanoseconds. It can effectively simulate VFTO pulses and assess the insulation damage to power equipment.

28.3.1 Pulse Source Discharge Branch Design

The discharge circuit is the fundamental unit of a megavolt-level nanosecond pulse device, and its electrical parameters and discharge characteristics determine the output waveform specifications of the device. To shorten the rise time, the fast Marx generator's discharge circuit adopts low-inductance discharge unit designs.

The discharge unit consists of one multi-gap gas switch and four low-inductance capacitors, as shown in Fig. 28.4. The four capacitors are divided into two groups: each pair is connected in parallel, and the two groups are symmetrically distributed on either side of the switch in series configuration. Two capacitors act as positive polarity and two as negative polarity capacitors. The capacitors are 80nF/100 kV low-inductance capacitors, with a maximum charging voltage of up to 100 kV. The use of multiple parallel capacitors effectively reduces the circuit's inductance.

The gas switch is a six-gap gas switch with a maximum operating voltage of DC $\pm$ 100 kV. The gas switch has a compact structure with a height of 126 mm, a diameter of 100 mm, and a gap distance of 5 mm for each gap, totaling 30 mm. This design achieves a compact and lightweight discharge circuit, with the circuit inductance

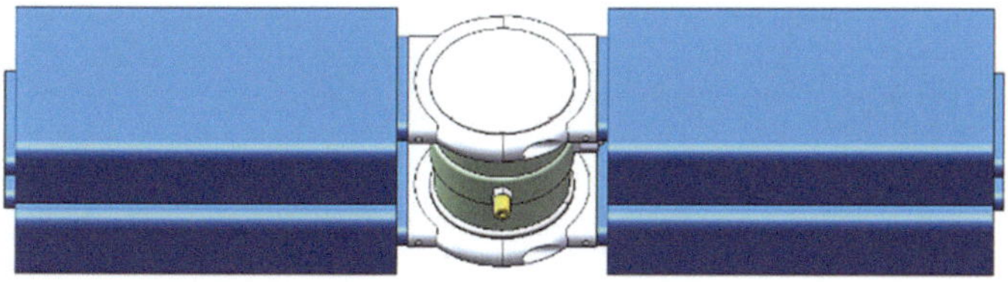

Fig. 28.4 Schematic diagram of discharge branch structure

reduced to only 150nH. This enables the pulse source to produce fast-rising electrical pulses, meeting the requirements for VFTO testing and evaluation.

During the DC withstand process and when the trigger pulse is applied, the axial electric field distribution of the switch is shown in Fig. 28.5. Under DC voltage, the electric field strength in the gaps near the positive and negative high-voltage electrodes is slightly lower, while the electric field strength across the other gaps is relatively uniform, with the maximum field strength reaching about 70 kV/cm. The uniform and relatively low electric field strength across the gas switch gaps ensures that self-discharge is avoided. When a trigger pulse is applied, the electric field distribution across the gaps changes significantly. The electric field strength in the trigger gap increases sharply, reaching up to 150 kV/cm, with a field distortion coefficient greater than 2. Following the application of the trigger pulse, the electric field strength in adjacent gaps 1 and 2 also increases noticeably, while the field strength in gaps 4, 5, and 6 decreases sharply. Once gap 3 reaches the breakdown thresh-old, it discharges first, followed by sequential breakdowns in gaps 2 and 1. The switch voltage is then applied to gaps 4, 5, and 6, where the electric field strength rapidly increases until breakdown occurs, leading to the full conduction of the switch.

The performance of the discharge circuit is stable, with a low probability of self-discharge and a wide operating range. The test results for the discharge circuit are shown in Fig. 28.6. As the gas pressure of the switch increases, the self-breakdown voltage rises approximately linearly. Under a charging voltage of $\pm$ 60 kV and a working coefficient of 50–80%, the discharge circuit operates reliably.

The effect of inconsistencies in the outer diameter of the insulation layer of the cable under measurement is similar to that caused by manufacturing and installation errors of the sampling electrode. Such inconsistencies alter the distance between the sampling electrode and the cable core, affecting the high-voltage arm capacitance. Similarly, an oversized insulation diameter decreases the high-voltage arm capacitance, resulting in an underestimated voltage measurement, while an undersized insulation diameter increases the capacitance, leading to an overestimated measurement. In engineering practice, efforts should be made to ensure that the measurement position of the cable aligns closely with the calibration position.

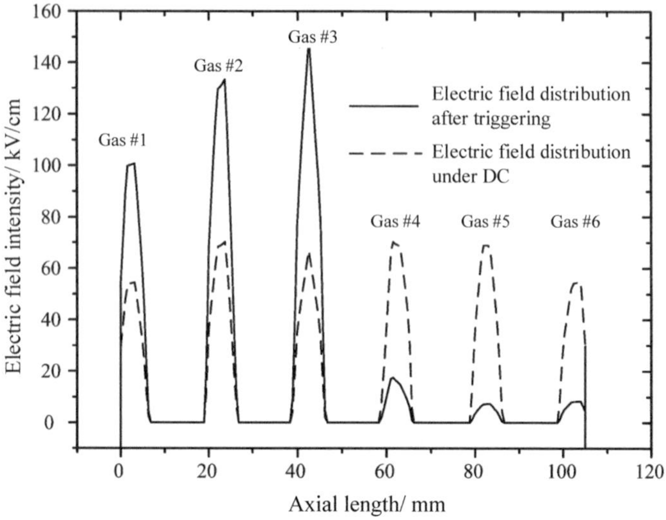

Fig. 28.5 Schematic diagram of switch gap electric field distribution

28.3.2 Pulse Source Overall Design

The overall structure of the nanosecond pulse source is illustrated in Fig. 28.7. It primarily consists of a fast Marx generator, charging power supply, trigger, control system, gas path system, load, and switching devices. The fast Marx generator can be controlled by a cylinder to switch between two output states: one from the upper end and the other from the lower end equipped with peak-shaping capacitors. The peak-shaping capacitors enable the generation of fast pulses with greater voltage amplitude and shorter rise times. The overall dimensions of the pulse source are approximately 2.5 × 1.2 × 1.3 m, making it significantly smaller than conventional lightning impulse voltage generators. Its compact structure allows for easy mobility.

The core of the fast Marx generator consists of seven discharge units connected in series, arranged in a linear configuration as shown in Fig. 28.8. Each discharge unit is equipped with a shielding ring to achieve uniform electric field distribution within the core. The shielding structure enhances insulation safety against ground faults. Additionally, the charging and triggering resistors employ bulk resistors, which mitigate the drawbacks of conventional water resistors, such as susceptibility to damage and instability in resistance values. This design significantly improves the reliability of the device.

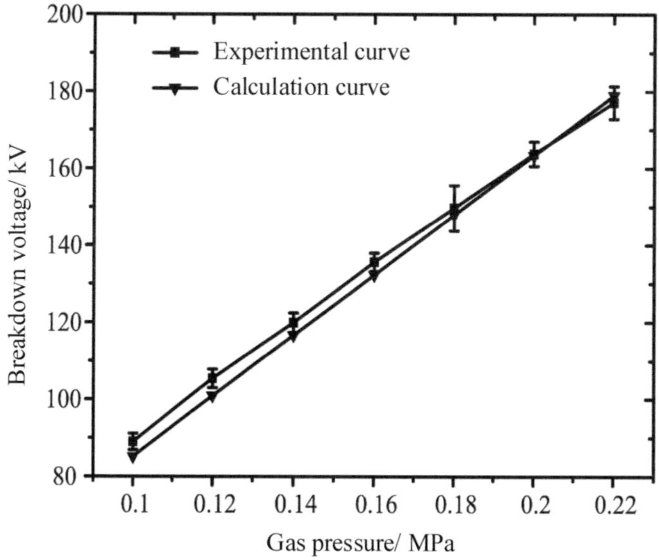

(a) Self-breakdown voltage curve

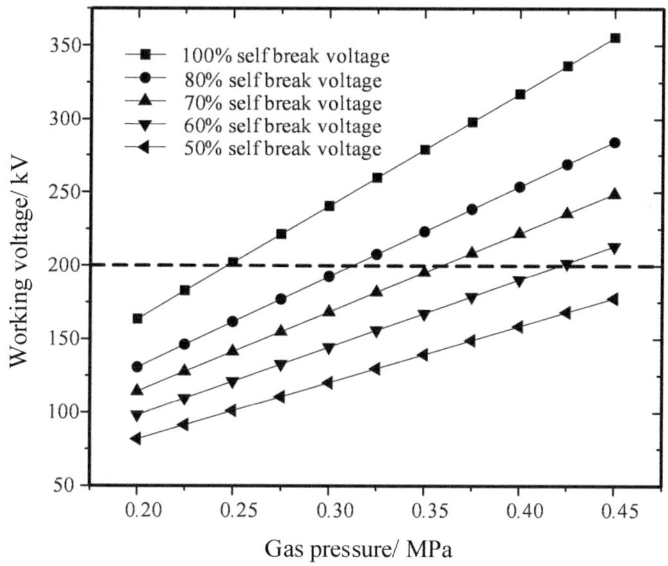

(b) Working coefficient curve

Fig. 28.6 Discharge branch test results operating principle and design of the spiral transformer

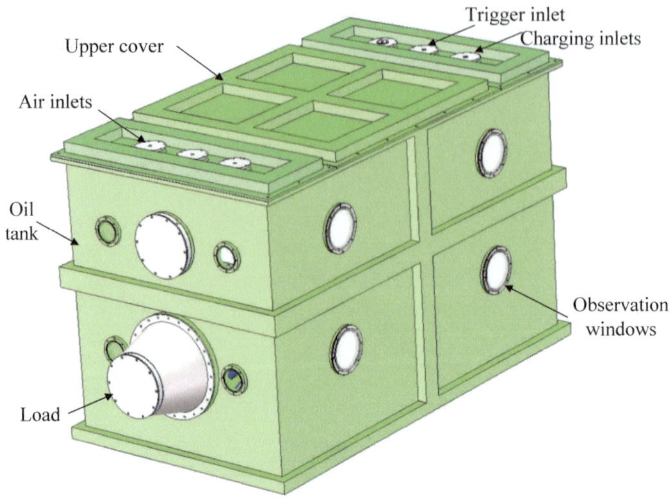

Fig. 28.7 Overall structure of pulse source

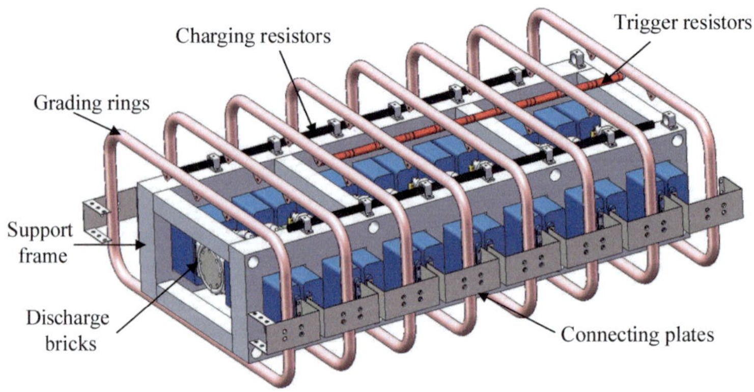

Fig. 28.8 Schematic diagram of the Marx generator movement structure

28.4 Nanosecond Pulse Source Output Wave-Form Test and Application

Through experimental testing of the nanosecond pulse source system, the actual output wave-form of the pulse source is obtained as shown in Fig. 28.9. Under a charging voltage of $\pm$ 60 kV and a load of 50 Ω, the output voltage in direct output mode is approximately 800 kV, with a rise time of about 80 ns. In the sharpening output mode, the output voltage reaches approximately 1.05 MV, with a rise time of about 20 ns.

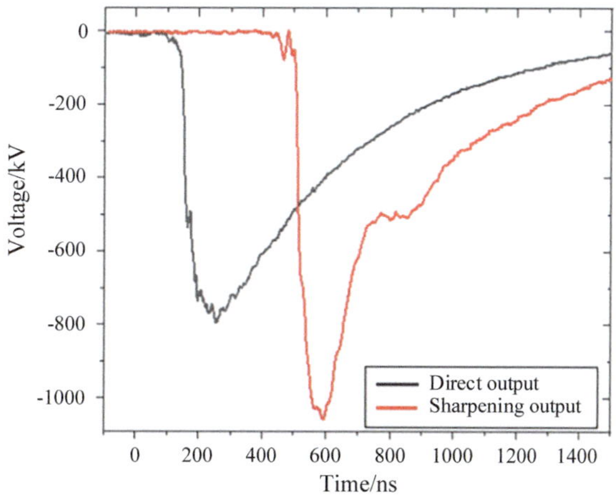

Fig. 28.9 Pulse source output voltage waveform

The experimental results indicate that the output pulse parameters obtained from both output modes are consistent with the simulation results. This confirms that the system can effectively simulate VFTO pulses for insulation testing of power equipment.

28.5 Conclusion

The nanosecond pulse device has broad application prospects in the testing of power equipment. Traditional lightning impulse voltage generators, due to their long rise times, can no longer meet the insulation testing requirements for nanosecond-level VFTO pulses in current power equipment. This paper presents the development of a fast pulse source system based on a fast Marx generator. The fast Marx generator features an innovative design with low-inductance, compact discharge branches and rapid switching loads. By utilizing a cylinder switch, it can switch between two output modes: direct output and peak-shaping capacitor output. This design offers advantages such as short rising time, high pulse intensity, compact structure, and enhanced operational safety, allowing for flexible adaptation to various testing requirements under different environmental conditions.

Based on the structural design, circuit simulation and experimental testing were conducted. The results indicate that the fast pulse source system designed in this paper can generate nanosecond-level pulse voltages ranging from several hundred kilovolts to megavolts. It can be used to simulate the insulation damage caused by VFTO pulses to power equipment such as GIS. Moreover, the system effectively mitigates the risks of VFTO-related failures, such as flashover of pin-type insulators and erroneous operations or failures of relay protection devices, thereby ensuring the safety of the power grid.

Funded Projects The Science and Technology Project of Zhejiang Huadian Equipment Testing and Research Institute Co., Ltd (Project Number: 5211HD230021).

References

1. Li X, Zhao H, Murphy A B.: SF6-alternative gases for application in gas-insulated switchgear. Journal of Physics D: Applied Physics 51(15), 153001 (2018).
2. YANG Xilong, MU Shuanglu, WANG Hao, et al.: Experimental study on VFTO suppression performance of magnetic ring damping busbar for UHV GIS. High Voltage Apparatus 60(08), 77-84 (2024).
3. HUANG kunpeng, KANG Liutao, ZHAO Ping, et al.: Research and development of lightweight 126 kV GIS equipment. High Voltage Apparatus 60(06), 89–98+106 (2024).
4. YANG Fan, GAO Youhua.: Simulation Analysis of Fast Transient Overvoltage in GIS Based on Mayr Arc Model. Electrical & Energy Management Technology (02), 28–32+38 (2024).
5. Li Y, Shang Y, Zhang L, et al.: Analysis of very fast transient overvoltages (VFTO) from onsite measurements on 800 kV GIS. IEEE Transactions on Dielectrics and Electrical Insulation 19(6), 2102-2110 (2012).
6. Ma G M, Li C R, Chen W J, et al.: A reliable wide-bandwidth VFTO sensor based on surface-mounted devices. IEEE transactions on power delivery 28(3), 1839-1846 (2013).
7. Pinguet S, Duperoux J P, Delmote P, et al.: Short-pulse Marx generator for high-power microwave applications. IEEE Transactions on Plasma Science 41(10), 2754-2757 (2013).
8. YAO Haowei, LI Zi, WANG Yonggang, et al.: Study on a Compact Solid-State Marx Generator. High Power Laser and Particle Beams 36(02), 39-45 (2024).
9. TIAN Yulu.: Research on Steep Front Optimization of Marx Pulse Generator Based on SiC MOSFET. Xi'an: Xidian University (2023).
10. Rao J, Liu K, Qiu J.: All solid-state nanosecond pulsed generators based on Marx and magnetic switches. IEEE Transactions on Dielectrics and Electrical Insulation 20(4), 1123-1128 (2013).
11. LI Saisai, MIAO Benjian, ZHOU Xiaomeng, et al.: Analysis of factors affecting lightning impulse voltage waveform generator and wavefront time measurement. Automation Application (11), 148-150 (2022).

Chapter 29
Development and Application of Compact Pulse X-ray Source Based on Spiral Transformer

Jiaming Liang, Zhiyao Zheng, Jialing Li, and Hanpeng Yang

Abstract Detection technology leveraging X-ray penetration characteristics has been widely applied as a key method in the field of non-destructive testing (NDT). Realizing more compact, portable, and lightweight X-ray sources has significant scientific and practical value. Based on an analysis of X-ray imaging principles and parameter requirements, this study developed a compact pulse X-ray source using a spiral transformer-based pulse high-voltage source and a vacuum-sealed rod-clamped diode. The source operates at approximately 100 kV, with a focal spot size of ~ 1 mm, capable of penetrating several millimeters of stainless steel. Experimental imaging of typical power system components demonstrated clear imaging capability for components such as fuses, lightning arresters, and composite insulators, indicating promising application potential.

Keywords Spiral line transformer · Pulse X-ray source · Radiation imaging · Power components

29.1 Introduction

Modern power systems are evolving toward larger grids, larger generating units, and ultra-high-voltage and high-capacity transmission. This evolution places increasingly stringent requirements on the operational reliability of power equipment, which necessitates advanced detection and diagnostic techniques. Particularly, there is great significance in developing online, non-destructive testing (NDT) technologies for live power equipment.

Funded Projects: the Science and Technology Project of Zhejiang Huadian Equipment Testing and Research Institute Co., Ltd. (Project Number: 5211HD230022)

J. Liang · Z. Zheng (✉) · J. Li · H. Yang
Zhejiang Huadian Equipment Testing and Research Institute Co., Ltd, Hangzhou, Zhejiang, China
e-mail: zzycau@qq.com

Y. Li et al. (eds.), *Proceedings of the 5th Chinese Conference on High Voltage and Discharge Plasmas*, Springer Proceedings in Physics 254,
https://doi.org/10.1007/978-981-96-5711-7_29

In recent years, X-ray technology has been introduced as a novel NDT method for fault detection in power equipment [1–3]. X-rays' strong penetration capability enables inspection without requiring disassembly of external structures and allows for direct visualization of internal defects such as loose components, foreign objects, and structural damage caused by prolonged operation. However, X-ray sources currently used for power equipment NDT primarily rely on direct current, resulting in large, expensive devices that pose significant challenges for field and high-altitude inspection, particularly in confined spaces.

Compared to Marx generators [4], spiral transformers offer the advantages of smaller size, compact structure, and the integration of energy storage and voltage multiplication within a single unit. First introduced by Fitch and Howell in 1964, these transformers are capable of producing output voltages of approximately 50 kV or higher through the use of spiral-wound strip conductors [5]. In 1980, Ruhl and GHerziger conducted a comprehensive theoretical analysis of the voltage multiplication mechanism in spiral transformers [6], employing coupled-wave analysis to derive analytical expressions for the output voltage. Their work accounted for factors such as switch impedance, load at the output terminals, and the finite conductivity of the spiral coil conductors. In 1977, Brau et al. proposed a simple and cost-effective design for an electron beam gun using a spiral generator [7]. In 2012, Pal'chikov and colleagues at the Russian Academy of Sciences [8] further refined the theoretical circuit model of spiral transformers, enhancing their design by incorporating an additional outer winding. This advancement formed the basis for the development and testing of a pulsed X-ray device. Most recently, in 2019, the team successfully utilized an innovative spiral transformer to achieve a compact, miniaturized X-ray source [9].

This paper reports the development of a compact 100-kV X-ray source, consisting of two main components: high-voltage generation using a spiral transformer and radiation conversion employing a vacuum diode.

29.2 Principles and Design of the Compact Spiral Transformer-Based X-ray Source

The operating principle of the helical transformer is based on the concept of vector inversion, where energy is stored in a strip line and released as needed. Its basic structure consists of a copper strip line and an insulating layer. Upon the application of voltage, closing the switch releases the energy stored in the strip line, generating a high-voltage pulse of short duration.

29.2.1 Operating Principle and Design of the Spiral Transformer

The spiral transformer is a high-efficiency pulse power device based on the concept of vector inversion. Its unique physical structure and electrical properties enable the generation of high-voltage pulses. The transformer primarily comprises a copper strip and a high-resistance insulating layer, forming a tightly wound spiral structure. The copper strip not only provides the conductive path but also serves as mechanical support, while the insulating layer ensures electrical isolation between the copper turns, preventing short circuits.

During operation, a DC voltage is first applied to the strip conductor to store electrical energy. When a pulse is needed, a precisely controlled switch is closed rapidly, instantly releasing the stored energy. This rapid release process causes the voltage to rise sharply within a very short time, generating a high-magnitude voltage pulse.

The structure and principle of the spiral transformer are illustrated in Fig. 29.1. Polypropylene (PP) insulating film is used to form the insulating layer. The spiral transformer includes a charged copper strip connected at one end, the power supply voltage, charging resistance at the input, and a self-breakdown gas switch, which generates a propagating voltage wave along the active conductor, forming the active layer. As the electromagnetic wave propagates, it converts the electrostatic field into an electromagnetic field; upon reaching the open end, the wave reflects back, converting the electromagnetic field back to an electrostatic field with reversed polarity, achieving voltage multiplication. Theoretically, the spiral transformer can generate an output voltage of $2nU_0$, where n represents the number of copper strip turns, and U_0 is the power supply voltage.

The output voltage of the spiral transformer is influenced by several factors, including switch resistance and inductance, transmission line resistance losses, and leakage inductance losses between adjacent coaxial capacitors [10]. These factors are closely related to the number of turns, width, and diameter of the copper strip. Therefore, when designing such transformers, key considerations include the thickness and width of the copper strip, as well as the selection and control of the switch. These parameters directly affect the energy storage efficiency and the performance of the pulse output. By optimizing these design parameters, the spiral transformer can achieve high-amplitude voltage output while ensuring the stability and reliability of the device.

Taking these factors into account, we developed a spiral transformer with 50 turns, utilizing an oil-insulated gas switch as the primary switch. Figure 29.2 presents the output voltage waveforms of the spiral transformer at various charging voltages. The results indicate that the step-up ratio of the spiral transformer can exceed 10, with a peak output voltage of 100 kV achieved when the charging voltage is 9.5 kV. Consequently, in subsequent experiments, the charging voltage was set to 10 kV to reliably achieve output voltages exceeding 100 kV.

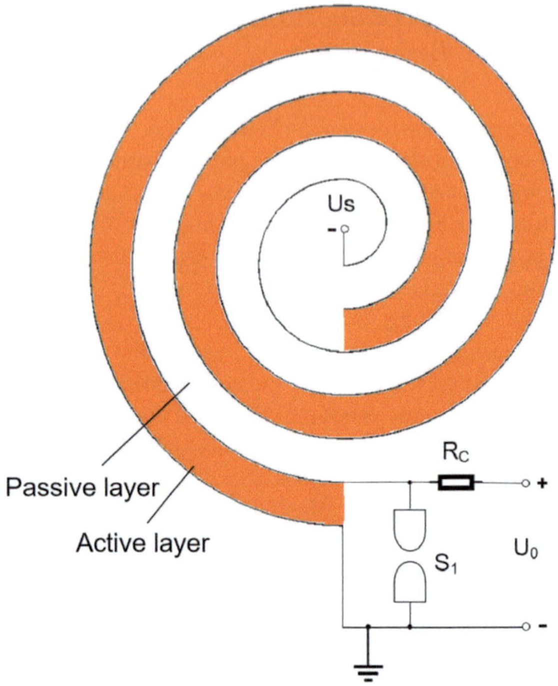

Fig. 29.1 Schematic diagram of the structure and working principle of the spiral transformer

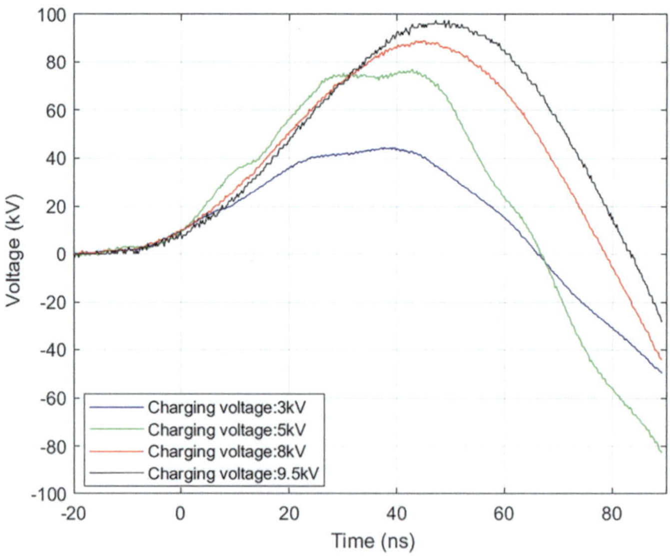

Fig. 29.2 Waveform of the first peak output voltage at different charging voltages

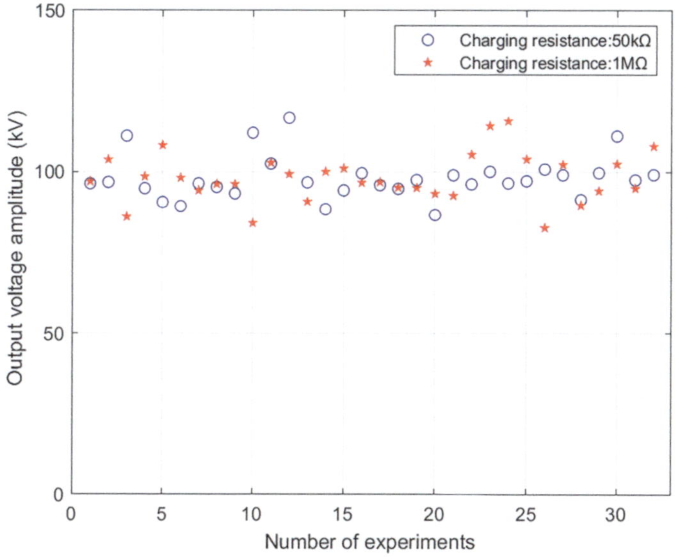

Fig. 29.3 Voltage stability test results

Additionally, stability testing was conducted on the spiral transformer. Figure 29.3 presents the voltage amplitude results from repeated experiments under varying charging resistances. The results indicate that the spiral transformer exhibits good output voltage stability, with amplitude deviations remaining within 10%. Although the discharge frequency of the spiral transformer varies with different charging resistances, there is no significant difference in the stability of the output amplitude. This suggests that the spiral transformer maintains consistent performance across different charging conditions, ensuring reliable high-voltage output suitable for practical applications.

Figure 29.4 presents the voltage rise rates across different loads, providing insights into the overall stability of the pulse source and diode. The test results demonstrate good consistency in the voltage rise rate of the X-ray source across various operating frequencies. Furthermore, the uniform rise rate observed across the loads suggests minimal performance degradation, highlighting the robustness of the spiral transformer design and its capability to efficiently generate high-voltage pulses. This stable performance under diverse loading conditions underscores the reliability of the system, making it well-suited for applications requiring precise and consistent high-voltage outputs.

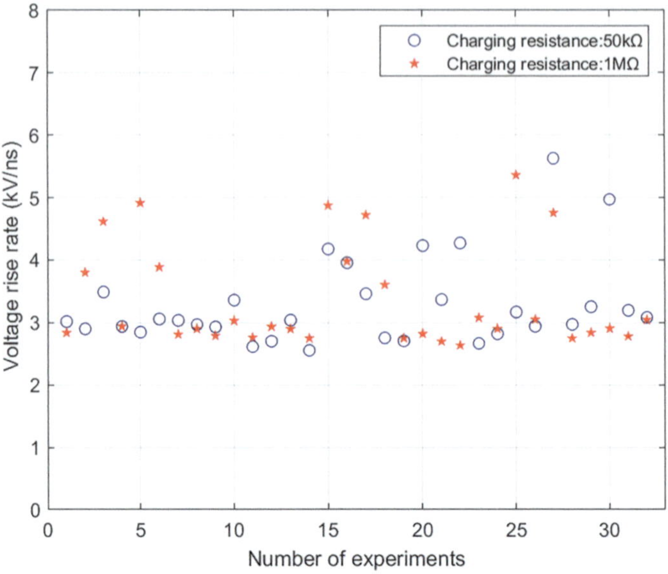

Fig. 29.4 Voltage rise rate test results

29.2.2 Design and Performance of the Compact X-ray Tube

The X-ray tube design can be categorized based on the method of electron generation, typically divided into cold cathode and hot cathode diodes. Cold cathode rod-clamped diodes are simpler in structure and do not require heating of the cathode, instead relying on field emission to generate electrons. This method provides greater stability in radiation quality. For the compact X-ray source operating at 100 kV, a cold cathode rod-clamped diode was employed for radiation conversion.

During the radiation conversion process in the cold cathode diode, a high-current electron beam (on the order of kiloamperes) strikes the target anode over an extremely short duration (tens of nanoseconds). This results in bremsstrahlung radiation as well as significant heat generation. Thus, the design and optimization of the radiation conversion target in the cold cathode diode is crucial for efficient performance.

Figure 29.5 shows the energy deposition profile on the surface of the anode for vertically incident electrons of varying energy levels. The results indicate that most of the energy from the electron beam is deposited within a few tens of microns of the metal surface, and lower-energy electron beams deposit their energy closer to the surface of the conversion target. This finding suggests that for lower-energy electron beams, using thinner conversion targets will not significantly impact the energy deposition process. Furthermore, to minimize photon attenuation within the target material during X-ray photon transport, thinner conversion targets are favorable for optimizing photon output.

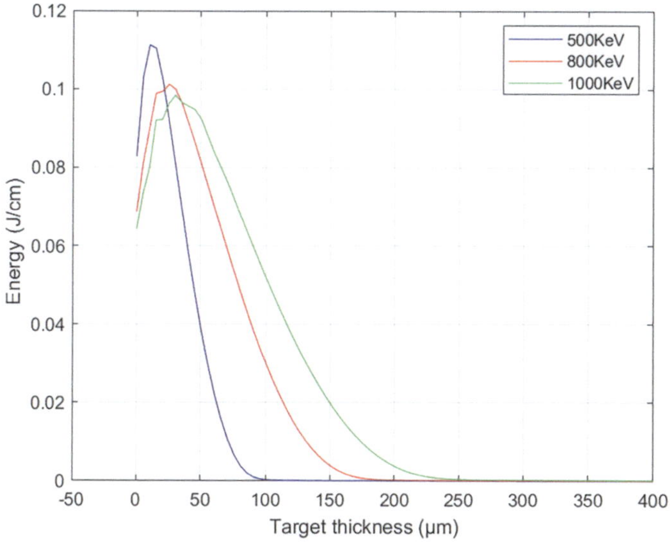

Fig. 29.5 Energy deposition profile for vertically incident electrons at different energy levels

Figure 29.6 illustrates the radiation field distribution directly in front of the diode following electron beam incidence on the conversion target. With appropriate radiation shielding, the strongest radiation dose is confined within a fan-shaped region of approximately 60°. This ensures that the radiation field is well-controlled, meeting both imaging and radiation safety requirements.

29.3 Experimental Evaluation of the Compact X-ray Source

29.3.1 Performance Testing of the Compact X-ray Source

Using the developed compact high-voltage source based on the spiral transformer to drive the cold cathode diode, we obtained the pulse voltage, pulse current, and pulse X-ray waveforms, as shown in Fig. 29.7. The diode operates at a peak voltage of approximately 100 kV, with a pulse current amplitude of 2.87 kA and an equivalent diode impedance of around 35 Ω. The pulse X-ray width is approximately 15 ns.

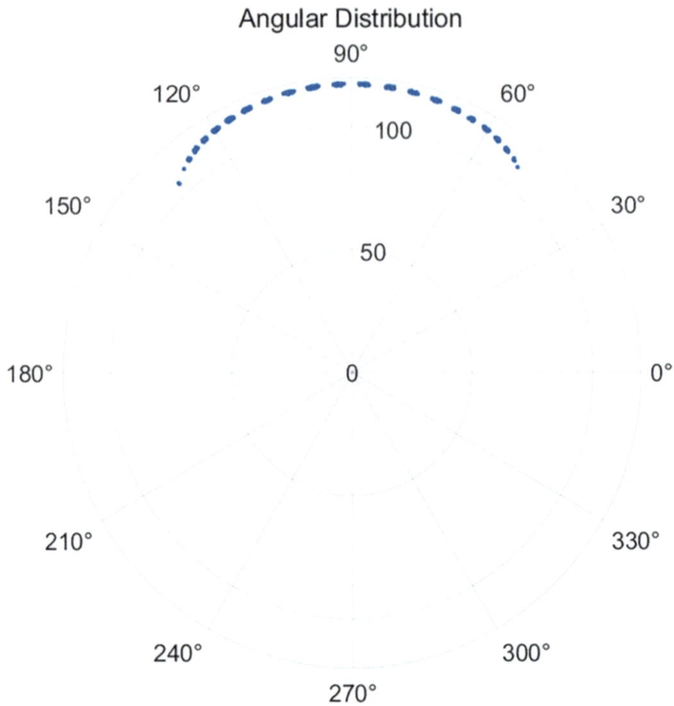

Fig. 29.6 Angular distribution of the radiation field

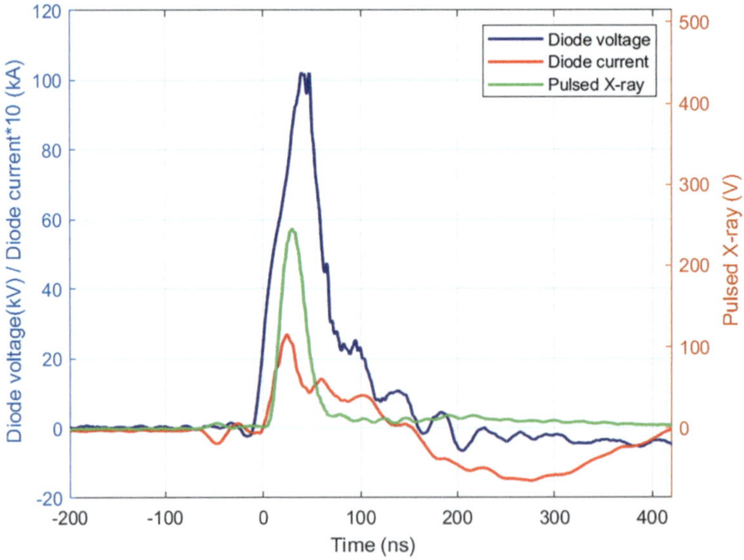

Fig. 29.7 Energy deposition profile for vertically incident electrons at different energy levels

29.3.2 *Imaging Experiments on Typical Power System Components*

The developed compact X-ray source, paired with a flat-panel imaging system, was used to conduct imaging experiments on three types of typical power system components: composite insulators, lightning arresters, and fuses.

(1) Imaging of Composite Insulators

Composite insulators consist of a fiberglass core encased in a layer of silicone rubber. Figure 29.8 shows the imaging results for a typical composite insulator. The test results reveal that the developed X-ray source can clearly distinguish the interface between the fiberglass core and the silicone rubber. This capability allows for intuitive diagnosis of defects such as delamination between layers in composite insulators within power systems. Notably, Fig. 8d highlights a manufacturing defect in the composite insulator's end: the presence of an air gap at the interface between the metal electrode, fiberglass core, and silicone rubber structure.

(2) Imaging of Fuses

Fuses are commonly used electrical protection devices designed to prevent circuit overload or short-circuit conditions. Key to their fault detection is the state of the fuse wire, as timely identification of abnormalities can prevent potential electrical failures. The imaging experiments using the developed X-ray source provide a clear visualization of the fuse wire's condition, including its breakage position, area, and deformation. Additionally, the imaging allows for the inspection of other critical internal components, such as contacts, insulation layers, and the housing structure. The condition of these components directly impacts the fuse's performance, and the imaging facilitates the assessment of contact wear, insulation aging, and possible damage or deformation to the housing.

In Fig. 29.9, the condition of the fuse wire inside the fuse tube is clearly displayed, allowing for effective diagnostics of whether the fuse wire has broken. Additionally, Fig. 9c provides a detailed imaging analysis of the metal and ceramic fusion, demonstrating excellent interface imaging quality. The clarity of the internal interface structure observed suggests superior fusion quality, which offers critical data for subsequent safety assessments and performance evaluations.

(3) Imaging of Lightning Arresters

Lightning arresters are extensively used in power systems to protect equipment from lightning strikes and over-voltage surges. By utilizing non-linear resistance materials, these devices quickly conduct excess current during surges, directing it safely to the ground, thereby preventing equipment damage. As power systems become more complex, the importance of lightning arresters continues to grow.

The developed X-ray source was employed to image the internal structure of lightning arresters, providing insights into the condition of internal components such as non-linear resistors, insulating media, and connecting parts. This non-destructive

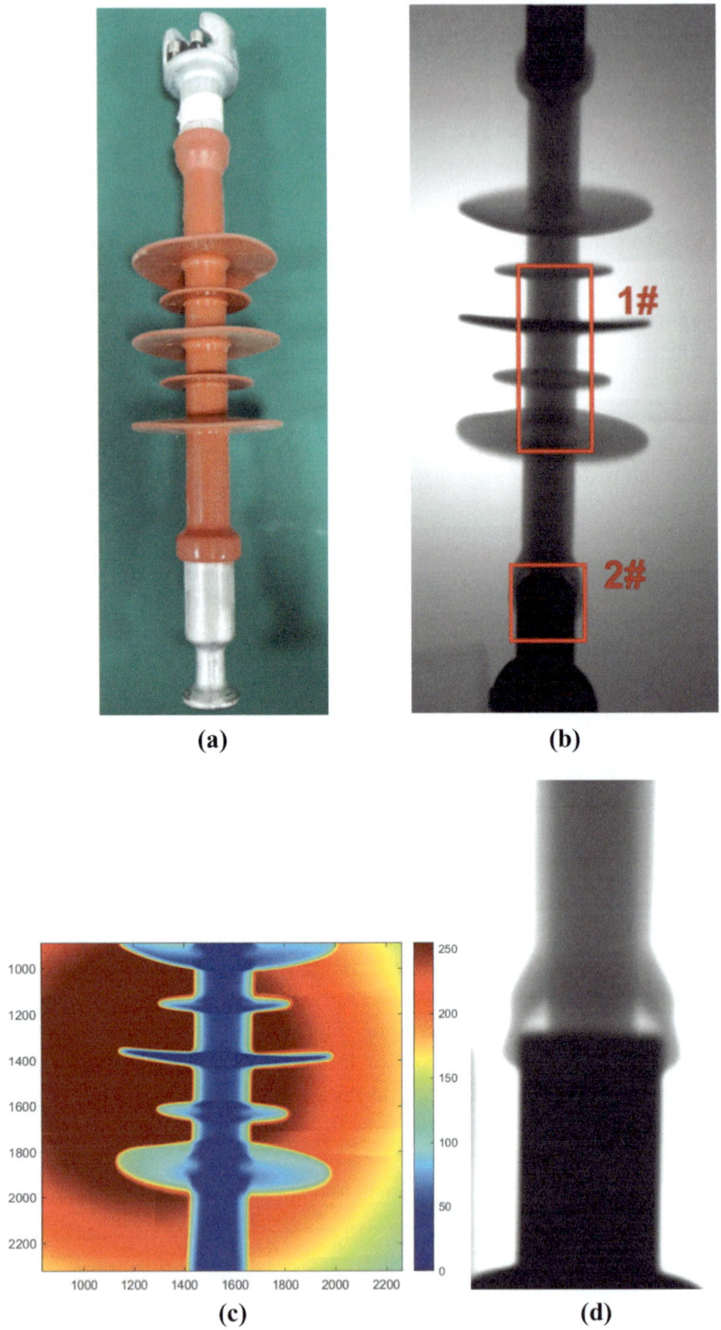

Fig. 29.8 Imaging results of a typical composite insulator. **a** Power equipment image, **b** Original imaging result, **c** Partial pseudo-color image, **d** Localized defect image

Fig. 29.9 Imaging experiment results of a typical fuse. **a** Fuse device image, **b** Original imaging result, **c** Partial imaging and pseudo-color image

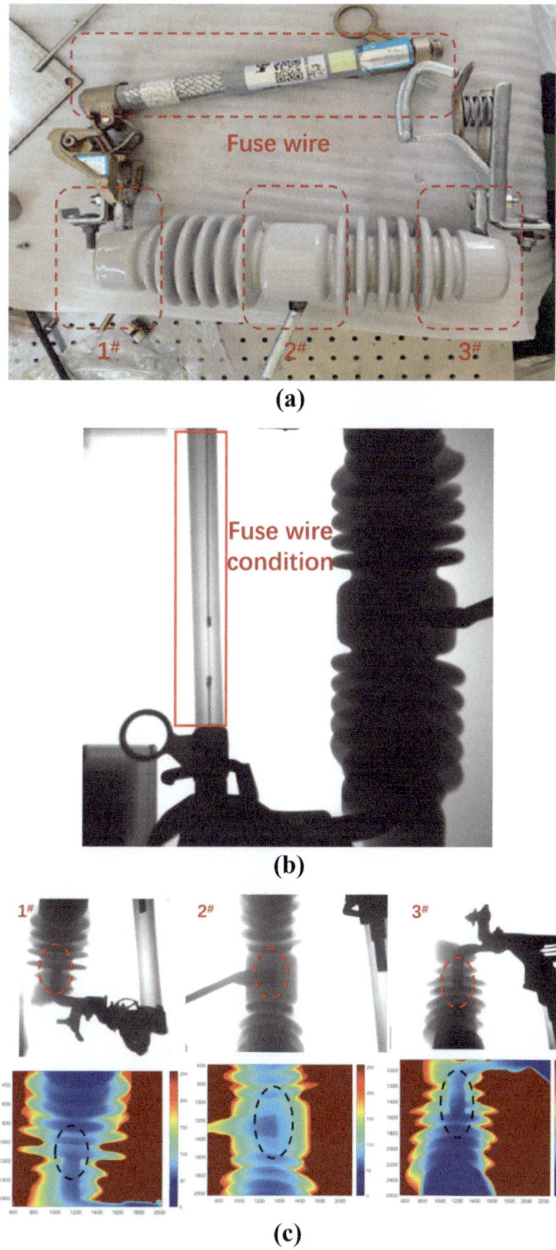

testing method eliminates the need for disassembly and provides real-time information about the internal condition of the arrester. The study reveals potential micro-defects in the arrester from prolonged usage, while also offering key data for assessing the reliability and performance of its insulating materials.

Figure 29.10 presents the imaging results of the lightning arrester, along with the corresponding pseudo-color transformation. Figure 10a shows the overall imaging of the arrester, while Fig. 10c offers an in-depth analysis of the specific positions of various connecting ends, allowing for clear observation of the internal connections. These imaging details not only support the functional assessment of the device but also lay the foundation for the evaluation of its long-term stability and reliability.

In summary, these imaging results fully demonstrate the excellent performance of the developed X-ray source. This advanced imaging technology not only enhances the visualization of internal structures and weld quality of electrical components but also aids in identifying potential defects and faults, providing a scientific basis for subsequent maintenance and improvement efforts. These findings offer a new perspective for research in the field, proving the significant value of high-quality imaging techniques in electrical safety and component reliability assessment.

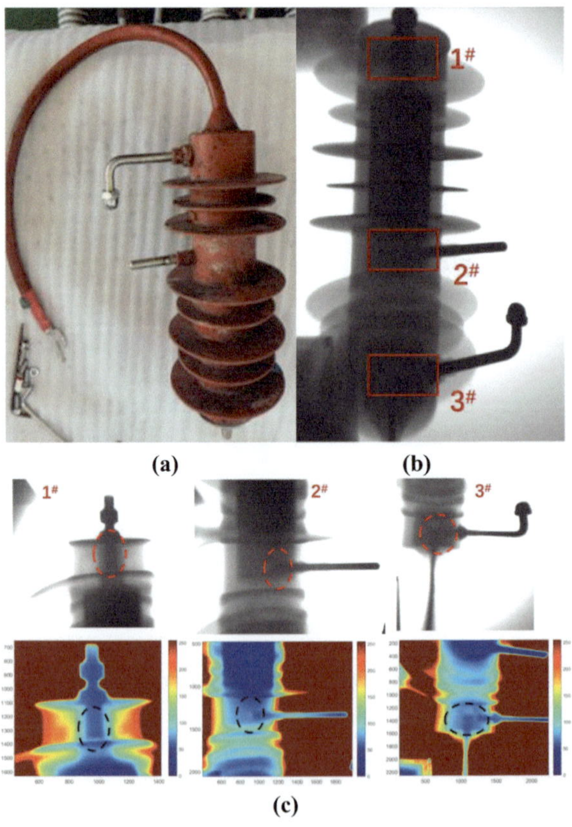

Fig. 29.10 Imaging results of a typical lightning arrester. **a** Lightning arrester device image, **b** Original imaging result, **c** Partial imaging and pseudo-color image

29.4 Conclusion

In this study, based on an analysis of X-ray imaging principles and parameter requirements, a compact pulsed X-ray source was developed using a spiral transformer-based pulsed high-voltage source and a vacuum-sealed rod-clamped diode. Application experiments were conducted on typical power system components. The results demonstrate that the X-ray source operates at approximately 100 kV and provides clear imaging of components such as fuses, lightning arresters, and composite insulators used in power systems, indicating promising application potential. Furthermore, the compact design and high imaging performance of the developed X-ray source suggest that, with appropriate modifications, it holds significant potential for live inspection of overhead transmission lines and non-destructive testing in confined spaces.

References

1. Li, Bo, Hu, Xiumin, He, Zhiqin, et al.: Research on defect detection of GIS equipment using X-ray nondestructive testing. Electronic Design Engineering 29(1), 78–82 (2021).
2. Li, Chenggang, Chen, Dabing, Zhang, Jianguo: Application of X-ray digital imaging technology in on-site inspection of power cables. Nondestructive Testing 37(2), 74–77 (2015).
3. Xu, Huanqing, Ma, Junpeng, Wang, Chengliang, et al.: X-ray digital imaging detection technology for typical defects in GIS equipment. Power System Technology 41(5), 1697–1702 (2017).
4. Geng, Lidong, Xie, Weiping, Yang, Qiang, et al.: Design and experiment of a 200 keV pulsed X-ray source based on PFN-Marx technology. High Power Laser and Particle Beams 34(8), 142–146 (2022).
5. Fitch, R.A., Howell, V.T.S.: Novel principle of transient high-voltage generation. Electrical Engineers, Proceedings of the Institution of 111(4), 849–855 (1964).
6. Rühl, F., Herziger, G.: Analysis of the spiral generator. The Review of Scientific Instruments 51(11), 1541–1547 (1980).
7. Brau, C.A., Raybun, J.L., Dodge, J.B., et al.: Simple, pulsed, electron beam gun. Review of Scientific Instruments 48(9), 1154–1160 (1977).
8. Pal'chikov, E.I., Ryabchun, A.M., Krasnikov, I.Yu.: Modified spiral high-voltage generator for feeding a pulsed X-ray apparatus. Technical Physics 57(2), 292–301 (2012).
9. Pal'chikov, E.I., Dolgikh, A.V., Klypin, V.V., et al.: Pulse X-ray device based on a combined spiral generator. Journal of Applied Mechanics and Technical Physics 60(3), 586–591 (2019).
10. Cheng, Liang, Zhang, Yongmin, Yao, Weibo, et al.: Loss parameters of helical voltage multiplier transmission lines. High Power Laser and Particle Beams 23(2), 555–558 (2011).

Chapter 30
Development of Nanosecond Pulse High Voltage Source Based on Spiral Transformer Principle

Jiaming Liang and Zhiyao Zheng

Abstract The spiral transformer is a type of vector inversion generator composed of a double-layer metal strip winding. By closing a switch, the energy stored in the strip is released, generating a high-voltage pulse. Compared with other generators, the spiral transformer has a simple structure and can generate voltage pulses with amplitudes ranging from tens to hundreds of kilovolts. This paper presents the development of a spiral transformer used to generate nanosecond pulse high voltage. Through theoretical analysis and modeling, the working principle and circuit structure of the spiral transformer are thoroughly examined. Based on this, a compact air-insulated spiral transformer pulse source with tens of kilovolts was designed and experimentally studied. The results show that a spiral transformer with an axial length of 80 mm and a maximum outer diameter of 74 mm achieves a voltage multiplication ratio of approximately 15 under air insulation conditions, realizing a nanosecond pulse high-voltage output of tens of kilovolts. This demonstrates significant potential for practical applications. Moreover, this paper systematically investigates the boosting efficiency of spiral transformers with typical parameters under different charging voltages and their output characteristics under varying loads.

Keywords Spiral transformer · Wave transmission process · Equivalent model · Telegraph equation · High-voltage pulse

30.1 Introduction

In recent years, X-ray inspection has been introduced as a novel non-destructive testing (NDT) method for fault diagnosis in power equipment. Due to the strong penetration capability of X-rays, inspections can be performed while equipment is energized and without dismantling the external structure. This technique produces

J. Liang (✉) · Z. Zheng
Zhejiang Huadian Equipment Testing and Research Institute Co., Ltd, Hangzhou, Zhejiang, China
e-mail: zzycau@qq.com

Y. Li et al. (eds.), *Proceedings of the 5th Chinese Conference on High Voltage and Discharge Plasmas*, Springer Proceedings in Physics 254,
https://doi.org/10.1007/978-981-96-5711-7_30

intuitive and visual images, making it effective for detecting internal defects such as loose components, foreign objects, and structural damage caused by prolonged operation. To enable on-site, high-altitude power transmission line inspections and X-ray NDT in confined spaces, there is significant value in developing compact, lightweight, reliable, and cost-effective X-ray sources. Currently, the primary X-ray sources for NDT in power systems are direct current (DC) sources and imported pulsed X-ray sources. However, DC sources are bulky and heavy, while imported devices are costly and challenging to maintain, presenting several inconveniences and risks for routine fault diagnosis. Against this backdrop, a pulsed X-ray source design based on a helical transformer is proposed. This system generates nanosecond pulsed high-voltage sources to drive cold cathode radiation converter devices, achieving a compact, lightweight, and reliable output. The helical transformer serves as the core component and key technology of the X-ray source system.

The helical transformer, first proposed by Fitch and Howell in 1964 [1], is a high-voltage generation method known for its simple structure, high step-up efficiency, and reliability. It has gained extensive attention and research in the field of nanosecond pulse generation. Compared to Marx or Linear Transformer Driver (LTD) generators [2–5], the helical transformer is simpler and more reliable in structure, consisting of only a low-inductance input switch and a helical strip line. The transformer's energy is capacitively stored within the strip line itself. When one side of the strip line is charged to 0 by a DC power source and the input switch is closed, an output pulse with an amplitude of approximately $-2\varepsilon ffNu0$ is generated across the load, where εff represents the voltage multiplication efficiency, and N denotes the number of turns in the helix. This method provides a straightforward and cost-effective approach for generating voltage pulses with amplitudes ranging from tens to hundreds of kilovolts from a relatively low initial charging voltage.

Ruhl and Gherziger [6] conducted a detailed analysis of the voltage step-up principle of helical transformers. Using Laplace transforms and an equivalent circuit model, they examined the impact of switching impedance, load at the output terminals, and the finite conductivity of the helical coil conductors on output voltage. However, they did not consider the discharge of the passive layer. Bichenkov et al. [7] improved upon the theoretical model by accounting for the effects of passive layer discharge. They modeled the helical transformer with relatively few turns as a variant of the Tesla transformer, effectively predicting the output voltage during oscillations within the helical transformer.

Helical transformers are widely used in applications such as triggers for high-current spark gap switches and power conditioning stages in ultra-compact high-voltage nanosecond pulse generation systems. Between 2003 and 2009, Rose et al. [8–12] conducted detailed experimental studies on helical transformers, examining failure characteristics and design criteria for their use as the power conditioning stage in explosive ultra-compact high-voltage nanosecond pulse generation systems. Helical transformers hold significant potential for generating high-voltage nanosecond pulse sources in a variety of applications.

30.2 Theoretical Analysis of the Helical Transformer

The operating principle of the helical transformer is based on the concept of vector inversion, where energy is stored in a strip line and released as needed. Its basic structure consists of a copper strip line and an insulating layer. Upon the application of voltage, closing the switch releases the energy stored in the strip line, generating a high-voltage pulse of short duration.

As shown in Fig. 30.1, which illustrates the schematic and electric field vector distribution of the helical transformer, the transformer consists of two layers of metal strips and two insulating layers wound around a supporting inner cylinder [13]. In this study, copper strips are used as the metal strips. Each copper strip layer is separated by insulation, which comprises an active layer and a passive layer. One end of the active layer is connected to the input switch S1, while the other end is open; both ends of the passive layer are open. Initially, with switch S1 open, the DC power source charges the helical transformer to voltage U_0. At this stage, the active layer holds an initial voltage, and, given the open state, the electric field vectors in the active and passive layers have equal magnitude but opposite directions. When switch S1 is closed at $t = 0$, a voltage wave of approximately $-U_0$ is generated in the active layer, propagating toward the open end. At $t = l/v$, the voltage wave reaches the open end, marking the completion of the line-charging process, with all electric field energy converted into magnetic field energy. This instant also initiates the reverse process, where the magnetic energy transitions back into electric field energy. Upon reaching the open end of the active layer, the wave reflects, reversing the direction of the electric field vector in the active layer. At this point, the electric field vectors in both the active and passive layers align, and at $t = tr = 2\ l/v$, the voltage wave completes a round-trip reflection, achieving maximum output voltage in the helical transformer. Figures 1a, b illustrate the cross-sectional electric field vector distributions of the helical transformer at $t = 0$ and $t = tr = 2\ l/v$, respectively, where tr denotes the rise time of the output voltage waveform, l is the total length of the active layer, and v is the propagation speed of the voltage wave in the active layer. Ideally, a peak triangular waveform output voltage of $2NU_0$ can be achieved across the load Z, where N represents the number of turns of the copper strips (or the number of layers in both the active and passive layers). However, due to practical constraints, the actual boost ratio is less than 2N, resulting in an output pulse amplitude approximately given by $Uout = -2\varepsilon ffNU_0$. The rise time of the output voltage, $tr \doteq 2N\pi D/c_{eff}$, corresponds to the round-trip propagation time of the voltage wave within the spiral, where D is the average winding diameter, and c_{eff} denotes the effective wave propagation speed within the spiral.

Overall, theoretical analysis of the spiral transformer necessitates consideration of factors including passive layer discharge, resistance losses across layers, output switch inductance, rise time, and the interdependence between load impedance and output voltage. Different load coupling conditions also warrant analysis. The spiral transformer is modeled by coupling input and output of the strip conductor, wherein

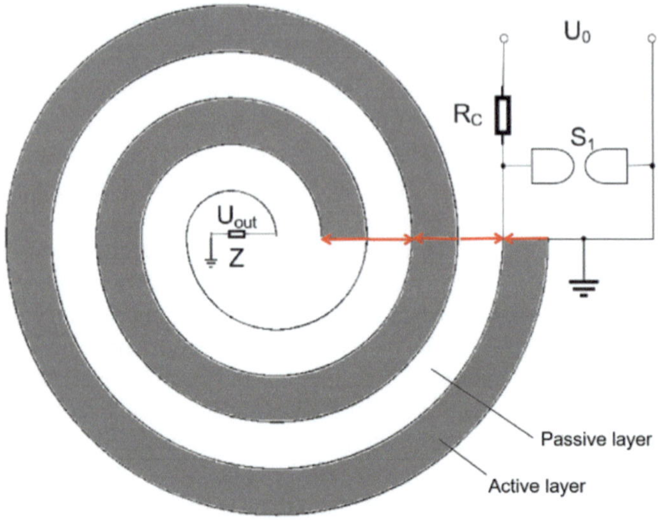

(a) t=0

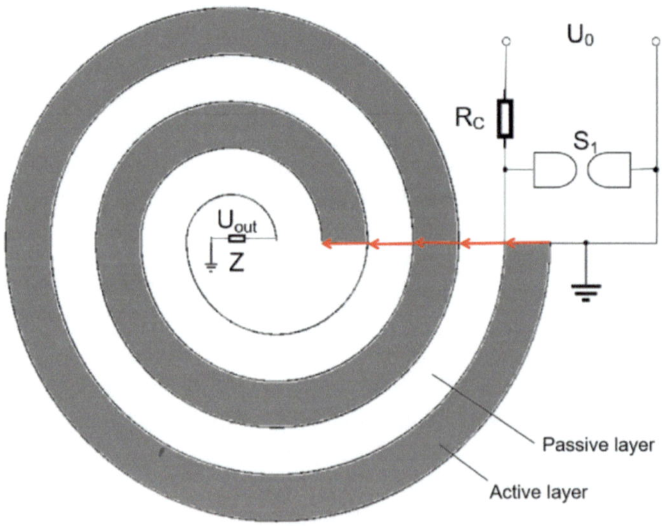

(b) $t=t_r$

Fig. 30.1 Spiral transformer structure diagram and principle

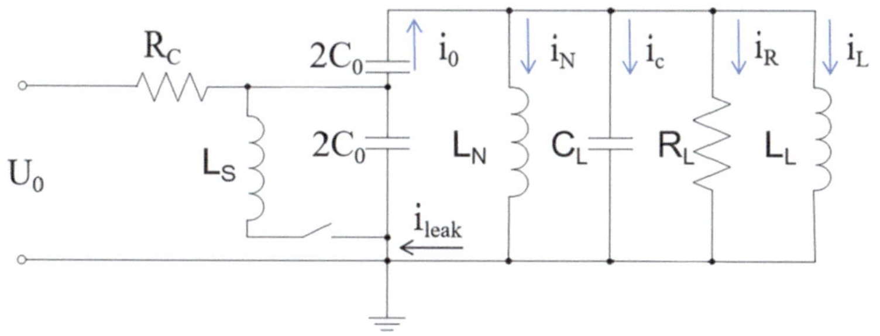

Fig. 30.2 Spiral transformer equivalent circuit

traveling waves within the active and passive conductors are described by the telegraph equation, and charge exchange between the strip, input switch, and load is implemented through an equivalent circuit. Crucially, variations in input switch inductance, resistance, and load characteristics influence wave propagation within the spiral.

The equivalent circuit of the spiral transformer is shown in Fig. 30.2. It consists of the input switch (including the inductance Ls), and a combination of inductance (L), resistance (RL), and capacitance (CL) that form the load, along with the spiral's output capacitance C0 and the inductance LN of the turns.

The output capacitance C0 is calculated as follows:

$$C_0 = \frac{\varepsilon_r \varepsilon_0 \pi D w}{2Nt} \tag{30.1}$$

where w is the width of the copper strip, t is the thickness of the copper strip, t_i is the thickness of the insulation layer, ε_0 is the permittivity of free space, and ε_r is the relative permittivity. It is noted that the spiral comprises two layers: the active and passive layers. When considering the total output capacitance of the spiral, these two layers are modeled in series, resulting in a total value of C_0. Thus, the capacitance of each layer is twice the output capacitance, $2C_0$, as illustrated in the equivalent circuit. The inductance LN is calculated using the following expression:

$$L_N = K\prime \frac{\mu_r \mu_0 \pi D^2 N^2}{2w} \tag{30.2}$$

where K' is a geometrical factor typically valued at 0.159, μ_0 is the permeability of free space, and μ_r is the relative permeability. According to the equivalent circuit, The circulating current iN flowing through the helical inductor can be calculated using Eq. (30.3). where i_0, i_L, i_C, and i_R represent the currents flowing through the spiral capacitance, inductance, load capacitance, and load resistance, respectively, and can be calculated using the Eqs. (30.4)-(30.7).

$$i_N = i_{leak} - i_0 - i_R - i_L - i_C \tag{30.3}$$

$$i_0 = \frac{1}{2N^2}\frac{du_s}{dt} \tag{30.4}$$

$$i_R = \frac{l}{2N^2cC_0R_Lu_s} \tag{30.5}$$

$$\frac{N^2L_L}{L_N}\frac{di_L}{dt} = \frac{\omega_0^2}{2}u_s \tag{30.6}$$

$$i_C = \frac{C_L}{2N^2C_0}\frac{du_s}{dt} \tag{30.7}$$

The established model takes into account the interaction between the active and passive conductors. To simplify the equations, a specific unit system is introduced, where the wave speed, conductor length, and initial voltage in both the active and passive conductors are normalized to one unit. To transition from the dimension system of the units x, t, u, i to the conventional unit system of "coordinate X, time T, voltage U, and current I," we define the following relation:

$$X = lx, T = \frac{l}{c_{eff}}t, U = u_0u, I = c_{eff}C\prime u_0i \tag{30.8}$$

In this expression, $l = N\pi D$ represents the total length of the unfolded strip line, and $C' = 2N^2C_0/l$ is the capacitance per unit length. Using this notation, the equations describing the propagation of electromagnetic waves in active and passive conductors are given by:

$$\begin{gathered}\frac{\partial u_a}{\partial t} = -\frac{\partial i_a}{\partial x} - Ni_s, \frac{\partial u_p}{\partial t} = -\frac{\partial i_p}{\partial x} - Ni_s,\\ \frac{\partial i_a}{\partial t} = -\frac{\partial u_a}{\partial x}, \frac{\partial i_p}{\partial t} = -\frac{\partial u_p}{\partial x},\\ u_s = N\int\limits_0^1 (u_a + u_p)dx\end{gathered} \tag{30.9}$$

Under the initial conditions:

$$\begin{gathered}u_a(x,0) = -1, u_p(x,0) = 1,\\ i_a(x,0) = i_p(x,0) = i_s = 0.\end{gathered} \tag{30.10}$$

where u_a and i_a are the normalized voltage and current in the active line, u_p and i_p are thenormalized voltage and current in the passive line; N is the number of turns; is is the inter-turn leakage current; $(i_a + i_p)$ is the current through the switch; and u_s is the output voltage of the generator.

Considering the boundary conditions of the helical transformer, since its two ends are not closed, we have $i_a(1, t) = i_p(1, t) = 0$(in our unit system, the unit in the parameter indicates x = l). The input current from the generator is specified as $i_{0a} = i_a(0, t)$ and $i_{0p} = i_p(0, t)$. Then, according to the boundary conditions, we can find the solution corresponding to the established model. Due to the close contact between the active and passive layers, k_{ap} approaches unity, and $k_{sa} \approx k_{sp} \ll 1$. In the model, we choose $k_{ap} = 0.9$ and $k_{sa} = k_{sp} = 0.1$. The relevant parameters of the designed helical transformer and the model calculation data are shown in Table 30.1.

Considering the interaction between currents i_{0p} and i_{0a}. the boundary conditions at the input are:

$$\begin{aligned} N^2\frac{di_s}{dt} - k_{sp}N\frac{di_{0p}}{dt} &= \frac{\omega_0^2}{2}u_s, \\ L_Z\frac{di_{0p}}{dt} - L_Z\frac{di_{0a}}{dt} &= \frac{\omega_0^2}{2}u_a(0), \\ k_{sp}N\frac{di_s}{dt} - (L_Z+1)\frac{di_{0p}}{dt} + L_Z\frac{di_{0a}}{dt} &= \frac{\omega_0^2}{2}u_p(0) \end{aligned} \tag{30.11}$$

where L_Z is the ratio of the switch inductance to the single-turn inductance of the helix, and ω_0 is the normalized oscillation frequency of the circuit.

$$L_Z = \frac{L_s}{L_N/N^2} \tag{30.12}$$

$$\omega_0 = \frac{1}{\sqrt{L_N C_0}}\frac{l}{c_{eff}} \tag{30.13}$$

Table 30.1 Typical parameters used in model calculations and experiments

Parameter name	Symbol	Values
Number of turns	N	20
The length of the helix	l	4.06 m
Width of copper strip	w	50 mm
The average diameter of the winding	D	64.6 mm
Thickness of winding	t_h	1.21 mm
Copper strip thickness	t	0.05 mm
Individual insulation thickness	t_i	0.2 mm
Wave velocity in a ribbon	c_{eff}	1.98 × 108 m/s
Output inductance	L_N	10.48 μh
Output capacitance	C_0	4.49 pF
Coupling coefficient	k_{sa},k_{sp}	0.1
Coupling coefficient	k_{ap}	0.9
Travel time	TO	20.51 ns

The electromagnetic wave velocity is calculated using the following formula, where the relative permittivity of the PP insulating film $\varepsilon_r \approx 2.3$ and the relative permeability $\mu_r \approx 1$. T_0 is the time required for the wave to propagate from one end of the helix to the other, and the inductance Ls of the input switch is estimated based on the length of the input switch circuit:

$$c_{eff} = c/\sqrt{\varepsilon_r \mu_r} \tag{30.14}$$

Based on the above model analysis and quantitative calculations, the structure of the helical transformer required to achieve the specified voltage multiplication and parameter requirements was determined.

30.3 Experimental Study on the Output Characteristics of the Helical Transformer

Based on the theoretical analysis results, an experimental setup for the helical transformer was designed and constructed. In this experiment, we used a copper strip that is 50 mm wide and 0.5 mm thick for the helical winding. The insulation material is PP insulating film with a thickness of 0.2 mm, and a multi-layer coaxial winding method was employed to create a 20-turn coil. An air-insulated self-breakdown low-inductance input switch was utilized. The output electrical pulse measurements were conducted using a North Star PVM high-voltage probe with a voltage divider ratio of 1000:1. The structure of the experimental system is shown in Fig. 30.3.

Figure 30.4 presents the output voltage waveforms of the helical transformer at different charging voltages. A comparison reveals that the output voltage of the helical transformer increases as the charging voltage rises. After multiple repetitions of the experiment, the statistics on the peak voltage amplitude of the first main pulse are illustrated in Fig. 30.5. It was observed that as the charging voltage increases, the ratio of output voltage to charging voltage slightly decreases. This indicates that there are some differences in the voltage multiplication factor of the helical transformer under varying charging voltages, with a slight downward trend in the voltage multiplication factor at higher charging voltages. The primary reason for this phenomenon is attributed to the changes in the inductance caused by switch breakdown. In this system, the primary switch utilizes a gas switch with an atmospheric pressure air gap. As the charging voltage increases, the switch gap widens. Under the working mode primarily dominated by single-channel breakdown, the discharge distance increases, leading to an increase in the inductance during the discharge process of the primary switch.

Figure 30.6 shows the output voltage waveforms of the helical transformer under a charging voltage of 4 kV for different resistive loads. A comparison reveals that as the load impedance increases, the output voltage amplitude significantly increases. This phenomenon can be attributed to the voltage division formed by the internal resistance

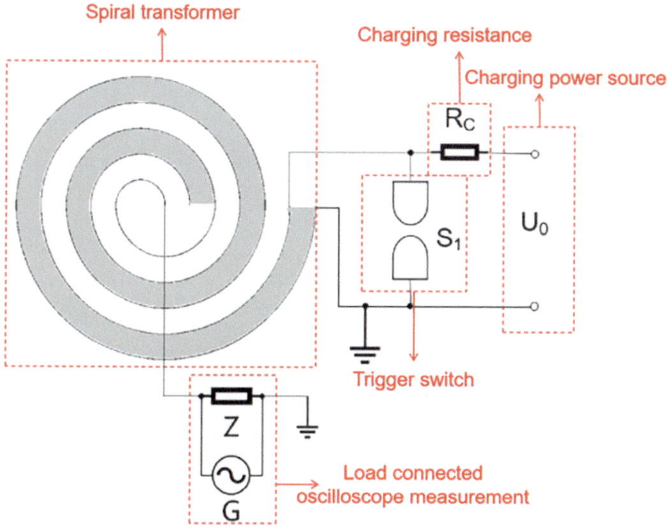

Fig. 30.3 Spiral transformer test system results schematic diagram

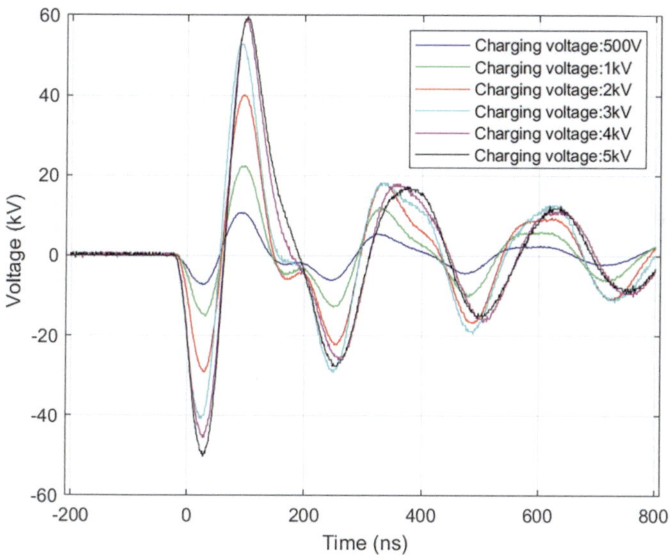

Fig. 30.4 Output voltage waveform at different charging voltage

of the pulsed source of the helical transformer and the load. Under high-resistance load conditions, it is easier to achieve high-voltage pulse outputs.

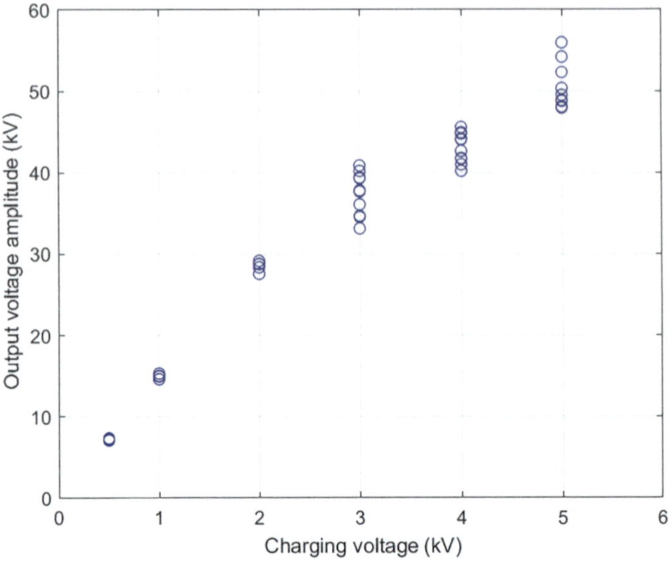

Fig. 30.5 The main pulse amplitude of the repeated experiment with different charging voltages was counted

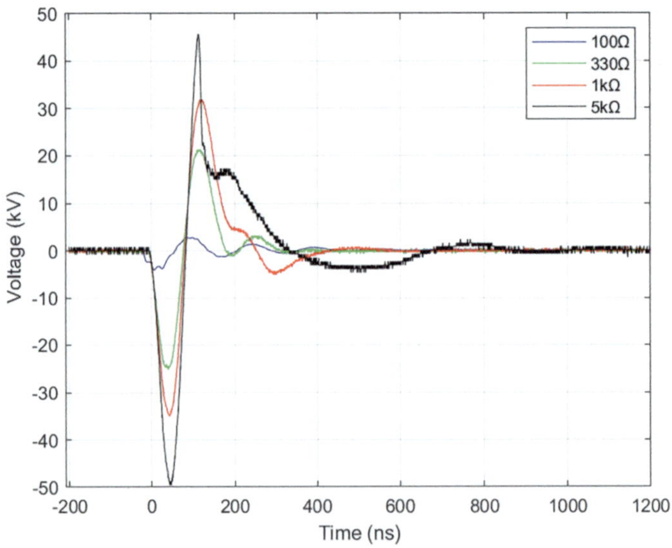

Fig. 30.6 Output voltage waveform under different resistive loads

This observation suggests that designing high-impedance loads is more advantageous for maximizing the voltage boosting capability of the helical transformer, thereby facilitating the generation of high-penetration pulsed X-rays.

30.4 Conclusion

This paper presents a systematic theoretical analysis of the helical transformer-based pulsed source and establishes a compact testing system with a high voltage multiplication factor for the helical transformer. Systematic experimental studies were conducted. When the outer diameter of the helical transformer is approximately 74 mm and the axial length is around 80 mm, a voltage multiplication factor of approximately 15 is achieved, allowing for the convenient generation of high-voltage nanosecond pulses in air on the order of tens of kilovolts. The experimental results validate the accuracy of the theoretical model and lay a solid foundation for the development of a compact pulsed X-ray source in the future.

Funded Projects The Science and Technology Project of Zhejiang Huadian Equipment Testing and Research Institute Co., Ltd. (Project Number: 5211HD230022).

References

1. Fitch, R.A., Howell, V.T.S.: Novel principle of transient high-voltage generation. Electrical Engineers, Proceedings of the Institution of 111(4), 849–855 (1964).
2. Han, Min, Zou, Xiaobing, Zhang, Guixin: Pulse Power Technology Basis. Tsinghua University Press, Beijing (2010).
3. Li, Fei, Song, Falun, Gan, Yanqing, et al.: Design and test on Marx trigger source with low output jitter. Journal of Terahertz Science and Electronic Information Technology 15(2), 323–327 (2017).
4. Sun, Fengju, Jiang, Xiaofeng, Wang, Zhiguo, et al.: Design and simulation of fast linear transformer driver with four stages in series sharing common cavity shell and mega-ampere current. High Power Laser and Particle Beams 30(3), 90–97 (2018).
5. Sun, Fengju, Qiu, Aici, Jiang, Xiaofeng, et al.: Twelve-stage linear transformer driver with one terawatt power on a sharing common cavity shell and internal in-situ triggering method. Modern Applied Physics 13(4), 57–65+204 (2022).
6. Rühl, F., Herziger, G.: Analysis of the spiral generator. The Review of Scientific Instruments 51(11), 1541–1547 (1980).
7. Bichenkov, E.I., Palchikov, E.I., Ryabchun, A.M., et al.: On the refinement of the theoretical model for a high-voltage spiral generator. Technical Physics Letters 52(12), 1597–1603 (2007).
8. Merryman, S.A., Rose, M.F., Shotts, Z.: Characterization and applications of vector inversion generators. In: 2003 IEEE Pulsed Power Conference, pp. 249–252 (2003).
9. Rose, M.F., Shotts, Z., Roberts, Z.: High efficiency compact high voltage vector inversion generators. In: 2005 IEEE Pulsed Power Conference, pp. 1207–1210 (2005).
10. Hanlon, J., Shotts, Z., Rose, M.F., et al.: High voltage solid state switched vector inversion generator. In: 2007 IEEE Pulsed Power Conference, pp. 918–922 (2007).

11. Shotts, Z., Roberts, Z., Rose, M.F.: Design principles for vector inversion generators. In: IEEE International Conference on Plasma Science, pp. 556–559 (2007).
12. Shotts, Z., Roberts, Z., Rose, M.F.: Limits and failure modes in high voltage vector inversion generators. In: 2009 IEEE Pulsed Power Conference, pp. 902–907 (2009).
13. Sun, Jingjing, Yang, Shuang, Sun, Yijie, et al.: Research on an improved spiral generator. High Power Laser and Particle Beams 35(10), 136–142 (2023).

Chapter 31
Development of a Capacitive Divider for Nanosecond Pulse Voltage Measurement in Flexible High-Voltage Cables

Dingfang Ke, Fangfang Wu, Hao Wu, Yibo Gao, Xuyang Zhao, Xue Lin, and Shengyi Xie

Abstract To obtain nanosecond-scale leading edge pulse voltage waveforms fed into high-voltage coaxial cables, a compact capacitive divider has been developed. This capacitive divider senses the differential signals of the pulse voltage, featuring a coaxial structure design that is highly integrated with the high-voltage cable. The sampling electrode forms a high-voltage arm capacitance with the inner conductor of the cable, while the outer conductor of the cable constitutes the low-voltage arm capacitance. The theoretical transformation ratio and frequency response of the capacitive divider have been calculated, and a high-voltage pulse testing platform has been established to calibrate the transformation ratio of the divider. The results indicate that the developed capacitive divider possesses the capability for nanosecond-scale leading edge pulse voltage measurement, meeting the insulation testing requirements for high-voltage coaxial cables under hundreds of kilovolts nanosecond pulse voltage conditions.

Keywords Capacitive divider · Coaxial cable · Nanosecond pulse · Frequency response · Insulation testing

Funded Projects: the Science and Technology Project of Zhejiang Huadian Equipment Testing and Research Institute Co., Ltd (Project Number: 5211HD230021)

D. Ke · F. Wu (✉) · H. Wu · Y. Gao · X. Zhao · X. Lin · S. Xie
Zhejiang Huadian Equipment Testing and Research Institute Co., Ltd, Hangzhou, Zhejiang, China
e-mail: 13819198545@163.com

Y. Li et al. (eds.), *Proceedings of the 5th Chinese Conference on High Voltage and Discharge Plasmas*, Springer Proceedings in Physics 254,
https://doi.org/10.1007/978-981-96-5711-7_31

31.1 Introduction

With the continuous development of power systems and high-voltage technology, the application of high-voltage coaxial cables has become increasingly widespread, resulting in higher requirements for insulation voltage levels and reliability. The insulation testing of high-voltage coaxial cables is particularly important [1–8]. Currently, there has been extensive research on the insulation characteristics of high-voltage cables under direct current, power frequency, and lightning pulse conditions, while testing under nanosecond leading edge pulse voltages is relatively lacking [9–11]. Under nanosecond pulse conditions, due to the narrow pulse width and high instantaneous power, the wave transmission process within the cable and the charge accumulation in the dielectric layer lead to significant differences in the insulation mechanisms compared to direct current, power frequency, and lightning tests. This results in noticeable degradation of the insulation capability of high-voltage cables under pulse conditions in power systems. Therefore, it is necessary to conduct research on the insulation characteristics of high-voltage coaxial cables under nanosecond leading edge pulse conditions. One of the technical challenges in this research is to convert transient high voltage into a low voltage that can be measured by recording instruments, such as oscilloscopes, without damaging the structure of the high-voltage coaxial cable, while also maintaining a nanosecond response capability.

The capacitive divider [12, 13] is a non-contact device used for measuring pulsed high voltage. Compared to resistive dividers [14, 15] and RC dividers [16, 17], it offers advantages such as good high-frequency characteristics, minimal impact on the original system, and a compact structure, making it widely used in engineering practice [18]. Capacitive dividers are further classified into self-integrating capacitive dividers and differential capacitive dividers. Among these, differential capacitive dividers have a higher signal-to-noise ratio and superior measurement performance, leading to their broader application compared to self-integrating capacitive dividers.

Capacitive dividers have been used since the 1970s to measure electromagnetic pulses from nuclear explosions, lightning environments, and space electric fields. They have gradually found applications in the measurement of overvoltages in power systems [19, 20], partial discharge signal measurement, and voltage measurement in pulsed power system transmission lines. Shu Yinbiao and colleagues designed a capacitive voltage probe for gas-insulated switchgear (GIS) [21–23], which successfully captured very fast transient overvoltage (VFTO) waveforms with leading edges of several tens of nanoseconds and amplitudes up to 2.5 times the rated ground voltage of the power system [24]. The low-voltage arm capacitance of the capacitive voltage probe for VFTO [25–27] measurement can be constructed using dielectric materials positioned between electrodes or by utilizing commercially available capacitive components. Huiskamp et al. developed a compact wideband coaxial differential capacitive divider with a bandwidth reaching several gigahertz [28]. Novac and colleagues further improved this divider by using the tip of a cable termination as the sensing electrode surface, conducting both theoretical and experimental research

[29]. Wei Bing and others developed a coaxial differential capacitive divider operating in the frequency range of 0.6–8 MHz [30, 31], capable of responding to fast pulse signals with 50 ns leading edges. This device was used for voltage measurement in the water transmission lines of the proton accelerator and in single-channel prototypes of the primary test stand (PTS).

To meet the measurement requirements of nanosecond, hundreds of kilovolts electrical pulses in power systems, this paper develops a differential capacitive divider for high-voltage nanosecond pulse measurements in flexible high-voltage cables, based on an analysis of the working mechanism and theoretical model of capacitive dividers. Calibration experiments were conducted under different operating conditions to obtain the transformation ratio of the capacitive divider, which was successfully applied in typical hundreds of kilovolts electrical pulse tests.

31.2 Fundamental Principles of the Differential Capacitive Divider for Nanosecond Pulse Measurement

The principle of the differential capacitive divider is illustrated in Fig. 31.1 and is generally used in conjunction with an integrator. In the figure, U_o represents the signal to be measured, U_i is the output signal of the differential capacitive divider, and U_s is the output voltage signal from the integrator. C_p is the high-voltage arm capacitance, formed by the distributed capacitance between the sampling electrode and the high-voltage conductor. C_c is the low-voltage arm capacitance, formed by the distributed capacitance between the sampling electrode and the ground electrode. R_p is the leakage resistance between the sampling electrode and the high-voltage electrode, while R_c is the leakage resistance between the sampling electrode and the ground electrode. R_j and C_j are the integration resistance and capacitance of the external integrator, respectively. To avoid distortion in the measurement waveform, the leakage resistance of the medium inside the transmission line at high frequencies should be significantly greater than the reactance of the distributed capacitance, that is $1/\omega C_p \ll R_P$, and $1/\omega C_c \ll R_c$.

The transformation ratio of the differential capacitive divider is given by $\beta_1 = U_0/U_i$, where the pulse voltage signal is extracted from the low-voltage arm capacitance.

$$U_i = (C_p\frac{dU_{C_p}}{dt} - C_c\frac{dU_{C_c}}{dt})Z_0 \tag{31.1}$$

When $1/\omega C_c \gg Z_0$ is satisfied, the capacitive divider functions as a differential type, and the low-voltage arm voltage can be simplified to:

$$U_i = -C_p Z_0 \frac{dU_0}{dt} \tag{31.2}$$

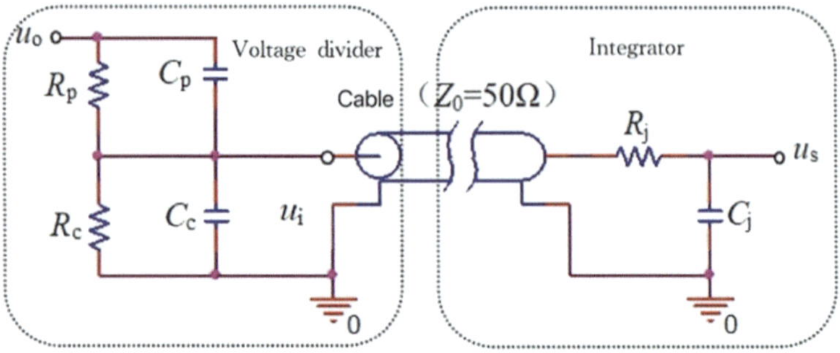

Fig. 31.1 Differential capacitor voltage divider principle circuit

The voltage signal across the low-voltage arm is proportional to the derivative of the source signal. To ensure accurate measurement of the target signal by the oscilloscope, integration of the output signal from the capacitive divider is required.

$$U_s = -\int (U_i/C_jR_j)dt \tag{31.3}$$

Thus, after using the integrator, the voltage division ratio of the signal measured by the oscilloscope is given by:

$$\beta = U_0/U_s = C_jR_j/C_pZ_0 \tag{31.4}$$

According to the voltage division ratio calculation Formula (31.4), the division ratio is inversely proportional to the high-voltage arm capacitance C_p, which itself is directly proportional to the area S of the sampling electrode facing the high-voltage electrode. Therefore, as S increases, the division ratio decreases, resulting in a higher output signal amplitude Us from the integrator for the same voltage signal under measurement. Assuming the noise interference amplitude in the measurement system remains constant, the signal-to-noise ratio of the measured signal decreases as the sampling electrode area S increases.

When the frequency of the pulse signal under measurement is excessively high or low, it becomes challenging to satisfy conditions such as $1/\omega C_p \ll R_P$ and $1/\omega C_c \gg Z_0$, resulting in a loss of linearity between the measured signal and the source signal, indicating that the capacitive divider possesses specific frequency response characteristics. It can be derived according to Kirchhoff's laws that:

$$(U_0 - U_i)/R_p + j\omega C_p(U_0 - U_i) - j\omega C_cU_i - I_1 = 0 \tag{31.5}$$

$$U_i = I_1 \cdot R_j + U_s \tag{31.6}$$

$$I_1 = j\omega C_jU_s \tag{31.7}$$

From the above equation, the frequency response characteristic H(ω), which represents the ratio of the output signal Us from the capacitive divider to the voltage signal UO under measurement, can be obtained as shown in Eq. (31.8).

$$H(\omega) = (1 + j\omega C_p R_p)/\left[(j\omega C_j R_j + 1)(j\omega C_p R_P - j\omega C_c R_p + 1) + j\omega C_j R_p\right] \quad (31.8)$$

31.3 Analysis of Factors Affecting the Working State of Capacitor Voltage Divider

In engineering applications, factors such as installation precision and environmental temperature can easily cause deviations in the capacitance parameters of the high- and low-voltage arms, thus affecting the measurement accuracy of the capacitive divider. Based on the above working principle of capacitive division, a preliminary analysis was conducted on the factors influencing the consistency of capacitive divider parameters.

31.3.1 Sampling Electrode Manufacturing and Installation Errors

Due to manufacturing errors in the sampling electrode or inaccuracies during installation, the lower end of the sampling electrode may not align perfectly with the insulation layer of the cable. This misalignment could result in the electrode either embedding into the insulation layer or not fully reaching it, ultimately affecting the measurement results. If the lower end of the sampling electrode extends beyond or falls short of the outer wall's lower edge, it will influence the high-voltage arm capacitance, while the low-voltage arm capacitance remains nearly unaffected. According to Eq. (31.4), if the sampling electrode extends beyond the outer wall's lower edge, the measurement result will be overestimated; if it falls short, the result will be underestimated. In engineering practice, efforts should be made to minimize manufacturing and installation errors.

31.3.2 Operating Principle and Design of the Spiral Transformer

The effect of inconsistencies in the outer diameter of the insulation layer of the cable under measurement is similar to that caused by manufacturing and installation errors of the sampling electrode. Such inconsistencies alter the distance between the sampling electrode and the cable core, affecting the high-voltage arm capacitance. Similarly, an oversized insulation diameter decreases the high-voltage arm capacitance, resulting in an underestimated voltage measurement, while an undersized insulation diameter increases the capacitance, leading to an overestimated measurement. In engineering practice, efforts should be made to ensure that the measurement position of the cable aligns closely with the calibration position.

31.3.3 Air Bubbles Formed During the Epoxy Resin Casting of the Low-Voltage Arm

When casting the low-voltage arm capacitance with materials such as epoxy resin, air gaps introduced during the casting process can lead to variations in the low-voltage arm capacitance. With air gaps present, the low-voltage arm capacitance can be considered a parallel combination of two capacitors. As the thickness of the air gap increases, the capacitance value of the low-voltage arm decreases, resulting in an increased capacitive impedance for the low-voltage arm, which may no longer satisfy the differentiation condition. Additionally, the air gap can lower the voltage withstand capability of the low-voltage arm, potentially leading to insulation breakdown. In engineering practice, efforts should be made to prevent air bubbles in the low-voltage arm casting. Provided that the low-voltage arm capacitance continues to satisfy the differentiation condition, any bubbles formed during casting will affect only the capacitance value and not the measurement result.

31.3.4 Eccentricity in the Machining of the Inner Cylinder of the Transmission Line

Eccentricity of the inner cylinder may occur during the machining process of the transmission line. Such eccentricity can impact the high-voltage arm capacitance, thereby affecting the measurement results. An analytical calculation and evaluation of the influence of inner cylinder eccentricity on the transmission line is conducted here.

The eccentricity of the inner cylinder in a transmission line is measured by the eccentricity length ΔL and the eccentricity angle θ. The eccentricity length ΔL represents the distance between the actual center and the theoretical center of the transmission line, while the eccentricity angle θ is the angle formed between the line connecting the actual and theoretical centers and the vertical axis of the transmission line cross-section.

Introducing the eccentricity coefficient:

$$n = \frac{\Delta L}{R_1} \tag{31.9}$$

At this point, the high-voltage arm capacitance can be expressed as:

$$C_{p}\prime = \frac{\pi \varepsilon_1 r^2}{\sqrt{(1 - n\cos\theta)^2 + (n\sin\theta)^2} R_1 \ln(\frac{\sqrt{(1-n\cos\theta)^2+(n\sin\theta)^2 R_1}}{R_2}} \tag{31.10}$$

When the measurement impedance of the cable and the time constant of the integrator remain constant, the measurement result is directly proportional to the high-voltage arm capacitance. In this context, the study focuses on analyzing the impact of eccentricity on the measurement results.

Let the ratio of the measurement result after eccentricity to the true measurement result be denoted as k.

$$k = \frac{\ln(\frac{R_1}{R_2})}{\sqrt{(1 - n\cos\theta)^2 + (n\sin\theta)^2} \ln(\frac{\sqrt{(1-n\cos\theta)^2+(n\cos\theta)^2 R_1}}{R_2})} \tag{31.11}$$

Use MATLAB to plot the relationship between k and n for the angles θ = 0, 0.25π, 0.5π, 0.75π, π, 1.25π, 1.5π, 1.75π, 2π, as shown in Fig. 31.2.

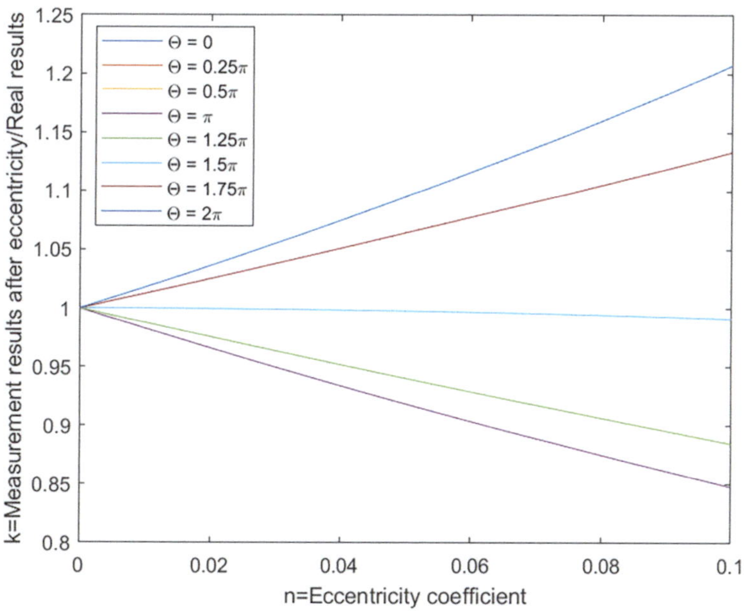

Fig. 31.2 Relationship between result ratio and eccentricity coefficient

Based on the analysis of the graph, when the eccentricity vector is symmetric about the vertical diameter of the transmission line cross-section, the values of R_1 are identical. This results in the two curves overlapping, as observed when $\theta = 0$ and $\theta = 2\pi$.

31.3.5 Fluctuations in Environmental Temperature

The ambient temperature of high-voltage cables fluctuates between −40 °C and 80 °C, while the calibration of various structural parameters is carried out in an environment with a room temperature of 20 °C. Temperature changes will cause thermal expansion and contraction, which will affect the structural parameters of the capacitive voltage divider. Since the thermal expansion coefficient of insulating materials (such as cross-linked polyethylene, etc.) is usually greater than that of metals, in order to ensure the structural stability of the measurement system and limit the impact of thermal expansion and contraction on the measurement effect, the capacitive voltage divider should be designed as a clamp-type structure, as shown in Fig. 31.4. Assume that the linear thermal expansion coefficient of the outer hoop material of the transmission line is α_1, the linear thermal expansion coefficient of the inner barrel material of the transmission line is $\alpha_{2,}$ and the thermal expansion coefficient of the probe material is α_3. The high-voltage arm capacitance becomes:

$$C_{p}\prime = \frac{\pi \varepsilon_1 (r + \alpha_3 r \Delta T)^2}{(R_1 + \alpha_1 R_1 \Delta T)\ln(\frac{R_1+\alpha_1 R_1 \Delta T}{R_2+\alpha_2 R_2 \Delta T})} \tag{31.12}$$

The ratio k of the measured result after the influence of thermal expansion and contraction to the true result is:

$$k = \frac{(r + \alpha_3 r \Delta T)^2 R_1 \ln(\frac{R_1}{R_2})}{r^2 (R_1 + \alpha_1 R_1 \Delta T)\ln(\frac{R_1+\alpha_1 R_1 \Delta T}{R_2+\alpha_2 R_2 \Delta T})} = \frac{(1 + \alpha_3 \Delta T)^2 \ln(\frac{R_1}{R_2})}{(1 + \alpha_1 \Delta T)\ln(\frac{R_1+\alpha_1 R_1 \Delta T}{R_2+\alpha_2 R_2 \Delta T})} \tag{31.13}$$

From the formula, it can be derived that the value of k is closely related to the ratio R_1/R_2. A simulation experiment was conducted using MATLAB, setting five value points for R_1/R_2 ranging from 5/4 to 25/4. The results are shown in Fig. 31.3. The findings indicate that as the value of R_1/R_2 decreases, the influence of thermal expansion and contraction on the results increases.

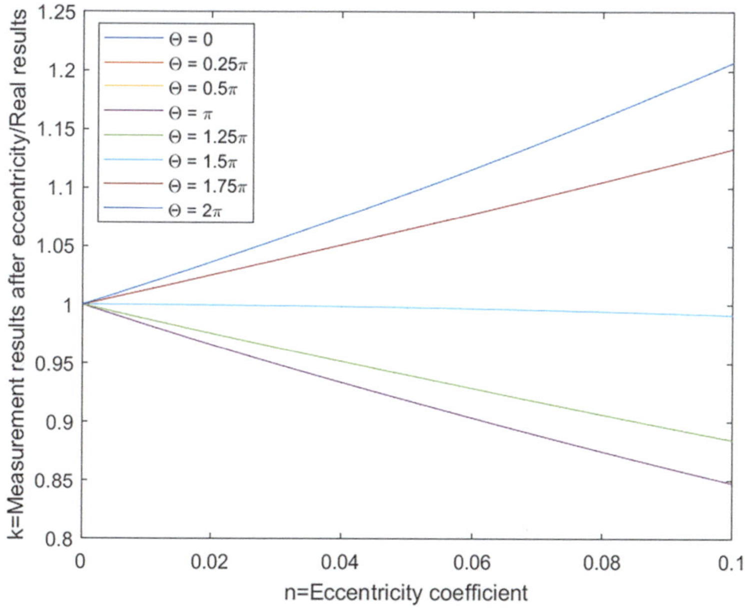

Fig. 31.3 Result ratio and temperature relationship diagram

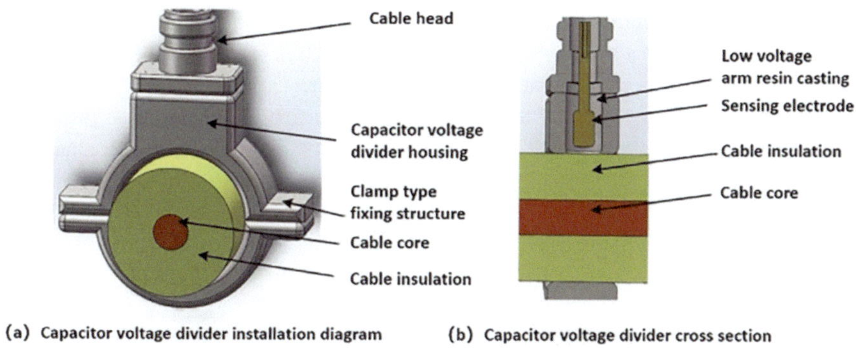

Fig. 31.4 Capacitor voltage divider installation diagram and typical cross-sectional view

31.4 Design and Calibration of Differential Nanosecond Pulse Capacitive Dividers

A capacitive divider has been designed for nanosecond pulse high-voltage measurements in flexible high-voltage cables. A typical flexible high-voltage cable consists of a tightly compressed copper conductor core and a super clean cross-linked polyethylene insulation layer. Depending on the manufacturing process and operating environment, some high-voltage cables may also include additional structural layers, such as semi-conductive non-woven fabrics, tinned copper wire shielding, and polyurethane. In the structural design of the capacitive divider, the complexity of the high-voltage cable is simplified by considering only the copper conductor core and the insulation layer.

Figure 31.2 presents a schematic diagram of the capacitive divider installation and a typical cross sectional view of its structure. The capacitive divider consists of several components, including the cable core, cable insulation layer, clamp type fixing structure, divider housing, sensing electrode, low-voltage arm resin, and cable termination. The overall design features a cylindrical coaxial structure, which offers advantages such as simple processing, high precision, and easy control of structural parameters.

The capacitive divider is fixed using a mounting flange, and the housing is grounded. When transmitting nanosecond pulse high voltage through the high-voltage cable, the cable core is at a high voltage, forming a high-voltage arm capacitance C_p between the sensing electrode of the capacitive divider and the high-voltage cable core. The insulation material used is super clean cross-linked polyethylene, which is the material of the high-voltage coaxial cable insulation layer. A low-voltage arm capacitance C_c is formed between the sensing electrode and the capacitive divider housing (which is grounded), with epoxy resin used for encapsulation. The high-voltage arm capacitance and the low-voltage arm capacitance create a voltage division, and there is a close correlation between the two.

The outer diameter of the insulation layer of the high-voltage cable to be measured is 30 mm, while the diameter of the cable core is 8 mm. The designed differential capacitive divider has a sensing electrode with a diameter of 10 mm and a length of 15 mm. The thickness of the insulation layer of the low-voltage arm is 0.25 mm.

According to the structure of the differential capacitive voltage divider, the high-voltage arm capacitance and the low-voltage arm capacitance can be calculated as follows:

$$C_p = \frac{2\pi\varepsilon_1 L}{\ln(\frac{R_1}{R_2})} \times \frac{\pi r^2}{2\pi R_1 L} = 0.081pF \tag{31.14}$$

$$C_p = \frac{2\pi\varepsilon_2 L}{\ln(\frac{r_1}{r_2})} = 58.152pF \tag{31.15}$$

After calculation, when the structural capacitor is used as the low-voltage arm capacitor, the conditions for forming a differential capacitor voltage divider are met, that is $\frac{1}{C_0} \gg Z_0$. After integration by an integrator with a time constant of, the oscilloscope measured voltage can be obtained:

$$U_s = C_0 Z_0 \frac{U_0}{\tau} \tag{31.16}$$

From the above formula, when paired with an integrator having a time constant of 2 μs, the voltage division ratio of the capacitive divider can be expressed as:

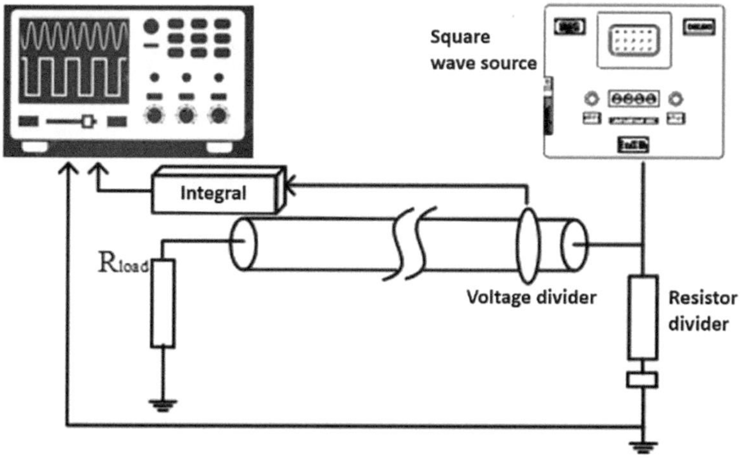

Fig. 31.5 Calibration circuit diagram

$$\beta = U_0/U_s = 4.96 \times 10^5 \quad (31.17)$$

A calibration experimental platform for nanosecond fast rise pulse measurement has been established to calibrate the voltage division ratio of the developed capacitive divider. Figure 31.5 illustrates the schematic diagram of the calibration circuit, where a standard square wave pulse source generates nanosecond fast rise square wave pulses that are fed into the cable under test. The amplitude of the square wave pulse is approximately 3 kV, with a rise time of less than 1 ns. Both a resistive divider and the capacitive divider to be calibrated are used to measure the voltage at the beginning of the cable. A load resistor, whose resistance value is equal to the wave impedance of the cable, is connected to the end of the cable to prevent reflections of the fast pulse signal at the cable's termination.

Figure 31.6 shows the calibration waveform of the capacitive divider. The measured calibration waveform allows for the determination of the voltage division ratio and response time of the divider. After calibration, the voltage division ratio of the developed capacitive divider is found to be \ (5.2 \times 10^5 \), which exhibits a small deviation from the theoretical calculations. The rise response time is approximately 5 ns. Assuming the amplitude of the pulse high voltage to be measured is 1 MV, the output voltage amplitude of the developed capacitive divider is approximately 2 V, which meets the testing requirements.

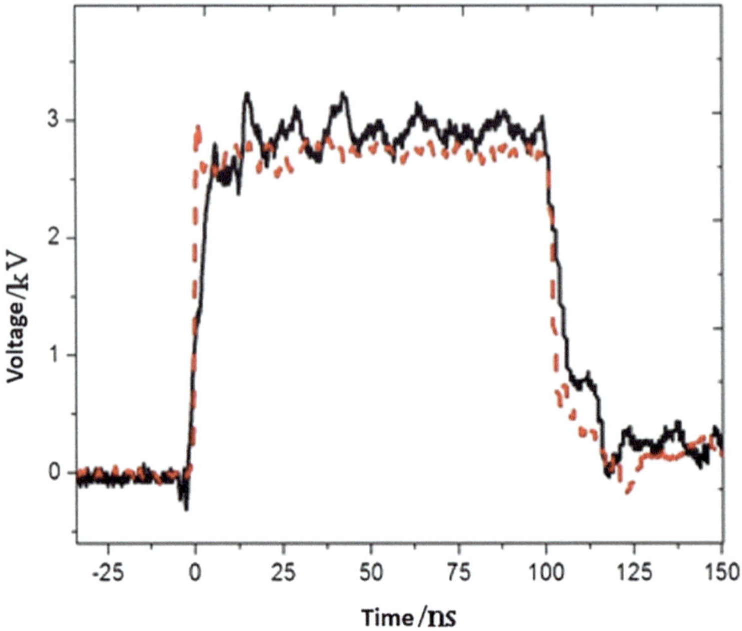

Fig. 31.6 Capacitor voltage divider calibration waveform

31.5 Nanosecond Pulse Testing Platform and Experimental Results

Building on the previous analysis, pressure testing assessments of flexible high-voltage cables under conditions of several hundred kilovolts and nanosecond-level rise time high-voltage electrical pulses were conducted. The experimental circuit is illustrated in Fig. 31.7. A classic Marx generator is used as the pulse source, which outputs high-amplitude, fast-rise electrical pulses through parallel charging and series discharging of multi-stage capacitive switches.

Typical voltage waveforms were obtained, as shown in Fig. 31.8. The capacitive divider accurately captured the electrical pulse waveform, reflecting key characteristics such as the double peaks of the pulse (located at −115 ns and 17 ns, with amplitudes of 88 kV and 457 kV, respectively) and the negative pulse superimposed after the main pulse (at a certain time, with an amplitude of −236 kV). The test results fully validate the reliability of the capacitive divider in nanosecond fast pulse testing.

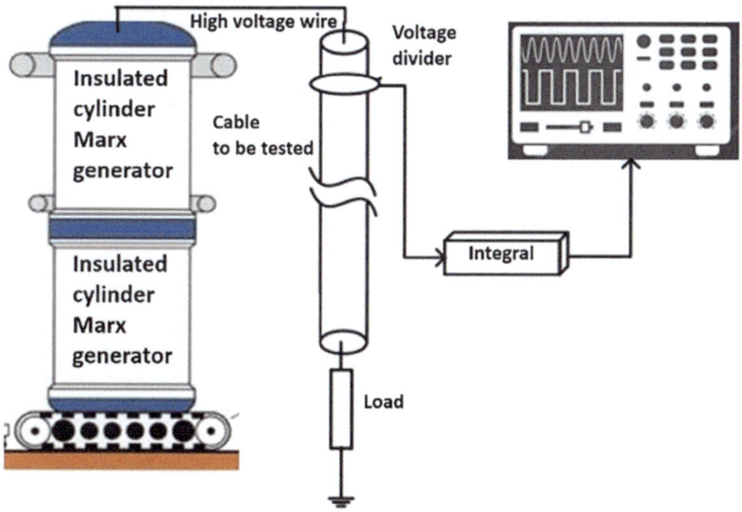

Fig. 31.7 Schematic diagram of coaxial cable withstand voltage test

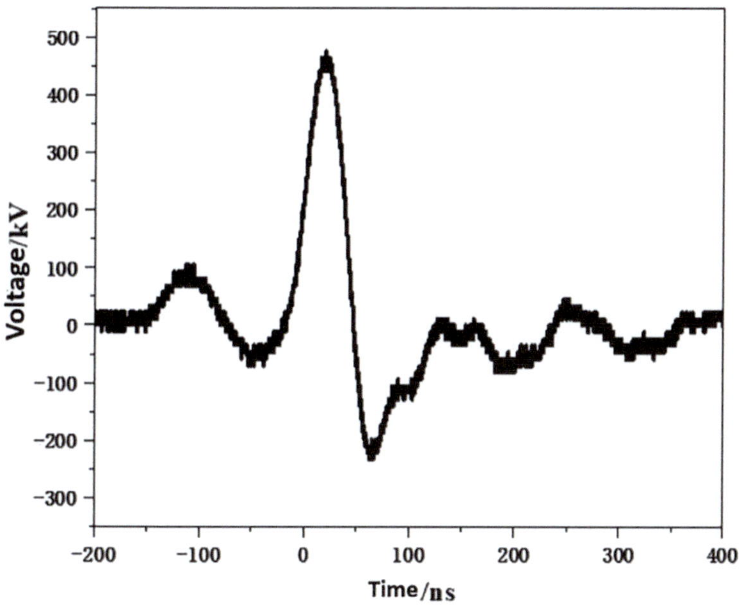

Fig. 31.8 Typical test waveform of capacitor voltage divider

31.6 Conclusion

This paper presents a theoretical analysis of the differential capacitive divider. Through parameter calculations and simulation experiments, a capacitive divider design that meets the requirements for measuring high-voltage electrical pulses in the hundreds of kilovolts range has been developed. The influences of factors such as air gaps in the insulating material of the low-voltage arm capacitor, processing and installation errors of the sampling electrodes on the measurement effectiveness of the differential capacitive divider have been analyzed. Using the capacitive divider developed in this study, kilovolt-level nanosecond pulse calibration tests and withstand voltage assessment experiments for several hundred kilovolts of nanosecond electrical pulses were conducted. Consistent electrical pulse waveforms were obtained, accurately reflecting the electrical parameter status on the high-voltage cable. The results indicate that the divider meets the technical requirements for nanosecond pulse withstand voltage testing of flexible high-voltage cables.

References

1. Yue Zongkun.: Design of 110kV and Above High-Voltage Cable Lines. Rural Electrification (2), 2 (2015).
2. Ye Ting, He Bangle, Li Ling, et al.: Design of a 20 kV High-Voltage Cable Insulation Tester Based on the K7G20 Touch Integrated Machine. Electrical Engineering Technology (08), 168–171+174 (2024).
3. Cheng Bin.: Research on Technology of Online Monitoring System for High-Voltage Cable Insulation. Electromechanical Information (08), 85–88 (2024).
4. Chen Yaxin.: Online Monitoring Technology and Application of High-Voltage Cable Lines. Tianjin University (2014).
5. Wang Kun, Qian Huoyun.: Development and Production of DC High-Voltage Pulse Cables for Linear Induction Accelerators. Proceedings of the 4th China Communications and Optoelectronic Cable Entrepreneurs Summit Forum (2011).
6. Yang Yuxin, Li Sheng, Xing Yaomin, et al.: Research on the Design of a Cable Insulation Diagnosis System Based on High-Frequency Partial Discharge Detection. Journal of Electric Power 35(03), 246–254 (2020).
7. Jia Chao, Wang Jinming, Yan Xifeng, et al.: Testing and Simulation of Insulation Materials for High-Voltage DC Cable Accessories. Electronic Technology & Software Engineering (17), 110–114 (2022).
8. Liu Xiaodong, Liu Peizhen.: Research on AC/DC Resistance Testing of the Buffer Layer in High-Voltage Cables and Insulation Shield Scorching. Electromechanical Engineering Technology 49(11), 142–144 (2020).
9. Wang Jufeng, Zhao Quan, Jin Haijun, et al.: Experimental Research on Countermeasures for Lightning-Induced Voltage in Secondary Cables of Power Systems. Guangxi Electric Power 27(2), 3 (2004).
10. Guan Genzhi, Yuan Congbo, Hu Shixiong.: Experimental Study on Induced Voltage of Secondary Lines by Simulated Lightning Currents and the Shielding Characteristics of Cables. High Voltage Technology 19(3), 19–25 (1993).
11. Le Yanjie, Xuan Yaowei, Yu Enke, et al.: Research on Lightning Impulse Testing of DC Cables Under Different Wiring Configurations. High Voltage Technology, (2015).

12. Cui Qiuju, Chen Guifeng, Peng Xiaoling, et al.: Design of Self-Integrating Capacitive Voltage Divider for Measuring Voltage in Ferrite Transmission Lines. High Voltage Apparatus 60(06), 107–113 (2024).
13. Wei Bing, Li Yong, Kang Junjun, et al.: Capacitive Voltage Divider for Measuring Microsecond-Order Pulse Width Square Wave Signals. High Power Laser and Particle Beams 35(02), 123–128 (2023).
14. Wang Chu, Zhang Lili, Zhang Guoshuai.: Research on the Nonlinear Mechanism of DC Resistance Voltage Dividers. Journal of Metrology 43(04), 521–525 (2022).
15. Wang Kun, Feng Jiaqi, Li Zhiyuan, et al.: Research on High-Voltage Capacitive Compensation Fast Front Pulse Resistive Voltage Dividers [J/OL]. High Voltage Technology, 1–9 (2024-08-15).
16. Lv Wei, Wu Guangning, Zhou Lijun, et al.: Parameter Setting and Simulation of RC-Type High-Voltage Impulse Dividers. Instrumentation Technology (05), 54–56 (2010).
17. Xu Lingjie, Zhou Yongjia, Zhou Qi.: Study on the Frequency Characteristics of Common High-Voltage RC Voltage Dividers. Zhejiang Electric Power 30(07), 13–15 (2011).
18. Liu Jinliang, Huai Wulong, Huo Zhe, et al.: A Capacitive Voltage Divider for Measuring Pulsed High Voltage. High Power Laser and Particle Beams, 2000, 12(1): 3.
19. Wang Meng, Zhang Wenzhao, Wang Ying, et al.: Control Strategies for Transient Overvoltages in Power Systems with High Proportions of Photovoltaic Integration. Journal of Solar Energy 44(10), 148–155 (2023).
20. Wang Weisheng, Li Guanghui, He Guoqing, et al.: Challenges and Prospects of Renewable Energy Grid Integration Control for New Power Systems. New Power Systems 1(02), 145–160 (2023).
21. Chen Gang, Guo Chong, Li Ping, et al.: Review of Research on Improving Insulation Reliability of Ultra/Extra High Voltage GIS [J/OL]. High Voltage Apparatus, 1–16 (2024-08-15).
22. Hou Shiying, Luo Ao, Yang Fan, et al.: Research on Steady-State Temperature Rise and Reduced-Order Model of Large Capacity GIS for Digital Twin Applications [J/OL]. High Voltage Technology, 1–11 (2024-08-15).
23. Huang Kunpeng, Kang Liutao, Zhao Ping.: Research and Design of Lightweight 126 kV GIS Equipment. High Voltage Apparatus 60(06), 89–98+106 (2024).
24. Shu Y, Chen W, Li Z, et al.: Experimental Research on Very-Fast Transient Overvoltage in 1100-kV Gas-Insulated Switchgear. IEEE Transactions on Power Delivery 28(1), 458–466 (2013).
25. Yang Fan, Gao Youhua.: Simulation Analysis of Fast Transient Overvoltages in GIS Under the Mayr Arc Model. Electrical Appliances and Energy Efficiency Management Technology 0(2), 28–32 (2024).
26. Li Chuang, Liao Xingrui, Yu Quanquan, et al.; Analysis of Conductive Interference of VFTO on Electronic Voltage Transformers and Research on Suppression Measures. Hunan Electric Power 44(02), 50–59 (2024).
27. Wang Yu, Xu Weidong, Lou Jie.; Simulation Study on VFTO Suppression Measures. Electrical Technology and Economy (02), 20–23 (2024).
28. Huiskamp T, Beckers F, Heesch E, et al.; B-Dot and D-Dot Sensors for (Sub) Nanosecond High-Voltage and High-Current Pulse Measurements. IEEE Sensors Journal 16(10), 3792–3801 (2016).
29. Novac B M, Xiao R, Huiskamp T, et al.; Theoretical and experimental studies of off-the-shelf v-dot probes. IEEE Transactions on Plasma science 46(8), 2985–2992 (2018).
30. Wei Bing, Fu Zhen, Wang Yujuan, et al.; Design and Application of Capacitive Voltage Dividers in Pulse Power Devices. High Voltage Technology (12), 39–43 (2007).
31. Wei Bing, Gu Yuanchao, Zhou Rongguo, et al.; Design, Calibration, and Experiment of D-Dot for Water Transmission Line in Linear Accelerators. High Power Laser and Particle Beams (05), 830–834 (2007).

Chapter 32
Seismic Performance Study of I-shaped Section Composite Column Insulators

Songsong Zhou, Le Li, Shuaichao Zhao, Yuqiao Yi, Tianfang Zhao, Jun Zhou, and Weixing Xu

Abstract This paper comprehensively applies seismic load calculations, finite element simulation, and modal analysis to explore the seismic performance of I-shaped and circular section column insulators under seismic dynamic loads. By comparing and analyzing, the study reveals the performance differences of insulators with different cross-sectional shapes in seismic de-sign, providing a scientific basis for the external insulation design of high-voltage DC converter stations.

Keywords I-shaped column insulators · Finite element simulation · Seismic performance · Modal analysis

32.1 Introduction

As the global energy structure transitions and new energy sources rapidly develop, ultra-high voltage (UHV) direct current transmission has become a key technology linking distant clean energy sources with consumer markets [1–3]. In this context, the external insulation design of UHV DC converter stations is crucial, directly affecting the safe and reliable operation of the entire power system [4, 5]. Column composite insulators, serving as vital components supporting flat wave reactors, disconnect switches, filter capacitors, and busbars, are extensively used in UHV converter stations [6–8].

Traditional research has primarily focused on the seismic performance analysis of solid or hollow insulators. International scholars, Selim Günay and Khalid M.

S. Zhou (✉) · J. Zhou
China Electric Power Research Institute, Beijing, China
e-mail: z18516979216@163.com

L. Li · Y. Yi · T. Zhao
North China Electric Power University, Baoding, China

S. Zhao · W. Xu
Henan Pinggao Electric Co., Ltd, Pinggao Henan, China

Y. Li et al. (eds.), *Proceedings of the 5th Chinese Conference on High Voltage and Discharge Plasmas*, Springer Proceedings in Physics 254,
https://doi.org/10.1007/978-981-96-5711-7_32

Mosalam [9], have specifically examined how the damping and stiffness of support structures affect the response of disconnect switches equipped with ceramic or composite column insulators under seismic loads. Sheng Li and others [10] have assessed the seismic performance of retrofitted composite insulators through full-scale shake table tests, confirming the safety performance of 1000 kV surge arresters under extreme seismic conditions and ensuring that the newly designed equipment is suitable for installation in any seismic risk area in China. Domestic researchers led by Gao Song [11] have explored the structural design and mechanical properties of new flat wave reactor composite column insulators through simulation analysis and experimental verification, particularly their performance under seismic loads. Li Yupeng and others [12] have conducted comprehensive studies on the design and seismic performance of new 126 kV disconnect switches through simulation and experimental methods, verifying their structural stability and insulation performance under extreme seismic conditions.

Despite extensive research, brittle failures of solid ceramic and composite column insulators have still been recorded in laboratory tests [13] and seismic monitoring [14–16]. Current research often overlooks the response of large-diameter composite insulators under high-intensity seismic impacts. Moreover, as the power grid evolves into an energy internet, the performance requirements for insulators continue to increase, underscoring the urgent need for research on their seismic design.

This paper introduces a type of I-shaped section composite column insulator and analyzes its seismic performance in a simulated seismic environment through finite element simulation. The study focuses on evaluating the different responses of I-shaped and circular section insulators to seismic dynamic loads.

32.2 Seismic Performance Analysis of Column Insulators

Common methods for analyzing seismic performance include the static method, response spectrum method, and dynamic time history method. This section utilizes the acceleration time history method to compare the seismic performance of I-shaped and circular section column insulators. The dynamic time history method, a commonly used frequency domain analysis method, is suitable for calculating the response of structures under seismic actions [17]. In this study, preliminary analysis will be conducted using the response spectrum method, which calculates the maximum response of each mode of the structure under a given load spectrum based on the earthquake response spectrum. By aggregating the maximum responses of each mode, the overall response of the structure can be obtained. Subsequently, a detailed analysis will be performed using the acceleration time history method, which will provide deep insights into the dynamic performance of I-shaped and circular section column insulators in actual earthquakes and compare their performance differences.

32.2.1 Finite Element Model of Column Insulators

The simplified ± 800kV column insulator finite element model is shown in Fig. 32.1. The column insulator consists of a core rod and flange, with the upper end of the core rod measuring 5530mm in length and the lower end 5590mm, and the flange baseplate thickness is 20mm. Solid elements are used for simulation, and binding constraints are applied between flanges, with the degrees of freedom at the bottom of the column insulator fixed.

The material properties of each component are presented in Table 32.1.

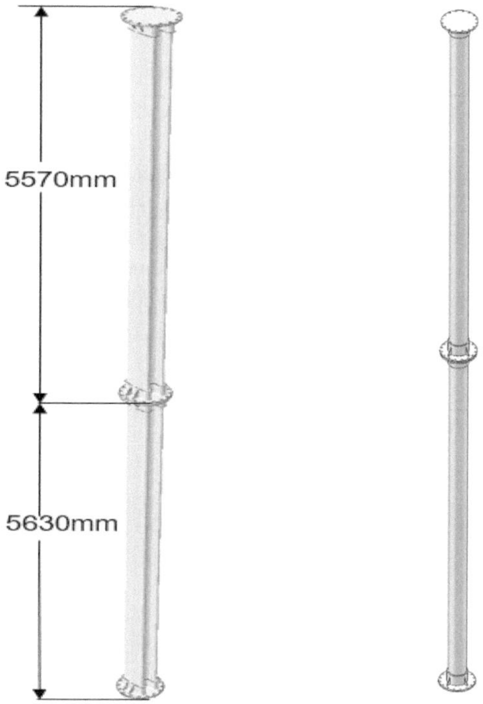

(a) I-shaped section column insulator **(b)** Circular section column insulator

Fig. 32.1 Finite element model of ± 800kv column insulators

Table 32.1 Material properties of each component

Material	Density (g/cm^3)	Elastic modulus (GPa)	Poisson's ratio
Fiberglass core rod	2.2	40	0.33
Metal flange	7.85	200	0.3

32.2.2 Calculation of Seismic Loads

The seismic influence coefficient is crucial for assessing the response of structures under seismic actions. This section will provide a detailed description of the seismic performance of the I-shaped section column insulator. According to relevant standard requirements, earthquake waveforms that meet the standards are selected, consisting of four segments as shown in Fig. 32.2: the linear rising segment, horizontal segment, curved descending segment, and linear descending segment. The calculation formula is shown in Eq. 32.1.

$$\alpha = \begin{cases} \left[0.40 + \frac{-0.40}{0.1}T\right]\alpha_{\max} & 0 \le T < 0.1 \\ \eta_2\alpha_{\max} & 0.1 \le T < T_g \\ \left(\frac{T_g}{T}\right)^{\gamma}\eta_2\alpha_{\max} & T_g \le T < 5T_g \\ \left[\eta_2 0.2^{\gamma} - \eta_1\left(T - 5T_g\right)\right] & 5T_g \le T \le 6.0 \end{cases} \quad (32.1)$$

where in formula (32.1): α—Seismic influence coefficient; α_{max}—Maximum seismic influence coefficient; T_g—Characteristic period; η_1—Linear Decline Section's Decline Slope Adjustment Coefficient; η_2—Damping Adjustment Coefficient.

The seismic fortification intensity in this study is set at level 8, with the characteristic period T_g typically being 0.4 s. The coefficient is adjusted by $\xi = 0.05$, resulting in an increase in the maximum seismic influence coefficient α_{max} by 1. The damping factor γ is initially estimated from Eq. (32.2) as 0.9. This factor is integral in modifying the seismic influence coefficient α by an additional factor of 0.02, as outlined in Eq. (32.3), concluding with an adjustment factor η_2 of 1.0 according to Eq. (32.4).

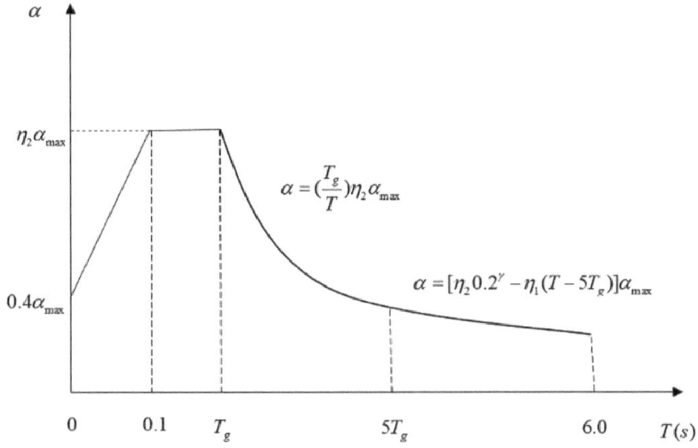

Fig. 32.2 Seismic coefficient influence curve

Table 32.2 Frequencies of the first four modes (Hz)

Mode	Circular section	I-shaped section
1	3.2823	1.6590
2	3.2825	1.9886
3	18.931	18.046
4	18.934	20.201

$$\gamma = 0.9 + \frac{0.05 - \xi}{0.3 + 6\xi} \tag{32.2}$$

$$\eta_1 = 0.02 + \frac{0.05 - \xi}{4 + 32\xi} \tag{32.3}$$

$$\eta_2 = 1 + \frac{0.05 - \xi}{0.08 + 1.6\xi} \tag{32.4}$$

32.2.3 Results of Seismic Analysis

Results of Modal Analysis

Modal analysis was performed on column insulators of both sectional types, obtaining the frequencies and mode shapes of each model. Table 32.2 displays the frequencies of the first four modal shapes of the circular section and I-shaped section column insulators. Figures 32.3 and 32.4 depict the mode shape contour plots of the column insulators.

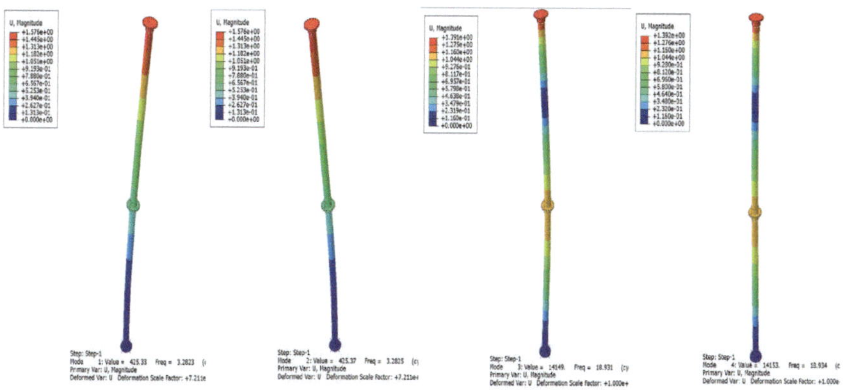

(a) First mode **(b)** Second mode **(c)** Third mode **(d)** Fourth mode

Fig. 32.3 Mode shape contour plots of circular section column insulator

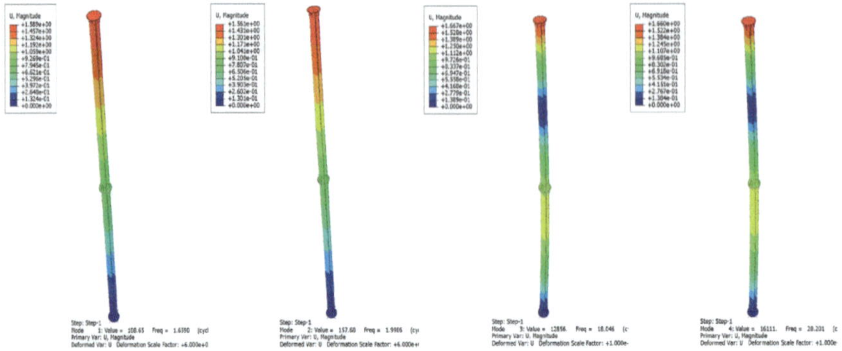

(a) First mode (b) Second mode (c) Third mode (d) Fourth mode

Fig. 32.4 Mode shape contour plots of I-shaped section column insulator

The frequencies of the first and second modes of the circular section are approximately 3.28Hz, while for the I-shaped section column insulator, the first mode is at 1.6590 Hz and the second mode at 1.9886 Hz, indicating that the I-shaped insulator is more likely to experience significant amplitude changes at the lowest frequencies. For the third and fourth modes, the frequencies for the circular section are 18.931 Hz and 18.934 Hz, respectively, whereas the I-shaped section displays frequencies of 18.046 Hz and 20.201 Hz. This shows that at higher modes, the frequencies of the I-shaped section surpass those of the circular section, which may indicate higher rigidity in specific vibrational modes of the I-shaped structure. Further analysis of the model stress under the acceleration time history spectrum is needed.

Results of Response Spectrum Analysis

The analysis was conducted on ± 800kV column insulators using the acceleration time history response spectrum calculated in Sect. 2.1 as the excitation. The seismic simulation results are displayed in Table 32.3, with corresponding contour plots in Figs. 32.5 and 32.6.

The seismic simulation results for the ± 800kV column insulators, as depicted in Table 32.3, are based on a detailed analysis of their seismic performance, comparing the stress and displacement responses of circular and I-shaped section column insulators under the same seismic excitation. As indicated by the simulation outcomes, the I-shaped insulators demonstrated superior seismic performance in terms of maximum stress and displacement compared to their circular counterparts. Specifically, the

Table 32.3 Seismic simulation results of ± 800kv column insulators

Cross-section type	Maximum flange stress (Mpa)	Maximum core rod stress (Mpa)	Maximum insulator displacement (mm)
Circular	556.5	122.4	43.02
I-shaped	402.7	69.49	27.17

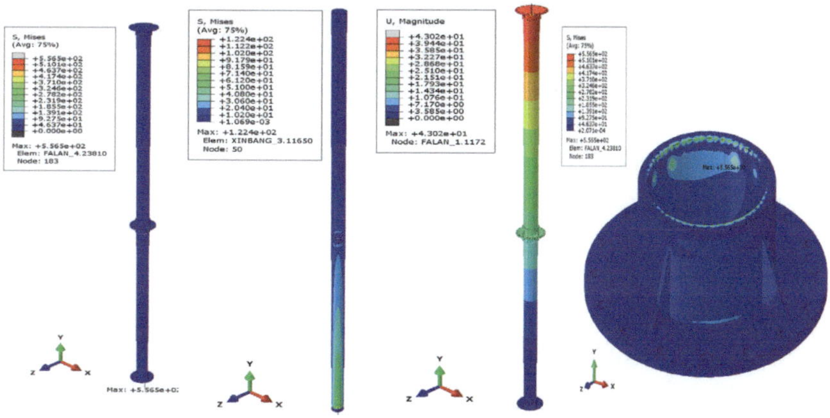

(a) Overall stress contour map **(b)** Column stress contour map **(c)** Overall displacement contour map **(d)** Flange stress contour map

Fig. 32.5 Seismic response of circular section column insulator

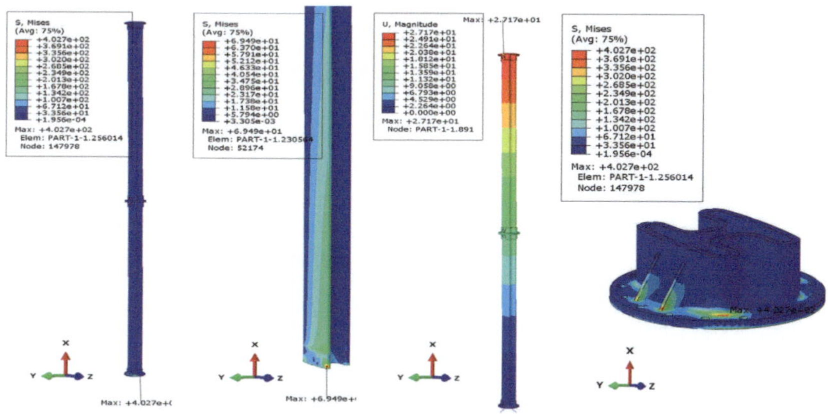

(a) Overall stress contour map **(b)** Column stress contour map **(c)** Overall displacement contour map **(d)** Flange stress contour map

Fig. 32.6 Seismic response of I-shaped section column insulator

maximum flange stress for the I-shaped insulators was 402.7 MPa, significantly lower than the 556.5 MPa observed for the circular insulators; similarly, the maximum core rod stress and maximum insulator displacement were also lower, at 69.49 MPa and 27.17 mm respectively. Such differences may stem from structural advantages inherent to the I-shaped section, whose unique shape provides enhanced mechanical properties and seismic capabilities. These findings underscore the importance of cross-sectional shape in enhancing the stability and safety of column insulators during seismic events.

32.3 Conclusions

In this study, detailed modal analysis and seismic simulations of I-shaped and circular section column insulators were conducted, leading to the following conclusions:

(1) I-shaped insulators exhibit larger amplitude variations at lower modal frequencies, while at higher modes, they show higher frequencies compared to circular sections, indicating greater structural rigidity in higher modes.
(2) In seismic simulations, I-shaped insulators performed better in terms of maximum stress and maximum displacement compared to circular sections, reflecting the potential advantages of the I-shaped section in seismic design. The structural shape contributes to improved overall stability and safety during earthquakes.

Acknowledgements This work is supported by the State Grid Corporation of China Headquarters Management Technology Project Funding (No. 5500-202255114A-1-1-ZN)).

References

1. Shu YB, Zhang ZL, Tang Y, et al.: Fundamental Issues of New-type Power System Construction. Proceedings of the CSEE 44(21), 8327–8341 (2024).
2. Zhuo ZY, Zhang N, Xie XR, et al.: Key Technologies and Developing Challenges of Power System with High Proportion of Renewable Energy. Automation of Electric Power Systems 45(9), 171–191 (2021).
3. Li H, Liu D, Yao D.: Analysis and Reflection on the Development of Power System Towards the Goal of Carbon Emission Peak and Carbon Neutrality. Proceedings of the CSEE 41(18), 6245–6259 (2021).
4. Li P, Gu C, Chen D, et al.: Development of Technologies in ±1500 kV UHV DC Transmission Research. High Voltage Technology 43(10), 3139–3148 (2017).
5. Gu C, Li P, He HW, et al.: Research on +1100kV and Above UHV DC Transmission Technologies. Proceedings of the CSEE 40(20), 6745–6754 (2020).
6. Sun BX, Cao MG.: Aseismic Design and Capability Assessment of DC Yard Equipment for UHV Xinsong Converter Station. High Voltage Electrical Appliances 54(12), 24–35 (2018).
7. Sun YH, Cheng YJ, Lu ZC, et al.: Experimental Research on Seismic Performance of UHV GIS Porcelain Bushing and Composite Bushing. High Voltage Technology 45(02), 541–548 (2019).
8. Zhang Q, Zhao WS, Li AR.: Status of Internal Insulating Mandrel for Insulators of UHV Composite Struts. Fiber Reinforced Plastics/Composites 0(9), 102–105 (2018).
9. uenay S, Mosalam MK.: Seismic Performance Evaluation of High-Voltage Disconnect Switches Using Real-Time Hybrid Simulation: II. Parametric Study. Earthquake Engineering & Structural Dynamics 43(8), 1223–1237 (2014).

10. Li, S, et al.: Full-scale testing on seismic performance of surge arrester with retrofitted composite insulators. Buildings 12(10), 1720 (2022).
11. Gao S, Gao G, Liu Y, et al.: Mechanical Properties of Composite Post Insulator for A New Type Smoothing Reactor. Insulation Material 53(06), 77–83 (2020).
12. Li YP, Guo HT, Li FY, et al.: Research on Design and Seismic Performance of New 126 kV Disconnector. High Voltage Electrical Appliances 58(10), 57–66 (2022).
13. Mosalam, K. M., et al.: Seismic performance of substation insulator posts for vertical-break disconnect switches. California Energy Commission Publication Number: CEC-500–2012 (2012).
14. Kumosa M, Kumosa L, Armentrout D.: Causes and potential remedies of brittle fracture failure of composite (nonceramic) insulators. IEEE transactions on dielectrics and electrical insulation, 11(6), 1037-1048(2004).
15. Kumosa M, Kumosa L, Armentrout D.: Failure analyses of nonceramic insulators. Part 1: Brittle fracture characteristics. IEEE Electrical Insulation Magazine, 21(3), 14–27(2005).
16. Kumosa M, Kumosa L, Armentrout D.: Failure analyses of nonceramic insulators: part II-the brittle fracture model and failure prevention. IEEE Electrical Insulation Magazine, 21(4), 28-41(2005).
17. Wu XF, Sun QG, Di JJ, etc.: A Numerical Simulation Comparison between Response Spectrum Analysis and Time His-tory Analysis. Northwest Seismological Journal, 33(03), 275–278+304(2011).

Chapter 33
A Self-Integrating Pulse Capacitive High-Voltage Divider Using a Simple Compensation Circuit

Xiaokuo Kou, Pei Guo, Donglei Zhang, and Tianyu Lin

Abstract To carry out experimental research on the breakdown characteristics of insulation materials to withstand system overvoltage, the characteristics of the breakdown voltage waveform should be observed and recorded, and a pulse capacitive divider with good dynamic response is a guarantee of the accuracy of the experimental data. A simple structure of low voltage arm compensation circuit is proposed for the waveform of breakdown process, and a self-integrating capacitive voltage divider is designed to take into account both high-frequency and low-frequency characteristics. By analyzing the self-integrating conditions and matching methods, the two-stage voltage division matching is combined with compensation matching, and the parameters of the compensation circuit are verified experimentally. The designed capacitive voltage divider has a high-voltage arm of 5pF, a low-voltage arm of 150pF, and a compensation resistance of 10kΩ. Through the step-response test, the capacitive voltage divider has a partial response time of 7.2 ns, which meets the demand of fast frontier test; after the whole casting, the capacitive voltage divider has a scale factor of 6389.26, which is in good consistency, and its linearity deviation is less than 1%.

Keywords Pulse voltage · Capacitive voltage divider · Match method · Compensation circuit; Step response

X. Kou · P. Guo
State Grid Henan Electric Power Company, Zhengzhou, China

D. Zhang · T. Lin (✉)
China Electric Power Research Institute Co, Ltd, Beijing, China
e-mail: ltyyy0927@163.com

Y. Li et al. (eds.), *Proceedings of the 5th Chinese Conference on High Voltage and Discharge Plasmas*, Springer Proceedings in Physics 254,
https://doi.org/10.1007/978-981-96-5711-7_33

33.1 Introduction

With the continuous expansion of power grid construction, the voltage level is increasing, the phenomenon of system overvoltage caused by the destruction of electrical equipment insulation occurs from time to time, so there is a need to continuously optimize the insulation design and improve the working field strength [1–3]. In order to meet this need, domestic and foreign scholars use the pulse voltage generator to simulate overvoltage, extensive re-search on breakdown characteristics of insulation materials have been conducted [4–6], pulse capacitive voltage divider matching the load carrying capacity and structure of the pulse voltage generator was developed [7–9]. Since the breakdown of insulation materials caused by overvoltage is often steep, with high amplitude and accompanied by oscillations, it is difficult to accurately record the waveform characteristics of the breakdown process, and the process from the application of excitation to breakdown is characterized by both long duration and short breakdown instant, which requires that the capacitive divider needs to take into account the low-frequency characteristics and high-frequency characteristics, and thus puts forward a higher requirement for the design of the pulsed capacitance voltage divider.

In this paper, overvoltage waveforms are simulated by a pulsed voltage source based on the closed-core pulsed transformer, the schematic diagram of the main discharge circuit is shown in Fig. 33.1, the insulation material specimen and load capacitance Cs are connected in parallel to withstand the microsecond pulse voltage, so the load capacitance is also used as a capacitive divider to measure the microsecond pulse voltage signals applied to the insulation material specimen, typical waveforms are: without breakdown the maximum irradiated value of 100 kV, wave front of tens of microseconds and du-ration of about 100 microseconds; with breakdown the maximum irradiated value of 100 kV, wave front of tens of microseconds and tens of nanoseconds at the falling edge at the end of the wave.

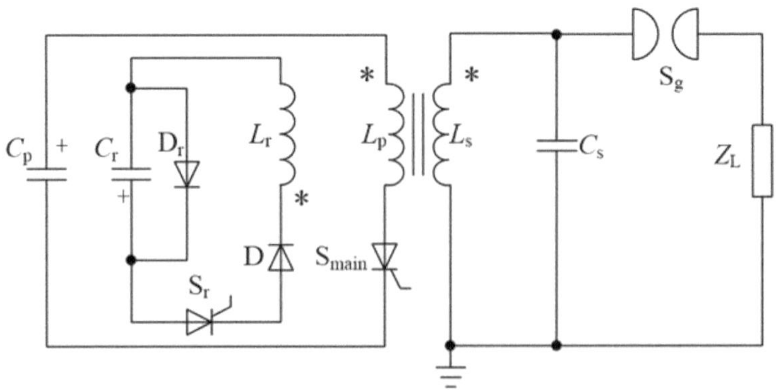

Fig. 33.1 Discharge circuit schematic of the pulsed voltage source

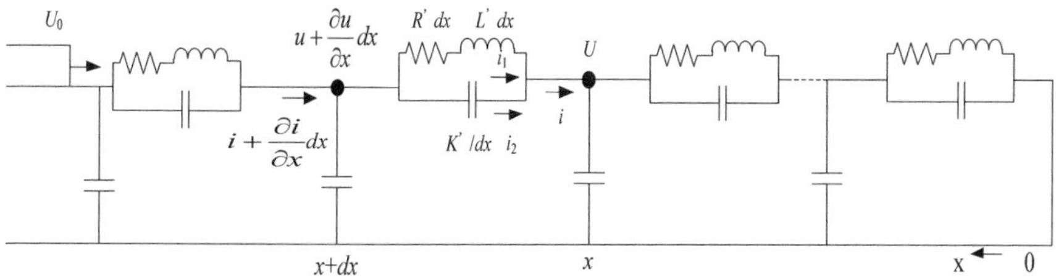

Fig. 33.2 Equivalent circuit of capacitive voltage divider distribution parameters

Let the total height of the capacitive divider be *l*, *R*', *L*', *K*', and C_p' are the insulation resistance per unit length, residual inductance, longitudinal capacitance, and distributed capacitance to ground, respectively. *R*, *L*, *K*, and C_p are then the total insulation resistance, residual inductance, and distributed capacitance to ground of the capacitive divider. If a unit square wave voltage U_0 is applied to the high-voltage side of the capacitive divider and the capacitance value of the capacitive divider is *C*, the capacitive divider can be represented by the equivalent circuit of the distribution parameters in Fig. 33.2 [10, 11].

The voltage output equation of the voltage divider built from the distributive parameter model of the voltage divider is shown in Eq. (33.1) [12] after approximating the high order term:

$$u_i(t) = u_0(t)\frac{x}{l}(1 - \frac{1}{6}\frac{C_p}{C + K}) \quad (33.1)$$

It can be seen that when the inductance is not taken into account, the output voltage of the capacitive voltage divider is independent of the frequency, and the error in the spoke value is determined by the stray capacitance to ground and the longitudinal stray capacitance. In practice, due to the inevitable existence of high-voltage leads, and the capacitance of the voltage divider itself is much larger than the stray capacitance, while the insulation resistance of the capacitor is also very large, so that the measurement waveforms appear to oscillate, and the response of steep wave is poor. This paper focuses on the waveform characteristics: a long duration, generally dozens or even hundreds of microseconds; with breakdown the waveform drop time is very short, generally tens of nanoseconds; totally speaking the waveform is characterized by a large span of waveform duration and rising (or falling) edge time, so it is necessary to take into account the characteristics both of the low-frequency response characteristics and high-frequency response characteristics. Not only is it necessary to improve the response characteristics of the capacitive divider when measuring steep waves, but also simultaneously satisfy the self-integrating conditions when measuring hundred-microsecond waveforms. Aiming at the measurement of nanosecond frontier voltage waveforms, scholars at home and abroad have carried out a large number of studies [13–15], when the measured signal pulse width is tens of nanoseconds or less, the use of small-capacity voltage dividers can measure fast frontal pulses and the self-integration condition is well satisfied, however, when the signal has a steep leading or falling edge and a long pulse width, experimental studies have shown that a capacitive voltage divider [16] needs to be used in conjunction with a matched integrator, making the design of a capacitive voltage divider complex and compromising the high-frequency characteristics. In addition, the Tektronix P6015A High Voltage Probe meets the requirements for voltage waveform measurements during breakdown, but has an upper limit of 40kV for pulse voltage spokes, R.E. Beverly [16] improved the withstand voltage of the P6015A by replacing the insulation, but the overall cost was high and the compensation circuit was complex. In this paper, a simple structure of low-voltage arm compensation circuit is proposed, and based on it, a self-integrating capacitive voltage divider with withstanding voltage not less than 90kV is designed, so that the measurement system can take into account both low-frequency and high-frequency characteristics, and completely and accurately record the sudden change of voltage before and after the breakdown of insulation materials.

33.2 Compensation Circuit Design

33.2.1 Matching Mode Analysis

The main factor affecting the low-frequency characteristics of the voltage divider is the time constant obtained by multiplying the equivalent sampling resistor with the capacitance of the low-voltage arm; the high-frequency characteristics are mainly affected by the stray inductance of the capacitor and the stray capacitance of the sampling resistor. Increasing the time constant to extend the low-frequency characteristics results in a deterioration of the high-frequency characteristics of the voltage divider due to an increase in the effect of the spurious parameter. First of all, to ensure that the voltage divider has sufficient high frequency response characteristics, the production of low-voltage arm should be as low as possible to select the capacitance value of the capacitor; secondly, the capacitor parameters are selected to satisfy the time constant requirement to ensure the low frequency response of the voltage divider. In order to make the capacitive voltage divider can work in the self-integrating mode, its time constant must be much larger than the duration of the measured waveform, generally take 5 to 10 times the order of magnitude [17]. Due to the measured signal duration of tens of microseconds, in order to meet the above self-integrating conditions, it is necessary to choose a capacitor with a large capacitance, which is contrary to meet the requirements of the high-frequency response characteristics of the voltage divider, in this paper the compensation circuit is put forward to solve this problem.

Meanwhile, for the measurement of pulsed voltage signals, the wave process of the signal transmission line must be considered, if no methods are taken, the signal will be reflected at both ends of the transmission line, affecting the quality of the

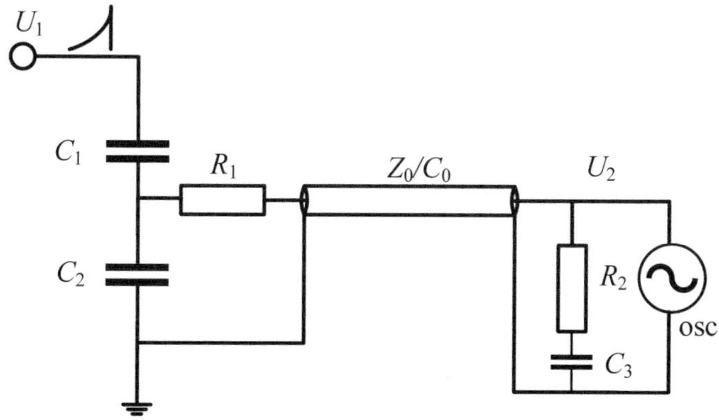

Fig. 33.3 The circuit topology of capacitive voltage divider with transmission line matching method

measurement waveform. The matching circuit topology of the capacitive voltage divider is shown in Fig. 33.3, in which C_1 and C_2 are the high-voltage arm and low-voltage arm capacitance of the capacitive voltage divider, respectively, and the characteristic impedance and the unit equivalent capacitance of the transmission line are Z_0 and C_0, respectively, and according to the different configuration strategies of R_1, R_2, and C_3, the matching method can be divided into the first-end matching, two-stage divider matching, and the compensated matching [18].

When the first-end matching method is used ($R_1 = Z_0$, R_2 and C_3 are 0), the end of the transmission line is connected to the oscilloscope, and its input impedance is usually 1 MΩ. Since the transmission line can be regarded as a capacitance under the condition of low frequency, the scale factor is $(C_1 + C_2 + C)/C_1$ when the steady state is reached. If the transmission line is shorter, and the signal transmission time is much shorter than the signal duration time, and the reflected wave can be superimposed on the header, the time constant is $C_2(Z_0 + R_{in})$; if the transmission line is longer, and the signal is equivalent to the input of an infinitely long transmission line before the reflected signal returns to the first end of the transmission line, the time constant is $C_2(Z_0 + R_1)$. It can be seen that the measurement signal is affected by the length of the transmission line when the first-end matching, and when it comes to the output signal of the voltage divider, after twice the transmission delay of the transmission line, the signal reflected back to the transmission line again back to the end of the transmission line, resulting in the output signal waveform distortion, so the first-end matching method in the case of long cables is not applicable.

When using two-stage divider matching method ($R_2 = Z_0$, $C_3 = 0$), the signal will not be reflected at the end of the transmission line under the fast pulse, R_1 and Z_0 constitute the second stage of voltage divider, and its scale factor is $(R_1 + Z_0)/Z_0$. When the measurement system reaches the steady state, the scale factor is $(C_1 + C_2 + C)(R_1 + Z_0)/(C_1Z_0)$, and the time constant is $C_2(Z_0 + R_1)$, which will not be reflected at the end of transmission line.

When using compensation matching method ($R_1 = R_2 = Z_0$, $C_1 + C_2 = C_3 + C$), in the high-frequency compensation capacitor C_3 is equivalent to a short-circuit, the signal is transmitted to the end of the transmission line does not produce refracted waves, at the steady state the scale factor is $(C_1 + C_2 + C_3 + C)/C_1$, equal to the initial scale factor, and high-frequency scale factor and low-frequency scale factor is the same as the time constant are $C_2(Z_0 + R_1)$. In order to balance the low-frequency characteristics with the high-frequency characteristics, while the parameters are chosen so that the capacitive voltage divider can operate under self-integrating conditions, in this paper, two-stage voltage divider matching is combined with compensation matching to ensure the high-frequency characteristics of the capacitive voltage divider due to the low-voltage arm of small capacitance value, and at the same time, the time constant of the capacitive voltage divider can be effectively improved by enhancing the compensation resistor at the first end to enhance the low-frequency characteristics of the capacitive voltage divider.

33.2.2 Verification Test of Compensation Circuit

In the compensation circuit, in order to make the end of the signal cable not to produce fold reflection, the end matching resistor $R_2 = Z_0 = 50\Omega$, the compensation capacitance meets $C_1 + C_2 = C_3 + C$, and the time constant and scale factor are adjusted by changing the first end compensation resistor R_1.

$R_1 = Z_0$ is usually selected in previous research work [13]. The characteristic impedance of the signal transmission line used in the experiment is 50Ω, so the first end of the compensation resistance is 50Ω. The high-voltage arm of the capacitance divider is composed of four 20pF high-voltage ceramic capacitors in series, with a capacitance value of 5pF, and the low-voltage arm is composed of five 30pF thin-film low-inductance capacitors in parallel, with a capacitance value of 150pF. The voltage divider is used to measure the breakdown voltage signal with the voltage amplitude of 40 kV and the duration of the 20 μs. The voltage waveform is shown as the red curve in Fig. 33.4a, and compared with the standard Tektronix P6015A measurement results, as shown by the black curve, it can be seen that the measured waveform of the pulse voltage signal shows obvious aberration in the falling edge stage of the breakdown process when the matching resistor at the first end is 50Ω.

When the first end resistance is adjusted to 500 Ω, the measured waveform shown in Fig. 33.4b shows that the distortion of the original breakdown process no longer appears at this time, however, the time constant does not satisfy the self-integrating condition due to the small compensation resistance, and the measured waveform is distorted in terms of the expression of the voltage duration; when increasing the first end resistance to 10kΩ, the consistency is good as the measured waveforms of the voltage divider are compared with the P6015A shown in Fig. 33.4c; When continuing to increase the first end resistance to 50kΩ shown in Fig. 33.4d, at this time the falling edge of the measured waveform has been slowed down, indicating that under this state the step response characteristics of the voltage divider has become worse, and

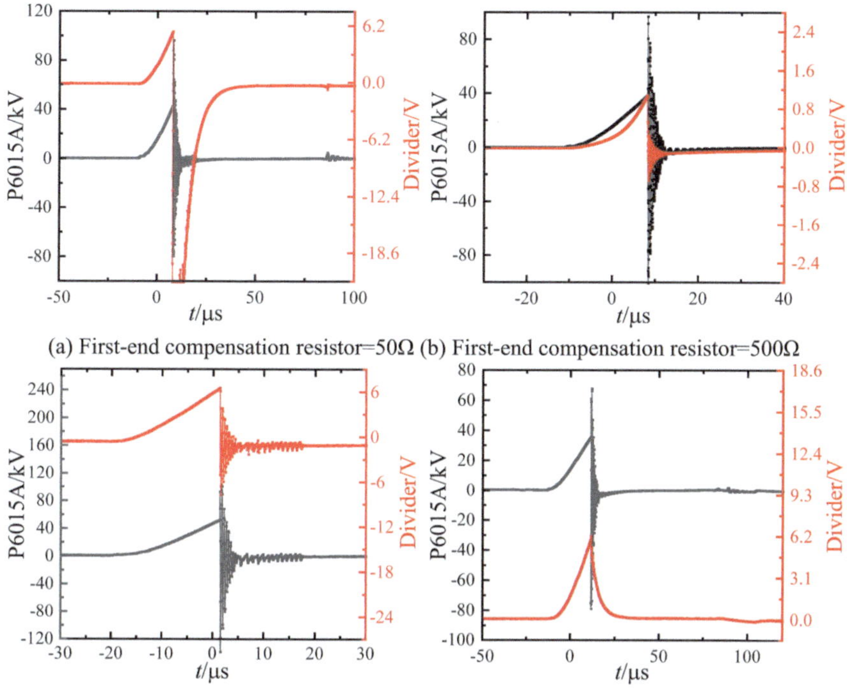

(a) First-end compensation resistor=50Ω (b) First-end compensation resistor=500Ω

(c) First-end compensation resistor=10kΩ (d) First-end compensation resistor=50kΩ

Fig. 33.4 Influence of the compensation resistance in the compensation circuit on the waveform of voltage measurement

it can't satisfy the measurement requirements of tens of nanoseconds level of fast signals.

The compensation resistor was finally determined to be 10kΩ and the pulse voltage signal is recorded. It can be seen from Fig. 33.5 that the measured waveforms compare well with Tektronix P6015A.

It is worth noting that the use of P6015A, there is an overshoot when the voltage waveform starts to fall and the oscillating spokes are larger, This is due to the introduction of additional high-voltage leads when the high-voltage probe is connected in parallel at both ends of the load capacitor, which makes the voltage measurement waveform distortion, and at the same time illustrates the necessity of using a pulsed-voltage-source load capacitor that also serves as a voltage divider, or else it would be necessary to connect a series of damping resistors to the high-voltage end of the connected sensors, which would be detrimental to the response characteristics of the voltage divider.

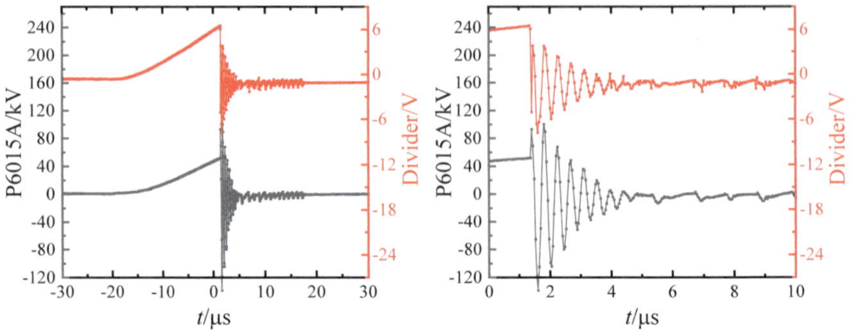

Fig. 33.5 Verification test of voltage measurement

33.3 Performance Test

33.3.1 Step Response

In order to perform a step response test of the capacitive voltage divider, a square wave source is used for calibration. Normally, the pulse voltage measurement system consists of a voltage divider, a digital oscilloscope, a signal transmission line, and a high-voltage lead between the voltage divider and the test divider (when fitted with a damping resistor r, the measurement system should be viewed as a cascade of two or three networks), as shown in Fig. 33.6. When calibrating a pulse voltage measurement system, it is assumed that the individual parameters have linear characteristics and do not change with voltage or current.

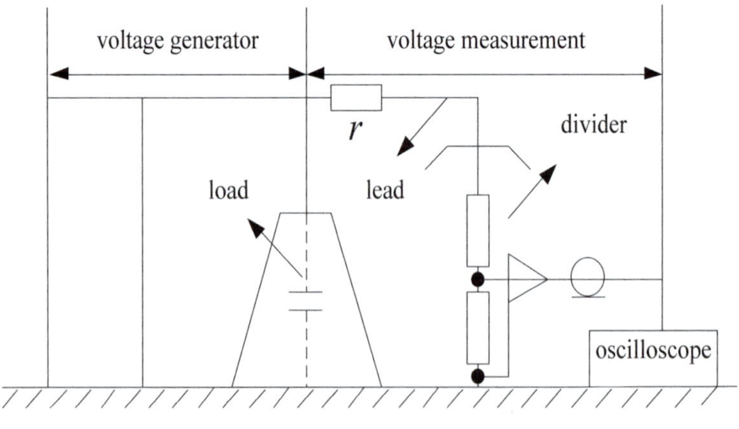

Fig. 33.6 Schematic diagram of pulse voltage measurement system

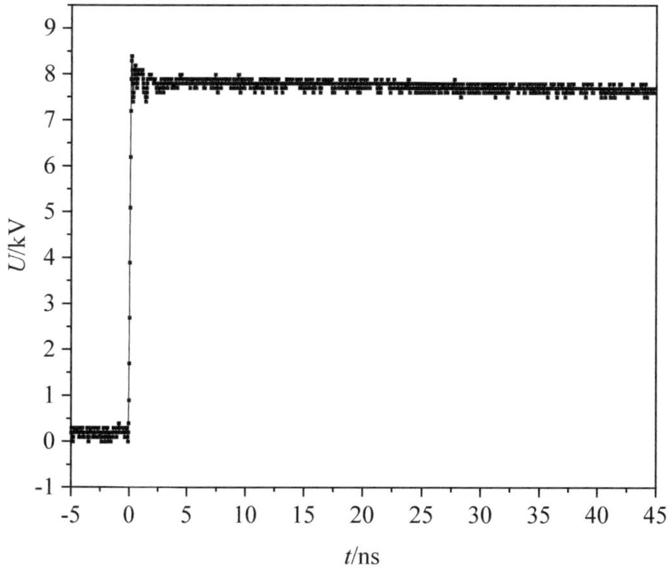

Fig. 33.7 The output waveform of square wave source

The charging capacitor of the square wave source is a low-inductance pulse capacitor with an withstand voltage of 15 kV and a capacity of 1.0μF. The components of the circuit are connected with thin copper skins, and the structure is compact to reduce stray inductance and stray resistance. A 10 kV high-voltage mercury relay is selected as the discharge switch, and the load resistance is a 400Ω non-inductive body resistor, which also serves as a voltage divider with a scale factor of 1000:1. The mercury switch is triggered to turn on when charging at 9 kV, and the output waveform of the square-wave voltage source is recorded by a Tek DPO3054 oscilloscope, as shown in Fig. 33.7.

When analyzing the rise time of the square wave voltage source output signal, it is also necessary to consider the rise time of the oscilloscope itself, according to the time composition of the measurement signal, the rise time of the measured signal and the actual rise time of the signal satisfy the equation:

$$t_{meas^2} = t_{sig^2} + t_{scope^2} \tag{33.2}$$

where t_{meas} is the rise time of the waveform recorded by the oscilloscope, t_{sig} is the actual rise time of the signal and t_{scope} is the rise time of the oscilloscope itself. Since the oscilloscope is 500M bandwidth, its own rise time is 3.5/5 = 0.7ns. Take ten consecutive test results as statistical samples, calculate the rise time of the waveform recorded by oscilloscope and the actual rise time of the signal, and calculate the average value, of which the average value of t_{scope} is the actual rise time of the square wave signal, i.e. 0.812ns.

When the voltage signal applied at the input of the voltage divider measurement system is a unit step waveform, a step response signal will be output at the low voltage end of the voltage divider as *G*(t), where the unit step signal is ε(t). Let the input voltage signal of the measurement system be u_i(t), the output voltage signal be u_0(t), and the ratio of voltage divider be *K*, then:

$$K = u_i(t) / u_0(t)|_{t \to \infty} \quad (33.3)$$

The step response *G*(t) of the measurement system will eventually stabilize as (1/*K*)ε(t). For measurement systems with good response characteristics, the response reaches its final value in a very short time. In order to calculate the response time conveniently, the output voltage signal is usually normalized, i.e., the value of the step response G(t) is multiplied by the value *K* of the ratio of voltage divider:

$$g(t) = KG(t) \quad (33.4)$$

The time at which the step response waveform begins to output is usually taken as the apparent origin, and a typical *g*(t) waveform is shown in Fig. 33.8.

For the above typical *g*(t) waveform with overshoot, it is more reasonable to use the overshoot β and the partial response time T_1 as the judgment criteria. β is the maximum value of the normalized step response beyond the value of the unit value, and t_1 is the maximum value of the step wave response integral *T*(t), so the step response integral is:

$$T(t) = \int_{o_1}^{t} \left[1 - g(t)\right] dt \quad (33.5)$$

As above let t_1 be the moment when *g*(t) first reaches unit 1, then $T_1 = T(t_1)$. Figure 33.9 shows the step response waveform of the capacitive voltage divider

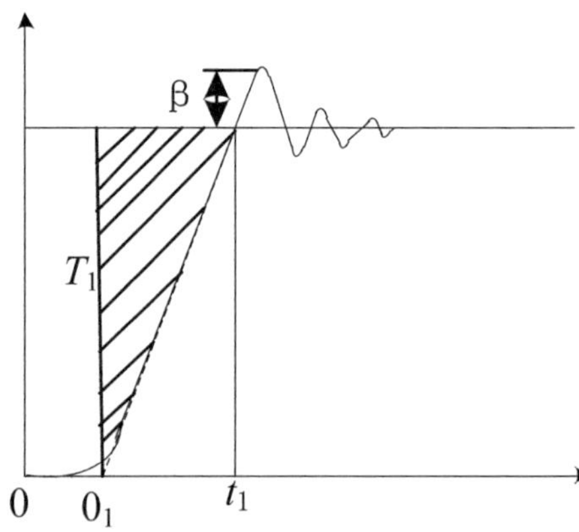

Fig. 33.8 Typical step response waveform

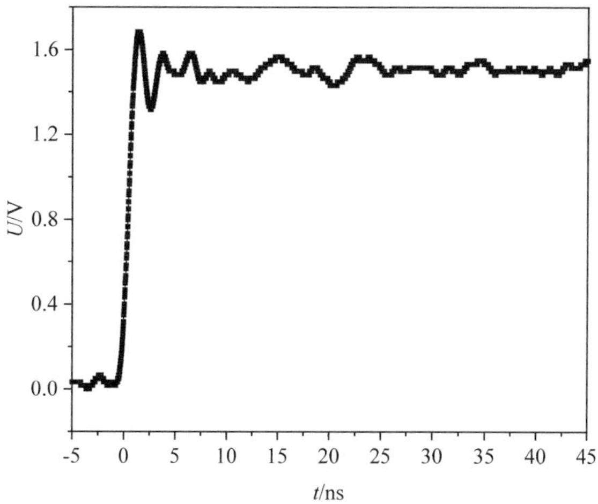

Fig. 33.9 The step response waveform of the capacitive voltage divider

designed in this paper. By integrating the corresponding area of the measured waveform, the partial response time of the measurement system is about 7.2 ns.

33.3.2 Scale Factor

The overall performance of the capacitive voltage divider designed in this paper is tested. First of all, in order to improve the overall insulation performance of the voltage divider, the solid insulation casting and potting is adopted to treat the high-voltage arm part, due to the higher requirements of voltage online monitoring on the measurement stability of the voltage divider, the same voltage divider before and after casting and potting is measured five times in succession at 90kV excitation voltage level, as shown in Fig. 33.10a, by comparing the scale factors of these five times, it was found that the stability of the measurements was significantly improved after casting, and its standard deviation decreased from 228.7 to 9.8, which is due to the silicone casting, the original air medium of the high-voltage arm of the voltage divider is replaced by silica gel, whose dielectric constant is closer to that of the high-voltage arm capacitor's original outer encapsulated epoxy material, which makes the difference in dielectric constant at the interface greatly reduced, thus weakening the electric field distortion at the interface and making the overall performance of the high-voltage arm more stable.

The scale factor and linearity of the capacitive voltage divider are tested after casting. The waveform to be measured adopts a pulsed voltage signal with a duration of 20μs, and the voltage amplitude in the range of 0 to 90 kV is selected from high to low at seven measurement points, and the same capacitive voltage divider after casting is compared with the standard Polaris high-voltage probe. Figure 33.10b shows the fitted curve of the scale factor of the capacitive voltage divider after casting,

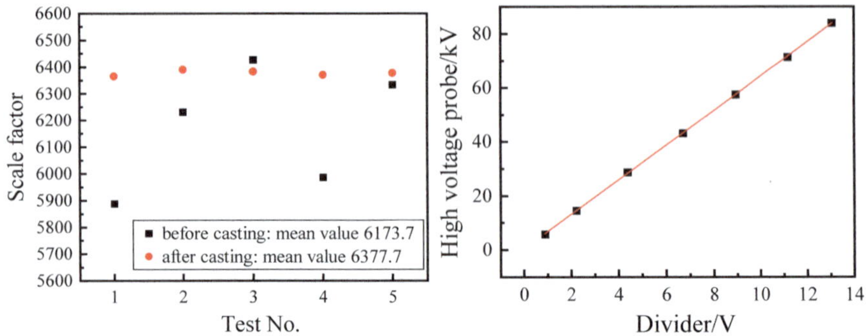

(a) Stability of scale factor (b) Linearity of scale factor

Fig. 33.10 Analysis of the effect of casting on scale factor

where the horizontal coordinate is the voltage amplitude measured by the voltage divider and the vertical coordinate is the amplitude measured by the standard Polaris high voltage probe. The slope of the fitted curve is the scale factor of the voltage divider, and the scale factor of the capacitive voltage divider after casting and potting is calculated to be 6389.26, with a linearity deviation of less than 1%.

33.4 Conclusions

In this paper, we carry out the experimental study of capacitive voltage divider for the typical waveforms in the breakdown characteristic test of insulation materials under impulse voltage, i.e., the duration is long, generally tens or even hundreds of microseconds, and the waveform drop time after breakdown is very short, generally tens of nanoseconds. The following conclusions were obtained:

(1) Through the analysis of the self-integrating conditions and matching methods, a low-voltage arm compensation circuit is proposed, including the first-end compensation resistor, end compensation capacitor and end matching resistor, and based on this, a self-integrating capacitive voltage divider is designed, which solves the problem of waveform aberration under the conventional first-end matching, and at the same time takes into account of the low-frequency characteristics and high-frequency characteristics, and is simple in structure, economic in cost, and convenient in use;
(2) Through the step response test, the capacitive voltage divider designed in this paper has a partial response time of 7.2ns, which meets the demand of fast frontier test;
(3) After the whole casting of the high-voltage arm, the withstand voltage of the capacitive voltage divider is not less than 90kV, scale factor is 6389.26 with good consistency, and its linearity deviation is less than 1%.

Acknowledgements This work was supported by the "Research and Application of IoT Perception System and Key Technologies for Digital Converter Stations" (5700-202324620A-3-2-ZN).

References

1. QIN Bingyang. Switchgear Partial Discharge Detection and Pattern Recognition Based on D-dot[D]. Jinan, China: Shandong University (2016).
2. Shu Y, Chen W, Li Z, *et al.* Experimental research on very-fast transient overvoltage in 1100-kV gas-insulated switchgear[J]. IEEE Transactions on Power Delivery, 28(1): 458-466 (2013).
3. Zhang L, Mao C, Lu P, *et al.* Conical voltage sensor for measuring very fast transient overvoltage up to 3 MV in ultra-high-voltage class gas-insulated switchgear[J]. IET Science Measurement & Technology, 12(3): 405-410 (2018).
4. YUE Gongcheng, LIU Weidong, CHEN Weijiang, *et al.* Calibration of Very Fast Transient Overvoltage Measurement System for Ultra High Voltage Gas Insulated Switchgear[J]. High Voltage Engineering, 38(2): 342-349 (2012).
5. YUAN Jie, LI Bin, WANG Ting, *et al.* Overview of transient over voltage mechanism and suppression strategies of high voltage direct current transmission grid[J]. Journal of Electric Power Science and Technology, 37(6): 3-16 (2022).
6. Lin TY, Zhang YP, Lu ZJ, *et al.* Dielectric Surface Flashover under Long-Term Repetitive Microsecond Pulses in Compressed Gas Environment[J]. Materials, 14(12): 3343 (2021).
7. LIN Tianyu, LI Zhou, LI Wei, *et al.* Flashover characteristics of stage insulators in the multi-gap gas switch under repetitive microsecond pulses[J]. High Voltage Engineering, 12(48): 342-349 (2022).
8. LI Dengshu, WANG Xin, WU Jianer, *et al.* XLPE cable insulation aging based on feature detection life prediction method[J]. Journal of Electric Power Science and Technology, 37(1): 168-177 (2022).
9. HUANG Jisheng, JIA Hongrui, LIU Xuezhong, *et al.* Simulation study on lightning overvoltage characteristic of equipment in the substation connection section[J]. Journal of Electric Power Science and Technology, 36(3): 148–156 (2021).
10. YE Bing. Investigations of Capacitive Divider to Measure the Pulse High Voltage of Electron Beam Accelerator with Water PFL[D]. Changsha, China:National University of Defense Technology (2006).
11. LIU Jinliang, YE Bing, ZHAN Tianwei, *et al.* Coaxial Capacitive Dividers for High-Voltage Pulse Measurements in Intense Electron Beam Accelerator with Water Pulse-Forming Line[J]. IEEE Transactions on Instrument Measurements, 58(1): 161-166 (2009).
12. Aidam M, Zorn T. Measurements with a 3D-D-dot-sensor in reverberation chambers[C]// 2016 International Symposium on Electromagnetic Compatibility - EMC EUROPE. Wroclaw, Poland: IEEE, 257–262 (2016).
13. WANG Qiwu, LI Yanxin, SHI Lihua, *et al.* Design and experimental research of D-dot probe for nuclear electromagnetic pulse measurement[J]. High Power Laser and Particle Beams, 27(11): 115004 (2015).
14. ZHANG Renyu, CHEN Changyu, WANG Changchang. High-voltage Testing Technology[M]. 3rd ed. Beijing, China: Tsinghua University Press (2009).
15. CHEN Xin. Research on the Measurement and Control System Using on 15kV Nanosecond Pulse Generator [D]. Chongqing, China: Chongqing University (2009).
16. LIU Xuandong, LI Dengyun, SUN Fengju, *et al.* A Coaxial Capacitive Voltage Divider for Nanosecond Pulse Measurement[J]. High Voltage Apparatus, 44(1): 32-36 (2008).
17. DU Jishi, LEI Yangjun, LIU Wei, *et al.* Research on a 10 kV voltage level contactless voltage transformer[J]. Journal of Electric Power Science and Technology, 38(3): 197-204 (2023).

18. WU Changfeng, JIANG Xingquan, HE Peng, *et al.* Analysis and Error Compensation of a Novel Resistance Divider[J]. High Voltage Apparatus, 44(2): 168–171 (2008).
19. Zhang Z, Zhang W, Cui X, *et al.* Measurement study of very fast transient overvoltage in GIS[C]//2009 International Conference on Sustainable Power Generation and Supply. Nanjing, China: IEEE, 1–5 (2009).
20. LAI Dingguo, ZHANG Yongmin, XIE Linshen, *et al.* Calibrating method of high voltage and current probe for fast pulse measurements[J]. High Power Laser and Particle Beams, 22(10): 2469–2472 (2010).
21. Yu B, Li R, Su J, *et al.* Analysis of the match problem of a capacitive voltage divider with a long measurement cable[J]. Measurement Science & Technology, 28(9): 095009 (2017).
22. Stanković K, Kovačević U. Combined measuring uncertainty of capacitive divider with concentrated capacitance on high-voltage scale[J]. IEEE Transactions on Plasma Science, 46(8): 2972-2978 (2018).
23. WEI Bing, GONG Boyi, WANG Zhi, *et al.* Frequency response characteristics of self-integralting capacitive divider[J]. High Power Laser and Particle Beams, 30(4): 045006 (2018).
24. Wang J C, Ding W D, and Qiu A C. Capacitive sensor for fast pulsed voltage monitor in transmission line[J]. Review of Scientific Instruments, 90(3): 035107 (2019).
25. R Wang, Y Gao, C Zhang, *et al.* Dynamics of Plasma Bullets in a Microsecond-Pulse-Driven Atmospheric-Pressure He Plasma Jet [J]. IEEE Transactions on Plasma Science, 44 (4): 393-397 (2016).

Chapter 34
Characteristic Gas Analysis of Ablative Defect in Buffer Layer of High Voltage XLPE Cable

Cheng Zhang, Ning Li, Qing Liu, Zhiguo Liu, Yuyang Jiao, Nan Chen, Shuyu Chen, Yu Tang, and Chengyan Ren

Abstract In recent years, operational failures in high voltage cables due to the ablation of cable buffer layers have significantly compromised power system safety. The ablation process generates characteristic gases from the buffer layer materials, which can serve as an early warning and diagnostic tool for ablation defects through gas detection methods. This study investigated the collection and analysis of gases emitted from buffer layers in decommissioned cables and those undergoing maintenance during power outages. The results show that the gas products during the buffer layer ablation process include CO_2, H_2, CO, and low molecular weight hydrocarbon gases. The characteristic gas detection method shows promise for identifying latent defects in areas with low electric field strength, such as those found in high-voltage cable buffer layers.

Keywords High voltage cable · Buffer layer ablation · Gas detection

34.1 Introduction

High voltage cross-linked polyethylene (XLPE) cables have become the mainstream for large-scale electricity transmission in urban areas due to their ease of installation, excellent electrical performance, and resistance to climate changes [1]. In recent years, several operational failures of cables have been reported due to discharges causing ablation in the buffer layer, which severely jeopardizes the safe operation

C. Zhang · N. Li · Q. Liu · Z. Liu · Y. Jiao · N. Chen
State Grid Beijing Electric Power Company, Beijing, China

S. Chen · Y. Tang · C. Ren (✉)
Beijing International S&T Cooperation Base for Plasma Science and Energy Conversion, Institute of Electrical Engineering, Chinese Academy of Sciences, Beijing, China
e-mail: rcy@mail.iee.ac.cn

University of Chinese Academy of Sciences, Beijing, China

Y. Li et al. (eds.), *Proceedings of the 5th Chinese Conference on High Voltage and Discharge Plasmas*, Springer Proceedings in Physics 254,
https://doi.org/10.1007/978-981-96-5711-7_34

of the cable systems [2]. Severe ablation defects in the buffer layer can damage the insulation shield, ultimately leading to breakdown of the cable itself, inducing power outages, and affecting the stable operation of the power system.

Moreover, high-voltage cable buffer layer ablation failures typically occur more than five years after the cables are put into operation. These defects have a long formation cycle, and their latent characteristics are not easily detectable using conventional techniques [3]. Currently, most buffer layer ablation defects in cables are only discovered during the dissection of the cable body or after a cable fault has been triggered during maintenance, leaving numerous hidden buffer layer ablation defects across the country. Thus, there is an urgent need for a reliable and rapid detection method that can facilitate early identification of latent defects within the buffer layer of operating cables.

Now research primarily focuses on gas generation analysis of cable buffer layers under laboratory conditions [4, 5], with a lack of onsite gas sampling analysis [6, 7]. This paper aims to collect gases from the buffer layer region of high-voltage XLPE cables under field conditions and analyze the gas components using gas chromatography. The goal is to obtain characteristic information about gases presently when buffer layer ablation defects occur, enabling early detection and warning of latent defects in cable buffer layers. The findings of this study are expected to provide significant guidance for fault diagnosis in cable buffer layers based on characteristic gases.

34.2 Gas Collection and Analysis of Cable Buffer Layer

34.2.1 Gas Collection and Analysis of Retired Cables

This subsection presented the collection and analysis of characteristic gases from the buffer layers of three retired cables. The gas sampling method involved drilling holes in the outer sheath to expose the buffer layer, thereby enabling the extraction of gases from the buffer layer region using a pump. The specific sampling conditions and status specifications of the cables are presented in Table 34.1. D1 and D2 referred to the same cable, with the distinction that a direct current voltage of 6 kV was applied between the conductor and the metal armor layer for a duration of 30 min for D2.

Table 34.1 The conditions of gas samples from retired cables

No	Ablation condition	Voltage application	Cable manufacturer/voltage level (KV)
D1	Yes	No	M1/110
D2	Yes	Yes	M1/110
D3	No	No	M2/110
D4	No	No	M3/220

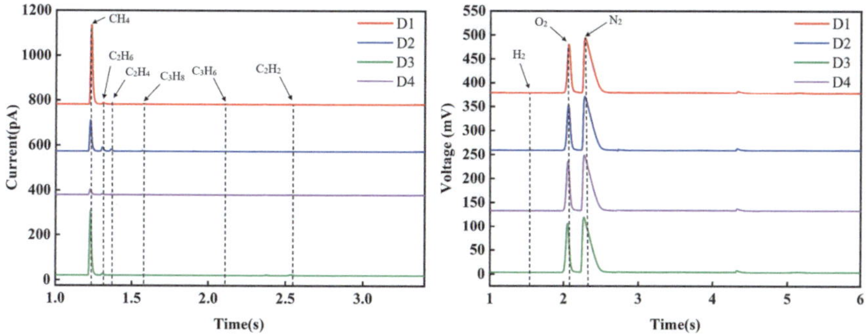

(a) Detection result of flame ionization detector (b) Detection result of thermal conductivity detector

Fig. 34.1 Gas chromatogram of the retired cable buffer layer

Figure 34.1a presents the gas chromatogram of hydrocarbon gases corresponding to the gas samples, while Fig. 34.1b displays the chromatogram of permanent gases. Both analyses were obtained using different sensors of the gas chromatograph [4].

Table 34.2 lists the type and percentage composition of gases for each gas sample. Comparing the detection results of the retired cables D1, D3, and D4, it is observed that due to being set for several months in an atmospheric environment, the composition of permanent gases primarily consists of a significant amount of N_2 and O_2, while the presence of CO_2 is not prominent in the chromatogram, and H_2 and CO are not detected. The D1 cable, which experienced ablation, shows higher CO_2, CH_4, C_2H_6 content compared to the other two normal retired cables. Under the conditions of applying a voltage significantly lower than the normal operating voltage of the cable, D2 exhibited considerable changes in the gas composition of the buffer layer compared to D1. D2 detected three additional gases, H_2, CO and C_2H_4. At the same time, the concentration of CO_2 increased significantly, while the concentration of O_2 decreased. Furthermore, the total amount of generated hydrocarbon gases rose, with C_2H_6 exhibiting the most substantial increase. This indicates that the condition of this cable is poor and the buffer layer ablation phenomena can occur even at lower voltages.

34.2.2 Gas Collection and Analysis of Cables During Power Outage Maintenance

The characteristic gases from the buffer layers of two segments of cables undergoing power outage maintenance were collected and analyzed in this subsection, using the same gas sampling method as described in Sect. 34.2.1. The specific sampling conditions and status specifications of the cables are presented in Table 34.3. The following location conditions indicate the distances of the sampling points from the cable terminal. P1 and P2 correspond to same locations but different phases of

Table 34.2 Gas phase composition of retired cable buffer layers

Gas type	D1 (%)	D2 (%)	D3 (%)	D4 (%)
H_2	N/A	0.0079	N/A	N/A
O_2	19.8241	18.5678	20.0421	20.0174
N_2	75.2261	71.5714	75.3953	74.9865
CO_2	0.0762	0.2822	0.0658	0.0608
CO	N/A	0.0301	N/A	N/A
CH_4	0.2904	0.3006	0.2243	0.0176
C_2H_6	0.0045	0.0142	0.004	0.0022
C_2H_4	N/A	0.007	N/A	N/A
C_3H_8	0.0006	0.0014	0.0012	0.0004
C_3H_6	0.0005	0.0001	0.0012	0.0004
C_2H_2	0.0034	0.0002	0.0046	0.0006

cables. The buffer layer ablation phenomena were observed in Phase C. Infrared detection technology was employed to measure the temperature of the cable outer sheath, revealing localized hot spots at P2, with a temperature difference of 0.5°C, suggesting the presence of buffer layer ablation at this point [8]. P3, and P4 serves as a comparison where buffer layer ablation phenomena were not observed, which are from the same cable manufacturer but different with P1 and P2.

The gas samples collected from the power outage maintenance cables and their respective gas components and concentrations are presented in Table 34.4. Compared to the retired cables that were cut and left idle for an extended period, the characteristic gases in the components extracted from the onsite cables exhibited less diffusion. In the former, the concentrations of N_2 and O_2 in retired cables are similar to those in the air, while the proportions of N_2 and O_2 in the cables undergoing power outage maintenance are significantly lower than those in the air.

P2 is ablation cable, which exhibits a markedly higher content of hydrocarbon gases compared to P1, P3 and P4, which do not experience buffer layer ablation. P2 detected localized hot spots in the outer sheath. Compared to P1, the concentration of oxygen decreased dramatically, while the levels of H_2, CO_2, CO, CH_4, and C_2H_6 significantly increased. Additionally, although no buffer layer ablation phenomena were observed, P3 exhibited higher concentrations of H_2, CO_2, CO, and hydrocarbon

Table 34.3 The conditions of gas samples from maintenance cables

No	Ablation condition	Location (m)	Phase	Outer sheath heating
P1	No	30	A	No
P2	Yes	30	C	Yes
P3	No	105	A	No
P4	No	105	B	No

Table 34.4 Gas phase composition of cable buffer layers during power outage maintenance

Gas type	P1 (%)	P2 (%)	P3 (%)	P4 (%)
H_2	30.8157	35.7937	28.1034	9.4602
O_2	10.4700	5.4356	11.1187	15.7046
N_2	36.7124	35.9901	41.9888	66.2599
CO_2	8.3040	11.4045	9.7232	3.5534
CO	5.0108	5.5471	4.7091	1.6082
CH_4	0.3342	0.7992	0.3214	0.7071
C_2H_6	0.1809	0.4777	0.1606	0.0463
C_2H_4	0.3802	0.1885	0.3481	0.1095
C_3H_8	0.0211	0.0161	0.0193	0.0058
C_3H_6	0.0290	0.0267	0.0218	0.0065
C_2H_2	0.1569	0.1440	0.0742	0.0251
C_4H_{10}	0.0020	0.0026	N/A	N/A
C_4H_8	0.0136	0.0153	0.0079	0.0063

characteristic gases compared to P4, suggesting that the condition of the buffer layer may harbor latent defects.

34.2.3 Analysis

When ablation occurs in the buffer layer, the primary gases that exhibit significant changes include CO_2, CO, H_2, CH_4, C_2H_6, C_2H_4, C_2H_2, C_3H_8, C_3H_6, C_4H_8, and C_4H_{10}. These 11 gases can serve as indicators for assessing buffer layer ablation defects and are referred to as the total generated gas. Table 34.5 presents the total generated gas amounts and the hydrocarbon gas generation amounts for each gas sample. Cables during power outage maintenance have their buffer layers subjected to electric fields for long-time running, resulting in hydrocarbon gas proportions exceeding 0.9%, compared to retired cables where the gas composition consists mainly of N_2 and O_2, with total generated gas amounts of less than 1%. For cables during power outage maintenance, hydrocarbon gases can serve as important indicators of buffer layer ablation defects. The hydrocarbon gas levels detected in P2, where overheating of the outer sheath was observed, are significantly higher than those in P1, P3, and P4.

Table 34.5 Characteristics of components in each gas sample

NO	Total generated gas (%)	Hydrocarbon gas (%)
D1	0.3756	0.2994
D2	0.6437	0.3235
D3	0.3011	0.2353
D4	0.0820	0.0212
P1	45.2484	1.1179
P2	54.4154	1.6701
P3	43.489	0.9533
P4	15.5284	0.9066

34.3 Conclusions

(1) Due to prolonged exposure to the atmospheric environment, the gas samples of retired cables show significant diffusion of internal characteristic gases. However, cables that have experienced buffer layer ablation still exhibit higher concentrations of characteristic gases. Furthermore, after applying lower voltage, the concentrations of characteristic gases in cables with buffer layer ablation demonstrate notable changes, with additional detections of H_2, CO, and various hydrocarbon gases.

(2) In the cables undergoing power outage maintenance, the concentrations of characteristic gases are significantly higher than those in idle retired cables, with H_2 showing the most substantial increase. In cables that have undergone ablation, both the total amount of characteristic gases and the amount of hydrocarbon gases are much greater than those in normally operating cables.

(3) H_2, CO, and various hydrocarbon gases were also detected in cables without buffer layer ablation defects. However, it is possible to determine whether ablation has occurred in the cable buffer layer based on indicators such as the total amount of characteristic gases and the quantities of hydrocarbon gases generated.

Acknowledgements This work was supported by the Science and Technology Project of Beijing Zhuoyue Electric Power Construction Co., Ltd. (ZYKCJS [2024]016).

References

1. Ren Chengyan, Niu Lei, Cao Xiangnan, et al: Trap characteristics and space charge behavior of high voltage XLPE cable insulation. IEEE Transactions on Plasma Science 52(10), 5094-5103 (2024).

2. Tian Yi, Zhu Guixin, Miao Qingqing, et al: Study on the ablation process and failure mechanism of the buffer layer in high-voltage XLPE cable. International Journal of Electrochemical Science 19(8), 100662 (2024).
3. Zhou Wenqing, Cheng Hao, Hui Baojun, et al: Key process of the defection of PET buffer layer in XLPE power cable by a case study: Thermo-oxidative degradation. Engineering Failure Analysis 146, 107131 (2023).
4. Li Senlin, Chen Shuyu, Chen Yikai, et al: Detection of buffer layer ablation defects for high voltage cable based on characteristic gases. Advanced Technology of Electrical Engineering and Energy 43(12): 73–82 (2024). (in Chinese).
5. Kong Jiamin, Zhou Kai, Chen Yidong, et al: A novel condition assessment method for corrugated aluminum sheathed XLPE cables based on evolved gas analysis. IEEE Transactions on Dielectrics and Electrical Insulation 30(2), 883-891(2023).
6. Wu Zhiheng, Lai Qingbo, Zhou Wenqing, et al: Analysis of influencing factors on buffer layer discharge for high-voltage XLPE cable. IET Generation, Transmission & Distribution 16, 4142–4157 (2022).
7. Zhou Wenqiang, Li Licheng, Hao Cheng, et al: Gas production analysis for the buffer layer of high-voltage cross-linked polyethylene cables. High Voltage 9(5), 1149–1158 (2024).
8. Zhang Haoran, Gao Jian, Wu Kangning, et al: Detection technology of buffer layer ablation failure in high voltage cables based on temperature measurement and gases analysis. High Voltage Engineering 49(12), 4929–4937 (2023). (in Chinese).

Chapter 35
Optical-Electrical Characterization of C_4F_7N/CO_2 Gas Mixture with Extremely Inhomogeneous Field Discharge

Junran Duan, Sijie Shen, Shihao Du, Jie Chen, Yuliang Wang, Haiquan Li, Dong Han, and Guoqiang Zhang

Abstract One of the most promising SF_6 replacement gases is the perfluoroisobutyronitrile (C_4F_7N)/carbon dioxide (CO_2) gas mixture. The experimental platform for C_4F_7N/CO_2 gas mixture in an extremely inhomogeneous field discharge is built, and a multi-source signal synchronization acquisition device is established to obtain optical and current pulses under different applied voltages. The discharges show a "hump" shape with the increase of voltage. The generation of optical and current pulses is synchronized with complementary phase distributions, i.e., the optical pulses are generated in the negative semiaxis and become stronger in the positive semiaxis; the current pulses are generated in the positive semiaxis and become stronger in the negative semiaxis. The Gaussian noise in the optical and current pulses is removed using the three-sigma principle, and different semiaxis root mean square (RMS) and pulse counts are calculated. At the early discharge stage, optical pulses at negative and current pulses at positive semiaxis have larger RMS and pulse counts. At the late discharge stage, the above phenomena are opposite. This process is related to the polarity effect of the extremely inhomogeneous field. This study provides a reference for detecting extremely inhomogeneous field discharges in C_4F_7N/CO_2 gas-insulated electrical equipment, promoting the application of combined optical-electrical detection techniques.

Keyword C_4F_7N/CO_2 gas mixture; Extremely inhomogeneous field; Optical-electrical characterization; Polarity effect

---

J. Duan · S. Shen · S. Du · D. Han (✉) · G. Zhang
Institute of Electrical Engineering, Chinese Academy of Sciences, Beijing, China
e-mail: donghan@mail.iee.ac.cn

University of Chinese Academy of Sciences, Beijing, China

J. Chen · Y. Wang · H. Li
Beijing Herosail Power Science and Technology Co.Ltd, Beijing, China

Y. Li et al. (eds.), *Proceedings of the 5th Chinese Conference on High Voltage and Discharge Plasmas*, Springer Proceedings in Physics 254,
https://doi.org/10.1007/978-981-96-5711-7_35

35.1 Introduction

The perfluoroisobutyronitrile (C_4F_7N)/carbon dioxide (CO_2) gas mixture is one of the main replacement gases for SF_6, offering notable advantages in electrical performance and environmental sustainability [1, 2]. Currently, C_4F_7N/CO_2 insulating medium is being researched, developed, and applied on a small scale [3, 4]. However, degradation and failure of the medium are the main concerns for researchers. Partial discharge, a primary contributor to the degradation of this new eco-friendly insulating gas during both production and commissioning, is unavoidable [5]. Consequently, partial discharge detection has become a crucial method for assessing insulation conditions.

Currently, partial discharge detection methods are primarily categorized into electrical and non-electrical. Researchers worldwide have made significant advancements in electrical detection techniques [5, 6]. They have analyzed the electrical characteristics of SF_6 and its gas mixtures, as well as clean air-insulated electrical equipment, and provided recommendations for using electrical detection in gas-insulated systems. However, due to electromagnetic interference, the effectiveness of these applications has not fully met expectations [7]. Non-electrical detection methods have attracted attention for their strong resistance to electromagnetic interference, making them valuable for identifying and assessing partial discharge in gas-insulated equipment. Nevertheless, these methods face challenges such as a limited detection range and issues with leakage detection [7]. To address these limitations, researchers have introduced a joint detection technique that combines both electrical and non-electrical methods. This approach allows for defect localization, discharge pattern recognition, and tracking discharge development in SF_6 and its gas-insulated equipment [8–10]. Studies have demonstrated the feasibility and effectiveness of this joint detection technique for use in gas-insulated electrical systems.

At present, the partial discharge detection and analysis of C_4F_7N/CO_2 gas-insulated electrical equipment is mainly based on SF_6 gas-insulated electrical equipment [11]. Due to the influence of electrical and physicochemical properties of insulating media, discharge characteristics are different between different insulating media [12]. Li Ze analyzed and diagnosed the optical properties under C_4F_7N/CO_2 gas mixture by silicon photomultiplier (SiPM), summarized the emission spectra under different discharge defects, and proposed a multispectral fault diagnosis method [13].

However, research on the combined optical-electrical detection technology for C_4F_7N/CO_2 gas mixture is rarely reported. Thus, we construct an optical-electrical pluses acquisition device consisting of an avalanche photon diode (APD) and a voltage detection unit to detect the optical and current pulses during partial discharge synchronously. Then, the phase distribution and intensity changes of the optical and current pulses are described. After denoising the data by the three-sigma rule of thumb, the root mean square (RMS) and pulse counts are analyzed.

35.2 Experimental

The experimental platform mainly consists of a discharge test platform and an optical-electrical synchronization acquisition device, as shown in Fig. 35.1.

35.2.1 Experimental Platform

The partial discharge test platform mainly consists of the YDTW-15/150 industrial frequency no partial discharge test transformer (output voltage of 150 kV, capacity of 15 kVA), a 10 KΩ protection resistor, a 500 pF capacitor voltage divider, and a stainless steel test tank. A stainless steel needle-plate electrode with 10 mm distance is used to simulate corona discharge.

35.2.2 Synchronization Acquisition Device

An oscilloscope, voltage detection unit, APD, and high-voltage probe are the main components of the synchronization acquisition device. The APD can transform

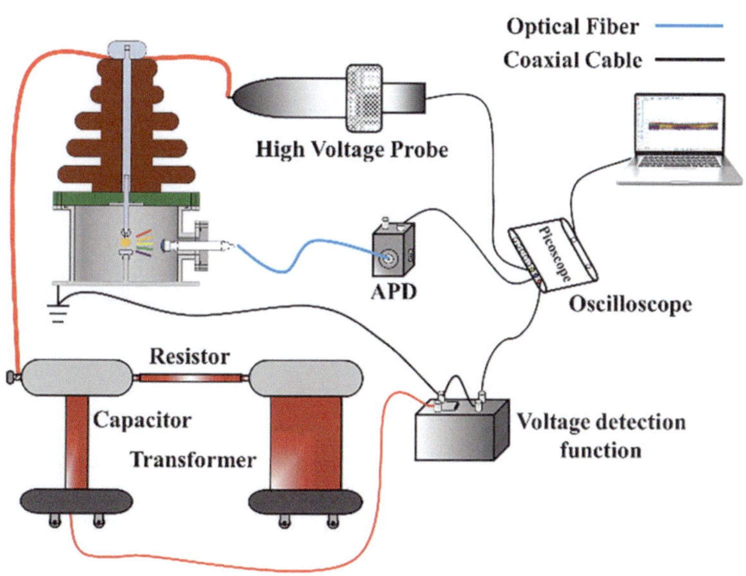

Fig. 35.1 Experimental platforms

optical impulses between 200 nm ~ 1000 nm into electrical signals, and the high-voltage probe (PVM-5) can detect voltages up to 60 kV. With a resolution of 12 bits, the oscilloscope (Picoscope 4824A) synchronizes the three signals at 80 MS/s.

35.2.3 Experimental Conditions

To prevent the influence of impurities on the experiment, the electrode surfaces are polished, and the interior of the tank is thoroughly wiped with anhydrous ethanol. After drying, the assembly of the test tank is completed. The assembled experimental tank is subjected to several vacuuming and gas-washing operations to eliminate the influence of impurities in the air. Finally, the stainless steel experimental tank is filled with 8% C_4F_7N/92% CO_2 gas mixture with an absolute pressure of 0.135 MPa according to the Daltons law of partial pressuresm [14]. Let it stand for 30 min, and start the discharge experiment after the gas is fully mixed. The discharge experiment is carried out in the way of equal differential voltage boosting, and the partial discharge signals are obtained sequentially.

35.3 Experiment Result and Analysis

35.3.1 Optical-Electrical Signal Variations

The range of the applied voltage is between 11 and 34 kV. Corona discharge emerges at 13 kV, and critical breakdown occurs at 34 kV. The partial discharge magnitude of the C_4F_7N/CO_2 gas mixture in a highly inhomogeneous field exhibits a "hump" structure, which is quite different from that of SF_6 gas. The amount of partial discharge first rises and then falls with an increase in applied voltage before abruptly increasing at the critical breakdown condition.

As shown in Fig. 35.2, the optical pulses in the APD first appear in the negative semiaxis with a broad phase distribution. As the discharge degree increases, the optical pulses appear in both the positive and negative semiaxes at the same time. In addition, the positive semiaxis' spectral intensity surpasses the negative polarity as the discharge develops. Strong pulses are present in the positive semiaxis at 21 kV and 34 kV. The former is a strong pulse in the critical breakdown state, while the latter is a strong pulse triggered by a concentrated burst of discharge initiation energy (Fig. 35.3).

The optical pulses exhibit a companion relationship with the current pulses. The generation of current pulses is often accompanied by optical pulses, and the detection limits of optical pulses and current pulses are basically the same (Fig. 35.4).

The optical and current pulses show complementary characteristics in the phase distribution during corona discharge. At the initial stage, the strong optical pulses are

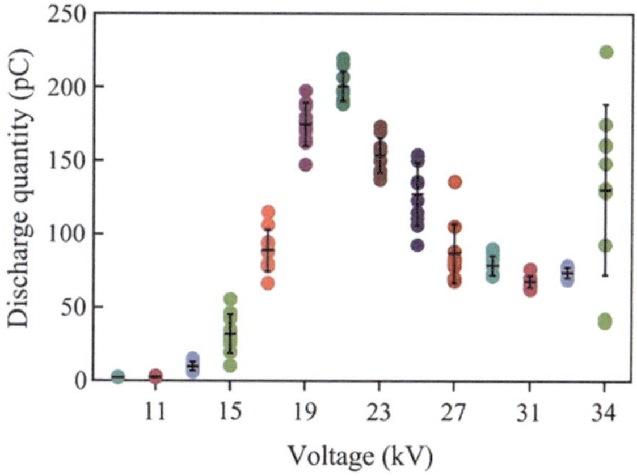

Fig. 35.2 Partial discharges at different voltages

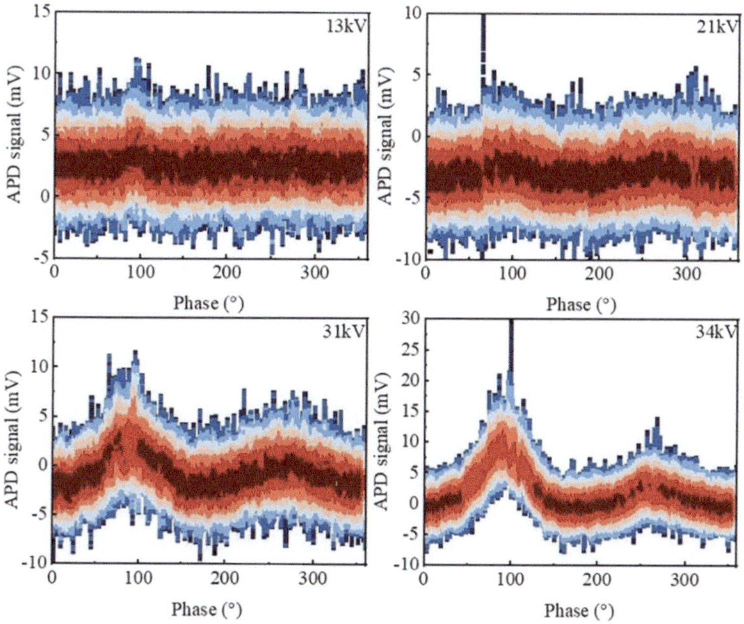

Fig. 35.3 Optical pulses of different voltages

concentrated near 270° of the phase, while the strong current pulses are distributed near 90°. As the discharge process advances, the signal intensity gradually moves toward the direction of the complementary phase. At the critical breakdown state, the strong optical pulses are uniformly distributed near 90°. In contrast, the strong current pulses are concentrated near 270°, forming a stable phase distribution pattern.

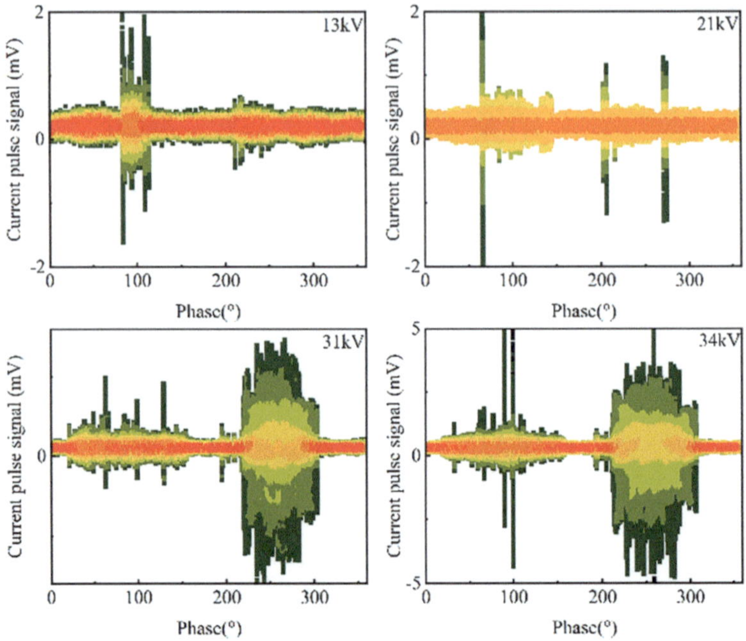

Fig. 35.4 Current pulses of different voltages

35.3.2 Character Analysis

1. Noise reduction

The three-sigma rule of thumb is utilized to denoise the data. The RMS and pulse counts of the denoised data are analyzed. The denoised APD and current pulses are shown in Figs. 35.5 and 35.6, respectively.

The effective signals are better conserved, data processing is made simpler, and the validity of the RMS and pulse counts statistics is improved by removing the Gaussian noise.

2. RMS

In Fig. 35.7, as the applied voltage increases, the RMS of optical pulses in the positive and negative semiaxes increases exponentially, with the overall positive semiaxis larger than the negative semiaxis. Between 9 and 15 kV, the RMS in the negative semiaxis is greater than that in the positive semiaxis.

Before 23 kV, the current pulses' RMS of the positive semiaxis is larger than that of the negative semiaxis. After that, the RMS of the positive semiaxis decreases, and the RMS of the negative semiaxis increases rapidly and exceeds that of the positive semiaxis. The trend of the RMS in the positive semiaxis is highly similar to the amount of partial discharges.

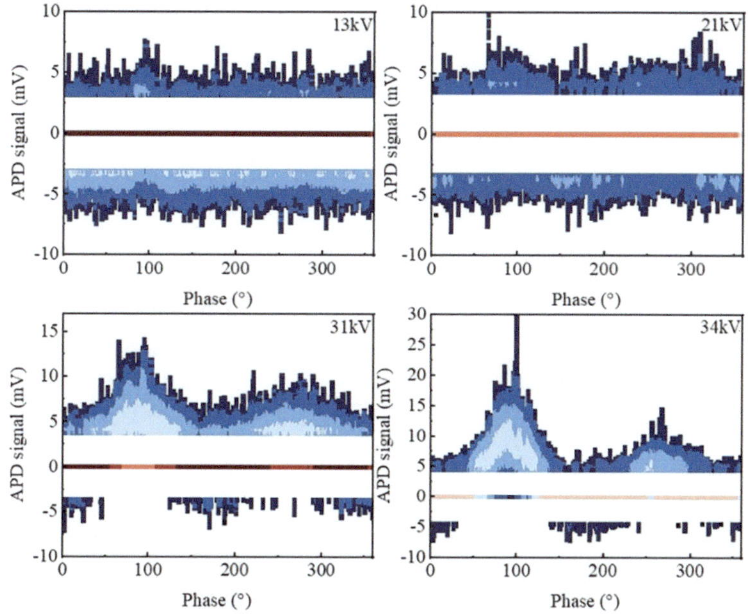

Fig. 35.5 Optical pulses of different voltages after denoising

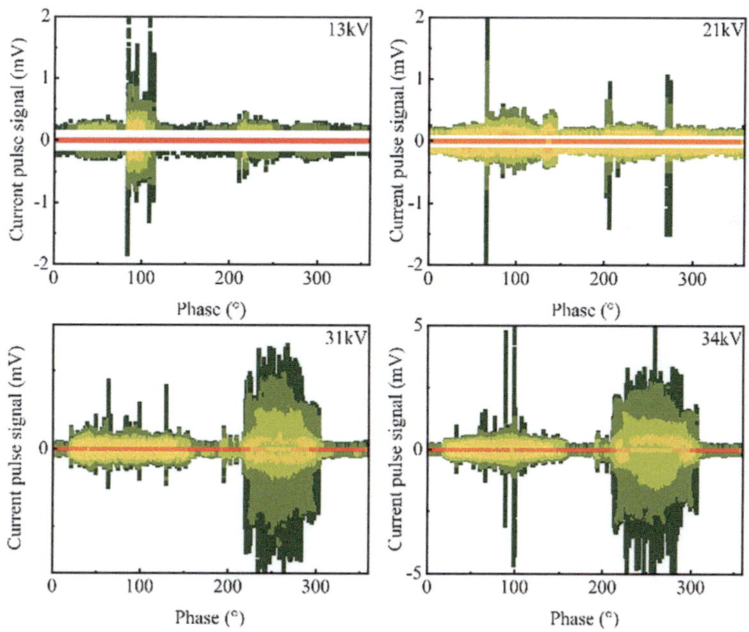

Fig. 35.6 Current pulses of different voltages after denoising

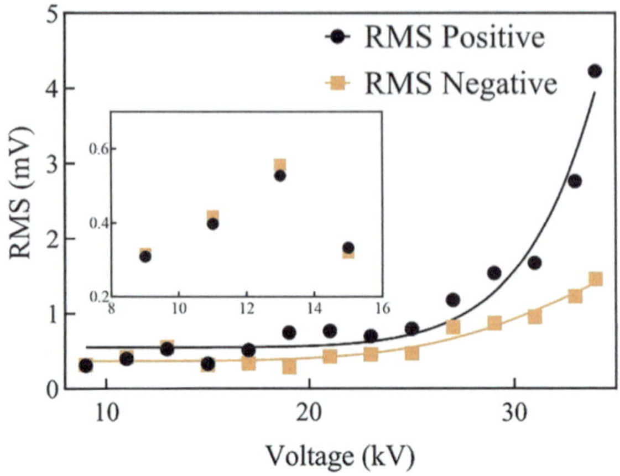

Fig. 35.7 RMS of optical pulses at different voltages

3. Pulse counts

As shown in Fig. 35.8, optical pulses count increase exponentially, with more pulses in the positive semiaxis than in the negative semiaxis. At the start of the discharge, however, there are more pulses in the negative semiaxis than in the positive semiaxis (Fig. 35.9).

When the applied voltage is less than 27 kV, as Fig. 35.10 illustrates, there are more current pulses in the positive semiaxis than in the negative semiaxis. Pulse counts in the negative semiaxis quickly surpass those in the positive semiaxis when the voltage exceeds 27 kV.

Polarity effects in the highly inhomogeneous field are linked to the RMS and pulse count variations. The negative polarity effect increases the field strength near the needle electrode, making it easy to produce electron collapse corona. In contrast, the positive polarity effect contributes to the generation of streamer discharge. When the applied voltage is low, the negative semiaxis shows electron-collapse discharge, and

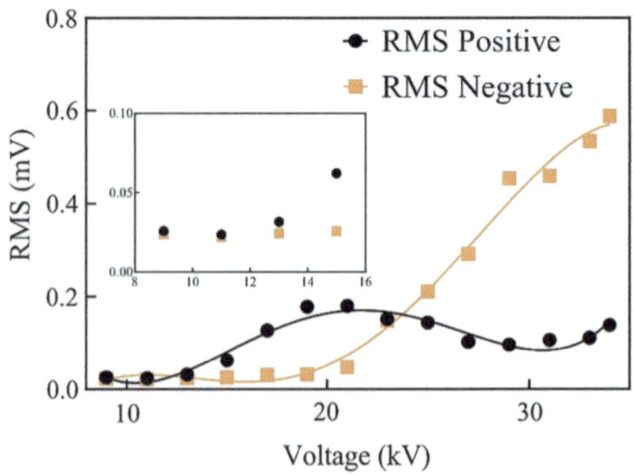

Fig. 35.8 RMS of current pulses at different voltages

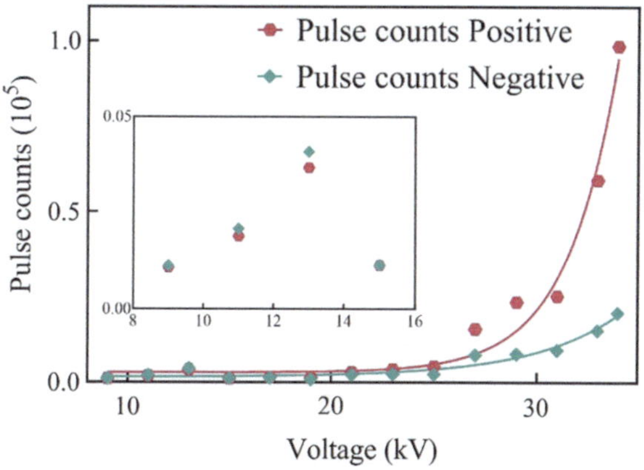

Fig. 35.9 Pulse counts of optical pulses at different voltages

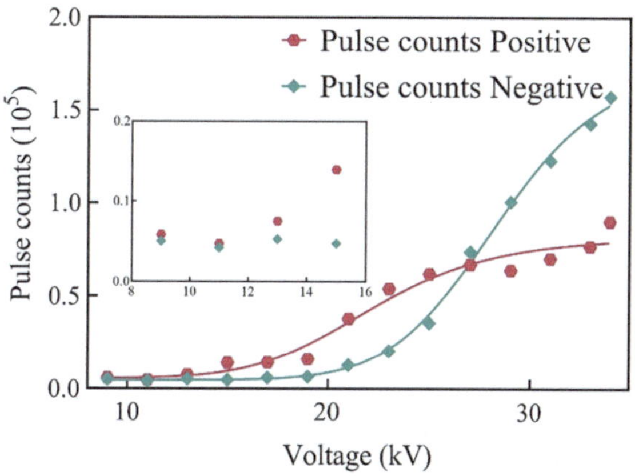

Fig. 35.10 Pulse counts of current pulses at different voltages

the positive semiaxis has no apparent corona discharge. With the increased applied voltage, electron-collapse discharge remains dominant at the negative semiaxis, and streamer discharge is dominant at the positive semiaxis. Streamer discharge produces the pulse current at the positive semiaxis, while a large number of positive charges and electrons form the pulse current at the negative semiaxis.

35.4 Conclusion

The optical and current pulses of the C_4F_7N/CO_2 gas combination are recorded by the acquisition device under extremely inhomogeneous field discharges. It is found that the discharge amount has a "hump" shape when the applied voltage changes. During partial discharge, the phase distributions show complementary characteristics. The optical and current pulses show synchronization, and the pulse strength distributions diverge by 180°. Because of their remarkable synchronization and complementary characteristics, the combination of optical and current pulses helps determine the discharge severity.

Regarding the RMS and pulse counts, a notable discrepancy exists between optical and current pulses. At the early discharge stage, due to the polarity effect, the optical pulse's RMS and pulse counts in the negative semiaxis are larger, and the current pulse at the positive semiaxis is larger. At the critical breakdown stage, the above phenomena are opposite. Overall, the RMS and pulse numbers show an upward trend.

This study analyzes the optical-electrical characteristics of the C_4F_7N/CO_2 gas mixture in an extremely inhomogeneous field discharge, which provides a reference for detection technology in C_4F_7N/CO_2 gas-insulated electrical equipment.

Acknowledgements This work is supported by a project supported by the Beijing Municipal Science and Technology Program (Z221100005822004).

References

1. ZHAO Mingyu, HAN Dong, RONG Wenqi, et al.: Decomposition Characteristics of Binary Mixtures of $(CF_3)_2CFCN$ Buffer Gases under Corona Discharge[J]. High Voltage Engineering, 2019, 45(4):1078–1085.
2. KIEFFEL Yannick, BIQUEZ Francois.: SF_6 Alternative Development For High Voltage Switchgears[C]. 2015 Electrical Insulation Conference (EIC), 2015: 379–383.
3. ZHAO Mingyue, HAN Dong, RONG Wenqi, et al.: Analysis of Decomposition By-products and Its Formation Mechanism of C_4F_7N/Air Mixed Gases Under AC Corona Discharge. High Voltage Engineering, 2018, 44(10): 3174-3182.
4. DUAN Junran, YAN Xianglian, GAO Keli, et al.: Compatibility Analysis of C_4F_7N/CO_2 and Its Gas Byproducts with Epoxy Resin[J]. Transactions of China Electrotechnical Society, 2024: 2–10.
5. CHAI Hua, Phung B.T. and Steve Mitchell.: Application of UHF Sensors in Power System Equipment for Partial Discharge Detection: A Review[J]. sensors, 2019, 19, 1029.
6. ARCONES Eduardo, ÁLVAREZ Fernando, ORTEGO Javier, et al.: Novel Reference Method for the Characterization of PD Measuring Systems Using HFCT Sensors[J]. sensors, 2024, 27, 3788.
7. KHAN Qasim, REFAAT Shady S., ABU-RUB Haitham, et al.: Partial Discharge Detection and Diagnosis in Gas Insulated Switchgear: State of the Art[J]. IEEE Electrical Insulation Magazine, 2019, 35(4): 16–33.
8. LI Xing, DING Dengwei, XU Yuan, et al.: Denoising and Localization for GIS/GIL Partial Discharge Based on Ultra-high Frequency and Ultrasound Method United[J]. High Voltage Engineering, 2024: 1–10.
9. SU Zhixiong, SUN Kang, DING Hao, et al.: Multi-source partial discharge detection method based on pulse current method and ultraviolet pulse method[J]. Insulating Materials, 2024, 57(1): 101-108.
10. GUO Wei, WANG Longzhen, ZHANG Suxin, et al. Analysis and Treatment of GIS Latent Defects Based on Detection of Acoustic-electric-gas[J]. High Voltage Apparatus, 2018, 54(1): 243–248.
11. THOMAS, BACKHAUS.: Optical-Electrical Investigation of the Partial Discharge Behaviour of a Protrusion in Gas-Insulated Systems Under AC and DC Voltage Stress in SF6 and Synthetic Air[C]. VDE High Voltage Technology 2020: 1–6.

12. LIU Yujie, REN Ming, WANG Kai, et al.: Comparative Analysis of the Difference in Partial Discharge Statistical Characteristics Between SF6/N2 Gas Mixture and Pure SF6 Gas[J]. High Voltage Engineering, 2024: 1–13.
13. LI Ze, QIAN Yong, ZANG Yiming, et al.: Optical Feature Analysis and Diagnosis of Partial Discharge in C $_4F_7N/CO_2$ Based on Multispectral Array[J]. Journal of Shanghai Jiao Tong University, 2023, 57 (9): 1176-1185.
14. HU Shizhuo, ZHOU Wenjun, ZHENG Yu, et al.: Power Frequency Breakdown Experiments and Synergistic Effect Analysis of C_4F_7N/CO_2 and C_4F_7N/N_2 Mixtures[J]. High Voltage Engineering, 2019, 45 (11): 3562–3570.

Chapter 36
Analysis of Armature-Orbit Ablation Mechanism and Multi-Physics Field Coupling During Electromagnetic Railgun Launch Process

Jingtong Feng, Luyao Liu, Pengfei Lu, Ali Mohammed Ali Abdo, Yisong Wang, and Hongshun Liu

Abstract Due to the potential burn damage between the armature and the rails during the launch process of the electromagnetic railgun, which can lead to a decrease in system performance or even launch failure, it is essential to investigate the forces on the armature and rails, as well as the distribution of multiple physical fields during the launch process. A modular modeling approach has been adopted to establish a mathematical model of the electromagnetic railgun, and a simulation analysis has been conducted considering the output characteristics related to the forces acting on the armature. Using COMSOL, a multi-physics coupling simulation was performed. By examining the distribution of current density, magnetic field strength, and temperature field within the electromagnetic railgun in conjunction with the force conditions of the armature during motion, it was found that in the initial motion phase, rail wear is primarily caused by armature-rail frictional heat and contact resistance Joule heat. In the later motion phase, severe rail ablation occurs mainly due to electric contact arcing between the armature and the rails. This research into the erosion mechanism of electric contact arcing under the influence of electromagnetic thermal multi-field coupling has significant implications for improving the reusability and lifespan of electromagnetic launching devices.

Keywords Electromagnetic railgun · Simulation modeling · Arc ablation · Multi-field coupling

---

J. Feng · L. Liu · P. Lu · A. M. A. Abdo · Y. Wang · H. Liu (✉)
Shandong Provincial Key Laboratory of UHV Transmission Technology, Shandong University, Jinan, China
e-mail: lhs@sdu.edu.cn

Y. Li et al. (eds.), *Proceedings of the 5th Chinese Conference on High Voltage and Discharge Plasmas*, Springer Proceedings in Physics 254,
https://doi.org/10.1007/978-981-96-5711-7_36

36.1 Introduction

The electromagnetic railgun is an advanced weapon system that accelerates projectiles via electromagnetic forces. As an emerging military technology, it exhibits significant advantages over conventional artillery, including higher muzzle velocity, extended operational range, and superior accuracy [1]. Additionally, its exceptional penetration capability and destructive potential render it strategically significant for naval warfare and missile defense applications. As a result, extensive research efforts are being directed toward its optimization and implementation.

A typical electromagnetic railgun consists of a pulsed power supply located at the breech, which provides energy through a centralized power feed mechanism. The rails, composed of high-conductivity and wear-resistant materials, serve as electrical conductors and structural guides for the armature. Positioned between two parallel rails, the armature conducts current and experiences a Lorentz force that accelerates the projectile along the launch trajectory [2–5]. The fundamental operating principle of the electromagnetic railgun is illustrated in Fig. 36.1.

During the launch process, extreme operating conditions—such as high power density, short-duration discharge, ultrahigh current, and hypervelocity—subject the armature-rail interface to various degradation mechanisms, including groove ablation, shear erosion, transition-phase ablation, and arc-induced ablation. In particular, localized electrical arcing at the contact interface results in intense thermal and electrical loading, leading to material vaporization and severe degradation of both the armature and rails. These effects contribute to system performance deterioration and may induce catastrophic launch failures. A comprehensive understanding of the force distribution on the armature, as well as the electric and magnetic field characteristics at the interface, is crucial for mitigating ablation-related failures and extending system lifespan.

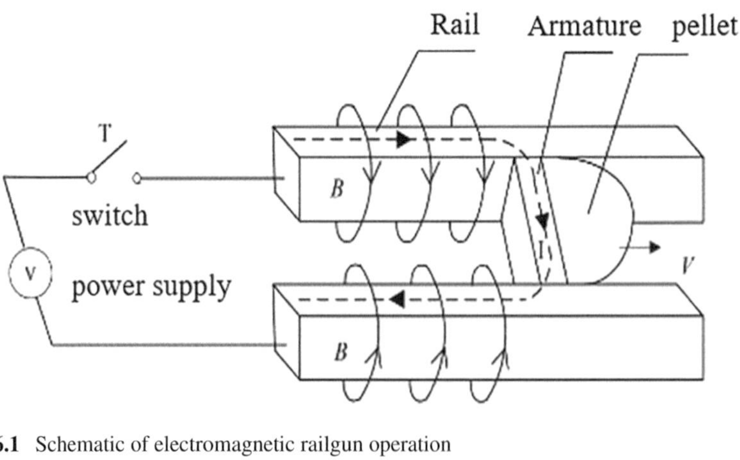

Fig. 36.1 Schematic of electromagnetic railgun operation

36.2 Introduction Simulation of Output Characteristics Considering Multi-Physics Interactions

36.2.1 A Subsection Sample Mathematical Model of the Electromagnetic Railgun

A simplified analytical model of the electromagnetic railgun consists of three interconnected sub-modules: the power supply circuit, the rail-armature impedance network, and the armature dynamic model.

Power Supply Circuit Model. The discharge process of the pulsed power supply follows two distinct stages: the discharge phase and the freewheeling phase.

The discharge phase. During the discharge phase, the freewheeling diode remains inactive due to reverse bias, and the governing circuit equation is formulated as:

$$L_1 C \frac{\mathrm{d}^2 i}{\mathrm{d}t^2} + R_1 C \frac{\mathrm{d}i}{\mathrm{d}t} + U = 0 \tag{36.1}$$

where U represents the initial capacitor voltage, L_1 denotes the total discharge circuit inductance, R_1 is the total circuit resistance, and C is the capacitor value.

Neglecting R_1, the transient discharge current i exhibits an underdamped oscillatory behavior:

$$i(t) = \frac{U}{\omega L_1} \mathrm{e}^{-\alpha t} \sin(\omega t) \tag{36.2}$$

Where α represents the damping coefficient, and ω denotes the oscillation frequency. The maximum discharge current $i_{\max}$ is given by:

$$i_{\max} = U \sqrt{\frac{C}{L_1}} \mathrm{e}^{\frac{-\alpha}{\sqrt{\omega_0^2 - \alpha^2}} \arctan\left(\frac{\sqrt{\omega_0^2 - \alpha^2}}{\alpha}\right)} \tag{36.3}$$

$$\omega_0^2 = 1/L_1 C \tag{36.4}$$

$$\alpha = R_1/2L_1 \tag{36.5}$$

The freewheeling phase. During the freewheeling phase, the governing circuit equation is:

$$L_2 \frac{di}{dt} + iR_2 = 0 \tag{36.6}$$

where L_2 and R_2 correspond to the total inductance and resistance of the freewheeling path, respectively. At this point the current can be expressed [6]:

$$i(t) = i_{\max} e^{-\frac{R_2}{L_2} t} \tag{36.7}$$

Rail-Armature Impedance Model. The mathematical representations of rail resistance and inductance components are given by:

$$R_r = R_0 + R_{r'} x \tag{36.8}$$

$$L_r = L_0 + L_{r'} x \tag{36.9}$$

where R_r and L_r denote the instantaneous rail resistance and inductance, whereas R_r' and L_r' represent their respective gradients per unit length.

The armature impedance model accounts for skin effect resistance and velocity-induced contact resistance [7, 8]:

$$R_a = \frac{s}{h_a} \sqrt{\frac{\pi \mu_a \rho_a}{2t}} \tag{36.10}$$

$$\rho_a = \rho_{a0} + \beta_a \frac{I}{h_a} \tag{36.11}$$

where ρ_a [9] is the instantaneous armature resistivity, and μ_a represents the armature permeability.

The velocity-dependent contact resistance is characterized by:

$$R_v = 2\rho_f v^{1.5} / A_c \tag{36.12}$$

Armature Dynamics Model. The governing equation for armature motion is given as [10–12]:

$$\frac{d^2x}{dt^2} = \frac{dv}{dt} = \frac{F_m - F_f - F_d}{m_a} \tag{36.13}$$

where F_m, F_f, and F_d correspond to the electromagnetic force, frictional resistance, and aerodynamic drag, respectively.

36.2.2 Simulation Analysis of Multi-Physics Interactions

A simulation duration of 2 ms is implemented, with an initial capacitor voltage of 5000 V and a capacitance of 3000 μF. The preloaded rail force F_{n0} is set at 500 N. The computed railgun output characteristics are presented in Fig. 36.2.

The armature acceleration initially rises before subsequently declining. At 1.85 ms, the armature attains a peak velocity of 1033.03 m/s, with a displacement

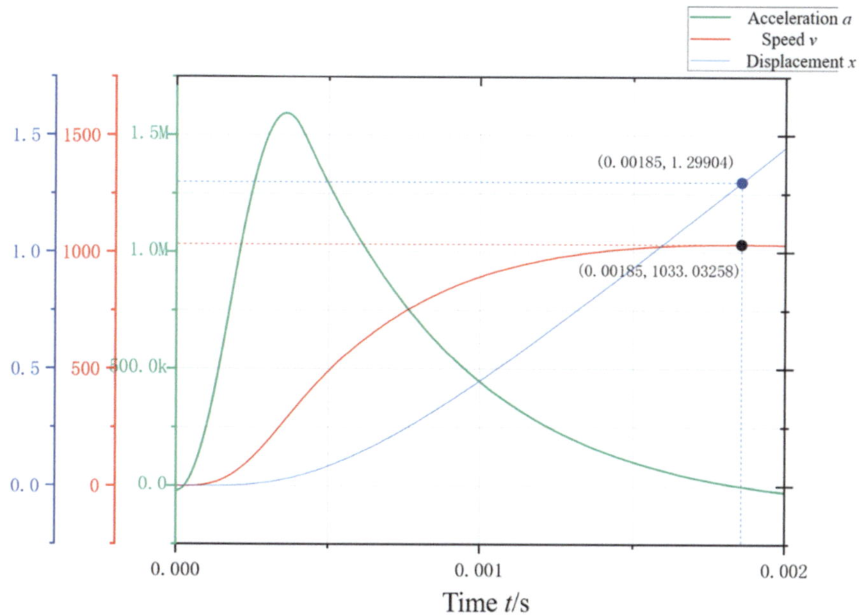

Fig. 36.2 Simulated output characteristics of the electromagnetic railgun

of 1.3 m. The temporal variations in electromagnetic force, frictional resistance, and aerodynamic drag are depicted in Fig. 36.3.

Simulation results reveal that the electromagnetic force dominates throughout the launch process, while frictional resistance decreases as velocity increases. In the later stages, aerodynamic drag becomes more pronounced, exerting a substantial influence on projectile motion.

36.3 Electromagnetic Field and Thermal Distribution Characteristics in the Railgun System

The distribution of current density and magnetic flux density significantly affects the thermal characteristics of the electromagnetic railgun system. Understanding these distributions helps in identifying areas susceptible to intense heating and potential failure due to excessive thermal loading.

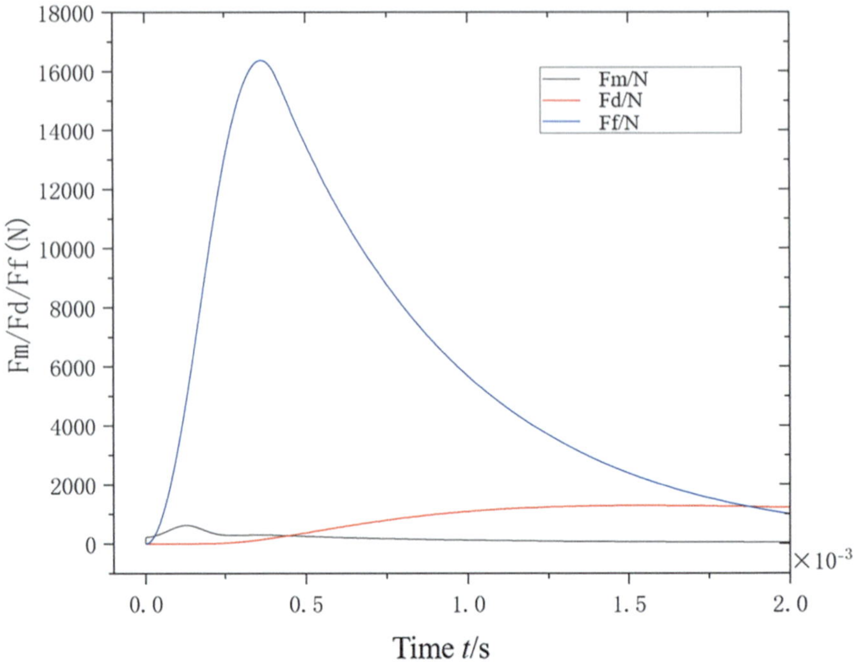

Fig. 36.3 Temporal evolution of forces acting on the armature

36.3.1 Current Density Distribution

The current density within the armature and rails dictates the extent of Joule heating, which in turn influences ablation patterns. Figure 36.4 presents the current density distribution at different time instants during the launch process.

Simulation results indicate that regardless of the time instance, current density is primarily concentrated in the rear groove of the armature. This localization results from the proximity effect and skin effect, which drive current toward the conductor edges, exacerbating heating and material degradation.

Further analysis of peak current density at the rear groove of the armature as a function of input pulse current amplitude is depicted in Fig. 36.5.

Results demonstrate a direct correlation between input pulse current and maximum current density. As the pulse current increases, the corresponding maximum current density also rises, peaking at 0.6 ms when the input pulse reaches 20 kV.

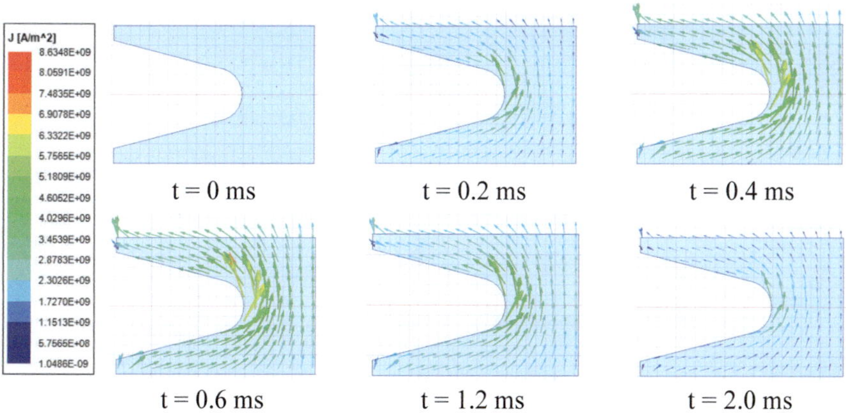

Fig. 36.4 Current density distribution in the armature at different moments

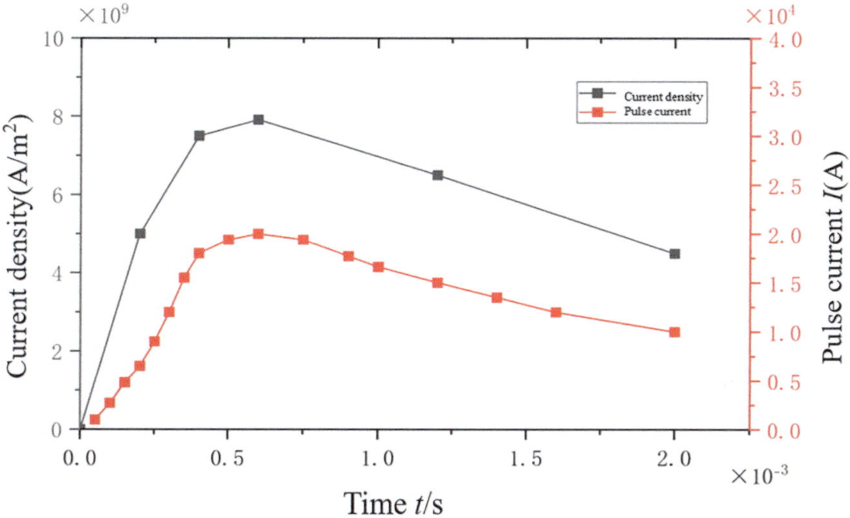

Fig. 36.5 Variation of peak current density with input pulse current amplitude

36.3.2 *Magnetic Flux Density Distribution*

Magnetic flux distribution follows a similar trend to current density due to the Ampère-Maxwell relationship. Figure 36.6 illustrates the distribution of magnetic induction intensity at different time instants.

The magnetic flux density is most concentrated at the rear groove of the armature and along the inner edges of the rails. At 0.6 ms, the peak magnetic flux density reaches 6.15 T, aligning with the highest current density region.

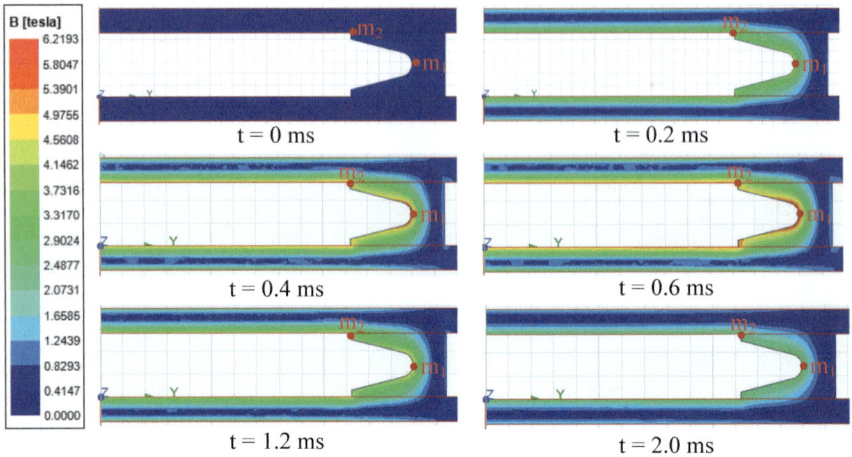

Fig. 36.6 Magnetic induction intensity in the railgun at different moments

A comparison of magnetic flux density at the armature groove and rail contact point over time is shown in Fig. 36.7.

Similar to current density, the magnetic flux follows the waveform of the input pulse current. The armature groove experiences consistently higher magnetic flux compared to the rail contact point, indicating significant localized heating effects.

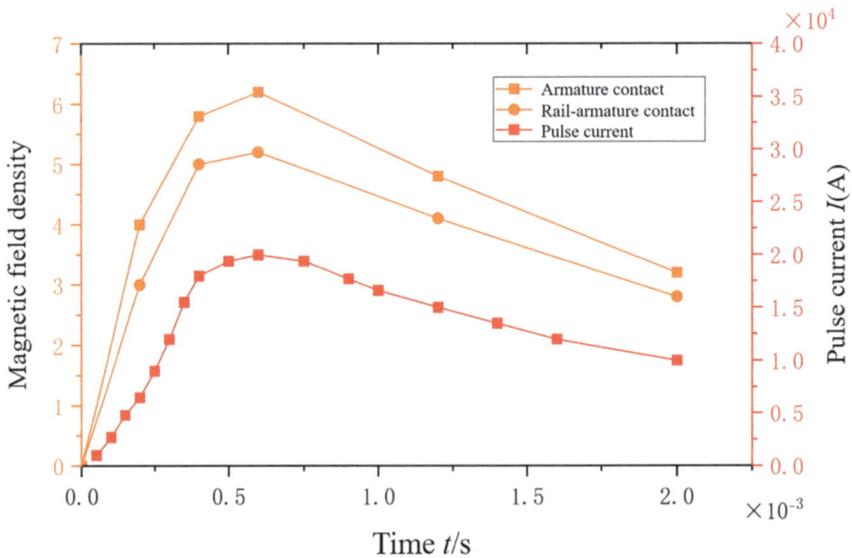

Fig. 36.7 Temporal evolution of magnetic flux density at key locations

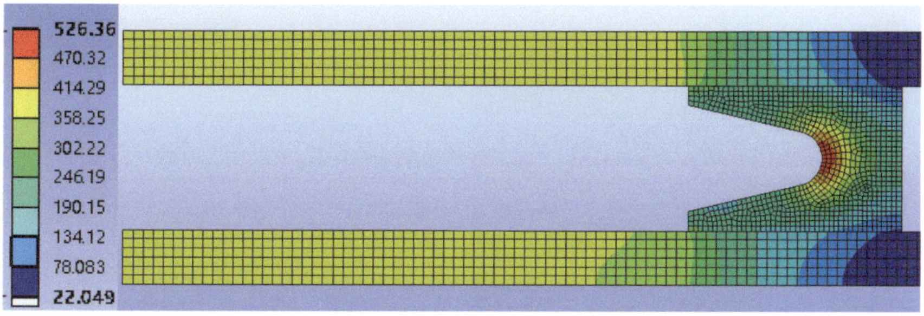

Fig. 36.8 Temperature field simulation results at 2 ms

36.3.3 Temperature Distribution in the Railgun System

The interplay between current density, magnetic flux density, and thermal effects determines the temperature distribution across system components. Figure 36.8 depicts the temperature field at 2.0 ms.

Results highlight that the highest temperature is concentrated in the rear groove of the armature, where Joule heating is most pronounced. Additionally, the rail-armature contact region experiences significant heating due to frictional effects and arc-induced energy dissipation.

A cross-sectional analysis of the temperature distribution at the armature-rail interface is provided in Fig. 36.9.

Temperature gradients align with the current and magnetic field distributions. The skin effect confines heat generation to the surface layers, exacerbating localized material degradation and potential failure.

36.4 Ablation Mechanism and Its Implications

Ablation in electromagnetic railguns is primarily driven by three mechanisms: Joule heating-induced surface erosion, frictional wear, and arc-induced material vaporization.

36.4.1 Early-Stage Ablation: Friction and Joule Heating

During initial motion, the armature overcomes static friction, generating heat due to mechanical contact. This stage is characterized by friction-induced wear and moderate Joule heating.

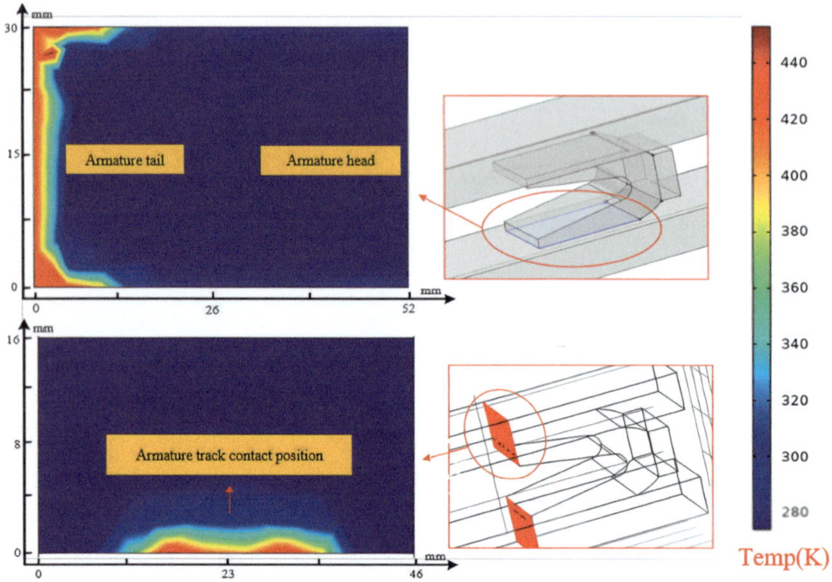

Fig. 36.9 Temperature distribution of the armature-rail interface

36.4.2 Late-Stage Ablation: Arc-Induced Degradation

As velocity increases, contact resistance at the rail-armature interface rises, leading to arc formation. These arcs concentrate energy within localized regions, causing material vaporization and severe erosion.

36.5 Conclusion

This study systematically investigates the electromagnetic, thermal, and mechanical interactions governing railgun performance. Key findings include:

Multi-physics coupling effects: Current density and magnetic flux distributions dictate thermal behavior, influencing ablation patterns.

Early-stage wear mechanisms: Frictional heating and Joule heating drive initial material degradation.

Late-stage degradation: Arc-induced energy concentration exacerbates surface erosion, reducing railgun lifespan.

Further research is required to refine late-stage arc dynamics and optimize rail materials for enhanced durability.

References

1. Harry, D.F.: Electromagnetic launch science and technology in the United States enters a new era. IEEE Trans. Magn. 41(1), 158–164 (2005).
2. Wang, G., Xie, L., Wang, Q., Song, S., Hu, X., Wang, W.: Electromagnetic mechanics analysis of electromagnetic railgun. J. Gun Launch Control 2011.
3. Yang, Y., Wang, J., Xue, W.: Analysis and simulation of velocity skin effect in railgun. High Power Laser Part. Beams 2011.
4. He, D.: Optimization design of interior ballistics for electromagnetic railgun. Master's thesis, Nanjing University of Science and Technology (2008).
5. Ellis, R.L., Poynor, J.C., McGlasson, B.T., et al.: Influence of bore and rail geometry on an electromagnetic naval railgun system. IEEE Trans. Magn. 41(1), 182–187 (2005).
6. Keshtkar, A., Maleki, T., Kalantarnia, A.: Determination of optimum rails dimensions in railgun by Lagrange's equations. IEEE Trans. Magn. 45(1), 594–597 (2009).
7. Mirzaei, M., Abdollahi, S.E.: Design optimization of reluctance synchronous linear machines for electromagnetic aircraft launch system. IEEE Trans. Magn. 45(1), 389–395 (2009).
8. Ramezani, E., Spahn, E., Bruderer, G.: A novel high current rate SCR for pulse power applications. In: 11th IEEE International Pulsed Power Conference, Baltimore, pp. 1016–1020 (1997).
9. Zhang, H.: Optimization design of key parameters for electromagnetic railgun. Master's thesis, Yanshan University, Hebei (2011).
10. Nearing, J.C., Huerta, M.A.: Skin and heating effects of railgun current. IEEE Trans. Magn. 25(1), 381–386 (1989).
11. Deadrick, F.J., Hawke, R.S., Scudder, J.K.: MAGRAC: a railgun simulation program. IEEE Trans. Magn. 18(1), 94–104 (1980).
12. Engel, T.G., Neri, J.M., Veracka, M.J.: Characterization of the velocity skin effect in the surface layer of a railgun sliding contact. IEEE Trans. Magn. 44(7), 1837–1844 (2008).

Chapter 37
Atmospheric Pressure Array Air Plasma Jet for Plasma Activated Water Generation and Its Application for *E. Coli* Inactivation

Xi Zhang, Hongji Li, Yikui Zeng, Shijie Hu, Jinping Zhong, Tiejian Liu, Keliang Wang, and Fada Feng

Abstract Plasma activated water (PAW) is produced through the interaction of water with plasma that contains a large number of active species. PAW has been widely used in sterilization, agriculture, and medical treatment. In this study, an array plasma jet composed of 6 DBD reactors was utilized to activate water. The pH value, conductivity, temperature, and total residual oxidants (TRO) were measured to characterize the PAW generated by the plasma jet under different treatment times, gas flows, and storage durations. Optical emission spectra (OES) of the plasma jet were recorded. The performance of *E. coli* inactivation was investigated. The results demonstrated successful generation of PAW using the array plasma jet. It was observed that the pH value decreased rapidly within several minutes and remained relatively stable during

Author Contributions: Conceptualization, X.Z., H.L. and S.H; methodology, S.H, T.L. and F.F.; validation, Y.Z., S.H, J.Z and F.F.; formal analysis, X.Z., S.H and T.L.; investigation, X.Z., H.L. and F.F.; data curation, Y.Z. and F.F.; writing—original draft preparation, X.Z. and H.L.; writing—review and editing, X.Z., H.L., Y.Z., S.H, J.Z, T.L., K.W. and F.F.; supervision, S.H, T.L., K.W. and F.F.; project administration, Y.Z. and F.F.; funding acquisition, F.F. All authors have read and agreed to the published version of the manuscript.

X. Zhang · K. Wang
Guangdong Polytechnic of Environmental Protection Engineering, Foshan, China

H. Li · T. Liu (✉)
Department of Neurosurgery, Third Affiliated Hospital of Southern Medical University, Guangzhou, China
e-mail: 183375865@qq.com

Y. Zeng · J. Zhong · F. Feng (✉)
School of Chemistry and Environment, Jiaying University, Meizhou, China
e-mail: fadafeng@jyu.edu.cn

S. Hu (✉)
Department of Laboratory Medicine, The Third Affiliated Hospital of Sun Yat-Sen University, Yuedong Hospital, Meizhou, China
e-mail: 15986494596@163.com

Y. Li et al. (eds.), *Proceedings of the 5th Chinese Conference on High Voltage and Discharge Plasmas*, Springer Proceedings in Physics 254,
https://doi.org/10.1007/978-981-96-5711-7_37

prolonged treatment periods. The conductivity and concentration of TRO increased significantly with the treatment time increased. The inactivation efficiency could be as high as 99.9999% when the PAW prepared by array plasma jet with 20 min treatment. Both the pH value of PAW and concentration of active species had a significant impact on its inactivation efficiency. The PAW had very high inactivation efficiency when it was used immediately after preparation, and it still had a certain sterilization ability with long storage time, which was of important for the application of PAW. The OES suggested that active species such as excited molecules, ions and radicals were produced in the air plasma jet. The active species were the origin of PAW generation, and the reactive oxygen and nitrogen species were responsible for the inactivation of *E. coli*.

Keyword Inactivation; Plasma activated water; Plasma jet; Non-thermal plasmas

37.1 Introduction

Plasma is characterized by the existence of electrons, ions, as well as highly excited atoms, radicals, and molecules in mixture with the other neutrals of gaseous systems [1]. In non-thermal plasma, the energy of electron can be as high as 1–20 eV, so the reactive processes that require extremely high activation energy can be realized. The chemical effects occurred in an electrical discharge are the consequence of energy injection into a gas stream through electron-impact processes under the influence of an electric field. When the species generated in the gas phase plasma interact with liquid, a series of reactions take place at the gas–liquid interface leading to the formation of new species in the water. Consequently, the plasma activated water (PAW) contains numerous active species such as NO_2^-, NO_3^-, H^+, H_2O_2, O_3, $ONOO^-$ and O_2^-, etc. which presents outstanding biological activities in agricultural and biomedical applications [2]. Plasma dose is very important for the characterization of the biological effect of plasma, but it is difficult to clear define the plasma dose in the application. Fortunately, the plasma dose defined by equivalent total oxidation potential (ETOP) is proposed by Cheng et al., which is based on the oxidation potential of reactive oxygen and nitrogen species (RONS) [3]. The ETOP provides a new insight into the fundamentals of plasma agricultural and biomedical applications. For PAW, the RONS can be detected by several qualitative and quantitative methods [4]. In the plasma activated process, the total residual oxidants (TRO) are mainly composed of H_2O_2, O_3, $\cdot OH$ radicals and other active oxygen particles such as HO_2^-, O_2^-, $HO_3\cdot$, $O_2^+H_2O$, etc., which are formed during the $\cdot OH$ chain reactions in water [5]. Therefore, the TRO can be used to represent the activity of PAW.

In recent years, PAW has attracted significant attention in the food and agricultural industries, primarily for its application in food preservation [6–8]. In terms of fruits and vegetables, PAW can inhibit the respiration of cells and inactivate microorganisms while degrading pesticide residues, and the soluble solids and VC content can be maintained as well. PAW leads to the completely close of the epidermal pores and change the physiological state of the fruits and vegetables after harvest, so that the nutrient consumption reduced, and water transpiration inhibited. It can effectively delay the decline of fruit firmness and induce the fruits or vegetables to produce various enzymes and other related substances that resist pathogenic microorganisms, thereby greatly enhancing the disease resistance of fruits and vegetables [9, 10]. Experimental evidence demonstrates that treating seeds with PAW can break seed dormancy, enhances germination rates, as well as promotes root growth and germ development [10]. Additionally, PAW can serve as a nitrogen fertilizer by providing ammonia-containing fertilizers for plant growth promotion [11].

Plasma medicine has been extensively studied for over five decades. Atmospheric plasma jets have found wide-ranging applications in skin disinfection, wound treatment including skin tumor therapy and caries treatment [12]. Water activated by plasma is also used in medical sterilization such as dentistry or wound care due to its high efficiency, rapid action on various bacteria and biofilms along with its eco-friendly nature when directly applied to the human body. For healing or tumor treatment procedures involving tissue exposure to aquatic fluid layers are common practice. The plasma-generated RONS activate and interact with the liquid before it can affect the tissue [13]. The RONS play an important role in the process of plasma inactivation of bacteria [14]. The uniformity of the solution and the good fluidity of the fluid allow the active components in the PAW to evenly cover the infected wound biofilm. In the sterilization process, the current and ultraviolet rays, radiation injure can be avoid by using PAW [13]. Moreover, PAW serves as an effective agent for surface decontamination and finds applications in disinfection and sterilization of medical equipment. PAW can also be combined with other disinfectants, for example, the combination of PAW and 3% H_2O_2 can effectively whiten teeth without damaging the oral mucosa, so it can be used for oral cavity sterilization [15]. In biomedicine, applications of PAW in antivirus attract attention. Guo et al. proposed that PAW can be used as an alternative disinfectant for S protein inactivation to prevent SARS-CoV-2 infection [16].

In the present work, a home-made PAW production reactor was introduced. The PAW was generated by an atmospheric pressure array air plasma jet composed of 6 DBD reactors. The variations of pH value, conductivity, TRO concentration, and temperature of the PAW were experimentally investigated immediately and up to 30 min after plasma jet treatment. The performance of inactivation efficiency of *E. coli* in water samples was evaluated. Based on the optical emission spectra (OES) of the air plasma jet, the RONS generation in PAW was initially described. Additionally, the possible mechanism of inactivation was discussed.

37.2 Experimental Section

Figure 1a shows the experimental setup used in this work. The PAW was generated by an atmospheric pressure array air plasma jet composed of 6 DBD reactors with electrode structure of cylindrical stainless-steel pipe-quartz tube-copper rod, in which the inner diameter of cylindrical stainless-steel pipe was about 4 mm, and the thickness and inner diameter of a quartz tube was about 0.75 mm and 0.5 mm, respectively. The stainless-steel pipes were connected to grounded electrode, and the inner copper rod high-voltage electrode was about 0.5 mm in diameter, and the discharge gap was 1 mm. A high frequency AC high-voltage power source was used to energize the DBD reactors. The AC high-voltage power source had been described in detail in our previous work [17], and the output voltage and frequency of the power supply were adjusted to about 5.2 kV and 16 kHz via a high voltage probe (HVP) (P5104, Tektronix, USA) combined with a digital oscilloscope (DS1102E, Rigol), respectively. And the input energy was about 35 W that was detected by a built-in dynamometer. The flow rate of air with relative humidity of 18% was controlled by a mass-flow controllers (MFC). The experiment was carried out at atmospheric pressure and room temperature. Figure 1b, c shows the DBD reactors without and with discharge, respectively. The plasma jet was generated well.

For PAW generation, the deionized water with volume of 100 ml in a beaker was placed below the array plasma jet and the outlet of DBD reactors were placed 10 mm below the water surface. In addition, according to the experimental requirements, the pH of deionized water was adjusted to 4, 7 and 10 by sulfuric acid and sodium hydroxide solution, respectively. The pH, conductivity and temperature were determined using pH (pH3-3C, INESA Instrument, China), and conductivity-temperature tester (DDB-303A, INESA Instrument, China), respectively. Total residual oxidants (TRO) concentration was tested by an UV/Vis spectrophotometer (Cary 300, Agilent, USA) with DPD method. The OES of the ionized and excited species in the plasma jet were recorded by a multichannel spectrometer (Avantes, Netherlands). The

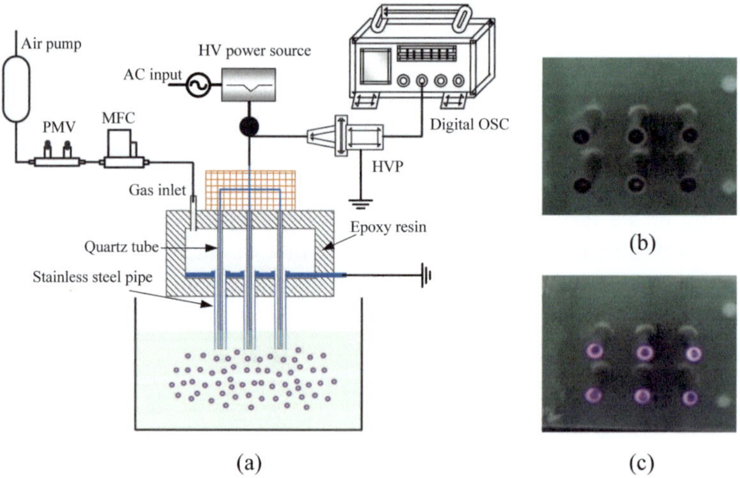

Fig. 37.1 Schematic diagram of experimental setup **a**, DBD reactors without discharge (**b**) and with discharge (**c**)

experiment was carried out in the air that before the plasma jet was placed into water.

E. coli (ATCC8739, Guangdong Huankai Microbial Technology Co., Ltd, China) was used to evaluate the inactivation efficiency of the PAW. The PAW prepared and bacterial solution were mixed in a 10 ml tube at a ratio of 1:9 for 10 min. After forming a mixed solution, it was diluted to an appropriate multiple according to the experimental requirements, 100 μl solution was sucked out with a pipette and inoculated on a solid agar plate using the coating method. And then the plate was put into a constant temperature incubator at 37 °C for 18–24 h. After culturing, the number of single colony cells were counted with a colony counter, and then the cell survival rate was calculated.

LB liquid medium and LB solid culture solution were prepared for the culture of *E. coli*. Preparation of *E. coli* suspension: Take out the prepared third-generation slanted strains, light the alcohol lamp to burn the inoculation ring in the ultra-clean workbench, and then gently scrape a ring of bacteria. The cells were inoculated into LB liquid medium, placed in a small-capacity vertical shaker at 37 °C, and the speed was adjusted to 200 rpm for 8–10 h, and then the cells were collected by high-speed centrifugation at 10,000 rpm for 2 min, and the cells were resuspended.

The enrichment culture of *E. coli* was carried out using LB liquid medium. A ring of bacteria from the third-generation slanted strains was scraped and inoculated into LB liquid medium, and then cultured in a vertical shaker at 37 °C with a rotation speed of 200 rpm for 10 h. The absorbance of bacterial solution was detected at a wavelength of 600 nm using a spectrophotometer every 1 h. Figure 2a shows the growth curve of *E. coli*. The OD_{600} of the bacterial solution was about 2.1 A over an 8 h culturation. As the density of *E. coli* in the bacterial solution was about 10^9 cfu/mL, it indicated that the growth of cells reached the log-phase and the *E. coli* were most active. Figure 2b shows a typical image of *E. coli* in the LB medium.

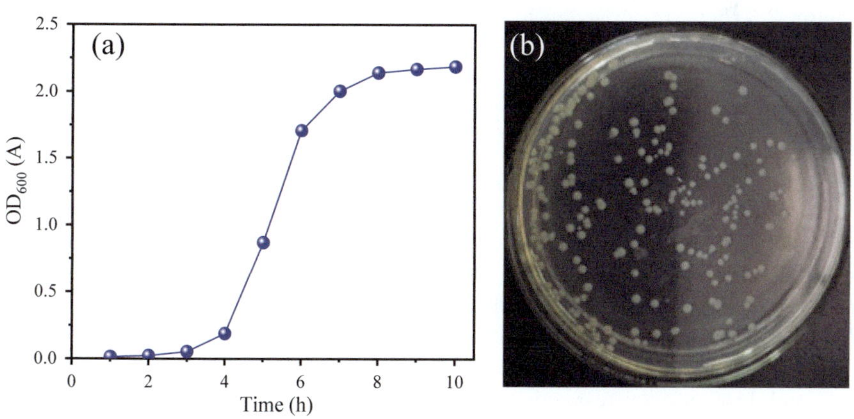

Fig. 37.2 Typical growth curve of *E. coli* (**a**) and image of *E. coli* in the LB medium (**b**)

Plate colony counting method was used in this experiment, which was an effective method to count the bacterial content of articles. By counting the number of colonies, the number of bacteria in the sample can be converted according to the dilution ratio of the bacterial solution and the sampling inoculum. The analysis of cell viability adopts the heterotrophic plate counting method (HPC). 100 μl of *E. coli* bacterial solution was introduced on the agar plate with a pipette and spreaded evenly with a coating rod, and then culturation in a biochemical incubator at 37 °C for 18–24 h. A colony with diameter of 1 mm or a single colony represents a single living cell in the original sample. In order to reduce the error of the experiment, three culture dishes were set for each concentration gradient, and then CFU was counted, and the average of the three dishes was taken to calculate the concentration of *E. coli* bacteria according to the dilution ratio. Agar plates with a colony count of 30–300 were available data. For the blank bacterial solution that was not mixed with PAW, the number of colonies in the plates was counted.

The inactivation efficiency of *E. coli* is calculated as:

$$h = \frac{(C_0 - C_1)/C_0}{100}\%$$

where C_0 and C_1 is the concentration of *E. coli* colony before and after PAW inactivation, respectively.

37.3 Results and Discussion

37.3.1 PAW Generation

Figure 37.3 shows the time development of pH, conductivity, concentration of TRO, and temperature in water with plasma jet and without plasma treatment. The pH value of PAW decreased quickly as the water activated by the array plasma jet, for example, decreased to 3.25 within 5 min and continued to decrease gradually over time due to acidification caused by the formation of hydrogen peroxides, nitric acid, and peroxynitrous acid as secondary products resulting from the interaction between NOx and water [18, 19]. Interestingly, it showed a reduction of pH in deionized water without plasma treatment, for example, the pH value decreased to 3.96 within 5 min, because of the CO_2 in the air dissolved into water. As the CO_2 in the air reaches equilibrium in the water, the pH value remains stable. The conductivity of PAW increased significantly with increasing treatment time, for instance, after 1 min and 30 min of array plasma jet activation, the conductivity reached approximately 28.7 and 1182 μS/cm, respectively. Previous studies have demonstrated that the active species generated during plasma discharge are highly soluble in water, and the formation of ROS and RNS will contribute to a significant increase in the conductivity of PAW [8]. As shown in Fig. 3c, the concentration of TRO in PAW was dependent on the duration of plasma activation. When the water was activated by array plasma jet, the active species produced during plasma process reacted with the solution and leaded to the formation of secondary active species. As the lifetime of ·OH radicals and other active oxygen radicals was very short, so most of the TRO detected in PAW were long-lived species such as H_2O_2 and O_3. The temperature of the PAW increased slightly with the increase of the treatment time, and it remained relatively stable after 20 min treatment time (Fig. 3d). In the process of the plasma generation by DBD, the discharge channel was heated by the electric current, and as the plasma jet entered the water, the energy carried was injected into the PAW, besides, the heat in the water was exchanged with its surroundings, resulted in the temperature to rise and remain relatively stable.

Figure 37.4 shows the change of pH, conductivity, concentration of TRO, and temperature in water with different gas flow by plasma jet and without plasma treatment. There was little change in pH with different gas flow because the pH was relatively stable when the pH was decreased to a certain level, which was agree well with the results in Fig. 3a. The conductivity decreased with the increase of gas flow because the residence time in the DBD region was short at higher gas flow and

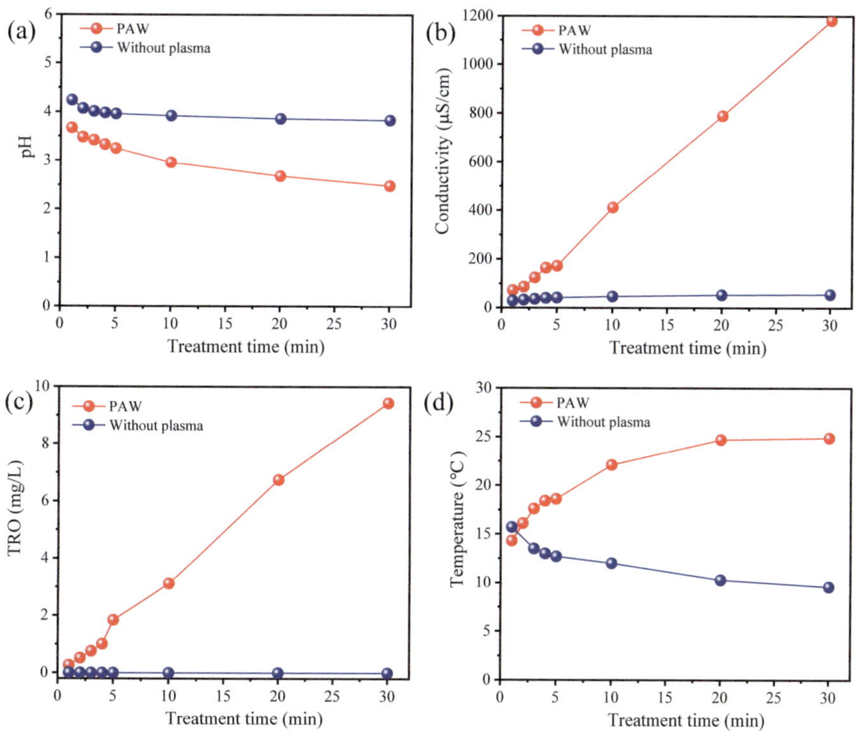

Fig. 37.3 The time development of pH **a**, conductivity **b**, TRO (**c**), and temperature (**d**) in water with plasma jet and without plasma treatment

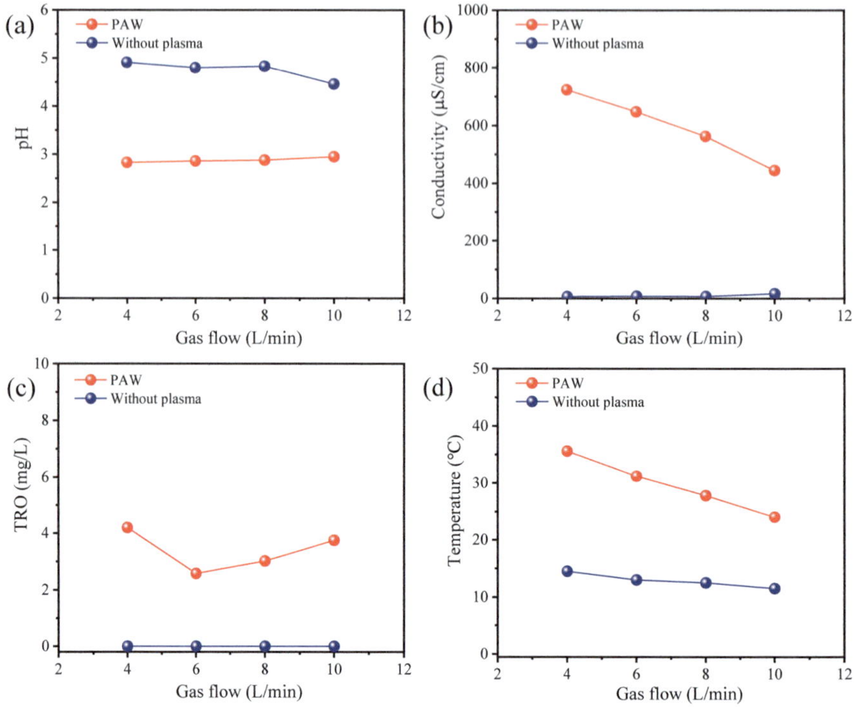

Fig. 37.4 The change of pH **a**, conductivity **b**, TRO (**c**), and temperature (**d**) in water with different gas flow by plasma jet and without plasma treatment

the concentration of active species generated by the plasma was relatively low. The concentration of TRO was detected with a maximum at a smaller flow rate of 6 L/min but changed very little with the variation of gas flow. The temperature of the PAW decreased as the gas flow rate increased because the heat generated in discharge would be taken away by the gas, resulting in a decrease in temperature.

Figure 37.5 shows the variation of pH, conductivity and TRO in PAW as the storage time increased. The change of pH in PAW was negligible. The acid–base ions in PAW were relatively stable chemicals, such as nitrogen ions, therefore, there was almost no change in PAW with time increased. The conductivity decreased with storage time, because some short-lived active species recombined and disappeared, and some compounds decomposed after the PAW prepared. As after the PAW was prepared, the concentration of TRO decreased quickly. For example, the TRO concentration decreased from 8.75 mg/L to 2.73 mg/L, and 68.8% of the TRO was dissipated in the water within 5 min. However, with the increased of storage time, the decreased of concentration of TRO was relatively slow. It indicated that once the PAW was prepared, it should be used as soon as possible. Furthermore, previous studies have reported that most of the characteristics such as pH, conductivity, NO_3^-, and H_2O_2 tend to remain stable even during long-term storage of PAW [20]. Therefore, it can

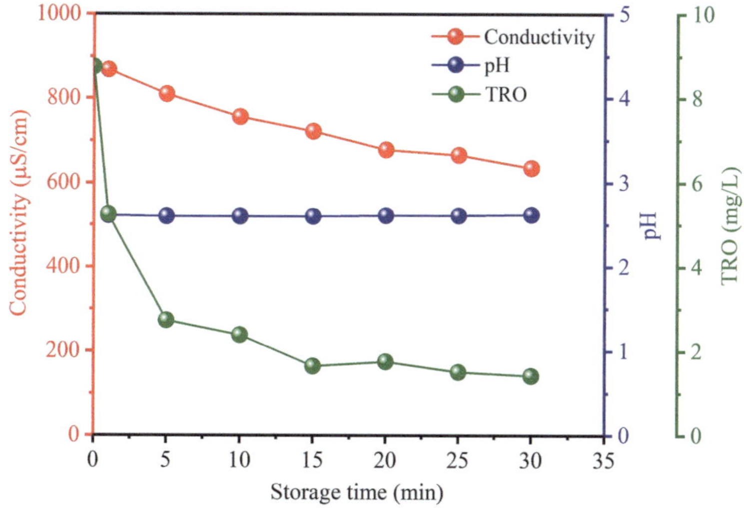

Fig. 37.5 The time development of pH, conductivity and TRO in PAW after preparation

be concluded that PAW remains effective for the situations that need to be used for a long time.

37.3.2 E. Coli Inactivation Using PAW

Figure 37.6 shows the inactivation efficiency of *E. coli* using PAW with different treatment time, gas flow, pH, and storage time. The treatment time for water activated significantly affected the inactivation efficiency of PAW, for example, when the treatment time increased from 30 s to 3 min, the inactivation efficiency increased from 2.0423% to 99.2767% (Fig. 6a). Furthermore, the inactivation efficiency of *E. coli* by PAW could be as high as 99.9999% when the water activated by array plasma jet for 20 min. The inactivation efficiency of *E. coli* decreased with the gas flow rate increased, for example, when the gas flow rate increased from 2 L/min to 10 L/min, the corresponding inactivation efficiency decreased from 99.9995% to 96.22% (Fig. 6b). The results obtained agreed well with the concentration of TRO with different gas flow in Fig. 4d. As shown in Fig. 6c, the pH value of the solution had a significant effect on the inactivation efficiency of *E. coli*. Especially, when the solution was in alkaline condition, the inactivation efficiency decreased significantly, because the active species in the PAW were neutralized by the base. As shown in Fig. 6d, when the PAW was prepared, the inactivation efficiency decreased as the storage time increased. It showed a reduction of inactivation efficiency from more than 99.99% to 86.1084% within 5 min due to the active species in PAW decreased with the storage time increased. The RONS in PAW with relatively short-lived such

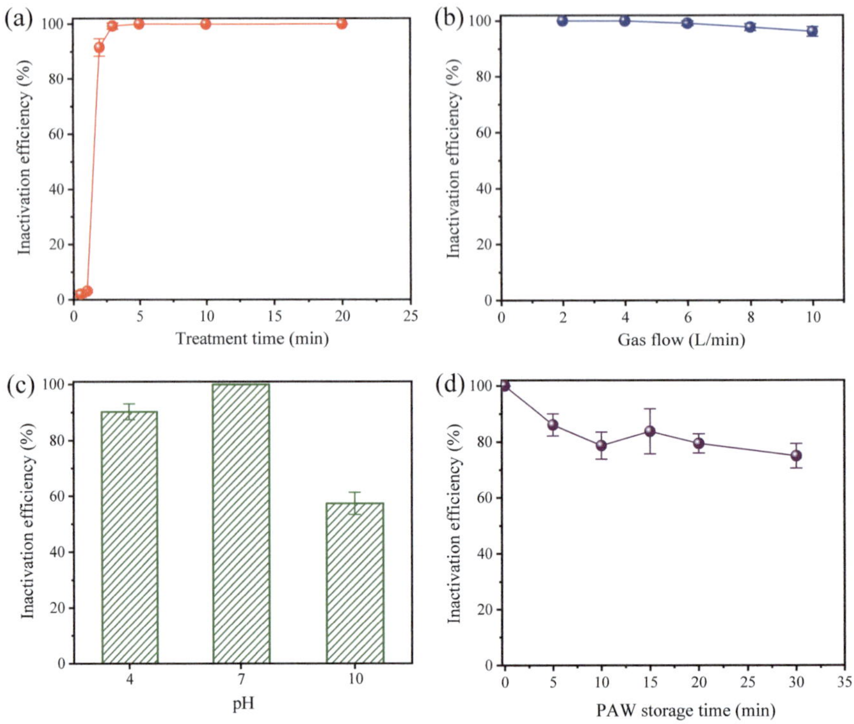

Fig. 37.6 The *E. coli* inactivation efficiency using PAW with different treatment time **a**, gas flow **b**, pH (**c**), and storage time (**d**)

as ·OH, NO·, O_2^-, $OONO_2^-$, and $OONO^-$, the half-lifetimes of which are proved to be ns to several seconds [21]. Interestingly, the storage time was increased from 5 to 30 min with only a little decrease in inactivation efficiency which agreed well with the concentration of TRO with different storage time in Fig. 5d. It is proposed that the concentration of H_2O_2 in PAW does not significantly decrease for dozens of days [20], which is the reason why PAW can still have a high sterilization effect long after it is prepared.

37.3.3 Mechanism of PAW Generation and Inactivation

The active species produced in the plasma jet are the origin of PAW generation [2]. Figure 37.7 shows the typical optical emission spectra of the plasma jet in air. Very strong intensity emission line of N_2 second positive system (SPS) (C-B) band between 315 and 405 nm was observed. In addition, the first negative system (FNS) N_2^+(B-X), N_2(B-A), N^+, and N bands were observed. All these N-based active species can be produced by direct electron impact excitations [13]. The O atom (446.8, 715.6,

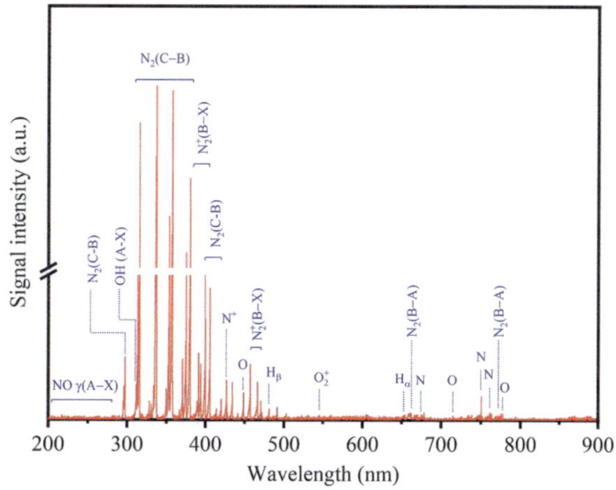

Fig. 37.7 Typical optical emission spectra of the plasma jet in air

and 777.4 nm) can be clearly observed in the emission spectra [22]. The bands of OH(A–X) around 295.1 and 309.2 nm are found [23]. Clearly, the emission spectrum of $OH(A^2\Sigma \rightarrow X^2\Pi, (0-0))$ is interfered by the emission spectrum of the $\Delta v = +1$ vibration transition band of $N_2(C^3\Pi u \rightarrow B^3\Pi g)$ [24]. The emission spectrum from NOγ(A–X), O_2^+, H·, and H· can also be observed in the plasma jet [25].

The discharge gases in plasma jet react with the water to form various ROS and RNS in PAW [26, 27]. The existence of N-based species, O, H and OH radical in the plasma-water interface region is very important for the PAW generation [20]. Nitrogen oxides, such as NO, NO_2, can be produced by discharge in air from several pathways (R1-R6) [22]. In the plasma process, most of the O_3 formed via the three-body reaction between O_2 and O, and the O_3 can also be generated in the bubble by the reaction between the O in the plasma jet and the O_2 in the bubble, and then the O_3 enters the liquid phase through the gas–liquid interface (R7) [28]. In the DBD reactors and plasma jet in air, the reaction of e and H_2O is responsible for the formation of OH (R8). The formation of H_2O_2 by OH radical recombination (R9) can occur in the gas, the liquid, or at the interface of gas–liquid [29]. O_3 in the gas and liquid has an important influence on the production of H_2O_2 (R10). Since the OH in the gas phase has a short lifetime and it is difficult to directly enter the liquid phase, the reaction between dissolved O_3 and H_2O_2 in water is an important source of OH in liquid (R11) [13]. The generated NOx and OH radical reacts with each other and H_2O in plasma jet and the gas–liquid interface, and then dissolved in water to form NO_2^- and NO_3^- ions in the form of nitric and nitrous acid (R12-R15) [21]. Noted that, the NO_2^- in liquid can be oxidized to NO_3^- by dissolved O_3 (R16) [22].

$$N_2 + e \rightarrow 2N + e \tag{R1}$$

$$O_2 + e \rightarrow 2O + e \tag{R2}$$

$$O + N \rightarrow NO \tag{R3}$$

$$O + N_2 \rightarrow NO + N \tag{R4}$$

$$NO + O \rightarrow NO_2 \tag{R5}$$

$$2NO + O_2 \rightarrow 2NO_2 \tag{R6}$$

$$O_2 + O \rightarrow O_{3g} \rightarrow O_{3aq} \tag{R7}$$

$$H_2O + e \rightarrow OH + H^+ + e \tag{R8}$$

$$2\,OH \rightarrow H_2O_2 \tag{R9}$$

$$O_3 + H_2O \rightarrow H_2O_2 + O_2 \tag{R10}$$

$$O_3 + H_2O_2 \rightarrow OH + OH_2 + O_2 \tag{R11}$$

$$NO_2 + OH \rightarrow H^+ + NO_3^- \tag{R12}$$

$$2NO_2 + H_2O \rightarrow 2H^+ + NO_2^- + NO_3^- \tag{R13}$$

$$NO_2 + NO + H_2O \rightarrow 2H^+ + 2NO_2^- \tag{R14}$$

$$4NO + O_2 + 2H_2O \rightarrow 4H^+ + 4NO_2^- \tag{R15}$$

$$O_3 + NO_2^- \rightarrow NO_3^- + O_2 \tag{R16}$$

It has been suggested that the acidic environment in PAW is not conducive to the survival of bacteria [6]. The synergistic effect of NO_2^- and H_2O_2 generated during plasma treatment, as well as the reduction of pH, has a significant bacteriostatic effect [30]. The highly oxidative environment in activated water induces oxidative stress in bacteria, leading to the inactivation of bacteria. Dissolved O_3 is recognized as a potent antimicrobial oxidant compound that plays a dominant role in cell inactivation [31]. Due to the long lifetime of O_3 and H_2O_2 in the liquid phase, the OH produced through their reaction within water also serves as an important sterilizing species [32]. The direct erosive impact of active species on bacteria results in damage to the cell structure of bacteria [33]. Active species can destroy the integrity of the outer cell membrane, causing a significant increase in the level of intracellular reactive oxygen species, and its mechanism of action may be related to the destruction of cell structure and enhancement of cell membrane permeability [34]. Therefore, it is proposed that reactive oxygen and nitrogen species play a crucial role in bacterial inactivation.

37.4 Conclusions

PAW was generated using an atmospheric pressure array air plasma jet equipped with 6 DBD reactors. The characteristics of PAW such as pH value, conductivity, temperature, and concentration of TRO were experimental investigated. And the *E. coli* inactivation performance of PAW was evaluated. The pH value of PAW rapidly decreased to 3.25 within 5 min and remained relatively stable throughout extended treatment periods. Both conductivity and TRO concentration exhibited significant increases with longer treatment times. For the water activated by plasma with 30 min, the conductivity and concentration of TRO could be as high as 1182 μS/cm and 7.40 mg/L, respectively. However, with an increase in the gas flow rate of the plasma jet, both conductivity and TRO concentration decreased, indicating that a shorter residence time had a negative impact on PAW production. The decrease in conductivity and TRO concentration immediately after PAW preparation can be attributed to recombination and disappearance of some active species. The inactivation efficiency was found to be as high as 99.9999% when water was activated by an array plasma jet for 20 min. Furthermore, the pH value and the concentration of active species in PAW had a significant impact on the inactivation efficiency. The inactivation efficiency of PAW was very low in the condition of alkaline solution. The PAW had very high inactivation efficiency when it was used immediately after preparation, but it also had a certain sterilization ability with long storage time, which was of important for the application of PAW. Based on the optical emission spectra of the air plasma jet, active species such as N_2 metastable, H metastable, NO metastable, N_2^+, OH, and O were identified, which were the origin of RONS in PAW. The RONS played an important role in the inactivation of *E. coli.*

Funding This research was funded by Guangdong Province Rural Revitalization Strategy Project, grant number KTP20210232 and Guangdong Basic and Applied Basic Research Foundation, grant number 2023A1515010769.

DataAvailability Data Availability Statement All data generated are included within this manuscript. For access to the files, please send an e-mail request to fadafeng@jyu.edu.cn (Fada Feng).

Conflicts of Interest The authors declare no conflicts of interest.

References

1. Tonks, L.; Langmuir, I. Oscillations in ionized gases. Phys. Rev. 1929, 33, 195-210.
2. Rathore, V.; Nema, S.K. The role of different plasma forming gases on chemical species formed in plasma activated water (PAW) and their effect on its properties. Phys. Scr. 2022, 97, 065003.
3. Cheng, H.; Xu, J.X.; Li, X.; Liu, D.W.; Lu, X.P. On the dose of plas ma medicine: Equivalent total oxidation potential (ETOP). Phys. Plasmas 2020, 27, 063514.
4. Hu, X.; Zhang, Y.; Wu, R.A.; Liao, X.; Liu, D.; Cullen, P.J.; Zhou, R.W.; Ding, T. Diagnostic analysis of reactive species in plasma-activated water (PAW): current advances and outlooks. J. Phys. D: Appl. Phys. 2021, 55, 023002.
5. Bai, M.; Zhang, Z.; Zhang, N.; Tian, Y.; Chen, C.; Meng, X. Treatment of 250 t/h Ballast Water in Oceanic Ships Using ·OH Radicals Based on Strong Electric-Field Discharge. Plasma Chem. Plasma Process. 2012, 32, 693-702.
6. Xiang, Q.; Fan, L.; Li, Y.; Dong, S.; Li, K.; Bai, Y. A review on recent advances in plasma-activated water for food safety: current applications and future trends. Crit. Rev. Food Sci. Nutr. 2022, 62, 2250-2268.
7. Hernández-Hernández, H.M.; Moreno-Vilet, L.; Villanue va-Rodríguez, S.J. Current status of emerging food processing technologies in Latin America: Novel non-thermal processing. Innov. Food Sci. Emerg. Technol. 2019, 58, 102233.
8. Rahman, M.; Hasan, M.S.; Islam, R.; Rana, R.; Sayem, A.S.M.; Sad, M.A.; Matin, A.; Raposo, A.; Zandonadi, R.P.; Han, H.; Ariza-Montes, A.; Vega-Muñoz, A.; Sunny, A.R. Plasma-Activated Water for Food Safety and Quality: A Review of Recent Developments. Int. J. Environ. Res. Public Health 2022, 19, 6630.
9. Krapp, A.; David, L.C.; Chardin, C.; Girin, T.; Marmagne, A.; Leprince, A.-S.; Chaillou, S.; Ferrario-Méry, S.; Meyer, C.; Daniel-Vedele, F. Nitrate transport and signalling in Arabidopsis. J. Exp. Bot. 2014, 65, 789-798.
10. Rashid, M.M.; Rashid, M.; Hasan, M.M.; Talukder, M.R. Rice plant growth and yield: foliar application of plasma activated water. Plasma Sci. Technol. 2021, 23, 075503.
11. Kaushik, N.K.; Bhartiya, P.; Kaushik, N.; Shin, Y.; Nguyen, L.N.; Park, J.S.; Kim, D.; Choi, E.H. Nitric-oxide enriched plasma-activated water inactivates 229E coronavirus and alters antiviral response genes in human lung host cells. Bioact. Mater. 2023, 19, 569-580.
12. Lu, X.; Naidis, G.V.; Laroussi, M.; Reuter, S.; Graves, D.B.; Ostrikov, K. Reactive species in non-equilibrium atmospheric-pressure plasmas: Generation, transport, and biological effects. Phys. Rep. 2016, 630, 1-84.
13. Machala, Z.; Tarabová, B.; Sersenová, D.; Janda, M.; Hensel, K. Chemical and antibacterial effects of plasma activated water: correlation with gaseous and aqueous reactive oxygen and nitrogen species, plasma sources and air flow conditions. J. Phys. D: Appl. Phys. 2018, 52, 034002.
14. Li, Y.; Zhang, T.; Song, Z.; Ding, C.; Chen, H. Molecular mecha nism study of Astragalus adsurgens Pall synergistically induced by plasma and plasma-activated water. Plasma Sci. Technol. 2022, 24, 075502.
15. Pan, J.; Li, Y.L.; Liu, C.M.; Tian, Y.; Yu, S.; Wang, K.L.; Zhang, J.; Fang, J. Investigation of Cold Atmospheric Plasma-Activated Water for the Dental Unit Waterline System Contamination and Safety Evaluation in Vitro. Plasma Chem. Plasma Process. 2017, 37, 1091-1103.
16. Guo, L.; Yao, Z.; Yang, L.; Zhang, H.; Qi, Y.; Gou, L.; Xi, W.; Liu, D.; Zhang, L.; Cheng, Y.; Wang, X.; Rong, M.; Chen, H.; Kong, M.G. Plasma-activated water: An alternative disinfectant for S protein inactivation to prevent SARS-CoV-2 infection. Chem. Eng. J. 2021, 421, 127742.
17. Liu, T.; Zeng, Y.; Xue, X.; Sui, Y.; Liang, Y.; Wang, F.; Feng, F. He-Plasma Jet Generation and Its Application for E. coli Sterilization. J. Spectrosc. 2021, 2021, 6671531.
18. Kaushik, N.K.; Ghimire, B.; Li, Y.; Adhikari, M.; Veerana, M.; Kaushik, N.; Jha, N.; Adhikari, B.; Lee, S.J.; Masur, K.; von Woedtke, T.; Weltmann, K.-D.; Choi, E.H. Biological and medical applications of plasma-activated media, water and solutions. Biol. Chem. 2019, 400, 39-62.

19. MedveckÁ, V.; Omasta, S.; Klas, M.; MoŠOvskÁ, S.; Kyzek, S.; ZahoranovÁ, A. Plasma activated water prepared by different plasma sources: physicochemical properties and decontamination effect on lentils sprouts. Plasma Sci. Technol. 2021, 24, 015503.
20. Vlad, I.-E.; Anghel, S.D. Time stability of water activated by different on-liquid atmospheric pressure plasmas. J. Electrostat. 2017, 87, 284-292.
21. Zhou, R.; Zhou, R.; Wang, P.; Xian, Y.; Mai-Prochnow, A.; Lu, X.; Cullen, P.J.; Ostrikov, K.K.; Bazaka, K. Plasma-activated water: generation, origin of reactive species and biological applications. J. Phys. D: Appl. Phys. 2020, 53, 303001.
22. Man, C.; Zhang, C.; Fang, H.; Zhou, R.; Huang, B.; Xu, Y.; Zhang, X.; Shao, T. Nanosecond-pulsed microbubble plasma reactor for plasma-activated water generation and bacterial inactivation. Plasma Process. Polym. 2022, 19, 2200004.
23. Lu, P.; Boehm, D.; Bourke, P.; Cullen, P.J. Achieving reactive species specificity within plasma-activated water through selective generation using air spark and glow discharges. Plasma Process. Polym. 2017, 14, 1600207.
24. Feng, F.; Zheng, Y.; Shen, X.; Zheng, Q.; Dai, S.; Zhang, X.; Huang, Y.; Liu, Z.; Yan, K. Characteristics of Back Corona Discharge in a Honeycomb Catalyst and Its Application for Treatment of Volatile Organic Compounds. Environ. Sci. Technol. 2015, 49, 6831-6837.
25. Li, Y.; Song, Z.; Zhang, T.; Xu, W.; Ding, C.; Chen, H.; Krishna Rao, K.S.V. Spectral Characteristics of Needle Array-Plate Dielectric Barrier Discharge Plasma and Its Activated Water. J. Spectrosc. 2021, 2021, 9771245.
26. Rathore, V.; Patel, D.; Butani, S.; Nema, S.K. Investigation of of Plasma Activated Water and its Bactericidal Efficacy. Plasma Chem. Plasma Process. 2021, 41, 871-902.
27. Rathore, V.; Nema, S. Optimization of process parameters to generate plasma activated water and study of physicochemical properties of plasma activated solutions at optimum condition. J. Appl. Phys. 2021, 129, 084901.
28. He, X.; Zeng, Y.; Chen, J.; Wang, F.; Fu, Y.; Feng, F.; Huang, H. Role of O3 in the removal of HCHO using a DC streamer plasma. J. Phys. D: Appl. Phys. 2019, 52, 465203.
29. Bruggeman, P.J.; Kushner, M.J.; Locke, B.R.; Gardeniers, J.G.E.; Graham, W.G.; Graves, D.B.; Hofman-Caris, R.C.H.M.; Maric, D.; Reid, J.P.; Ceriani, E.; Fernandez Rivas, D.; Foster, J.E.; Garrick, S.C.; Gorbanev, Y.; Hamaguchi, S.; Iza, F.; Jablonowski, H.; Klimova, E.; Kolb, J.; Krcma, F.; Lukes, P.; Machala, Z.; Marinov, I.; Mariotti, D.; Mededovic Thagard, S.; Minakata, D.; Neyts, E.C.; Pawlat, J.; Petrovic, Z.L.; Pflieger, R.; Reuter, S.; Schram, D.C.; Schröter, S.; Shiraiwa, M.; Tarabová, B.; Tsai, P.A.; Verlet, J.R.R.; von Woedtke, T.; Wilson, K.R.; Yasui, K.; Zvereva, G. Plasma–liquid interactions: a review and roadmap. Plasma Sources Sci. Technol. 2016, 25, 053002.
30. Laurita, R.; Barbieri, D.; Gherardi, M.; Colombo, V.; Lukes, P. Chemical analysis of reactive species and antimicrobial activity of water treated by nanosecond pulsed DBD air plasma. Clin. Plasma Med. 2015, 3, 53-61.
31. Pavlovich, M.J.; Chang, H.W.; Sakiyama, Y.; Clark, D.S.; Graves, D.B. Ozone correlates with antibacterial effects from indirect air dielectric barrier discharge treatment of water. J. Phys. D: Appl. Phys. 2013, 46, 145202.
32. Sun, P.; Wu, H.; Bai, N.; Zhou, H.; Wang, R.; Feng, H.; Zhu, W.; Zhang, J.; Fang, J. Inactivation of Bacillus subtilis Spores in Water by a Direct-Current, Cold Atmospheric-Pressure Air Plasma Microjet. Plasma Process. Polym. 2012, 9, 157-164.
33. Chen, Z.; Cui, Q.; Chen, C.; Xu, D.; Liu, D.; Chen, H.L.; Kong, M.G. Inactivation of myeloma cancer cells by helium and argon plasma jets: The effect comparison and the key reactive species. Phys. Plasmas 2018, 25, 023508.
34. Joh, H.M.; Choi, J.Y.; Kim, S.J.; Kang, T.H.; Chung, T.H. Effects of the pulse width on the reactive species production and DNA damage in cancer cells exposed to atmospheric pressure microsecond-pulsed helium plasma jets. AIP Adv. 2017, 7, 085106.

MIX
Papier aus verantwortungsvollen Quellen
Paper from responsible sources
FSC® C105338

If you have any concerns about our products,
you can contact us on
ProductSafety@springernature.com

In case Publisher is established outside the EU,
the EU authorized representative is:
Springer Nature Customer Service Center GmbH
Europaplatz 3, 69115 Heidelberg, Germany

Printed by Libri Plureos GmbH
in Hamburg, Germany